GLENIS HEATH HEATHER MCKENZIE LAUREL TULLY

FOOD by DESIGN

5e

years 7–10

Food by Design
5th Edition
Laurel Tully
Glenis Heath
Heather McKenzie
ISBN 9780170495714

Product Manager: Caroline Williams
Content Developer: Chantelle Bryant
Associate Content Project Manager: Garima Singh
Project Designer: Mariana Maccarini
Cover Design: Mariana Maccarini
Cover Photography: Mark Fergus Photography
Text Design: Jennai Lee Fai and Mariana Maccarini
Editor: Natasha Broadstock
Proofreader: Lauren McGregor
Permissions/Photo Researcher: Lumina Datamatics
Cover: Stocksy United/Wizemark
Typeset by Lumina Datamatics

Any URLs contained in this publication were checked for currency during the production process. Note, however, that the publisher cannot vouch for the ongoing currency of URLs.

For product information and technology assistance,
in Australia call **1300 790 853;**
in New Zealand call **0800 449 725**

For permission to use material from this text or product, please email **aust.permissions@cengage.com**

National Library of Australia Cataloguing-in-Publication Data
A catalogue record for this book is available from the National Library of Australia

Cengage Learning Australia
Level 7, 80 Dorcas Street
South Melbourne, Victoria Australia 3205

For learning solutions, visit **cengage.com.au**

Printed in China by 1010 Printing International Limited.
1 2 3 4 5 6 7 27 26 25

Stock illustrations used in chapter opener infographics: anna/Adobe Stock Photos, dyachenkopro/Adobe Stock Photos, alysantwanet/Adobe Stock Photos, Artem/Adobe Stock Photos, A.Selawi/Adobe Stock Photos, RavenL/Adobe Stock Photos, MITstudio/Adobe Stock Photos, viktorijareut/Adobe Stock Photos, kaisorn/Adobe Stock Photos, Cena Studio/Adobe Stock Photos, colorcocktail/Adobe Stock Photos, yime/Adobe Stock Photos, Igor/Adobe Stock Photos, setory/Adobe Stock Photos, Natalia/Adobe Stock Photos, NikAndr/Adobe Stock Photos, yatate/Adobe Stock Photos, paitoonpati/Adobe Stock Photos, 24hours/Adobe Stock Photos, OlgaStrelnikova/Adobe Stock Photos, Olesia_g/Adobe Stock Photos, Abhinash/Adobe Stock Photos.

CONTENTS

9780170495714

ABOUT THIS BOOK

Working with *Food by Design*

Food by Design for Years 7–10 (5th edition) has been written to inspire a lifelong love of food, nutrition and cooking in students. This contemporary resource provides comprehensive coverage of the curriculum, with pages that are loaded with activities and recipes designed to support student engagement and learning.

Drawing on the latest research and developments, *Food by Design* provides an insight and understanding of the topics in the Food Technology curriculum, including sustainability, food security, food sovereignty, food citizenship, and Australian First Nations' food and culture. Food by Design offers a wealth of learning experiences designed to engage students and to foster design thinking, in creating and applying solutions to real-world problems. It is crafted to spark students' interest in Food Technology and equip them with the knowledge and skills necessary for VCE-level studies.

Features

Each chapter has been carefully constructed to ensure that students are meeting the achievement standards of the Victorian Curriculum.

CHAPTER OPENER

Visually engaging **infographics** scaffold key information contained within each chapter.

Key terms in each chapter are highlighted, and definitions are provided at the start of the chapter.

Lists of the student **resources available on Nelson MindTap** are provided; these include videos, worksheets, templates and weblinks to support the chapter content.

RECIPES

Clear and extensively tested **recipes** provide enriching practical experiences for students, and help them to develop the necessary skills to prepare heathy and nutritious food.

Each recipe includes a set of evaluation questions that encourage students to record their observations and make connections with the theory covered in the chapter.

DESIGN ACTIVITIES

Design activities provide scaffolded opportunities for students to follow a design process to investigate, generate, produce and evaluate solutions to design problems.

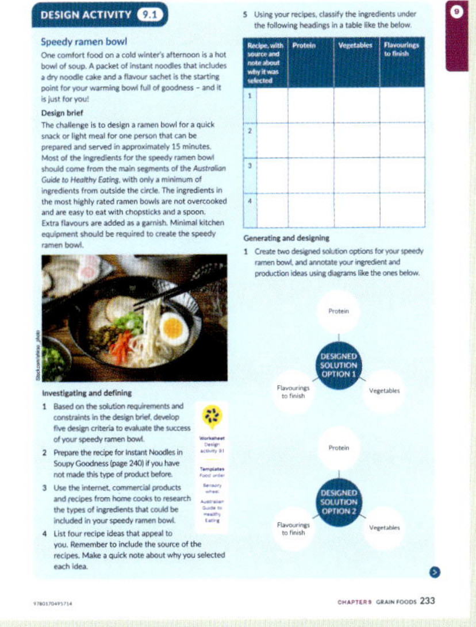

CASE STUDIES

With a focus on local businesses and initiatives, **case studies** include real-life examples that broaden student understanding, and link to the wider context of the topics covered in each chapter. They include questions and activities that allow students to reflect and make connections.

PRACTICAL ACTIVITIES

New in this edition, **practical activities** are designed to provide students with the opportunity to engage in the practical and hands-on application of key skills and knowledge.

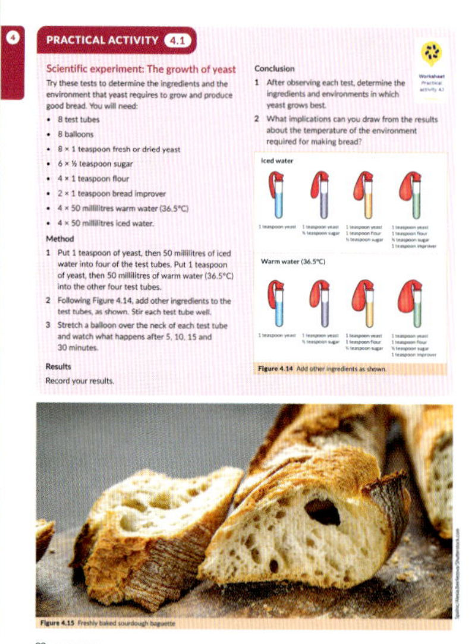

PRACTICAL ACTIVITY 4.1

Scientific experiment: The growth of yeast

Try these tests to determine the ingredients and the environment that yeast requires to grow and produce good bread. You will need:

- 8 test tubes
- 8 balloons
- 8 × 1 teaspoon fresh or dried yeast
- 6 × ½ teaspoon sugar
- 4 × 1 teaspoon flour
- 2 × 1 teaspoon bread improver
- 4 × 50 millilitres warm water (36.5°C)
- 4 × 50 millilitres iced water.

Method

1 Put 1 teaspoon of yeast, then 50 millilitres of iced water into four of the test tubes. Put 1 teaspoon of yeast, then 50 millilitres of warm water (36.5°C) into the other four test tubes.
2 Following Figure 4.14, add other ingredients to the test tubes, as shown. Stir each test tube well.
3 Stretch a balloon over the neck of each test tube and watch what happens after 5, 10, 15 and 30 minutes.

Results

Record your results.

Conclusion

1 After observing each test, determine the ingredients and environments in which yeast grows best.
2 What implications can you draw from the results about the temperature of the environment required for making bread?

Figure 4.14 Add other ingredients as shown.

Figure 4.15 Freshly baked sourdough baguette

82 FOOD BY DESIGN 9780170495714

ACTIVITIES

Activities provide learning opportunities that allow students to build food skills and address general capabilities in a fun and engaging way.

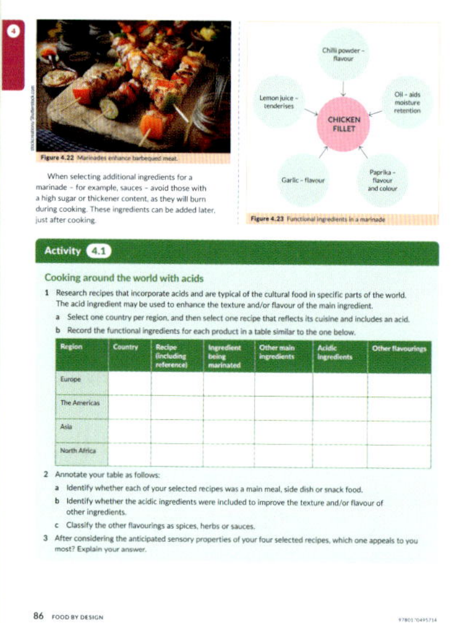

Figure 4.22 Marinades enhance barbequed meat.

When selecting additional ingredients for a marinade – for example, sauces – avoid those with a high sugar or thickener content, as they will burn during cooking. These ingredients can be added later, just after cooking.

Figure 4.23 Functional ingredients in a marinade

Activity 4.1

Cooking around the world with acids

1 Research recipes that incorporate acids and are typical of the cultural food in specific parts of the world. The acid ingredient may be used to enhance the texture and/or flavour of the main ingredient.
   a Select one country per region, and then select one recipe that reflects its cuisine and includes an acid.
   b Record the functional ingredients for each product in a table similar to the one below.

Region	Country	Recipe (including reference)	Ingredient being marinated	Other main ingredients	Acidic ingredients	Other flavourings
Europe						
The Americas						
Asia						
North Africa						

2 Annotate your table as follows:
   a Identify whether each of your selected recipes was a main meal, side dish or snack food.
   b Identify whether the acidic ingredients were included to improve the texture and/or flavour of other ingredients.
   c Classify the other flavourings as spices, herbs or sauces.
3 After considering the anticipated sensory properties of your four selected recipes, which one appeals to you most? Explain your answer.

86 FOOD BY DESIGN 9780170495714

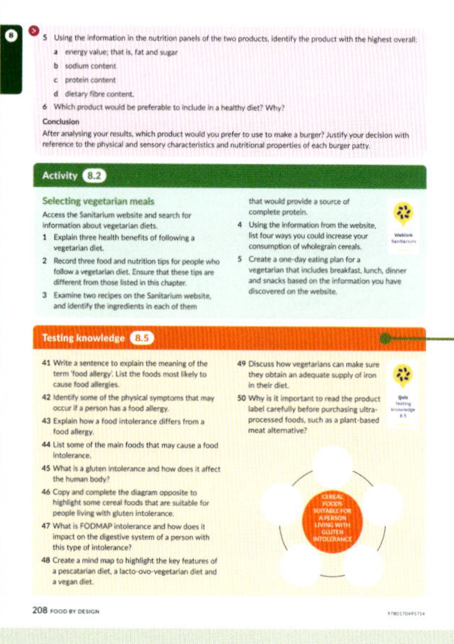

5 Using the information in the nutrition panels of the two products, identify the product with the highest overall:
   a energy value; that is, fat and sugar
   b sodium content
   c protein content
   d dietary fibre content.
6 Which product would be preferable to include in a healthy diet? Why?

Conclusion

After analysing your results, which product would you prefer to use to make a burger? Justify your decision with reference to the physical and sensory characteristics and nutritional properties of each burger patty.

Activity 8.2

Selecting vegetarian meals

Access the Sanitarium website and search for information about vegetarian diets.

1 Explain three health benefits of following a vegetarian diet.
2 Record three food and nutrition tips for people who follow a vegetarian diet. Ensure that these tips are different from those listed in this chapter.
3 Examine two recipes on the Sanitarium website, and identify the ingredients in each of them that would provide a source of complete protein.
4 Using the information from the website, list four ways you could increase your consumption of wholegrain cereals.
5 Create a one-day eating plan for a vegetarian that includes breakfast, lunch, dinner and snacks based on the information you have discovered on the website.

Testing knowledge 8.5

41 Write a sentence to explain the meaning of the term 'food allergy'. List the foods most likely to cause food allergies.
42 Identify some of the physical symptoms that may occur if a person has a food allergy.
43 Explain how a food intolerance differs from a food allergy.
44 List some of the main foods that may cause a food intolerance.
45 What is a gluten intolerance and how does it affect the human body?
46 Copy and complete the diagram opposite to highlight some cereal foods that are suitable for people living with gluten intolerance.
47 What is FODMAP intolerance and how does it impact on the digestive system of a person with this type of intolerance?
48 Create a mind map to highlight the key features of a pescatarian diet, a lacto-ovo-vegetarian diet and a vegan diet.
49 Discuss how vegetarians can make sure they obtain an adequate supply of iron in their diet.
50 Why is it important to read the product label carefully before purchasing ultra-processed foods, such as a plant-based meat alternative?

208 FOOD BY DESIGN 9780170495714

TESTING KNOWLEDGE

Testing knowledge questions are checkpoints for students' understanding and comprehension. They appear at regular intervals throughout each chapter. These questions are also available through Nelson MindTap.

9780170495714

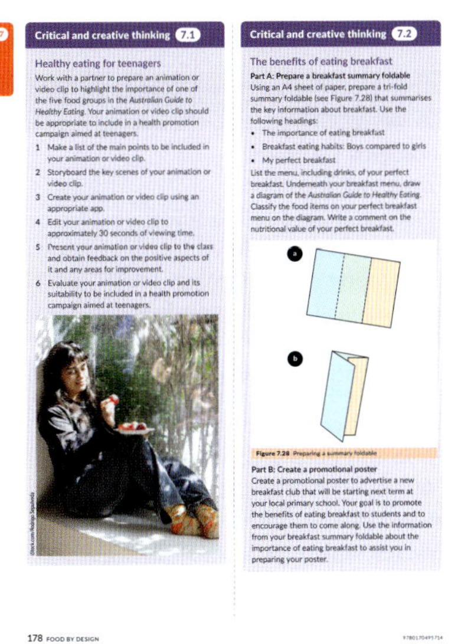

CRITICAL AND CREATIVE THINKING

Critical and creative thinking activities are designed so that students can practise and demonstrate higher-order critical and creative thinking in the Food Technology context.

RECIPE INDEX

The **recipe index** is an important tool for both students and teachers, enabling them to search for specific ingredients and/or recipes. The recipe index includes all recipes in the student book, and there is an archive of past recipes on Nelson MindTap.

GLOSSARY INDEX

Curriculum Grid

<table>
<tr><th>Chapters</th><th>Victorian Curriculum 2.0</th><th>Australian Curriculum</th></tr>
<tr><td>1 – Designing with food</td><td rowspan="4">Technologies contexts:
• Food specialisations
Creating designed solutions:
• Investigating and defining
• Generating and designing
• Producing and implementing
• Evaluating
• Planning and managing</td><td rowspan="4">Technologies contexts:
• Food specialisations
Processes and production skills:
• Investigating and defining
• Generating and designing
• Producing and implementing
• Evaluating
• Collaborating and managing</td></tr>
<tr><td>2 – Preparing food safely</td></tr>
<tr><td>3 – Recipe basics</td></tr>
<tr><td>4 – The food lab</td></tr>
<tr><td>5 – Indigenous foods</td><td>Technologies contexts:
• Food and fibre production
• Food specialisations
Creating designed solutions:
• All (see above – Ch 1 - 4)
Cross-curriculum priorities:
• Aboriginal and Torres Strait Islander histories and cultures
• Sustainability</td><td>Technologies contexts:
• Food and fibre production
• Food specialisations
Processes and production skills:
• All (see above – Ch 1 - 4)
Cross-curriculum priorities:
• Aboriginal and Torres Strait Islander histories and cultures
• Sustainability</td></tr>
<tr><td>6 – Sustainable food systems</td><td>Technologies contexts:
• Food and fibre production
• Food specialisations
Creating designed solutions:
• All (see above – Ch 1 - 4)
Health and PE:
• Personal, social and community health
Cross-curriculum priorities:
• Sustainability</td><td>Technologies contexts:
• Food and fibre production
• Food specialisation
Processes and production skills:
• All (see above – Ch 1 - 4)
Health and PE:
• Personal, social and community health
Cross-curriculum priorities:
• Sustainability</td></tr>
<tr><td>7 – Eat well, be well</td><td rowspan="2">Technologies contexts:
• Food specialisations
Creating designed solutions:
• All (see above – Ch 1 - 4)
Health and PE:
• Personal, social and community health</td><td rowspan="2">Technologies contexts:
• Food specialisations
Processes and production skills:
• All (see above – Ch 1 - 4)
Health and PE:
• Personal, social and community health</td></tr>
<tr><td>8 – Eating well for lifelong health</td></tr>
<tr><td>9 – Grain foods</td><td rowspan="5">Technologies contexts:
• Food and fibre production
• Food specialisations
Creating designed solutions:
• All (see above – Ch 1 - 4)
Health and PE:
• Personal, social and community health
Cross-curriculum priorities:
• Sustainability
• Asia and Australia's engagement with Asia</td><td rowspan="5">Technologies contexts:
• Food and fibre production
• Food specialisations
Processes and production skills:
• All (see above – Ch 1 - 4)
Health and PE:
• Personal, social and community health
Cross-curriculum priorities:
• Sustainability
• Asia and Australia's engagement with Asia</td></tr>
<tr><td>10 – Vegetables and legumes</td></tr>
<tr><td>11 – Meat, poultry, fish and eggs</td></tr>
<tr><td>12 – Fruit</td></tr>
<tr><td>13 – Milk, yoghurt and cheese</td></tr>
<tr><td>14 – Only sometimes!</td><td>Technologies contexts:
• Food specialisations
Creating designed solutions:
• All (see above – Ch 1 - 4)
Health and PE:
• Personal, social and community health
Cross-curriculum priorities:
• Sustainability</td><td>Technologies contexts:
• Food specialisations
Processes and production skills:
• All (see above – Ch 1 - 4)
Health and PE:
• Personal, social and community health
Cross-curriculum priorities:
• Sustainability</td></tr>
</table>

9780170495714

Nelson MindTap

An online learning space that provides students with tailored learning experiences.

- Access tools and content that make learning simpler yet smarter.
- **Flexible formats:** choose how you navigate using either the online eText or offline PDFs.
- Margin links in the student book signpost multimedia student resources found on Nelson MindTap.

Worksheet Practical Activity 5.1

For students:

- Watch **videos** to enhance your understanding of key concepts
- Check your understanding with **testing knowledge quizzes**
- Complete all practical and design activities with downloadable **worksheets** and **templates**
- Access a curated list of **weblinks** to videos, articles and other online resources associated with the content to build on your understanding of the concepts through real-world applications

For teachers*:

- Gain access to all the student digital resources, including **quizzes**, **worksheets** and **templates** and to the online and offline **eText**
- Enjoy much-loved recipes from past editions of Food by Design in our **recipe archive**
- Use **scope and sequence planners** to support easy program and assessment planning across Years 7-10.
- Access **solutions** to all testing knowledge questions in the student book
- Save time and money and avoid unnecessary food waste, using the **recipe ingredients calculator** for all recipes in the student
- Access a curated list of up-to-date **weblinks** to videos, articles and other online resources associated with the content to enhance student engagement and learning
- Use Nelson MindTap's **course customisation** tools for complete flexibility with the ability to add links, resequence, or hide any content based on your school's approach and students' needs.

* *Food by Design* **Nelson MindTap** *gives you access to an eText with additional resources. Nelson MindTap products are not for individual use; they are for adoption, class set or book hire. Nelson MindTap is teacher-led courseware and requires an active course for access. This is a student activation code that can only be used to access a Nelson MindTap course as part of a school's adoption. Please contact your learning consultant to discuss purchasing options.*

Security & privacy: Nelson MindTap joined the Safer Technologies 4 Schools (ST4S) Product Badge Program in 2024. The annual ST4S assessment is part of our commitment to supporting the online security and safety of students and schools. Learn more at st4s.edu.au

THE DESIGN PROCESS

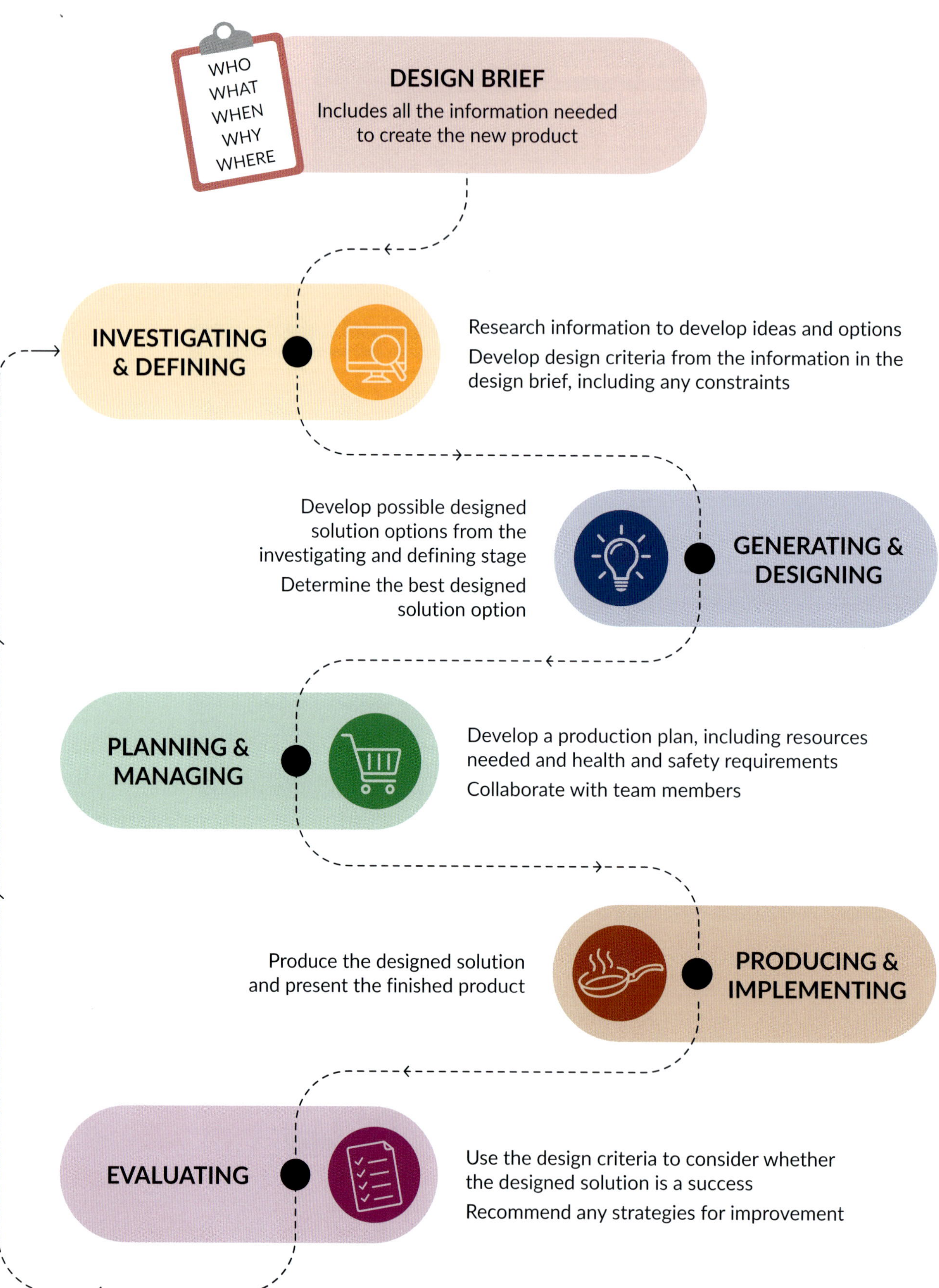

9780170495714

OUR SENSES & FOOD

Our five senses – sight, smell, taste, touch and hearing – enable us to enjoy, evaluate and describe food.

Appearance
When looking at food, we notice its colour, shape and texture

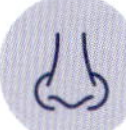

Aroma
The nose detects the smell or aroma released from food, e.g. spicy, sweet or burnt

Flavour
When we put food into our mouth we note whether the taste is sweet, sour, bitter, salty or umami

Texture
Foods have different textures or mouth feel when eaten; they may be smooth, lumpy or granular

Sound
Foods make different noises when eaten – apples sound crunchy; noodles sound slurpy

1 DESIGNING WITH FOOD

KEY TERMS

constraints factors in the design brief with which the product must comply

design brief a statement that defines the need or opportunity to be resolved and identifies the users of the product; it contains information about the type of product, any constraints, the resources needed and a time frame

design criteria criteria drawn from the constraints and solution requirements in the design brief; they are used to determine if the designed solution is appropriate

design process the process of investigating and defining, generating and designing, planning and managing, producing and implementing, and evaluating

designed solution a product created for a specific purpose that has been outlined or defined in the design brief

qualitative or sensory analysis the evaluation of the sensory properties of food, such as appearance, aroma, flavour, texture and sound

quantitative measures ways to measure the physical, chemical or nutritional properties of food

sensory properties the appearance, aroma, flavour, texture and sound of food

Templates
- Sensory wheel
- Food tasting table
- Recipe map
- SWOT analysis table
- Decision table
- PMI table
- Food order
- Production plan

Worksheet
- Design activity 1.1

Quizzes
- Testing knowledge 1.1
- Testing knowledge 1.2

To access resources above, visit **cengage.com.au/nelsonmindtap**

9780170495714

1

The role of food

Food plays a vital part in our lives. It is the essential fuel that keeps us alive. Food is frequently the focal point of our social lives – we share food with family and friends in our homes, in cafés and restaurants, at school, at sporting events, at festivals, and in a variety of other settings. Because food is fundamental to life, it is important to understand how to prepare it so that it will provide us with the essential nutrients we need to maintain good health, as well as be appealing to eat. Consideration must also be given to providing sustainable diets that consider the environmental issues associated with food systems.

iStock.com/Drazen

Figure 1.1 Sharing food with family and friends

Describing food

To work successfully with food, it is necessary to develop an understanding of its characteristics or **sensory properties**: its appearance, aroma, flavour, texture and sound. To discuss food with other people, you need to have a specific language that enables you to describe and explain the properties of food. This is distinctly different to a subjective statement of your opinion about how much you like or dislike a particular food. We all have certain foods that we particularly enjoy, but what is it about these foods that makes us think of them as our 'favourites'?

Weblink How the five senses affect food choices

The body's senses of sight, smell, taste, touch and even hearing are important in building up our knowledge of food. The eyes, nose, tongue, ears and skin send messages to the brain to tell us which foods give us pleasure and which foods we find unpleasant. The sweet, smooth taste of chocolate, the aroma of a roasting chicken or baking bread, or the sound of a crisp apple being bitten can give us a sense of excitement and anticipation.

The sensory wheel (see Figure 1.2) is designed as a strategy to assist individuals to objectively describe food; for example, your eye looks at the appearance of food and observes its colour, shape and texture. The adjectives allow you to describe the food to others.

Template Sensory wheel

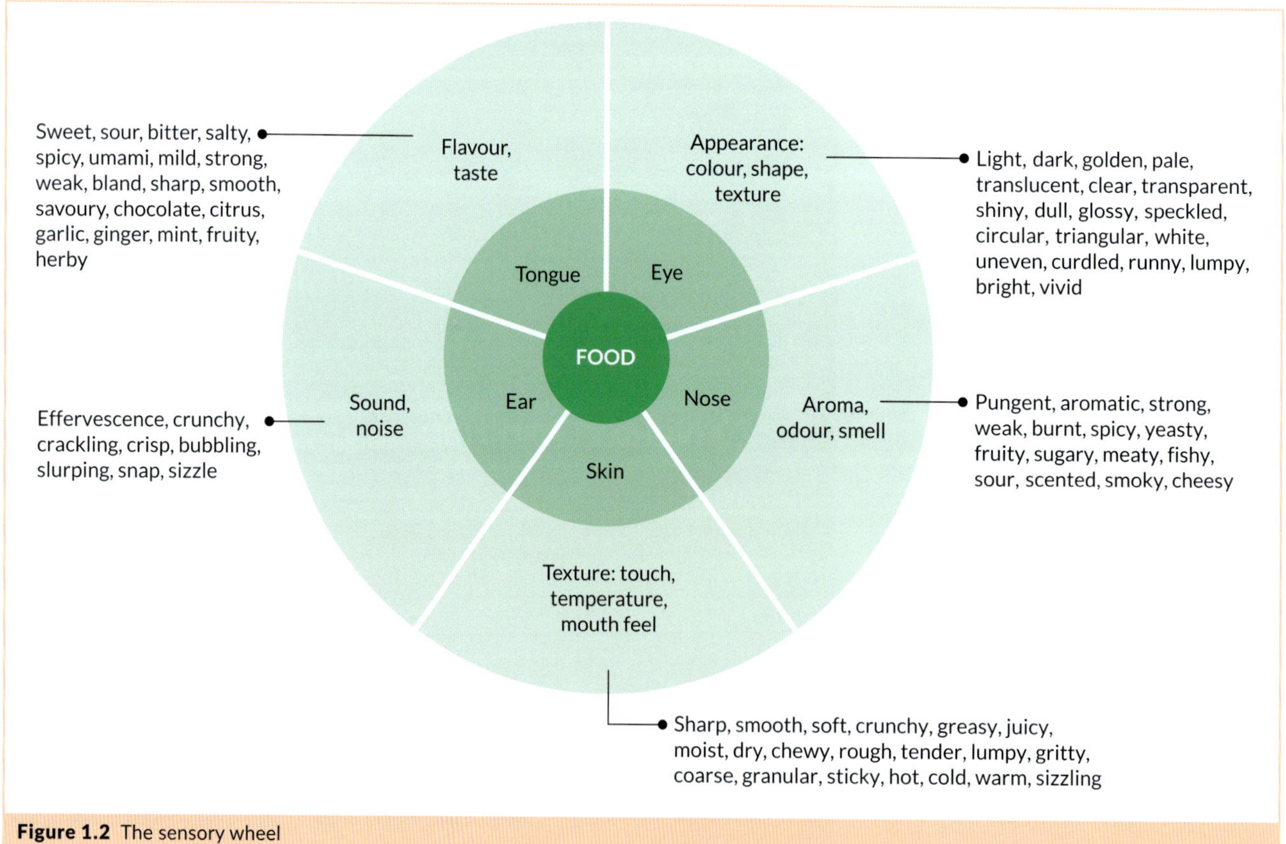

Figure 1.2 The sensory wheel

9780170495714

Activity 1.1

Favourite foods

1 a Sketch four of your favourite foods, leaving space around each sketch for annotations.
  b Annotate each sketch to describe the food's appearance, aroma, flavour, texture and sound. Refer to the sensory wheel to help you with descriptive words for each property.

2 Look at Figures 1.3 and 1.4.
  a Write down as many words as you can to describe the appearance of each food item, and how you imagine the aroma, flavour, texture and sound of each would be.
  b Which of the properties of each of these two food items appeals to you most? Why?

Template Sensory wheel

Syed/Adobe Stock Photos

Figure 1.3 Chicken and salad wrap

Tasneem Ali Khamaiseh/Shutterstock.com

Figure 1.4 Acai bowl

Tasting food

The sensory impression food creates every time we eat builds our memory bank of the foods we enjoy. Think about the aromas of hot chips frying, a freshly baked cake or lamb chops cooking on a barbecue. Many chemicals combine in a complex way to create different foods, and we react as our senses experience every bite or sip we take.

After we have looked at a food, our senses of taste and smell work together to determine its flavour. As the food enters our mouths, the taste buds recognise the basic flavours of sweet, sour, bitter, salty and umami. The taste of *umami* can be described as 'delicious and savoury', and one that deepens the flavour of a food.

You will have noticed that when you have a cold and a blocked nose, your food seems tasteless because your sense of smell is not working as well as it could be.

Every food falls within one or more of the basic taste categories:

- Sweet foods, which most people love, include chocolate, strawberries, honey and ice-cream.
- Sour foods include vinegar and citrus fruits, such as lemons and grapefruit.
- The pith of an orange, bitter melon, rocket and radicchio are examples of bitter foods.
- Bacon, potato crisps, soy sauce, feta cheese and olives all have a salty flavour.
- Tomatoes and parmesan cheese are good examples of umami foods.

Activity 1.2

Taste test

For this taste test, you will need the following equipment:

- two trays, labelled 'Tray A' and 'Tray B'
- 12 different food samples, six on each tray – the samples should be cut into small pieces and the trays should be covered so that the foods cannot be seen
- enough blindfolds for half of the class
- enough small plates and spoons for every member of the class
- a copy of the food tasting table on the following page for each class member to use to record the flavour, texture and identification of the food samples.

Template Food tasting table

Table 1.1 Food tasting table

Sample	Flavour	Texture	What I think the food is	What the food really is
Sample 1				
Sample 2				
Sample 3				
Sample 4				
Sample 5				
Sample 6				

1 Work with a partner to complete the taste test.
2 One partner should be blindfolded.
3 Begin with six food samples from Tray A. The blindfolded partner must not be told which foods are on the tray.
4 Use a spoon to feed the blindfolded partner one of the six samples.
5 Ask them to describe the sample's flavour and texture, and to state what they think the food is.
6 Record their answers on the table.
7 Feed the blindfolded partner the remaining samples on Tray A and record their answers.
8 Remove the blindfold and share the results.
9 Change places, and repeat the process using Tray B.

Analysing the properties of food

After a food product has been prepared and served, it is important to examine it closely to determine whether it was successful; that is, did people enjoy eating it, and did it meet the needs identified in the design brief (see page 9)?

When evaluating the success of a food product, food manufacturers, chefs and home cooks identify the ingredients used and how they worked together in a recipe. This helps them determine whether improvements should be made to the processes and equipment used in production. Measuring the size, weight, volume and nutrient content of a food product enables comparisons to be made between similar and different products.

Most importantly, what people remember about food is whether eating it was a pleasurable experience; this hinges on the sensory analysis of appearance, aroma, flavour, texture and sound.

iStock.com/Elisaveta Ivanova

Figure 1.5 The sensory experience of eating an apple is shaped by the appearance, aroma, flavour and texture of the apple, and the sound of the crunch as you eat it.

Qualitative or sensory measures

Qualitative or sensory analysis is used to evaluate the sensory properties of food: appearance, aroma, flavour, texture and sound.

As consumers, we use all of our five senses – sight, smell, taste, touch and hearing – to form opinions about food products. This form of sensory analysis is subjective, since some foods are appealing to some people but not to others. In the food industry, specially trained testers carry out controlled sensory analysis tests on food to ensure that the end product will appeal to a wide range of consumers.

Several types of sensory analysis tests are used to collect this data and collate it in the form of descriptors, which can be visual images or may involve the use of words or sentences. Descriptors describe specific characteristics of the food – look at the sensory wheel on page 4 for some examples.

Facial hedonic descriptors enable the consumer to rate how much they like or dislike a particular food by using a hedonic scale graded from a very unhappy to an extremely happy face.

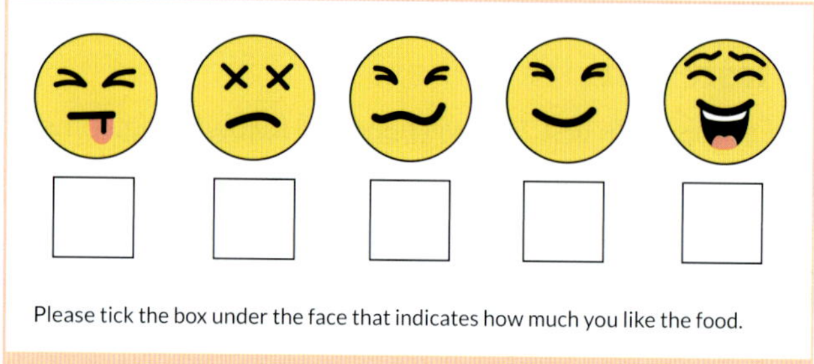

Figure 1.6 A hedonic scale

9780170495714

Attitudinal descriptors indicate how people feel about a particular food. These descriptors provide more detailed statements about the food to allow the consumer to state their attitude to the product.

Table 1.2 Attitudinal descriptors

Tick the statement that best describes your attitude to this product.	
I would eat this at every opportunity I had.	
I would eat this very often.	
I would eat this frequently.	
I like this and would eat it now and then.	
I would eat this if it was available but would not go out of my way for it.	
I don't like this, but I would eat it occasionally.	
I would hardly ever eat this.	
I would eat this only if there were no other food choices.	
I would eat this only if I was forced to.	

Quantitative measures

Quantitative measures are ways of measuring the physical, chemical or nutritional properties of food. They enable consumers to compare similar food products. Features of food products that can be measured accurately using quantitative measures include the:

- ratio of ingredients
- weight of the finished product
- microbiological content to ensure that the food is safe to eat and to determine the shelf life of the product
- colour
- nutrient content
- volume
- consistency
- texture.

The recording of this type of data is an important part of the quality control process for food manufacturers. Products are analysed to determine whether they meet the Australian Standards, which are set out in the Food Standards Code and managed by Food Standards Australia New Zealand. This gives consumers confidence about the food they purchase – that it is what the label says it is and that it is safe to eat. It also provides a basis upon which consumers can compare competitors' products.

Testing knowledge 

1.1

Quiz Testing knowledge 1.1

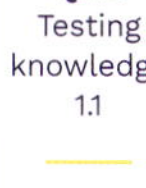

Template Sensory wheel

1. In one sentence, explain why food plays such an important part in our lives.
2. List the sensory properties of food.
3. Discuss how understanding the sensory properties of food enables people to share their views about the qualities of a particular food.
4. For each of the food items listed in the table below, provide one descriptor of each sensory property. Use the sensory wheel on page 4 to help you.

Food item	Sensory descriptors
Red apple	
Full-cream milk	
Wholemeal bread	
Potato crisp	

5. Explain how we determine the flavour of food.
6. What is the meaning of 'umami'?
7. Explain why facial hedonic descriptors are a useful feedback tool for food manufacturers.
8. Discuss the type of feedback consumers provide about particular foods when using attitudinal descriptors.
9. List four properties of food that can be accurately measured using quantitative measures.
10. Explain the difference between qualitative analysis and quantitative analysis of food products.

Figure 1.7 Measuring the consistency of sauces is a quantitative measure

Designing with food

Creating **designed solutions** involves thinking creatively to develop a product for a specific purpose that has been outlined/defined in the design brief. That means that a person or a group of people have spent time, effort and money to design, create and produce something functional – in this case, we are focusing on food.

New food products and meal ideas seem to appear in the supermarket on an almost weekly basis. Food manufacturers respond to consumers' needs by designing and producing new snack foods, instant meals, frozen foods, convenience foods and beverages.

Whenever a new food product is designed and produced, the manufacturer works through the **design process** to create a designed solution. The design process involves:

- investigating and defining
- generating and designing
- planning and managing
- producing and implementing
- evaluating.

Table 1.3 Using the design process to create a designed solution

Design brief		
Year 9 Food Studies students will be holding a lunchtime fundraiser in the main school courtyard. They will be selling a bread-style product that contains a protein food and salad ingredients, and that can be easily eaten in the hand.		
The stages in the design process	Design thinking in each stage	Examples of each stage
Investigating and defining	• Development of design criteria • Collection of information from a wide range of resources • Use of creative thinking to analyse the information gained and develop design options	• Take suggestions from student focus groups • Research trending hand-held bread-style products • Consider examples of suitable protein foods, salad ingredients and bread products
Generating and designing	• Use of the ideas from the 'investigating and defining' stage to develop viable design options • Testing of the design options to determine which one best meets the constraints in the design brief; this becomes the designed solution	• Consider possible solutions: – burger – souvlaki – burrito – steak sandwich • Use PMI (Plus, Minus, Interesting) to determine the preferred designed solution • Source recipe for the designed solution – souvlaki
Planning and managing	• Identification of the steps to plan and manage production: – food order – production plan – work individually or collaborate as part of a team	• Complete a food order for ingredients for souvlaki • Prepare a production plan • Allocate tasks to team members
Producing and implementing	• Production of the designed solution • Use of appropriate tools and equipment • Presentation of the finished product	• Collect ingredients • Prepare the souvlaki following health and safety guidelines • Follow recipe accurately to maximise the sensory properties of the souvlaki
Evaluating	• Sensory analysis • Response to design criteria	• Analyse the sensory properties of the souvlaki • Respond to the design criteria to evaluate the success of the designed solution • Make recommendations for improvement

9780170495714

Design brief

A **design brief** is a concise statement that sets out the project or task to be undertaken. It becomes a 'road map' for the creative thinking that will result in the designed solution. It includes information about the problem to be solved, the need to be resolved or the opportunity to be explored. A design brief may also include information about the end user or the target market for the new product, as well as any constraints.

Outlining the 'who, what, when, why and where' – the '5 Ws' – is a strategy to help determine and identify different aspects of the problem to be solved. When designers are developing a design brief for a new food product, they need to investigate information about the '5 Ws' of the product. For example, the design brief will include information about for whom the new product will be designed, such as children. It also will also provide information about what type of product is to be developed, such as a new healthy snack food. These factors are often called **constraints**, because they are factors with which the product must comply, and the designer must include them in their thinking when creating designed solutions.

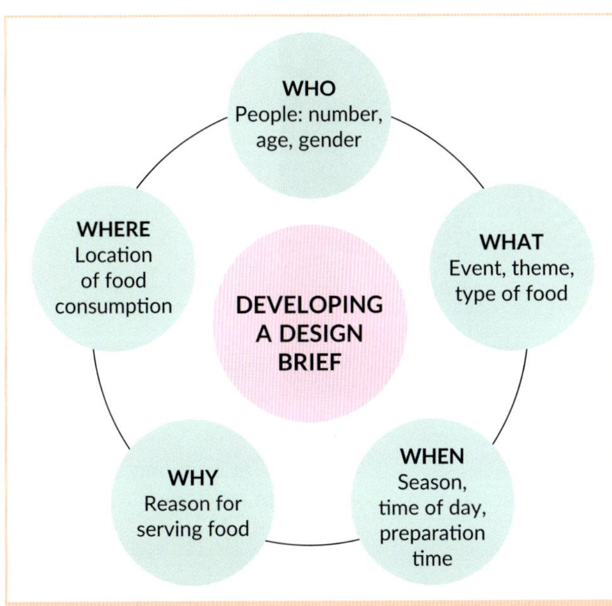

Figure 1.8 Things to consider when developing a design brief

Just like a food manufacturer, you too probably had to work through the development of a design brief this morning without realising it. For example, after waking up in the morning you had to decide about breakfast:

- Who – do I have to make my own breakfast?
- What – what food that I could eat for breakfast is in the pantry or refrigerator?
- When – how much time do I have to make and eat breakfast before leaving for school?
- Why – why is it important to eat breakfast after waking up?
- Where – will I eat breakfast at home or on the bus going to school?

Investigating and defining

The first stage of the design process involves investigating and defining the requirements of the task. This may involve researching and exploring a wide range of ideas and information, such as new food trends, any new ingredients or flavours that have come onto the marketplace, and the current consumer demand for ethically or sustainably produced foods. The designer will also need to investigate whether the company will need to upskill its staff on the latest production methods, as well as the safe use of tools and equipment required.

This research can be undertaken using a wide range of sources such as the internet, magazines, technical journals, recipes, primary producers and competing manufacturers. A food manufacturer may begin with a consumer focus group, where the needs and wants of consumers are explored. Methods of packaging the new product may require investigation to identify the material used and the impact of its disposal on the environment.

Developing design criteria

During the investigating and defining stage, it will be necessary to define or establish the **design criteria** based on the information in the design brief. The design criteria are a set of guidelines that the designer will respond to, following production, to determine if a proposed designed solution meets the requirements of the design brief. They are developed from the requirements of the design brief, including any constraints. The design criteria may take the form of questions.

Design criteria questions that may be asked about the food product could include:

- Was the food suitable for the people consuming it?
- Did the food reflect the event or theme?
- Were the quality and sensory properties of the food – appearance, aroma, flavour, texture and sound – appropriate to the needs of the design brief?
- Were the food items appropriate for the season?
- Was the food prepared and served within the time frame specified in the design brief?
- Did the food meet the needs of the design brief? Was the overall product a success?

Figure 1.9 shows a sample annotated design brief, including design criteria, for a project: a food truck at the Battle of the Bands competition.

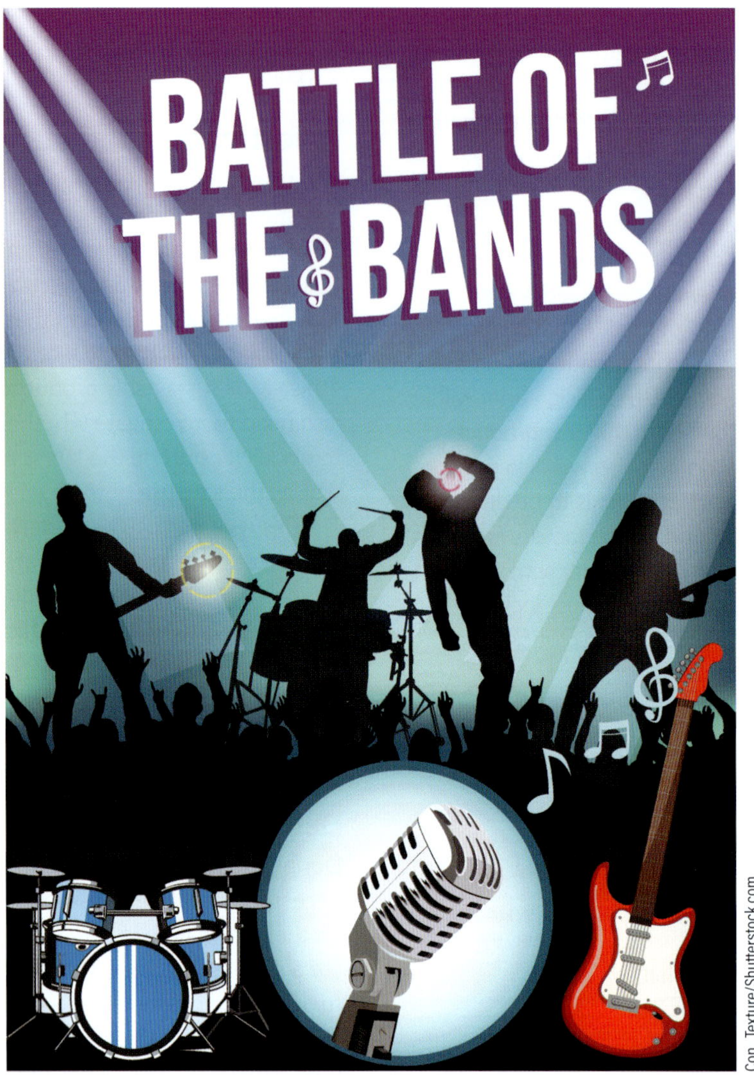

Was all of the food for sale in the food truck based on Mexican-style cuisine?

Were there three savoury snack foods available for customers to purchase?

Was there at least one vegetarian snack food on the menu?

Battle of the Bands

The Battle of the Bands is a competition for young musicians and is sponsored by the Youth Stage. It is held once a year to showcase the talent of local musicians. Local bands will compete for a contract with a recording studio to produce an album of their music. A temporary stage will be constructed in parklands close to the Yarra River, and music fans are encouraged to bring a picnic rug and purchase food to share during the event.

Within the parkland, several food trucks will sell a variety of food to the concert goers. One of the food trucks will focus its menu on Mexican-style foods. Its menu must include three savoury snack foods, and at least one of these must be suitable for vegetarians. The snack foods must have appealing sensory properties. All the food must be sold in individual serves and packaged in environmentally friendly materials.

Were of all the snack foods packaged in environmentally friendly materials?

Did the snack foods have appealing sensory properties?

Were all of the snack foods sold in individual serves?

Figure 1.9 An annotated design brief, with design criteria

Generating and designing

The next stage of the design process involves generating and designing to create ideas or options for the new food product. The final designed solution is the result of creative design thinking, and working through the design and production processes.

Design thinking and ideas can be documented in a concept sketch, an annotated diagram including images, or a recipe map as shown in Figure 1.10. The creation of a range of design ideas allows the designer to weigh up and consider each option to enable the best designed solution or choice to be made.

The designed solution options are carefully analysed and evaluated to ensure that the chosen designed solution meets the solution requirements and constraints identified in the design brief. The final decision can be guided by testing a range of recipe ideas or creating a SWOT (Strengths, Weaknesses, Opportunities, Threats) analysis table, decision table or PMI (Plus, Minus, Interesting) table. Once the design thinking and ideas have been developed, the designer may outline the advantages and disadvantages of each idea, showing how it meets the requirements of the design brief. Finally, each of the design options is evaluated, and the designer decides on the best designed solution to the problem outlined in the design brief.

SWOT analysis

The SWOT analysis is a useful tool for evaluating design ideas by identifying strengths, weaknesses, opportunities and threats. Justification of the selected designed solution is also important.

Template SWOT analysis table

Table 1.4 A SWOT analysis table

Analysis of designed solution options	
Designed solution 1:	
Strengths:	Weaknesses:
Opportunities:	Threats:
Designed solution 2:	
Strengths:	Weaknesses:
Opportunities:	Threats:
Preferred designed solution:	
Justification:	

Meat type

Accompanying dips

Marinade ingredients

Salad ingredients

Bread to wrap

Your designed solution: souvlaki recipe

Template Recipe map

derGriza/Shutterstock.com

Figure 1.10 A recipe map for a souvlaki to develop designed solution options

Decision table

A decision table allows the designer to compare the advantages and disadvantages of various designed solution options.

Table 1.5 A decision table

Comparison of designed solution options		
Designed solution 1:	Designed solution 2:	Designed solution 3:
Advantages:	Advantages:	Advantages:
Disadvantages:	Disadvantages:	Disadvantages:
Preferred designed solution:		
Justification:		

PMI

PMI is a tool to examine and consider the positives, negatives and any interesting features or characteristics of a possible designed solution.

Table 1.6 A PMI table

Plus	Minus	Interesting

Planning and managing

The ability to plan and effectively manage resources is critical for people who make decisions about the food we eat every day. In fact, working systematically through the design process when creating a meal or catering for a major event requires the same logical thinking and project management skills as are used in the wider food industry. In some stages of the process – for example, generating and designing – people may be working individually, while at other stages – for example, producing and implementing – they may work collaboratively with a team. Remember, good communication is essential when working for or with others.

Some aspects to consider are:

- knowledge of the relationship between nutrition and health
- skills and knowledge in food preparation
- access to appropriate ingredients
- tools and equipment needed to prepare and cook food
- health and safety requirements for food production
- skills in managing time efficiently to meet deadlines
- the ability to manage a food budget.

Before producing the designed solution, the home cook or food manufacturer must undertake considerable planning. Food orders and production plans are important steps. These may involve collaborative planning if working in a group or team.

Food orders

A food order must be created for all the ingredients required to make the product so that they will all be available at the time of production. It is also essential to list any specialist equipment that is required.

Table 1.7 A food order for the home cook

Food order		
Name:		
Production date:		
Recipes:		
Supermarket	Fresh fruit and vegetables	Meat/chicken/fish
Specialist equipment:		

9780170495714

Production plans

A production plan (or project plan) is an important tool that the home cook or food manufacturer can use to ensure that the product they plan to make is completed successfully and made within the time available. The plan focuses on the 'how to' part of the designed solution. Production plans manage efficiency by eliminating time-wasting. They identify health and safety issues in the production process and areas of risk when working with food.

At school or at home, preparing a production plan prompts the cook to read the recipe carefully, think about which step in the recipe needs to be done first, and consider ways to organise tasks so that there will be a smooth workflow in the kitchen.

A production plan should contain:

- information about materials or ingredients needed
- a list of all the tools and equipment required
- the sequence of steps or processes or techniques involved
- an estimate of the time it will take to produce the product – it is important to note the time it will take to complete each of the steps involved (you should break up these steps into 5–10-minute intervals) as well as the time needed to clean up
- information about all necessary health and safety requirements to be followed during the producing and implementing stage.

During the producing and implementing stage, the cook, chef or food manufacturer will record any modifications or changes they have made to the production plan, so that they can consider these in future productions.

Table 1.8 is an example of a simple production plan suitable for the home cook.

Activity 1.3

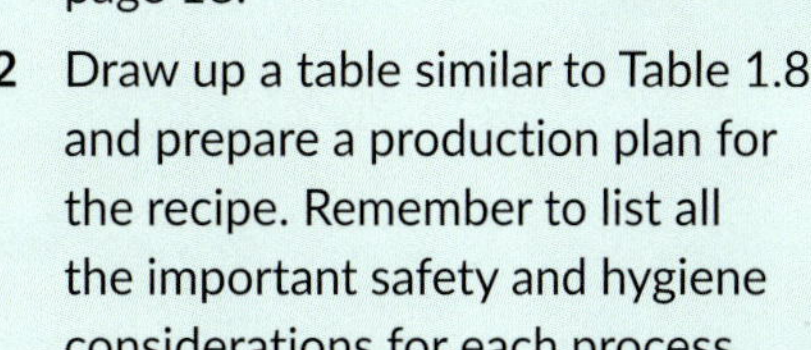

Preparing a production plan

Template Production plan

1. Read the recipe for Salad Roll-up on page 18.
2. Draw up a table similar to Table 1.8 and prepare a production plan for the recipe. Remember to list all the important safety and hygiene considerations for each process.
3. Share your production plan with a classmate and ask them to give you ideas for steps, equipment or safety/hygiene considerations you might have overlooked. Add these to your production plan.

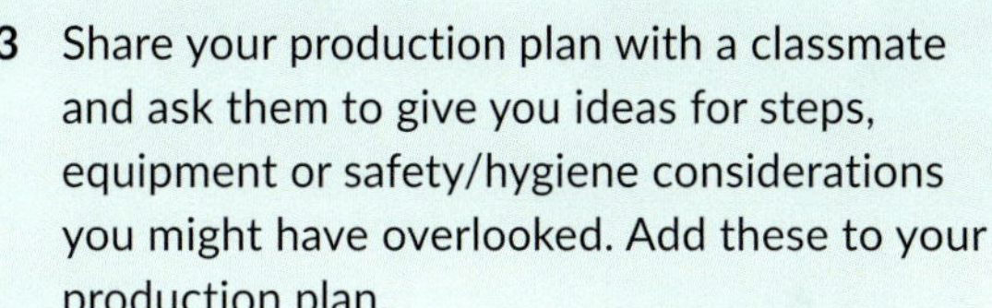

Table 1.8 A production plan for the home cook

Name of recipe:		**Ingredients:**	
Cheese Omelette (see recipe on page 95) *Serves one*		2 eggs 1 tablespoon water 2 shakes pepper 1 teaspoon butter 1 tablespoon parsley, finely chopped ¼ cup cheddar cheese, grated	
Time (a.m.)	**Important steps**	**Equipment required**	**Safety and hygiene considerations**
9.00–9.10	Collect ingredients	Tray, bowls, spoons	Thoroughly wash hands.
9.10–9.20	Chop parsley; grate cheese	Cook's knife, grater	Keep fingers away from blade of knife while chopping.
9.20–9.25	Beat eggs and water	Bowl, whisk/fork	Make sure all equipment is clean and dry.
9.25–9.30	Prepare the omelette pan	Omelette pan	Use an oven mitt to handle the omelette pan.
9.30–9.40	Cook the omelette	Omelette pan, spatula	Do not leave the omelette pan unattended on the cooktop.
9.40–9.50	Serve and garnish	Clean serving plate	Place the hot omelette pan on a chopping board on the bench so that it cools before cleaning.
9.50–10.00	Clean up	Washing-up equipment	Do not put a sharp knife into a sink filled with soapy water.

Producing and implementing

It is now time to undertake the producing and implementing stage of the design process. Food production processes, practical skills, tools, equipment and techniques are applied to food ingredients to produce the selected designed solution. Consideration must be given to safe work practices undertaken during the production.

Odua Images/Shutterstock.com

Figure 1.11 Food production processes, practical skills, tools, equipment and techniques are applied to food ingredients to produce the selected designed solution.

Evaluating

The designer will continually evaluate as they move through each stage of the design process. After the completion of the designed solution, the final evaluation is documented. This stage involves reviewing the designed solution using the design criteria. The designer will make judgements about whether the product is an effective solution to the problem – including the constraints – outlined in the design brief.

The evaluation of the product will also assess the sensory properties of the food. Descriptive words can be used to describe the characteristics of food. For example: 'The bread roll had a golden crispy crust and a slight yeasty aroma; inside, the roll was soft and fluffy.' Everyone has an opinion about how much they like a particular food and this is a subjective evaluation, where food is rated using a 5-star scale. It will help the designer to determine if any modifications are needed to the product if it is made again at home, or if it will be something that the general public will buy.

In addition, the evaluating stage may involve assessing the time taken and considering other modifications that need to be made to the planning and managing stage. Suitability of the equipment and processes used during the producing and implementing stage is also evaluated. If you collaborated with others, it is worth reflecting on how effectively you worked as a team to solve the design problem, and then recommend strategies for improvement when you next undertake a co-designed solution to a problem.

This stage is also an opportunity for you to reflect on your growth as a designer of food products and identify your strengths in the design process.

Testing knowledge 1.2

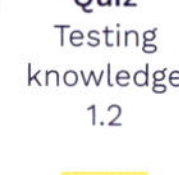

11 When creating new products, explain why the design brief is an important part of the process.

12 Outline the types of information that food manufacturers are likely to investigate before they begin to design new food products.

13 Explain why it is important for the designer to develop design criteria.

14 Discuss strategies the designer could use to share their design ideas with others.

15 Explain how the designer determines the final design option they will take into the producing and implementing stage.

16 When planning and managing the production of a new food product, what are the key resources the designer should consider?

17 Create a mind map of the type of information a production plan should contain.

18 Why is it critical that the designer uses their established design criteria when evaluating their final designed solution?

19 What is the purpose of evaluating the sensory properties of the food in the designed solution?

20 Explain why it is important to reflect on the effectiveness of working together as a team when creating new food products.

9780170495714

Critical and creative thinking 1.1

Compare the sensory properties of two popular pasta dishes

Describe in detail the sensory properties of both spaghetti bolognese and lasagna using a table like the one below. In your descriptions, consider all elements of the food product; for example, spaghetti bolognese has pasta and a meat sauce, while lasagna has pasta, a meat sauce and a cheese sauce.

1 Discuss the similarities and differences of the sensory properties of spaghetti bolognese and lasagna.

2 Identify the food product that you like most and justify your answer.

Template Sensory wheel

Sensory properties	Spaghetti bolognese (Jaded Art/Shutterstock.com)	Lasagna (ndsign/Shutterstock.com)
Appearance		
Aroma		
Flavour		
Texture		
Sound		

DESIGN ACTIVITY 1.1

Asian-style meatballs to share for lunch

Meatballs appear throughout the world as a lunchtime favourite. They can be skewered and grilled, pan-fried, baked in the oven or cooked in a sauce. Meatballs can be wrapped in a lettuce cup or rice paper, used to fill a baguette such as in bành mi, served on top of a bowl of rice noodles or tossed through a green leaf and herb salad.

Design brief

You are hosting a summer lunch for friends during the school holidays. Design a casual, Asian-style, shared lunch featuring meatballs that is set out on the table so everyone can help themselves and create their own meal. Meatballs work well when served in a wrap, in a baguette or on top of a bowl of noodles. The other elements of your lunch should include fresh green vegetables, tasty herbs, and a sauce that packs a flavour punch to finish off the selection.

Figure 1.12 Bánh mi

iStock.com/JoeLeeFoto

Investigating and defining

1 Based on the solution requirements and constraints in the design brief, develop five design criteria to evaluate the success of the shared lunch featuring meatballs.

2 Use the internet and recipes from home cooks to research the types of ingredients for the meatballs and the accompanying items – wrap/baguette/noodles, fresh green vegetables, tasty herbs and a flavoursome sauce – that all complement the other elements of the lunch.

Worksheet Design activity 1.1

	Idea 1	Idea 2	Idea 3	Idea 4
Recipe and source				
Type of meatballs				
Wrap/baguette/noodles				
Fresh green vegetables				
Flavoursome sauce				
Thinking notes on why the idea was selected				

3 Make the recipe for Pork Meatball Bành Mi on page 20 if you have not tried this type of product before.

4 Record four recipes that appeal to you and meet the needs of the design brief in a table like the one above. Remember to include the source of the recipes.

Generating and designing

1 Create two designed solution options for your meatball lunch to share with friends and annotate the ideas for ingredients in diagrams like the ones below.

2 Select the designed solution that you would prefer and justify your choice. Explain why you did not select the other option.

3 Write up new recipes for all the elements of the lunch ready for production.

Planning and managing

1 Prepare a food order.

2 Annotate a copy of your recipes for the lunch to indicate the tools and equipment needed in production and to highlight key safe work practices.

Producing and implementing

1 Produce all the elements of your Asian-style meatball lunch to share.

2 Take a photo after plating the meal and before taste testing. Include it with your evaluation.

Evaluating

1 Respond to the design criteria you developed.

2 Describe the sensory properties – appearance, aroma, flavour, texture and sound – of your Asian-style meatball lunch.

3 Discuss the effectiveness of your time management for this production.

4 What recommendations would you make to improve the quality of the lunch, if you were to make it again?

5 Classify the ingredients of your Asian-style meatball lunch on a diagram of the *Australian Guide to Healthy Eating*. Comment on how well the meal meets the recommendations of this food selection model.

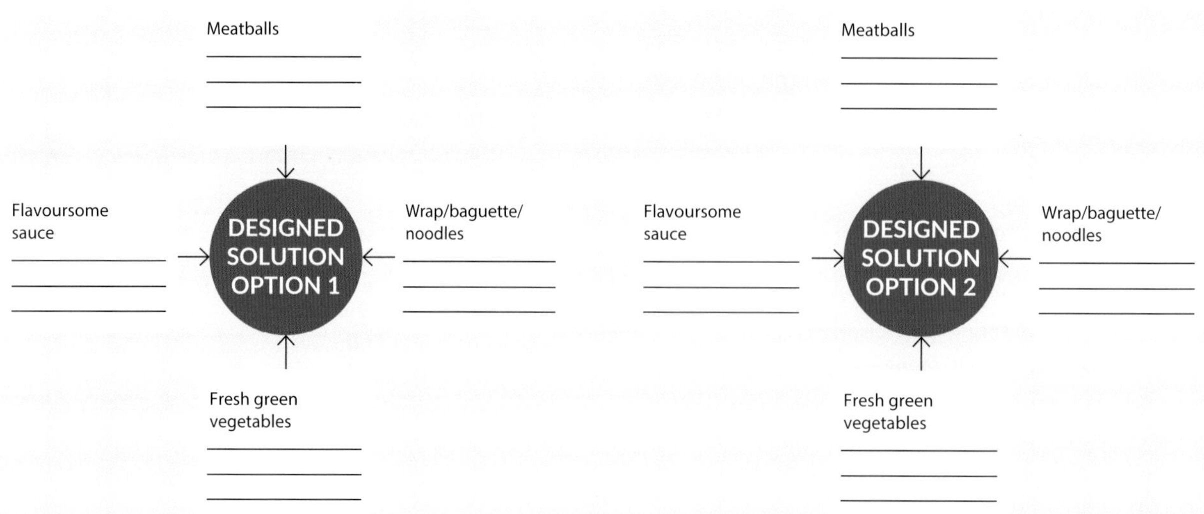

9780170495714

Open Fish Finger Sandwich with Quick Red Onion Pickle

INGREDIENTS

Quick Red Onion Pickle

1 small red onion

⅓ cup white vinegar

1 tablespoon caster sugar

2 teaspoons salt

½ teaspoon fennel or coriander seeds

¼ teaspoon black peppercorns

Dressing

4 cornichons or 2 gherkins, finely diced

¼ cup Greek yoghurt

1 teaspoon lemon juice

pinch salt and pepper

Open Fish Finger Sandwich

6 fish fingers

4 cherry tomatoes or 1 tomato

2 slices multigrain bread (toast thickness)

2 small cos lettuce leaves

SERVES ONE

Mark Fergus Photography

METHOD

Quick Red Onion Pickle

1. Peel the onion and cut in half lengthways. Place the flat side of the onion on a chopping board and slice thinly into half-moons.
2. Place the sliced onion in a medium bowl.
3. Place the vinegar, caster sugar, salt, seeds and black peppercorns in a small saucepan. Bring to boil over high heat. Pour the hot liquid over the onion slices.
4. Allow to cool for 15 minutes on the bench.
5. Store in a sterilised glass jar in the refrigerator.

Dressing

Prepare the dressing by combining all the ingredients in a small bowl. Taste test to check the seasoning. Set aside.

Open Fish Finger Sandwich

1. Preheat oven to 220°C.
2. Line a baking tray with baking paper. Lay out the fish fingers on the lined baking tray in a single layer, allowing space between them.
3. Bake for approximately 18 minutes. Gently turn the fish fingers over halfway through the cooking time. They will be ready when they are golden brown and crisp to the touch.
4. Cut the cherry tomatoes in half or slice the tomato thinly, using a serrated knife.
5. When the fish fingers are nearly cooked, lightly toast the bread and place it on a serving plate.
6. Spread the dressing onto the toast, then arrange the fish fingers, tomatoes and lettuce on top.
7. Top with some drained Quick Red Onion Pickle and serve.

EVALUATION

1. Describe the sensory properties – appearance, aroma, flavour, texture and sound – of your Open Fish Finger Sandwich with Quick Red Onion Pickle.
2. Describe the physical changes to the onion slices as they soak in the pickling solution. Explain why these changes occur in the physical properties of the onion.
3. What are the health benefits of baking the fish fingers rather than frying them?
4. Why is a serrated knife the most effective tool to slice fresh tomatoes?
5. Classify the ingredients of the Open Fish Finger Sandwich with Quick Red Onion Pickle on a diagram of the *Australian Guide to Healthy Eating* and discuss how well it meets the recommendations of this food selection model.

Salad Roll-up

INGREDIENTS

1 egg, or 1 slice ham, or 1 slice turkey loaf

1 teaspoon mayonnaise (optional)

1 large lettuce leaf

½ tomato

¼ cucumber

¼ carrot

30 grams cheese

1 pita bread or 2 slices mountain bread

2 tablespoons dip, such as hummus or tzatziki

SERVES ONE

METHOD

1 Hard-boil the egg by placing it in a small saucepan with enough warm water to just cover it. Bring to the boil, then simmer for 8 minutes. Shell and mash the egg, with mayonnaise if desired.
2 Finely shred the lettuce and slice the tomato and cucumber very thinly.
3 Grate the carrot and cheese.
4 If using pita bread, split it into halves, so that you are left with two thin rounds of bread.
5 Spread the bread with the dip.
6 Place half of the filling ingredients over each slice of bread.
7 Roll up the bread tightly in plastic wrap. Allow to rest in the refrigerator for 10–15 minutes so that the roll-up will hold its shape when cut.
8 Cut into serving portions.

EVALUATION

1 Explain why it is important to finely shred the lettuce and slice the tomato and cucumber very thinly in step 2.
2 Describe how you would safely use the grater.
3 List three other ingredients you could use to spread over the bread instead of hummus or tzatziki.
4 Make a list of other ingredients you could use in the Salad Roll-up filling.
5 Write a paragraph to explain why the Salad Roll-up is a healthy snack or lunch food.

Mark Fergus Photography

9780170495714

Crustless Quiches

INGREDIENTS

2 slices wholemeal bread (preferably not fresh)

2 rashers bacon, diced

½ onion, grated

1 small clove garlic, crushed

1½ cups vegetables, such as sweet potato, carrot, pumpkin, zucchini, capsicum, celery or sweet corn

2 eggs

⅓ cup milk

¼ cup self-raising flour

½ cup (60 grams) cheese, grated

2 tablespoons oil

pepper

SERVES TWO

METHOD

1. Preheat oven to 180°C.
2. Grease two small ovenproof dishes or foil takeaway containers.
3. Use a food processor to process the bread into breadcrumbs.
4. Place the breadcrumbs in the bottom of the ovenproof dishes or takeaway containers.
5. Heat the oil in a small frying pan.
6. Place the bacon, onion and garlic in the frying pan and sauté gently until softened and lightly browned. Cool.
7. Grate or finely dice the vegetables into a medium bowl.
8. Add the sautéed bacon, onion and garlic. Add the eggs, milk, flour, grated cheese, oil and pepper. Stir until well combined.
9. Pour the mixture over the breadcrumbs in the ovenproof dishes or takeaway containers.
10. Place the ovenproof dishes or takeaway containers on a baking tray.
11. Bake in the preheated oven for approximately 20–25 minutes or until the quiches are set.

Mark Fergus Photography

EVALUATION

1. Refer to the words for describing food on page 4. Use some of these words to help you describe the sensory properties – appearance, aroma, flavour, texture and sound – of your Crustless Quiches.
2. If you were to make this recipe again, would you change it in any way to alter the flavour?
3. Make a list of the health and safety issues you needed to consider in the production of your Crustless Quiches.
4. Do you think this recipe is high in dietary fibre? What ingredients are included in this recipe that increase the dietary fibre content of this dish?
5. Classify the ingredients for the Crustless Quiches on a diagram of the *Australian Guide to Healthy Eating*. Explain how well this recipe meets the recommendations of this food selection model.

Pork Meatball Bành Mi

INGREDIENTS

Pork Meatballs

250 grams minced pork

½ small onion, grated

½ cup (50 grams) cooked rice

1 clove garlic, crushed

2 centimetres fresh ginger, finely grated

2 tablespoons coriander leaves, finely chopped

1 tablespoon mint leaves, finely chopped

1 tablespoon fish sauce

pinch salt and black pepper

1 tablespoon oil

spray olive oil

Nuoc Cham Sauce

1 tablespoon fish sauce

1 tablespoon caster sugar

1 tablespoon rice wine vinegar

1 tablespoon lime juice

¼ long red chilli, finely chopped

1 garlic clove, finely chopped

To serve

½ Lebanese cucumber

2 white long rolls or 1 baguette

1 tablespoon Japanese or whole egg mayonnaise

½ small carrot, finely julienned

shredded iceberg lettuce

coriander leaves, picked

⅓ long red chilli, finely sliced

METHOD

Pork Meatballs

1. Add the minced pork, onion, rice, garlic, ginger, coriander, mint and fish sauce to a large mixing bowl. Season with salt and pepper.
2. Mix the ingredients together for at least 3 minutes to allow the protein in the pork to bind the ingredients together.
3. Roll a heaped tablespoon of pork mixture into a ball and place on a plate. Repeat with the remaining mixture to make 10 meatballs. (Hint: wet your hands lightly before rolling the meatballs as the mixture is sticky.) Refrigerate the meatballs for 10 minutes.
4. Heat 1 tablespoon oil in a frying pan and cook the meatballs over a medium heat for approximately 10 minutes. Shake the pan occasionally as they are cooking until they are well browned and cooked through.
5. Drain on paper towel.

Nuoc Cham Sauce

Mix all of the ingredients together and stir well to dissolve the sugar. Set aside for 10 minutes.

To serve

1. Use a vegetable peeler to slice the cucumber lengthwise into ribbons.
2. Cut the rolls or baguette in half lengthwise – almost, but not quite, all the way through.
3. Spread both sides with a little mayonnaise.
4. On half of the bread, layer the carrot, cucumber, meatballs and lettuce. Add the coriander and red chilli.
5. Drizzle with a little Nuoc Cham Sauce or serve the Nuoc Cham Sauce separately.
6. Top with the other half of the bread. If using a baguette, cut it into two portions to serve.
7. Wrap in baking paper if serving the Pork Meatball Bành Mi as a takeaway.

9780170495714

EVALUATION

1 Describe the sensory properties – appearance, aroma, flavour, texture and sound – of your Pork Meatball Bành Mi.

2 Create a flow chart to demonstrate how to chop herbs safely.

3 Identify the process that occurred to the physical properties of the protein in the pork when it was mixed for 3 minutes in step 2. Explain how this impacted on the physical properties of the raw meat.

4 When the meatballs are fried, they develop a golden brown colour. Identify the process that creates the colour change in the meatballs.

5 Classify the ingredients for the Pork Meatball Bành Mi on a diagram of the *Australian Guide to Healthy Eating*. Using this data, write a paragraph to explain whether you think this recipe would make a healthy lunch.

Mark Fergus Photography

Chicken Souvlaki

INGREDIENTS

125 grams chicken thigh fillet

2 teaspoons olive oil (for frying)

Souvlaki Marinade

1 tablespoon lemon juice

2 teaspoons soy sauce

1 clove garlic, crushed

2 teaspoons olive oil

Tzatziki Dip

¼ cup plain yoghurt

1 clove garlic, crushed

½ Lebanese cucumber, coarsely grated and drained

½ teaspoon lemon rind, finely grated

To serve

1 pita or naan bread – you could use commercial bread or make your own using the recipes on page 242

2 lettuce leaves, finely shredded

SERVES ONE

METHOD

1 Preheat the oven to 100°C.
2 Combine the ingredients for the Souvlaki Marinade in a medium bowl.
3 Add the chicken thigh fillet and allow to marinate for 15 minutes.
4 Prepare the Tzatziki Dip by combining all the ingredients and mixing well.
5 Chill the dip for 15 minutes to allow the flavours to develop.
6 Warm the pita or naan bread in the preheated oven for 5–10 minutes. Remove and wrap in a clean tea towel to keep warm.
7 Heat the olive oil (for frying) in a frying pan over a moderate heat.
8 Remove the chicken from the marinade and discard the remaining marinade.
9 Place the chicken in the hot frying pan and brown lightly. Turn and brown on the other side. Fry the chicken until golden brown and cooked through.

To assemble the dish

1 Slice the cooked chicken thinly.
2 Place the lettuce in the centre of the warmed pita or naan bread, and top with the sliced chicken. Spoon some of the Tzatziki Dip over the chicken.
3 Roll the pita or naan bread over to enclose the filling.
4 Serve with more Tzatziki Dip.

Mark Fergus Photography

EVALUATION

1 Describe the sensory properties – appearance, aroma, flavour, texture and sound – of your Chicken Souvlaki.
2 Outline two health and safety steps that you followed to prevent cross-contamination when cooking with raw chicken.
3 Identify the role of the lemon juice, soy sauce and oil in the Souvlaki Marinade.
4 What rules are important to remember when heating and cooking with oil?
5 Classify the ingredients for the Chicken Souvlaki on a diagram of the *Australian Guide to Healthy Eating*. Explain whether you think this recipe would be suitable to serve as a healthy evening meal.

9780170495714

Pancakes

INGREDIENTS

¾ cup self-raising flour
2 teaspoons caster sugar
1 egg, lightly beaten
½ cup milk
extra milk (if required)
10 grams butter

METHOD

1. Sift flour and sugar into a large bowl.
2. Make a well in the middle. Pour in the egg and half of the milk. Stir.
3. Add the remainder of the milk and mix into a lump-free batter.
4. Allow the batter to rest for 10–15 minutes. Adjust the consistency with extra milk if required.
5. Heat a frying pan. With a piece of paper towel, grease lightly with the butter.
6. Spoon the batter into the hot pan or use a ¼ cup measure. Allow space to turn the pancakes.
7. When bubbles appear on each pancake's surface, turn it over.
8. Stack the cooked pancakes on a wire rack, inside a tea towel, to cool.

EVALUATION

1. Describe the sensory properties – appearance, aroma, flavour, texture and sound – of your Pancakes.
2. Why is it important to add only half of the milk with the egg in step 2 of the recipe?
3. Explain the importance of allowing the batter to rest for 10–15 minutes before making it into pancakes.
4. Briefly explain how you would use the frying pan safely.
5. Do you think that the Pancakes would be suitable to include as part of a healthy diet? Justify your answer.

Mark Fergus Photography

PERSONAL HYGIENE IN THE KITCHEN

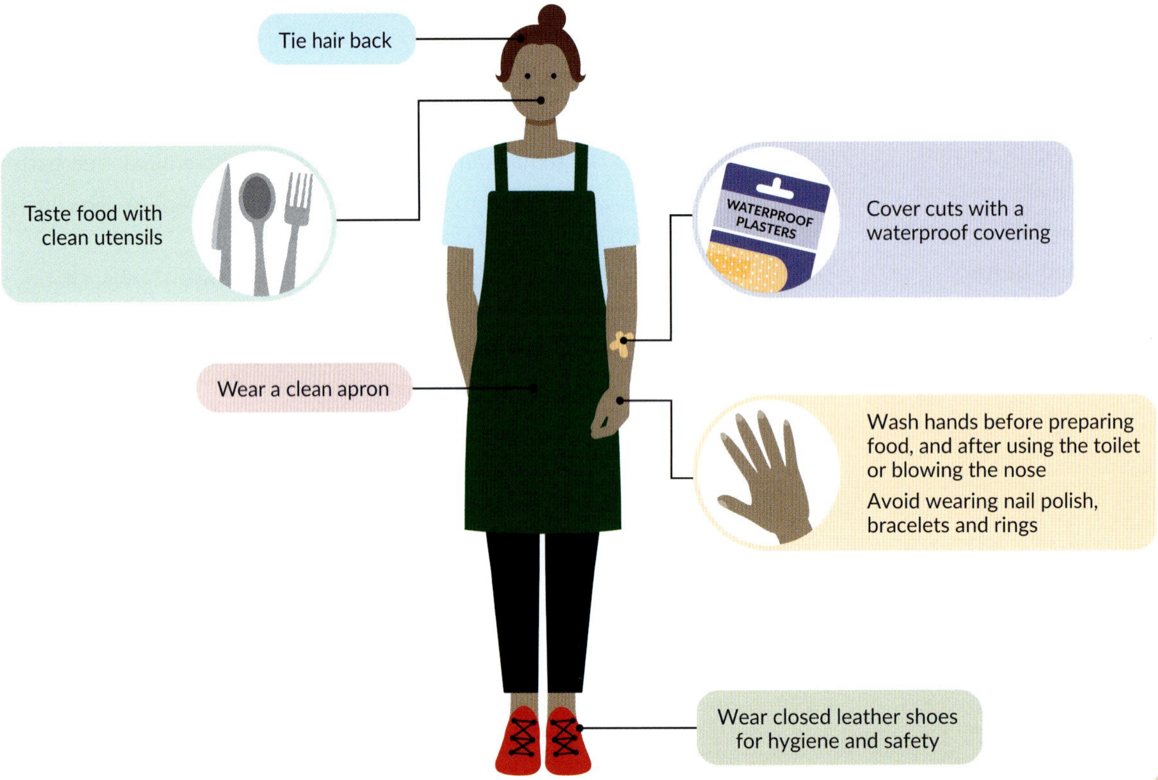

Tie hair back

Taste food with clean utensils

Cover cuts with a waterproof covering

WATERPROOF PLASTERS

Wear a clean apron

Wash hands before preparing food, and after using the toilet or blowing the nose

Avoid wearing nail polish, bracelets and rings

Wear closed leather shoes for hygiene and safety

USING SMALL APPLIANCES SAFELY

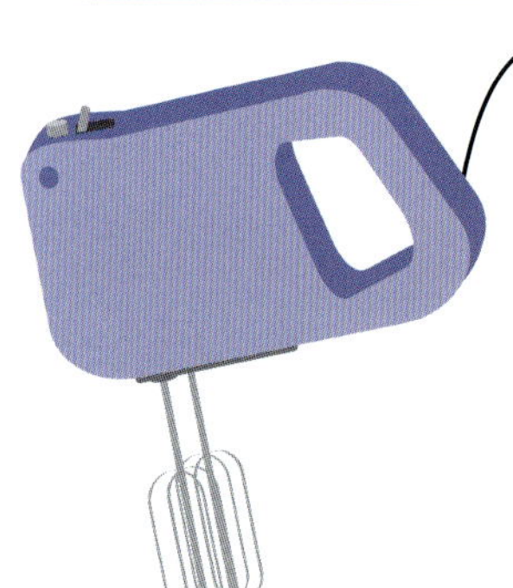

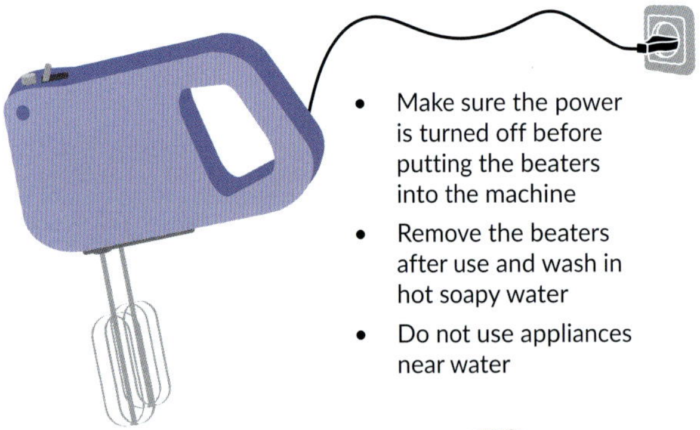

- Make sure the power is turned off before putting the beaters into the machine
- Remove the beaters after use and wash in hot soapy water
- Do not use appliances near water

- Check the cord is not damaged
- Use wooden tongs to remove toast that doesn't pop up
- Turn appliances off when not in use

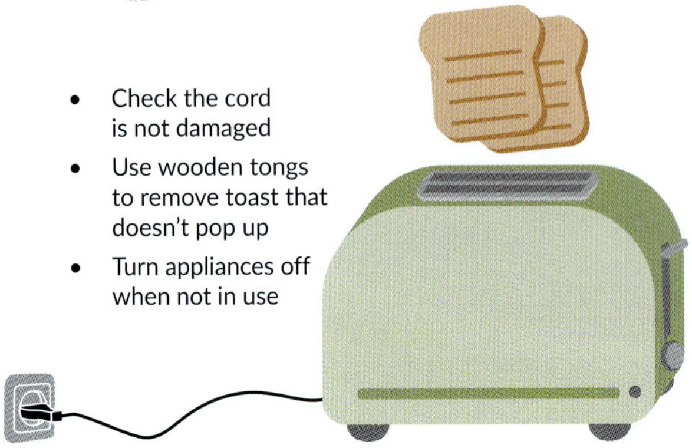

USING KNIVES SAFELY

- Use the 'claw' position when cutting
- Don't leave knives in the sink
- Always cut food on a chopping board made from wood or polyethylene
- Make sure knives are away from the edge of the bench

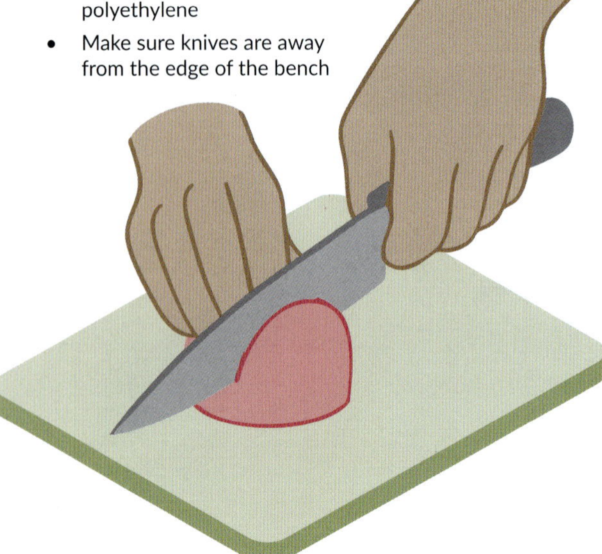

9780170495714

STEPS TO PREVENT FOOD POISONING

1. STORING

Separate raw and cooked foods; refrigerate food below 5°C; keep food out of the danger zone (between 5°C and 60°C)

2. PREPARING

Wash hands and clean chopping board after cutting raw meat

3. COOKING

Cook most foods to at least 75°C; cook poultry until juices run clear

4. SHOPPING

Transport high-risk foods such as raw chicken in an insulated bag

2 PREPARING FOOD SAFELY

KEY TERMS

cross-contamination the transfer of harmful bacteria from uncooked food to food that has been cooked or prepared

danger zone the temperature – between 5°C and 60°C – at which bacteria can multiply very quickly

electric oven an oven that uses radiant and convection heat produced by electricity to cook food

fire blanket an insulated blanket used to extinguish small fires in the kitchen

food poisoning an illness caused by eating food that has been contaminated with harmful bacteria

food spoilage the deterioration in the physical, sensory and chemical properties of food over time; however, the food will not usually be harmful to eat

gas oven an oven that uses radiant and convection heat produced by gas to cook food

small appliances pieces of equipment such as toasters, food processors, hand-held beaters or blenders

Worksheets

- Practical activity 2.1
- Design activity 2.1

Quizzes

- Testing knowledge 2.1
- Testing knowledge 2.2
- Testing knowledge 2.3

To access resources above, visit **cengage.com.au/nelsonmindtap**

2

Safety in the kitchen

Just like the home kitchen, the school kitchen is usually a very busy place. Whether you are preparing food at home or at school, you use a range of tools and equipment that – if not handled with care – have the potential to cause accidents. Like many workspaces, kitchens can be dangerous places because of the equipment used in them, as well as the conditions created by the use of heat and liquids.

Activity 2.1

Risk areas in the kitchen

1 With a partner, make a list of the main risk areas you can see in your school kitchen.

2 Draw up a risk and safety table for your school kitchen, using the one below as an example. Suggest a simple safety rule that should be followed to avoid accidents occurring in each risk area that you identify.

Risk area	Safety rule
Floor	Wipe up all spills immediately.

3 Using the information from your risk and safety table, design and produce a safety poster for one of the risk areas. Use a digital device to develop the poster.
   - Your poster should be bright and eye-catching.
   - Make sure that the appropriate safety rule for the risk area is highlighted on the poster.
   - Develop a symbol for use on your safety poster to highlight the risk area.

4 As a group, look at all the safety posters produced by the class.
   - Discuss the key features of each poster and the way each highlights a specific risk area.
   - Does the safety symbol on each poster stand out?
   - How could these symbols be used in other areas of the school?

5 Make a digital display or print the posters and display them in your school kitchen, near the relevant risk areas, so that they can be part of an ongoing health and safety campaign at school.

Fire safety

The possibility of fire is a major safety concern in the kitchen. The most common cause of kitchen fires is the oil or fat in a piece of equipment – such as a frying pan or wok – catching alight.

If a fat fire occurs in the kitchen, it is important to smother the fire. This will stop oxygen from feeding the fire. To do this, you can use a **fire blanket**, which is extremely effective in extinguishing small kitchen fires.

Following these simple rules is important when installing and using a fire blanket:

- The fire blanket should be attached to the kitchen wall just above waist height, so that it is easy to access.
- Pull down firmly on the tabs of the fire blanket to remove it from its cover.
- If possible, gently place the fire blanket over the fire. Throwing it on the fire may spread the fire further.
- Turn off the gas or electricity heat source under the fire.
- Leave the fire blanket over the source of the fire until it has been extinguished and the equipment that contained the fire has cooled.
- A fire blanket can also be used to extinguish fire on a person's clothing. Place the blanket over the person and wrap it around them; the person should then roll on the ground to extinguish the fire as soon as possible.
- Replace a fire blanket after it has been used. You should never reuse a fire blanket.

In addition, small fire extinguishers can be purchased for the kitchen. These utilise a dry chemical powder and can be used to extinguish a fat fire, which, as noted above, is the most common fire to occur in a kitchen.

If you do not have a fire blanket or a fire extinguisher, the best method of putting out a fat fire is to cover it with a large saucepan lid, or pour flour or sand onto it. You should never pour water on a fat fire, as this will only cause the fire to spread.

Weblink How to use a fire blanket

Mark Fergus Photography

Figure 2.1 A fire blanket and a fire extinguisher

9780170495714

Activity 2.2

Fire safety in the kitchen

1 Check to see if your school kitchen has any fire safety equipment, such as a fire blanket or a fire extinguisher. If it does:
   a Where are they located?
   b Are the instructions on how to use them easy to read?
2 Ask your teacher to demonstrate how you should use a fire blanket.
3 Look at the fire evacuation plan for your classroom. Is it clear and easy to follow?
4 Write a short article for your school bulletin or website about the effectiveness of your current school fire evacuation drill.

Personal hygiene

Preparing food hygienically to avoid food contamination is just as important as taking care when working with equipment in the kitchen. Bacteria that can cause **food poisoning** thrive on the warmth and moisture that the human body produces:

- Bacteria live on and in all parts of the body – the hands, fingernails, skin, hair, nose, mouth and even ears all provide a wonderful home for bacteria to thrive.
- The clothes you wear are contaminated by the bacteria that naturally occur in the environment. When you sit down at home or when you are out, bacteria transfer from the environment onto your clothes.
- Bacteria are found in all sorts of places, including on the handles of school bags, on lockers, on the handrails of buses, trains and trams, and on desks, tables and chairs.

Following the personal hygiene rules below helps to ensure that the food you prepare is safe to eat.

Personal hygiene rules

1 Thoroughly wash hands with soap and water before preparing food. This prevents disease-carrying bacteria from transferring to the food.
2 Make sure that you wash your hands after using the toilet. Harmful bacteria live in faeces, and they can be passed onto toilet paper and hands when you use the toilet.
3 Make sure you wash your hands after you blow your nose. Bacteria can pass onto your hands from the tissue or handkerchief.

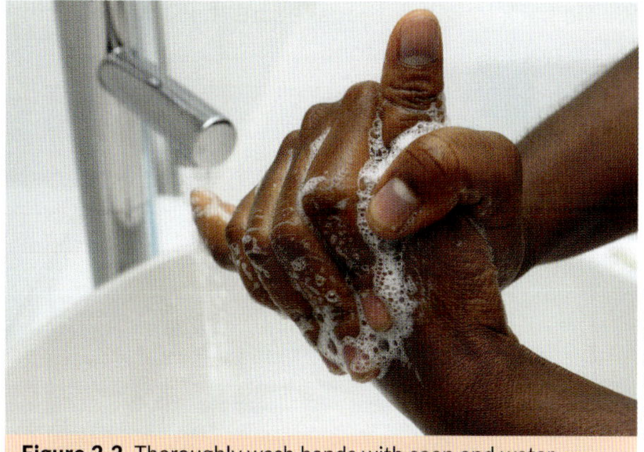

Figure 2.2 Thoroughly wash hands with soap and water.

lostinbids/E+/Getty Images

4 Do not sneeze or cough over food. Bacteria can easily be transferred through the air.
5 Do not wear nail polish or false nails while you are preparing food. Chips of nail polish or the false nails themselves can fall into the food.

Figure 2.3 Finger nails should be clean and without nail polish.

iStockphoto.com/solidcolours

6 Take off rings and bracelets before you begin to prepare food. These can be hiding places for bacteria.
7 Cover cuts with a clean, waterproof covering. If you cut yourself while preparing food, remember to sanitise the cut with disinfectant before you cover it.
8 Wear a clean apron to cover your street clothes. Clothes pick up dirt and dust, and can therefore transfer bacteria onto food.

Figure 2.4 Wear a clean apron to cover your street clothes.

SolStock/E+/Getty Images

9 Tie back long hair to make sure that any loose strands do not fall into the food.

Figure 2.5 Tie back long hair.

iStockphoto.com/Kiwis

10 Don't 'double dip'. Always use a clean spoon to taste food. Bacteria that live in your mouth can easily be passed onto the food from a used spoon.

Figure 2.6 Always use a clean spoon to taste food.

Mixetto/E+/Getty Images

Kitchen hygiene

Working in a clean environment and using hygienic practices in the kitchen are equally as important as following the rules for personal hygiene.

Kitchen hygiene rules

1 Make sure that you use clean tea towels when drying the dishes. Tea towels should not be used to wipe up spills from the floor, flung over your shoulder while you are cooking or used for drying hands.

2 Like tea towels, dishcloths must be kept clean. Replace your dishcloth regularly, or soak and sanitise it if it is reusable.

3 Keep the food preparation area clean and tidy while you work. Clean up the work area after each process of production by stacking and washing the dishes you have just used and wiping over the bench area.

Figure 2.7 Sanitise the bench after food preparation.

iStock.com/undrey

Washing-up techniques

Thoroughly washing and drying dishes is another important aspect of kitchen hygiene. Today, many families wash most of their dishes in a dishwasher and hand-wash only a few items, such as saucepans.

Tables 2.1 and 2.2 summarise some key points about using dishwashers and hand-washing dishes.

Table 2.1 Advantages of using a dishwasher versus hand-washing dishes

Advantages of washing dishes in a dishwasher	Advantages of washing dishes by hand
Alena Matrosova/Shutterstock.com	iStock.com/SolStock
Dishwashers can save the family time.	You can use an eco-friendly detergent.
Dishwashers dry the dishes at a very high temperature, and therefore thoroughly disinfect the dishes.	Hand-washing dishes is often quicker than using a dishwasher, which can have a very long washing and drying cycle.
Dishwashers are very convenient and allow you to wash a large number of dishes at once.	You can wash the dishes as you use them, and therefore you may not need as much crockery and glassware.
Using a dishwasher once a day can be more water-efficient than washing dishes by hand after every meal.	Hand-washing dishes with family members provides opportunities for family conversation.

Table 2.2 Steps for using a dishwasher and hand-washing dishes

Steps for washing dishes in a dishwasher	Steps for washing dishes by hand
1 Scrape the food scraps from the dishes and rinse the dishes well.	1 Scrape the food scraps from the dishes. Rinse the dishes and carefully stack them into piles of similar kinds.
2 Stack the cutlery container so that the eating surface of the knives, forks and spoons are facing up.	2 Fill the sink with hot, soapy water. Grease will not come off in cold or lukewarm water.
3 Make sure glasses are carefully stacked and unlikely to tip over during the washing cycle.	3 Wash the glassware first, and then rinse it in hot water. Allow to drain and air-dry, if possible.
4 Place heavy items on the lower shelf. Stack all plastic items so that they cannot flip over and fill with water.	4 Cutlery should be washed after the glassware. Rinse carefully. Remember not to put sharp knives into a sink filled with soapy water, as they may become hidden from view.
5 Do not over-stack the dishwasher – if you do, some items may not be properly washed.	5 Next, wash all the crockery, and then rinse in hot water.
6 Regularly clean the filter. Secure the detergent's childproof lock and store the detergent away from children.	6 Finally, wash the mixing bowls and saucepans. Rinse well.

Testing knowledge 2.1

1. Describe the correct procedure for using a fire blanket.
2. Outline the best procedure for extinguishing a fire in a frying pan or wok if you do not have a fire blanket.
3. Explain why it is important to wash your hands before preparing food.
4. Why should jewellery be removed before preparing food?
5. At school, before production begins, everyone puts on a clean apron. Why?
6. Discuss why 'double-dipping' is considered to be a health hazard.
7. Explain one environmental advantage of using a dishwasher to wash dishes and one environmental advantage of washing dishes by hand.
8. List what you think are the three most important rules when using a dishwasher.
9. Explain why a dishwasher may be considered a hazard for children.
10. Explain why you should wash glassware first when washing dishes by hand.

Food poisoning

Food poisoning is an illness caused by eating food that has been contaminated with harmful bacteria. It can occur when bacteria, which are invisible to the naked eye, are transferred onto food – often unknowingly – because of poor personal hygiene or poor food handling. One of the most common causes of food poisoning is the **cross-contamination** of food. This occurs when harmful bacteria from uncooked food are transferred to food that has been cooked or prepared. For example, if raw chicken was cut on a chopping board, and then salad ingredients were cut on the same board without it being washed, harmful bacteria could be transferred. Washing the chopping board in hot, soapy water between cutting the raw chicken and cutting the vegetables will prevent the transfer of bacteria.

Food spoilage, on the other hand, is the deterioration in the physical, sensory and chemical properties of food over time. Food that has 'spoiled' may be attacked by yeast, moulds or the natural enzymes present in the food, such as when a cut apple begins to brown. However, while the appearance, aroma, flavour or texture of the food might have deteriorated during storage, spoiled food will not usually be harmful to eat.

To minimise the risk of food poisoning it is very important to keep hot food hot (above 60°C) and cold food cold (below 5°C) so that it is out of the temperature range in which bacteria thrive. Between 5°C and 60°C is described as the **danger zone** because in this temperature range, one bacterium can multiply into approximately 17 million bacteria within eight hours. Bacteria that cause food poisoning also need a moist environment, time to grow, a food supply and a low-acid environment. Many bacteria also need oxygen.

It is important to note that poisoned food may look, smell and taste just like normal food. Symptoms of food poisoning can appcar almost as soon as food is eaten (that is, within one hour) or take up to 36 hours to develop. Food poisoning symptoms can include diarrhoea, stomach cramps and vomiting. In severe cases, food poisoning can cause death.

Figure 2.8 The conditions bacteria need for growth

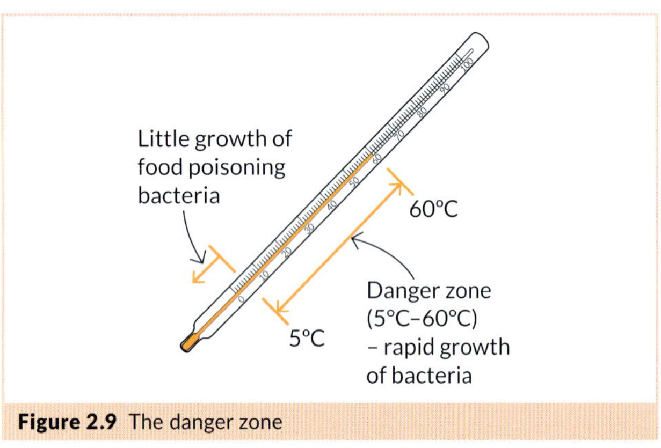

Figure 2.9 The danger zone

Preventing food poisoning

When working with food, it is vital to do all you can to prevent food from being contaminated in the first place, therefore reducing the risk of bacteria in the food growing and multiplying. When storing, preparing, cooking and shopping for food, there are some strategies you should follow that can minimise the risk of food poisoning.

When storing food

- Keep raw food separated from cooked food, and store raw food at the bottom of the refrigerator to avoid juices dripping onto and contaminating other food.
- Check that the temperature of the refrigerator is below 5°C.
- Allow cooked foods to cool to room temperature (about 21°C) before storing in the refrigerator. This should not take more than two hours. Note that cooling will occur more quickly if you put the hot food into a number of smaller containers rather than leaving it in one large one. This prevents the refrigerator temperature from rising, and reduces the risk of bacterial growth in all food stored in the refrigerator.

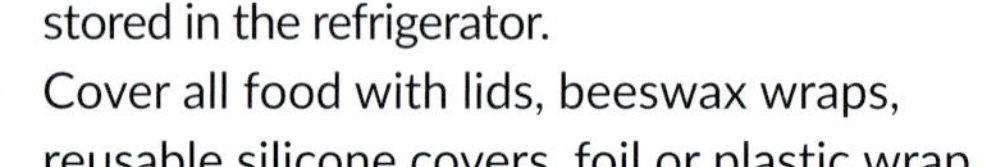

- Cover all food with lids, beeswax wraps, reusable silicone covers, foil or plastic wrap.
- Do not store food in opened tin cans.

Weblinks
Four Golden Rules of food safety
Avoid food poisoning – Shop Safe, Store Safe, Cook Safe, Eat Safe

When preparing food

- Always wash your hands with soap in warm water before touching food.
- Do not cut salad ingredients on the same chopping board as raw meat without first washing the board in hot, soapy water. This rule also applies for chopping boards that are used to cut raw meat before being used for cooked meat. Thoroughly cleaning the board between uses reduces the chances of cross-contamination.

When cooking food

- Most food should be cooked to a temperature of at least 75°C.
- If you do not have a cooking thermometer, make sure that you cook poultry until the meat is white, particularly near the bone. When cooking hamburgers, mince and

sausages, the juices will run clear when they are ready. White fish will easily flake apart with a fork when cooked.

When shopping for food

- Keep potentially high-risk foods outside the temperature danger zone.
- Keep hot foods and cold foods separate and buy them at the end of your shopping trip.
- Always check labels and do not buy food that is past its use-by or best-before date.
- Avoid buying food that is in swollen, dented, leaking or damaged cans, containers or other packaging.
- Check that serving staff use separate tongs when handling separate food types, such as meats and vegetables.
- Take your shopping home quickly and store it immediately.

Testing knowledge 2.2

Quiz Testing knowledge 2.2

11 Explain the meaning of the term 'food poisoning'.

12 List the causes of food poisoning.

13 What is the relationship between food poisoning and food spoilage?

14 What physical symptoms might a person with food poisoning display?

15 Explain the meaning of the term 'cross-contamination' and outline two strategies that can be used to avoid it.

16 Define the term 'danger zone' in the context of food safety.

17 Copy the diagram below and fill it in to highlight the conditions that bacteria need for growth.

18 Outline two strategies to follow to ensure that food is stored safely.

19 When cooking food, how can you tell if it is out of the danger zone?

20 List two safety strategies to follow when shopping for food.

Safe use of knives

Knives are among the most frequently used pieces of kitchen equipment, and are essential for cutting, slicing, dicing and peeling a wide range of foods. Although knives seem simple to use, it is important to use them safely to minimise the risk of injury.

- Select the most appropriate knife for the food you are preparing; for example, a cook's knife for large pieces of food or a vegetable knife for small pieces of fruit or vegetables.
- Keep your fingertips tucked under while using the 'spider' or 'claw' position.

Figure 2.10 Using the 'spider' position for cutting, slicing and dicing

- Make sure you keep knives sharp. Sharper knives are safer because they cut through the food more easily and require less pressure to be used.
- Never run your finger along the cutting edge of the knife to test its sharpness.
- Make sure the handle of the knife is clean and dry – not greasy – so that the knife does not slip.
- Always cut food on a chopping board made from wood or polyethylene – this will help to protect the sharp edge of the blade. Do not use knives on glass boards, metal or plates, as these materials will blunt the knives.
- Make sure knives are kept away from the edges of benches and out of reach of small children.

- When passing a knife to someone else, remember to pass the handle of the knife, not the blade.
- If you need to move around the kitchen with a knife, hold it close to the side of your body with the blade pointing down.
- Do not put knives in a sink filled with hot, soapy water, as they may become hidden from view.
- Store knives in a knife block or a wall-mounted magnetic rack, not in a drawer with other kitchen utensils.

Stoves, cooktops and ovens

Traditionally, a stove is a freestanding appliance that includes an oven and a cooktop.

ppart/Shutterstock.com

Figure 2.11 A stove with a cooktop that incorporates a wok burner and an oven griller

Cooktops

A cooktop can be used to cook food in saucepans, woks or frying pans, sitting on an electric hob, a gas burner or an induction hob. An electric or gas cooktop works by heating up the saucepan that then heats the food. An induction cooktop works by sending an electric current through a copper coil underneath the cooktop surface that creates a magnetic field and, if the saucepan is made of magnetic material – such as cast iron or some types of stainless steel – an electric current is produced in the saucepan. This causes electrons in the saucepan to move rapidly and generate heat to cook the food.

Douglas Sacha/Moment/Getty Images

Figure 2.12 An induction cooktop

Ovens

The oven is a very versatile appliance that is used to cook a wide variety of recipes. The dry heat of the oven can produce beautiful cakes, biscuits and scones, crisp roast potatoes, golden roast chicken and delicious pastries.

There are many types of oven, including freestanding ovens, ovens built into a wall or under a bench, and ovens with internal or external grillers. Multi-function electric ovens are useful because they allow the cook to use a combination of oven features; for example, the cook can regulate the heat on the top and bottom shelves, or use the griller function or fan to achieve the best results from their cooking. Combination ovens – which have convection heat and microwave options – have also become popular in some homes. A steamer oven is another option for home cooking.

Ovens are used to bake, roast, casserole and reheat food. Traditional ovens use radiant heat and convection heat to cook food, and include a thermostat that is used to measure the temperature in the oven. Convection ovens have a fan that moves the heat around the oven, using convection currents to ensure an even temperature throughout.

Cooking with electricity

An **electric oven** has a coil that heats when the oven is turned on. The switch or dial is turned to the required temperature setting, and an indicator – such as a buzzer or light – will show when the oven has reached the desired temperature. Most electric ovens take between five and 10 minutes to heat to the set temperature. Preheating the oven before cooking ensures that food does not dry out, and that cakes, scones, muffins and bread rise to maximum size during cooking, to achieve a light texture.

9780170495714

Cooking with gas

When using a **gas oven**, the oven must be lit according to the instructions, which are usually on the oven's doorplate. The oven should be set to the correct temperature, and when the fog has cleared from the glass, the oven is ready to use. Preheating a gas oven takes about 10 minutes.

Oven temperatures

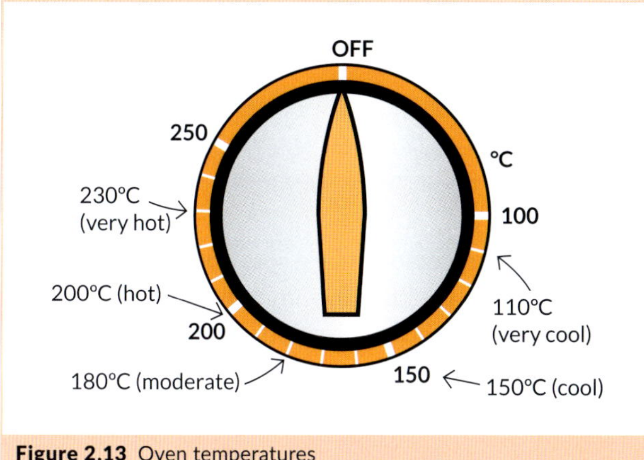

Figure 2.13 Oven temperatures

Small appliances

Small appliances are used in the preparation of many food products. Most households have some small appliances, such as a toaster, a food processor, hand-held beaters, an electric juicer, a blender, a stick mixer, a sandwich maker, an electric kettle, an electric wok, a crepe maker, an air fryer or a coffee maker.

Using small appliances safely

Most small appliances are electrically powered, and many have sharp components. Therefore, considerable care is needed when using, cleaning and storing them.

Food processors

Food processors are popular small appliances that come in many styles, mainly because they can be used in a wide variety of ways; for example: to chop, slice or shred vegetables; to puree soups; to make pastry; to prepare breadcrumbs; to make a quick-mix cake batter; or to blend ingredients. Different brands of food processors usually have slightly different features, so each one will operate in a slightly different way.

It is important to follow the manufacturer's safety instructions when using a food processor. It is particularly important to make sure that you use the food plunger that is supplied with the machine for pushing food through the feeding tube (see page 34), rather than a knife or your fingers. Food processors are designed so that they will not turn on unless the lid switch is in the safety position.

B Calkins/Shutterstock.com

Figure 2.14 A food processor

Hand-held beaters

Hand-held beaters are another useful small appliance. They are generally simple to use, since they are light to hold and have a series of speeds that can be easily adjusted, even while running. Hand-held beaters can be used to cream butter and sugar when making cakes or biscuits, to beat egg whites to a stiff foam for meringues, or to make batter for pancakes.

Observe the following safety precautions when using hand-held beaters:

- Always make sure that the power is turned off when putting the beaters into the machine.
- Make sure the beaters are securely pushed into the machine.
- Do not operate the beaters near water.
- Remember to turn off the power and remove the beaters from the machine before washing them. Wash the beaters in hot, soapy water. Thoroughly wipe the machine, wind up the cord and store it with the beaters in a clean, dry place.

George Dolgikh/Shutterstock.com

Figure 2.15 Hand-held beaters

Electric juicers

Electric juicers have grown in popularity since people have become more aware of the importance of fresh fruit and vegetables juices. Juicing one variety of fruit, such as apples or oranges, can make a beautifully refreshing drink for breakfast. Some people prefer to make an exotic 'fruit combo' by juicing a range of their favourite fruits, such as oranges, pineapple and mangoes. Equally delicious is vegetable juice made by combining a variety of vegetables, such as carrot, celery and capsicum. You can also combine fruits and vegetables.

When using an electric juicer, remember to follow the same safety procedures as you do with a food processor.

Proxima Studio/Shutterstock.com

Figure 2.16 An electric juicer

Rules for the safe use of small appliances

1 Carefully read through the instructions manual so that you understand how to correctly use each small appliance.

2 Do not use appliances near water or a stove.

View Stock/Getty Images

Figure 2.17 Ensure that you follow the rules for the safe use of small appliances.

3 Ensure that you have dry hands before plugging in, unplugging or operating appliances.

Antonio Guillem/Shutterstock.com

Figure 2.18 Ensure that you have dry hands before using appliances.

4 With food processors, make sure you use the plunger supplied with the appliance for pushing food through the feed tube – do not use your fingers, a knife or another utensil.

Mark Fergus Photography

Figure 2.19 Make sure you use the plunger supplied with the food processor.

5 Carefully wash the blades of a food processor after use. Do not put them in a sink filled with water.

6 An air fryer will get hot very quickly, so stay nearby while it is in use. Check that there is no smoke coming from the air fryer.

fcafotodigital/E+/Getty Images

Figure 2.20 Take care when using an air fryer.

9780170495714

7 Never try to remove toast from a toaster with a knife, fork or other metal utensil.
8 Make sure the appliance is unplugged before you begin to clean it.
9 Store appliances in a clean, dry place away from moisture and dust.
10 Check appliances frequently to make sure that the electrical cords do not become damaged. A frayed electrical cord makes an appliance unsafe to use.

PRACTICAL ACTIVITY 2.1

Making a Strawberry and Banana Smoothie

Worksheet
Practical activity 2.1

Aim

To use a small electrical appliance such as a stick mixer safely.

Ingredients

½ very ripe banana

4 large strawberries

½ cup milk, chilled

1 tablespoon vanilla yoghurt

1 teaspoon honey

Method

1 Peel and slice the banana.
2 Rinse and pat dry the strawberries. Remove the green leaves, then slice each strawberry.
3 Combine the sliced banana, sliced strawberries, milk, yoghurt and honey in a tall jug or measuring container. Blend the ingredients with a stick mixer until smooth.
4 Pour the smoothie into a glass and serve immediately.

Analysis

1 Why is it important to read the stick mixer's operating instructions or have them explained before using this appliance?
2 Explain why it is important to fit the cutting blade onto the stick mixer before plugging the appliance into the power point.
3 Why is it important to use a tall jug or measuring container to blend the smoothie?
4 Identify the risks of operating the stick mixer near water or submerging it in the washing up water when cleaning it.
5 After you have made the smoothie, describe the steps you will undergo when you are cleaning the stick mixer.

Mark Fergus Photography

Conclusion

After using the stick mixer, list three safety points that you could teach another person who is new to working in the kitchen about using this small appliance.

Testing knowledge 2.3

Quiz
Testing knowledge 2.3

21 Describe the best method of safely dicing an onion.
22 Explain why it is recommended to store knives in a knife block or a wall-mounted magnetic rack, rather than in a drawer with other kitchen utensils.
23 Outline the best way to safely wash a cook's knife.
24 Identify two safe work practices you should follow when cooking food in a saucepan on a cooktop.
25 Describe the steps involved in safely removing food from an oven once it is cooked.
26 List two important safety rules to observe when using the griller function.
27 Describe how to safely clean and store hand-held beaters.
28 The electric toaster is one of the most commonly used small appliances. Briefly explain three important rules for safely using a toaster.
29 Identify one of the most important safety rules to follow when using a food processor.
30 List two rules for safely using an electric juicer.

Critical and creative thinking 2.1

Food poisoning causes and prevention

Complete two summary frames:

- one focusing on the causes of food poisoning
- one focusing on strategies to prevent food poisoning.

Definition of food poisoning	Conditions for growth of bacteria
Cross-contamination	Danger zone

DRAW A PICTURE TO HELP YOU REMEMBER FACTS ABOUT THE CAUSES OF FOOD POISONING

Avoiding cross-contamination	Preparing food
Cooking food	Shopping for and storing food

DRAW A PICTURE TO HELP YOU REMEMBER HOW TO PREVENT FOOD POISONING

Critical and creative thinking 2.2

Spot the food safety mistakes

Zan wants to make her favourite cousin a quick and easy lunch. Her cousin will be arriving in about two hours, so Zan has to do a refrigerator and pantry raid and then create something tasty. She has all the ingredients she needs in the pantry to throw together a chicken Caesar salad.

2 chicken fillets
2 tablespoons olive oil
2 slices sourdough bread, cubed
spray olive oil
1 rasher bacon, diced
6 cos lettuce leaves
2 anchovies, chopped
4 tablespoons Caesar mayonnaise dressing
parmesan cheese, shaved

Zan didn't have a recipe to follow, so this how she made the salad. She:

1. preheated the oven to 200°C
2. cut the chicken on a chopping board into 1-centimetre strips
3. heated the oil and fried the chicken strips until golden brown, and then removed the pan from the heat and cooled the chicken strips on a plate
4. cut the bread into cubes on the same chopping board, transferred the cubes to an oven tray, coated the cubes of bread with spray oil, and then baked them for 10 minutes until golden
5. dusted the crumbs off the chopping board, diced the bacon, and then cooked it in the same pan used for the chicken until it was crunchy

9780170495714

6 coarsely chopped the cos lettuce on the chopping board and then transferred it to a serving bowl
7 added the chicken strips, bread cubes, bacon and anchovies to the bowl and gently mixed the ingredients together with her hands
8 drizzled the Caesar mayonnaise dressing over the salad and scattered the shaved parmesan on top.

Zan placed the finished bowl of salad on the table and still had 45 minutes to tidy up before her cousin arrived. Zan and her cousin enjoyed the salad, but later in the day both of them became unwell.

Task

1. Use your knowledge of how to work safely and hygienically when preparing food to prepare a list of actions where food poisoning could occur during the production of a chicken Caesar salad.
2. Identify the possible cause of food poisoning with the salad Zan prepared.
3. Recommend strategies that Zan should follow when preparing and serving food to ensure that it is safe to eat.

DESIGN ACTIVITY 2.1

Australian Food Safety Week

Australian Food Safety Week is an annual event to encourage all Australians to practise good food safety when purchasing food and preparing it at home.

Design brief

The Food Safety Information Council (FSIC) has decided to run a promotional campaign on social media as part of its Australian Food Safety Week campaign: 'Look, Think, Act'. It is seeking ideas to include on a poster to promote awareness of food safety for Australian consumers and reduce the risk of food poisoning.

The FSIC has identified three key areas of food safety that will be the focus of the campaign:

- personal hygiene
- preparing and cooking food safely
- storing food safely.

Investigating and defining

1. Investigate the number of Australians affected by food poisoning each year.
2. Research the personal hygiene practices that are important for Australians to follow when working with food.
3. Explain why it is necessary for consumers to understand the importance of use-by and best-before dates.
4. Investigate the key risks Australians are likely to face when preparing, cooking and storing food.
5. Compile a list of 'high-risk' foods that should be included in the campaign.
6. Research guidelines for storing, preparing and cooking foods that are considered 'high risk'.

Generating and designing

Worksheet Design activity 2.1

Create a table like the one below.

1. Develop a list of two mistakes that Australians are likely to make for each of the FSIC's focus areas of food safety.
2. For each mistake, propose a simple safe work practice that consumers could implement to avoid the mistake.

Focus area	Mistake	Safe work practice
Personal hygiene	• •	• •
Preparing and cooking food safely	• •	• •
Storing food safely	• •	• •

3. Make a decision about which one of the mistakes and safe work practices in each area identified in your table will most effectively promote awareness of food safety for Australian consumers and reduce the risk of food poisoning.
4. Explain why the other mistakes and safe work practices you identified would not reduce the risk of food poisoning as effectively.

Producing and implementing

1. Use a digital device to create a poster for the 'Look, Think, Act' campaign that will highlight the mistakes and safe work practices you have identified.

Seasonal Fruit Kebabs with Yoghurt and Coconut Dip

INGREDIENTS

Seasonal Fruit Kebabs

4 or 5 seasonal fruits:

summer fruits: peach, nectarine, grapes, melon, pineapple, strawberries

or

winter fruits: orange, mandarin, apple, pear, kiwifruit, banana

1 teaspoon lemon juice

bamboo skewers

Yoghurt and Coconut Dip

1 tablespoon shredded coconut

1 tub (200 grams) fruit yoghurt

METHOD

Seasonal Fruit Kebabs

1 Cut the fruit into bite-sized pieces.

2 Sprinkle with a small amount of lemon juice to prevent browning.

3 Thread onto the bamboo skewers.

4 Serve with Yoghurt and Coconut Dip.

Yoghurt and Coconut Dip

1 Place the coconut on an oven tray.

2 Toast the coconut in the oven at 200°C for approximately 2–3 minutes. Remove and check if it is pale golden in colour. Alternatively, toast the coconut under a griller or in a dry, non-stick frying pan.

3 Allow the coconut to cool.

4 Stir the coconut through the yoghurt.

EVALUATION

1 Explain how you would safely use a vegetable knife when cutting the fruit.

2 Describe two safety rules you observed when using the oven, griller or frying pan to brown the coconut.

3 If you needed to substitute canned fruit for some of the fresh fruit in the kebab, list two canned fruits that would be most suitable to use.

4 If you were to make this product again, what changes would you make to the ingredients you selected or the processes you used to improve the finished product?

5 Fruit is one of the main food groups identified in the *Australian Guide to Healthy Eating*. Research how many serves of fruit per day are recommended in this food model and discuss how well your Fruit Kebabs with Yoghurt and Coconut Dip rated.

Mark Fergus Photography

9780170495714

Seeded Health Bars

These Seeded Health Bars make an appealing, healthy snack to add to a lunch box or to enjoy as an after-school treat. The chia seeds are very nutritious, as they are high in omega-3 fatty acids, protein and fibre. This delicious snack is also nut-free and gluten-free.

INGREDIENTS

½ cup shredded coconut
¼ cup sultanas
¼ cup currants
¼ cup pepita seeds
¼ cup sunflower seeds
1 tablespoon black chia seeds
1 tablespoon white sesame seeds
2 tablespoons tahini
3 tablespoons honey
2 tablespoons coconut oil
½ teaspoon ground cinnamon
½ teaspoon vanilla extract
¼ teaspoon sea salt

METHOD

1 Preheat oven to 160°C.
2 Grease and line the base and sides of a 10 × 20-centimetre bar tin with baking paper.
3 Place the coconut, sultanas, currants, pepita seeds, sunflower seeds, chia seeds and sesame seeds in a medium bowl and stir well to combine.
4 In a small saucepan, combine the tahini, honey, coconut oil, cinnamon, vanilla and salt. Using a wooden spoon or silicone spatula, stir the ingredients over a low-medium heat until the mixture is smooth and well blended and starts to boil.
5 Pour the hot ingredients over the seed and fruit mixture and stir thoroughly until the seeds are well coated.
6 Press the mixture firmly into the lined tin.
7 Bake for 20–25 minutes or until golden brown. Turn the tin once or twice during the cooking time to ensure that the mixture browns evenly.
8 Place the tin on a wire rack and allow to cool completely before cutting.
9 When cold, slice into eight bars approximately 2.5 centimetres wide.

Mark Fergus Photography

EVALUATION

1 Describe the sensory properties – appearance, aroma, flavour, texture and sound – of your Seeded Health Bars.
2 Predict a possible outcome if you were to stir the wet ingredients over a high heat rather than a low-medium heat in step 4.
3 Describe one safety rule to observe when using the oven to bake the slice.
4 Identify another type of dried fruit or a type of nut that could be added to the recipe if you did not want to include the currants.
5 Classify the ingredients for the Seeded Health Bars on a diagram of the *Australian Guide to Healthy Eating*. Justify why eating a Seeded Health Bar as a snack is a more nutritious option than a chocolate-coated biscuit.

2 RECIPES

Basic Scones

INGREDIENTS

2 cups self-raising flour

1 tablespoon (20 grams) butter

1–1¼ cup milk (approximately)

1 tablespoon milk, for glazing

MAKES 12 SCONES

METHOD

1. Preheat oven to 230°C. Grease an oven tray.
2. Sift the flour into a large bowl.
3. Rub the butter into the flour using your fingertips until the mixture resembles fresh breadcrumbs.
4. Add 1 cup of milk all at once. Mix with a spatula until a soft dough is formed. Add a little extra milk if the dough is too dry.
5. Turn onto a lightly floured board and lightly knead for 30 seconds. Handle the dough as little as possible to prevent it becoming tough.
6. Gently pat out the dough to 2.5-centimetre thickness and cut scones out.
7. Place scones on the oven tray and glaze with milk.
8. Bake for 10–12 minutes or until golden brown.
9. Remove from oven and wrap in a clean tea towel to cool.

Mark Fergus Photography

EVALUATION

1. How did you know when the butter was sufficiently rubbed into the flour?
2. Why are the scones patted out rather than rolled with a rolling pin?
3. Why are the scones glazed before going into the oven?
4. Identify two safety rules you followed when baking your scones in the oven.
5. Classify the ingredients for the Basic Scones on a diagram of the *Australian Guide to Healthy Eating* and comment on whether you would rate them as very healthy, healthy or not very healthy. Explain how the rating you gave them might change if you were to serve them with jam and cream for afternoon tea.

9780170495714

Pumpkin Noodle Soup with Herb and Garlic Bread

INGREDIENTS

Pumpkin Noodle Soup

400 grams butternut pumpkin (approximately 300 grams after peeling and seeding)

1 onion, diced

1 medium potato, peeled and diced

2 cups water

½ packet chicken noodle soup mix

ground pepper

½ tablespoon chopped parsley

Herb and Garlic Bread

1 bread roll

1 teaspoon parsley

1 teaspoon chives

1 clove garlic

15 grams butter, softened

METHOD

Pumpkin Noodle Soup

1. Carefully peel the pumpkin and remove the seeds. Cut into small pieces, about 2 centimetres in size.
2. Place the pumpkin, onion and potato in a saucepan with the water and bring to the boil.
3. Add the chicken noodle soup mix.
4. Simmer for approximately 30 minutes or until the pumpkin is tender.
5. Puree the soup in a blender or by using a stick mixer.
6. Season with pepper to taste.
7. Sprinkle with a little chopped parsley to garnish.
8. Serve hot with Herb and Garlic Bread.

Herb and Garlic Bread

1. Preheat oven to 200°C.
2. Cut the bread roll into slices nearly all the way through.
3. Finely chop the parsley and chives.
4. Crush the garlic.
5. Mix the softened butter with the herbs and garlic.
6. Spread one side of each bread slice with the herb and garlic butter.
7. Wrap the bread in aluminium foil.
8. Heat through in the oven for approximately 10 minutes until warm.

Mark Fergus Photography

EVALUATION

1. Describe the sensory properties – appearance, aroma, flavour, texture and sound – of your Pumpkin Noodle Soup.
2. Describe how to peel the pumpkin safely.
3. How else could you puree the soup if you did not have a blender or stick mixer?
4. Outline two side dishes other than the Herb and Garlic Bread that you could serve with the soup to make it a more substantial meal.
5. Classify the ingredients for the Pumpkin Noodle Soup with Herb and Garlic Bread on a diagram of the *Australian Guide to Healthy Eating*. Explain how well this meal meets the recommendations of this food selection model.

Food-processor Sweet Shortcrust Pastry

INGREDIENTS

1 cup plain flour

1 cup self-raising flour

2 tablespoons caster sugar

125 grams butter, directly from the refrigerator

1 teaspoon lemon juice

⅓ cup cold water (approximately)

METHOD

1. Place the flours and sugar in the bowl of the food processor and pulse five times.
2. Chop the butter into small pieces and add to the dry ingredients in the food processor.
3. Process until the mixture resembles fine breadcrumbs.
4. Add the lemon juice and water and process for a further minute or until the mixture just comes together.
5. Remove the mixture from the food processor and place on a lightly floured board. Bring together into a ball.
6. Wrap in plastic wrap and refrigerate for 20 minutes.

EVALUATION

1. Why is a mixture of plain flour and self-raising flour used to make this pastry?
2. Explain why it is important to pulse the flour and sugar before adding the butter.
3. Why is lemon juice added to this recipe?
4. What is the purpose of wrapping the pastry in plastic wrap and resting it in the refrigerator for 20 minutes?
5. Explain why pastry is classified as an 'only sometimes and in small amounts' food in the *Australian Guide to Healthy Eating*.

Anastasiia Nurullina/Shutterstock.com

9780170495714

Apple and Cinnamon Turnovers

INGREDIENTS

½ quantity Food-processor Sweet Shortcrust Pastry

⅔ cup pie apples, or 1 apple, peeled and sliced

2 teaspoons caster sugar

¼ teaspoon cinnamon

2 tablespoons sultanas

1 tablespoon milk

1 tablespoon icing sugar

METHOD

1 Preheat oven to 200°C.
2 Roll out the pastry to a square of approximately 24 × 24 centimetres. Cut the pastry into half so that you have two rectangles each of 24 × 12 centimetres in size. Place on a baking tray.
3 Mix the pie apples or peeled and sliced apple, caster sugar, cinnamon and sultanas together in a small bowl.
4 Divide the apple mixture into half.
5 Place one portion of the apple mixture on the lower half of each rectangle of pastry.
6 Brush the edges of the pastry with milk.
7 Turn the top half of the pastry over the apple mixture and press the edges together firmly. Trim the edges of the pastry if necessary.
8 Use a fork to decorate the edges of the pastry. Cut a steam vent in the top of the turnover.
9 Glaze the pastry with the milk.
10 Bake for 15 minutes.
11 Remove from oven and dust lightly with the icing sugar. Return to oven and continue to bake for a further 5 minutes.

EVALUATION

1 Describe the sensory properties – appearance, aroma, flavour, texture and sound – of your Apple and Cinnamon Turnovers.
2 Explain why it is necessary to make a vent in the top of the turnovers.
3 Identify two safety rules you followed when baking your turnovers in the oven.
4 Explain why the pastry is glazed with milk before baking.
5 Classify the ingredients of your Apple and Cinnamon Turnovers on a diagram of the *Australian Guide to Healthy Eating* and decide if they are a healthy dessert. Justify your answer.

Mark Fergus Photography

9780170495714

Cherry Tomato, Feta and Egg Surprise

Cherry tomatoes, along with all other types of tomato, are classified as fruit. The sweet flavour, thin skin and slightly dense flesh of cherry tomatoes means they are ideal to use in cooking. This recipe for Cherry Tomato, Feta and Egg Surprise makes a nutritious breakfast or brunch dish that is surprisingly delicious.

Note: You will need a 24-centimetre frying pan with a tight-fitting lid to cook the Cherry Tomato, Feta and Egg Surprise.

INGREDIENTS

8 cherry tomatoes, halved
2 eggs
⅛ teaspoon salt and pepper
1 spring onion
40 grams tasty cheese
1 teaspoon olive oil
1 wheat tortilla or soft white wrap
2 teaspoons sweet chilli sauce
1 cup baby spinach leaves
30 grams feta cheese, crumbled

SERVES ONE

METHOD

1. Preheat oven to 200°C. Line a small oven tray with baking paper.
2. Place the cherry tomatoes on the lined tray and roast in the preheated oven for approximately 15 minutes until just cooked.
3. Break the eggs into a small jug and beat until well combined. Season with salt and pepper.
4. Finely slice the spring onions and grate the tasty cheese.
5. Place the oil in the frying pan and turn the heat to medium-high.
6. Add the tortilla or wrap and cook for 1 minute.
7. Spread the roasted tomatoes over the tortilla and drizzle with the sweet chilli sauce.
8. Pour the egg mixture evenly over the tortilla and tomatoes.
9. Sprinkle over the chopped spring onions and grated cheese. Top with the spinach leaves and crumbled feta.
10. Cover with a tight-fitting lid and cook for approximately 7–8 minutes or until the egg mixture is set and the spinach has wilted.
11. Slide the Cherry Tomato, Feta and Egg surprise onto a serving plate and serve while warm.

Mark Fergus Photography

EVALUATION

1. Describe the sensory properties – appearance, aroma, flavour, texture and sound – of your Cherry Tomato, Feta and Egg Surprise.
2. Describe how the physical and sensory properties of the eggs change when they are cooked in the frying pan.
3. Outline one safety rule to follow when cooking on a cooktop.
4. Suggest another vegetable you could add to this recipe if you wanted to increase the number of serves of vegetables it provides.
5. Classify the ingredients for the Cherry Tomato, Feta and Egg Surprise on a diagram of the *Australian Guide to Healthy Eating* and discuss how well this meal meets the recommendations of this food selection model.

9780170495714

Chicken and Vegetable Stir-fry

INGREDIENTS

2 teaspoons lemon juice
1 tablespoon soy sauce
2 tablespoons oil
1 chicken fillet
½ onion, cut in wedges
1 clove garlic, sliced
1 centimetre fresh ginger, diced
¼ red capsicum, sliced
4 baby sweet corn, halved lengthwise
6 green beans, sliced
1 bok choy, sliced coarsely
160 grams egg noodles
¼ cup chicken stock
1 tablespoon sweet chilli sauce
1 tablespoon soy sauce (extra)

SERVES TWO

METHOD

1. In a small bowl, combine the lemon juice, soy sauce and 1 tablespoon oil to make a marinade.
2. Slice the chicken fillet across the grain into thin strips and mix into the marinade. Cover and refrigerate.
3. Prepare the vegetables. (Remember to keep them separate, since they have different cooking times and will not be added to the stir-fry at the same time.)
4. Boil water in a medium saucepan and pour over the noodles to soak.
5. Heat the remaining oil in a wok. Add the onion, garlic and ginger. Stir-fry for 30 seconds.
6. Drain the chicken and add to the wok. Stir-fry for 2 minutes or until the chicken changes colour.
7. Add the vegetables separately to the wok, starting with the vegetable that takes the longest time to cook and finishing with the vegetable that takes the shortest time to cook.
8. Pour in the chicken stock and cook for a further 2 minutes.
9. Drain the noodles and toss through the chicken and vegetables. Add the sweet chilli sauce and extra soy sauce.

Mark Fergus Photography

EVALUATION

1. Describe the sensory properties – appearance, aroma, flavour, texture and sound – of your Chicken and Vegetable Stir-fry. Rate the success of your product using a hedonic face (see page 6).
2. What is cross-contamination? How did you prevent cross-contamination from occurring when preparing the ingredients for this recipe?
3. How did you decide on the order in which the vegetables were added to the stir-fry during the cooking process?
4. Describe how you worked safely when cooking the stir-fry in the wok.
5. Classify the ingredients for the Chicken and Vegetable Stir-fry on a diagram of the *Australian Guide to Healthy Eating* and comment on the nutritional value of this meal.

9780170495714

WHY WE COOK FOOD

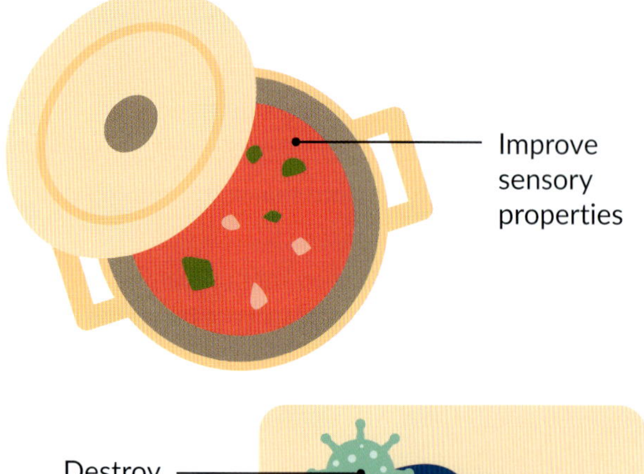

HOW WE COOK FOOD

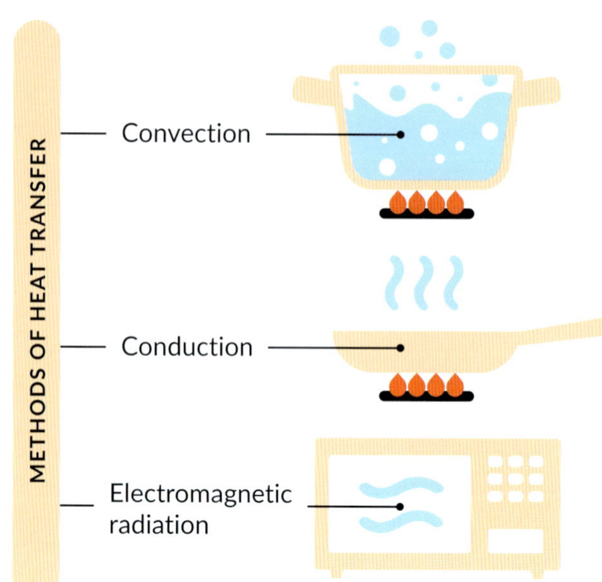

DRY METHODS OF COOKING FOOD

MOIST METHODS OF COOKING FOOD

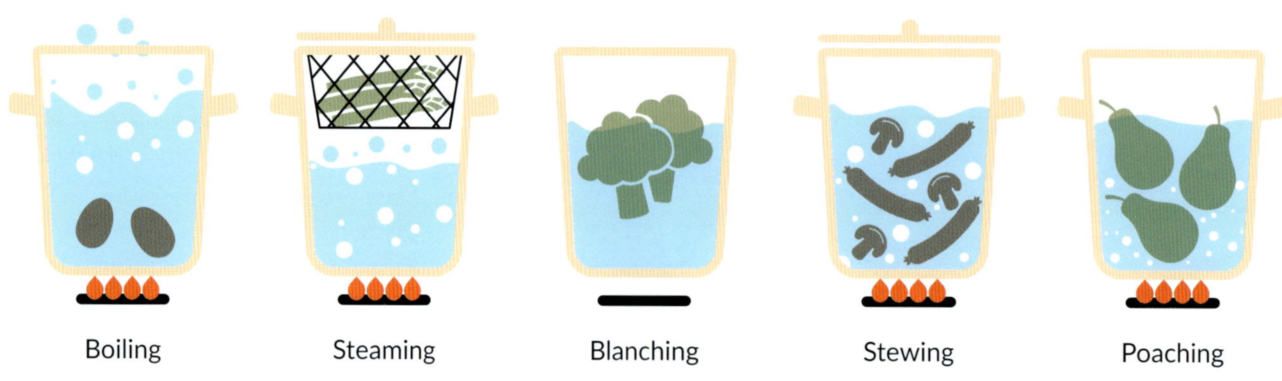

9780170495714

PROCESSES AND TOOLS FOR COOKING

CUTTING AND PEELING

MEASURING

GRATING AND CRUSHING

MIXING AND BEATING

SERVING AND STRAINING

LIFTING

3 RECIPE BASICS

KEY TERMS

baking a method of cooking food in an oven without the addition of fat or oil

boiling a method of cooking food in water at boiling point (100°C)

conduction when heat is transferred from one molecule to another by collision or movement

convection when the molecules in liquids or gases move from a warmer area to a cooler one

cooking the transfer of energy from a heat source to food

frying a method of cooking food by total or part immersion in fat or oil that is heated to temperatures between 150°C and 220°C

grilling a fast, dry method of cooking food that uses intense heat radiated by an electrical element, a gas flame, glowing charcoal or an open wood fire

metric measuring tools measuring spoons, cups, jugs and scales that have been calibrated to accurately measure ingredients by volume or weight using the metric system

poaching a method of cooking delicate foods in liquid at a temperature just below simmering point (85°C)

radiation the transmission of heat energy in the form of rays

recipe a list of ingredients and instructions for preparing food

roasting a method of cooking food in an oven using a minimal amount of fat or oil

steaming a method of cooking food in the steam from boiling water

Templates
- Food order
- Sensory wheel
- Australian Guide to Healthy Eating

Worksheets
- Practical activity 3.1
- Practical activity 3.2
- Design activity 3.1
- Design activity 3.2

Quizzes
- Testing knowledge 3.1
- Testing knowledge 3.2

Nelson MindTap

To access resources above, visit **cengage.com.au/nelsonmindtap**

9780170495714

Tools of the trade

Every skilled tradesperson has their own special tools that are specifically designed for working with particular materials, such as wood, fabric, metal or clay. Working with food requires specialist tools too.

Many of the tools used for preparing food are small pieces of equipment called utensils, and each utensil is used for a specific task. Utensils are often grouped with other pieces of equipment that perform similar functions, such as cutting, peeling or measuring. Other equipment used for preparing food includes small appliances, such as food processors.

The table below will help you to identify some of the tools needed for working with food.

Figure 3.1 Cutting vegetables

iStockphoto.com/Pattanaphong Khuankaew

Table 3.1 Processes and tools for cooking

Process	Tools of the trade	Use and safety
Cutting and peeling	• Knives: – cook's – vegetable – serrated • Peeler	• Always cut downwards with a knife onto a chopping board. • Wash knives separately – never put knives into a sink of soapy water. • Carry knives close to your side, with the tip down. • Pass the handle of the knife, never the blade.
Measuring	• Spoons • Cups • Jugs • Scales	• Measure accurately: – level dry ingredients with a spatula – measure liquid ingredients at eye level – reset scales between ingredients.
Grating and crushing	• Meat mallet • Garlic crusher • Potato masher • Grater • Ricer	• Grater: – keep your fingertips away from the cutting edge on the grater – remove grated food with a pastry brush. • Rinse the equipment immediately after use.
Mixing and beating	• Wooden spoon • Whisk • Hand-held beaters • Blender • Stick mixer • Food processor	• Electrical equipment: – follow the manufacturer's safety instructions when using – always use the plunger supplied when pushing food through the feed tube of a food processor – switch off the power supply and unplug before cleaning – wash the equipment thoroughly in hot soapy water immediately after use.
Sieving and straining	• Colander • Sieve • Slotted spoon	• When straining hot food or boiling water: – use oven mitts to hold the equipment – hold the equipment away from the body. • Wash and thoroughly dry the equipment before storing.
Lifting	• Tongs • Egg lifter • Wire skimmer	• Do not rest the handles of the tools onto hot saucepans or frying pans – heat may be transferred and cause burns.

9780170495714

Activity 3.1

Using tools safely

1 Read the recipe for Spaghetti Bolognese on page 154. List all the equipment that would be needed to produce this recipe.

2 Draw up the Spaghetti Bolognese recipe as a flow chart, highlighting the main stages in the recipe. Annotate the flow chart with the safety issues regarding the equipment involved in each stage.

Making sense of a recipe

A **recipe** is a list of ingredients and instructions for preparing food. A recipe has several components:

- a name
- a list of ingredients, including quantities and – sometimes – details about preliminary preparation
- a method that explains how to prepare the ingredients and the order in which the processes should be completed.

A recipe also gives the following information:

- an indication of cooking temperature and time
- any special equipment required
- an indication of the number of serves the recipe makes.

Ideas for garnishes or decorations are sometimes included in the recipe, or other foods that will be complementary to the finished product may be suggested. Photographs can indicate how the finished product will look.

iStock.com/PeopleImages

Figure 3.2 A recipe is a list of ingredients and instructions for preparing food.

Activity 3.2

Recipe formats

1 Collect two examples of recipes from product labels, magazines or online. Draw up a table like the one below and identify the features of each recipe.

2 After comparing the features of each recipe, identify which one you would be most likely to make, and explain why.

	Recipe 1	Recipe 2
Name		
Photograph		
Format of recipe		
Number of ingredients		
Number of steps to complete recipe		
Time it will take to prepare and cook recipe		
Equipment required		
Number of serves		
Garnishes or decorations		
Serving suggestions		

Compare the formats of the following two soup recipes (for Minestrone Soup and Chicken and Sweet Corn Soup) and then answer the following questions.

3 If you were preparing a shopping list in a hurry, which recipe format would be easier to work from? Why?

4 What are the advantages of working from the recipe format used for Minestrone Soup?

5 What are the advantages of working from the recipe format used for Chicken and Sweet Corn Soup?

6 Reorganise and rewrite the Chicken and Sweet Corn Soup recipe so that it is in the same format as the Minestrone Soup recipe.

Minestrone Soup

Recipe name

Minestrone soup goes well with fresh, crusty bread.

Suggestion for complementary food

Mark Fergus Photography

Ingredients

Quantities of ingredients

Some processes that need to be completed before starting the method

½ onion, diced
⅓ carrot, diced
½ potato, diced
½ stick celery, sliced
4 green beans, sliced
⅓ zucchini, sliced
½ cup cabbage, shredded
1 tablespoon oil
1 clove garlic, crushed
1 tablespoon tomato paste
100 grams canned diced tomatoes
1 cup beef stock
1 ½ cups water
1 tablespoon canned cannellini beans
1 tablespoon small pasta (for example, rigatoni)
1 tablespoon grated parmesan cheese, to serve
salt and pepper

Serves two

Number of serves

Method

Order in which the ingredients should be put together

1. After cutting, place each vegetable in a separate pile, as they will be cooked at different times.
2. Heat the oil in a large saucepan. Add the onion and cook over medium heat until transparent.
3. Add the carrot and cook for 1–2 minutes. Stir occasionally and take care not to brown the vegetables.
4. Repeat this process, adding the celery, then the green beans, zucchini and potato.
5. Add the cabbage and cook until it wilts.
6. Add the garlic, tomato paste, diced tomatoes, stock and water, and bring to the boil.
7. Reduce the heat to a simmer and cover the saucepan with a lid. Cook for 15–20 minutes or until the vegetables are soft.
8. Add the cannellini beans and pasta and cook for a further 10 minutes.
9. Adjust the amount of liquid if necessary. Season with salt and pepper.
10. Serve topped with parmesan cheese.

Cooking time

Mark Fergus Photography

Chicken and Sweet Corn Soup

Finely dice half an onion and one-quarter of a green capsicum. Then slice one stick of celery into thin pieces. Take a large saucepan and sauté the onion, capsicum and celery in one tablespoon of vegetable oil over medium heat. Do not brown. Open a 220-gram can of creamed corn and add to the saucepan. Stir in two cups of water and half a packet of chicken noodle soup. Simmer for 15 minutes. Serve topped with a finely sliced spring onion.

Serves two

9780170495714

Abbreviations in recipes

Measurements are usually given in recipes to ensure that a successful product can be made, eaten and enjoyed. To make recipes easier to read and quicker to write, some aspects are often abbreviated. Some of the most common abbreviations are shown in Table 3.2.

Table 3.2 Common abbreviations in recipes

Abbreviation	Meaning
g	gram
kg	kilogram
mL	millilitre
L	litre
°C	degrees Celsius
tsp	teaspoon
tbs	tablespoon
c	cup
SR flour	self-raising flour
cm	centimetre
min	minutes

Measurement in recipes

Accurate measurement in food preparation is important to ensure success in recipes and to allow the same product to be made again. Correctly calibrated **measuring tools** are essential for accurate measurement: see Tables 3.3 to 3.5. In Australia, we use the metric measurement system, so it is important to ensure that your tools are labelled according to this system.

Table 3.3 Dry ingredients

<table>
<tr><td rowspan="9">Measuring tools</td><td rowspan="4">Spoons</td><td>1 tablespoon = 20 mL</td></tr>
<tr><td>1 teaspoon = 5 mL</td></tr>
<tr><td>½ teaspoon = 2.5 mL</td></tr>
<tr><td>¼ teaspoon = 1.25 mL</td></tr>
<tr><td rowspan="4">Cups</td><td>1 cup = 250 mL</td></tr>
<tr><td>½ cup = 125 mL</td></tr>
<tr><td>⅓ cup = 80 mL</td></tr>
<tr><td>¼ cup = 60 mL</td></tr>
<tr><td colspan="2">The holding capacity of the item should be written on its handle.</td></tr>
<tr><td>Method to accurately measure ingredients</td><td colspan="2">Dip the spoon or cup into the ingredient, slightly overfilling it, then level it off with a spatula.</td></tr>
<tr><td>Examples of dry ingredients measured by volume (spoons and cups)</td><td colspan="2">Flour, sugar, cocoa, desiccated coconut, spices.</td></tr>
</table>

Table 3.4 Liquid ingredients

<table>
<tr><td>Measuring tools</td><td>Measuring jugs</td><td>Those with both cup and millilitre measurements are the most useful.</td></tr>
<tr><td>Method to accurately measure ingredients</td><td colspan="2">Sit the jug on a level surface, pour in the liquid and read the quantity at eye level.</td></tr>
<tr><td>Examples of liquid ingredients measured by volume (jugs)</td><td colspan="2">Milk, stock, water, cream</td></tr>
</table>

Figure 3.3 Measuring liquid ingredients

iStockphoto.com/Chua Aik Wui

Table 3.5 Other ingredients

<table>
<tr><td>Measuring tools</td><td>Kitchen scales</td><td>These are graduated in either 1-gram or 5-gram measures.</td></tr>
<tr><td>Method to accurately measure ingredients</td><td colspan="2">Check that the scales are set on zero before starting to measure ingredients.</td></tr>
<tr><td>Examples of ingredients measured by weight</td><td colspan="2">Butter, cheese, meat, whole nuts, fresh fruit and vegetables</td></tr>
</table>

Figure 3.4 Measuring using digital scales

iStockphoto.com/Tsyb_Oleg

9780170495714

Activity 3.3

Measurement revision

1 Describe the method to accurately measure dry ingredients using a measuring spoon or cup.

2 List three liquid ingredients that could be measured using a measuring jug other than those included in the table.

3 List four dry ingredients that could be measured using a measuring spoon or cup other than those included in the table.

4 Describe the method used to accurately measure liquid ingredients using a measuring jug.

5 What are the benefits of using scales rather than measuring spoons and cups to measure meat or cheese?

6 When measuring dry ingredients, why is it more accurate to dip and level off with a spatula than to pack the ingredients into the measuring spoon or cup?

7 Describe one important rule to follow when using scales.

8 Explain how you would accurately measure the following ingredients:

a ⅓ cup wholemeal flour

b 150 millilitres milk

c 1½ teaspoons curry powder

d 100 grams mushrooms

iStock.com/happyfoto

9 Copy and complete the following table, filling in the equivalent measures.

1 tablespoon = ___ millilitres
1 teaspoon = ___ millilitres
½ teaspoon = ___ millilitres
¼ teaspoon = ___ millilitres
1 tablespoon = ___ teaspoons
1 cup = ___ millilitres
1 litre = ___ cups
1 kilogram = ___ grams

10 Identify two ingredients that would be easier to measure with scales than with cups.

Testing knowledge 3.1

Quiz
Testing knowledge 3.1

1 Identify the piece of equipment you would use if you wanted to cut or chop:

a stewing steak

b celery

c apples

d a loaf of sourdough bread.

2 When preparing butternut pumpkin to cook, identify the tools used to:

a remove the skin

b scoop out the seeds

c chop the flesh.

iStockphoto.com/Longfin Media

3 Identify the equipment you could use to cream together butter and sugar for a cake mixture.

4 List the information you would expect to find in a recipe.

5 List the abbreviations that are sometimes used in recipes for gram, litre, teaspoon, tablespoon, cup and degrees Celsius.

6 Describe the method you would use to measure 1 tablespoon of cocoa for a biscuit recipe.

7 List how many millilitres there are in 1 tablespoon, 1 teaspoon, 1 cup and 1 litre.

8 Describe the process of accurately measuring liquid ingredients.

9 Summarise the reasons why accurate measurement is important when following a recipe.

10 Describe how to accurately use scales to measure ingredients.

Commonly used food preparation terms

The following table lists food preparation terms that you will frequently find in recipes, along with their descriptions and information on equipment and food relevant to each.

9780170495714

Table 3.6 Food preparation terms

Food preparation term	Description of process	Appropriate equipment	Food examples
Bake	Cook food using dry heat in an oven without the addition of fat or oil.	Baking tray, cake tin	Bread, biscuits, cakes, pastry
Beat	Vigorously mix ingredients to incorporate air or combine.	Bowl, wooden spoon, hand-held beaters, whisk	Cream, egg whites
Bind	Stir ingredients to combine.	Bowl, wooden spoon, spatula	Hamburger mixture
Blanch	Plunge food into boiling water for 30 seconds. Drain and refresh in iced water.	Saucepan, sieve, colander	Almonds, snow peas, broccoli
Blend	Mix a dry ingredient with a moist or wet ingredient until it forms a smooth paste.	Bowl, wooden spoon	Cornflour and water
Boil	Heat a liquid to 100°C or to boiling point.	Kettle, electric jug, saucepan	Water, stock
Chop	Roughly cut food into small pieces.	Chopping board, cook's knife	Vegetables
Cream	Beat together sugar and butter until they resemble lightly whipped cream. The mixture will become lighter in colour.	Bowl, wooden spoon, hand-held beaters	Butter cakes, biscuits
Dice	Cut food into small, even-sized cubes.	Chopping board, cook's knife	Onions
Fold	Gently combine a light, airy mixture into a heavier mixture, e.g. beaten egg white into custard sauce. Use a metal spoon or spatula in short strokes to prevent loss of air or volume.	Bowl, metal spoon, spatula	Fluffy omelette, sponge
Fry	Cook food in hot fat or oil. Food may be deep-fried, shallow-fried or stir-fried.	Frying pan, wok, lifter	Bacon and eggs, potato chips
Garnish	Add edible decoration to a dish to enhance its appearance.	Chopping board, vegetable knife	Fresh herbs such as parsley
Glaze	Brush a thin liquid such as milk or beaten egg over food before baking to create a shiny, golden brown surface.	Pastry brush, jug	Scones, pies, tarts
Grate	Reduce a piece of food into thin shreds by rubbing it against the serrated metal surface of a grater.	Box grater, microplane grater	Cheese, vegetables
Grill	Cook small pieces of tender food by dry, radiant heat, e.g. on a griller or a barbecue.	Griller, barbecue	Small, tender cuts of meat or poultry, kebabs, satay sticks
Julienne	Cut food into thin, matchstick-sized pieces.	Chopping board, cook's knife	Carrot, celery, capsicum
Knead	Mix and shape a flour dough by hand. In bread making, this process strengthens the gluten.	Hands, floured board	Bread, pastry
Marinate	Soak foods such as meat or poultry in a seasoned liquid to improve its flavour and sometimes to tenderise.	Bowl	Meat strips for a stir-fry, tandoori chicken pieces
Mix	Combine ingredients so that they are evenly incorporated.	Bowl, spoon, spatula	Flour and sugar
Poach	Gently cook food in a simmering liquid.	Saucepan	Eggs, pieces of fresh fruit
Purée	Make food into a smooth paste by passing through a sieve or by blending.	Sieve, food processor, blender, stick mixer	Stewed apple, vegetable soup, tomato sauce
Roux	Mix melted butter or margarine and flour, blend, then cook. A roux is used to thicken a sauce.	Saucepan, wooden spoon	White sauce, gravy

Food preparation term	Description of process	Appropriate equipment	Food examples
Rub in	Mix butter or margarine through dry ingredients with the fingertips until the mixture looks like breadcrumbs.	Hands, bowl	Scones
Sauté	Lightly toss food in fat or oil in a frying pan over direct heat. The process assists in flavour development but does not brown.	Frying pan, saucepan, wooden spoon	Soups, casseroles
Sear	Brown food quickly over a high heat to seal in juices.	Frying pan	Steak, lamb chops
Shred	Cut food into thin strands using a knife, a grater or a shredding disc in a food processor.	Chopping board, cook's knife, grater, food processor	Lettuce, cabbage, carrot
Sift	Pass dry ingredients through a fine mesh sieve to mix, aerate and remove lumps.	Sieve	Sponges, cakes
Simmer	Bring liquid to just below boiling point so that small bubbles appear on the surface of the liquid.	Saucepan, sometimes covered with a lid	Stock, soup
Slice	Cut food into thin pieces.	Cook's knife or serrated knife, chopping board	Processed meats, salad vegetables
Steam	Cook food over boiling water on a rack or in a special basket in a covered pan. This retains the food's shape and minimises nutrient loss.	Saucepan with a tight-fitting lid, steaming basket made of metal or bamboo, steaming oven	Pork buns, dim sims, vegetables
Stew	Simmer food covered in liquid for a long time. This is used in dishes containing tough cuts of meat with vegetables.	Saucepan with a tight-fitting lid	Lamb, root vegetables, fruit
Stir	Lightly mix ingredients, using a wooden spoon.	Bowl, saucepan, wooden spoon	Custard sauce, gravy
Toss	Mix ingredients by lightly lifting and folding several times.	Wok and chuan, or bowl and salad servers	Vegetables (stir-fried in a wok), salad ingredients
Whisk	Incorporate air into ingredients such as cream, egg whites and sauces.	Bowl, whisk	Souffle, sponge cake

Activity 3.4

Recipe terms

1 Read the recipe for Mini Quiches on page 67. Make a list of the terms from Table 3.6 that are used in this recipe.

2 List any other terms from the recipe that you think are important and write a definition for each of them.

3 Make a list of the ingredients in the recipe and write down the best method of storing each of them.

4 Select six other terms from Table 3.6 and, for each one, find a recipe in this textbook that uses the term. Using a table like the one below, write down a list of the terms and the matching recipe and page number. The first has been done for you, as an example.

Term	Recipe name	Page number
Shred	Salad Roll-up	18

9780170495714

Cooking *destroys harmful microorganisms* that can spoil food, and therefore extends its life

Cooking *improves the sensory properties of food*, e.g. brightening the colour of vegetables or forming a golden crust on baked potatoes

Cooking *aids digestion and nutrient absorption*, as well as making food easier to chew and swallow

Figure 3.5 Reasons for cooking food

How food is cooked

Cooking food involves the application of heat, which brings about a range of physical and chemical changes to the food. We cook food for a number of reasons, as shown in Figure 3.5.

Different cooking methods achieve their results by transferring heat to the food using different mediums such as water, oil or air.

Conduction

Conduction is a process of cooking food by direct heat. Heat is produced when energy is transferred from one molecule to another by collision or movement.

The metal base of the saucepan comes into direct contact with the heat source and heat is absorbed by the saucepan. This heat is transferred to the surface of the food and then by conduction to the inside of the food, cooking it.

Figure 3.6 The metal base of the saucepan conducts heat to the food.

Convection

Convection is the transfer of heat in liquids and air that happens when the molecules from a warmer area move to a cooler area. For example, when an oven is heated, the hot air rises and forces the colder air down to the base of the oven, where it is heated. This also occurs when you cook vegetables – such as potatoes – in boiling water or in steam. As the water in the saucepan is heated, the water molecules rise as they become hot, and then sink again as they reach the surface and start to cool, creating convection currents.

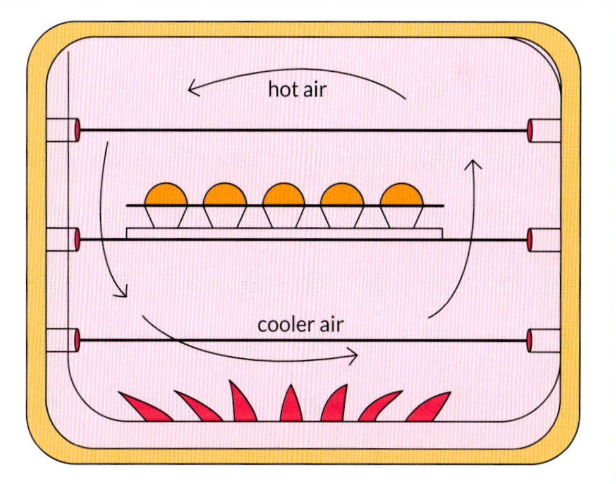

Figure 3.7 Convection currents in a traditional oven

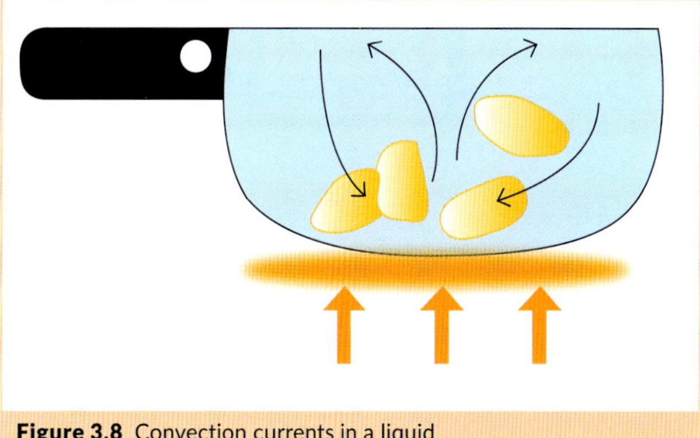

Figure 3.8 Convection currents in a liquid

Methods of cooking food

Food can be cooked in a variety of ways using dry heat, moist heat and electromagnetic radiation.

Dry methods of cooking food

Dry methods of cooking include roasting, baking, grilling, frying and wok cooking. When food is cooked using dry methods, heat is transferred to the food by convection or radiation.

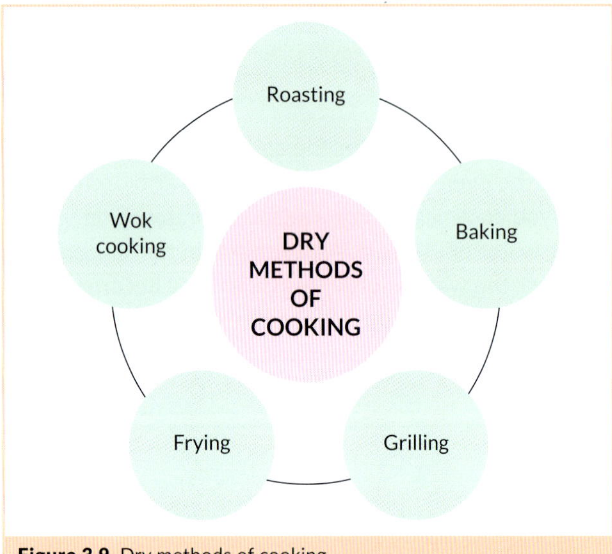

Figure 3.9 Dry methods of cooking

Roasting

Roasting is a method of cooking food in the oven with a small amount of fat or oil. Food is cooked by convection heat and heat radiating from the oven walls. Roasting is used to cook meats such as chicken or pork and vegetables such as potatoes and pumpkin.

Figure 3.10 Roasting

Baking

Baking is similar to roasting, in that it is a method of cooking food in the oven with heat provided by a combination of convection and radiation. However, baked food is cooked without the addition of fat or oil. Baked foods include cakes, breads, pastries, egg custards and lasagna.

Figure 3.11 Baking

Grilling

Grilling is a fast, dry method of cooking that uses intense radiated heat from a wood fire, gas or electric element. The surface of the food is cooked by radiation and the interior by conduction. It is a suitable method for cooking foods that are tender and require a short cooking time, such as meat (for example, steak) fish, vegetables and fruit (for example, tomatoes and pineapple).

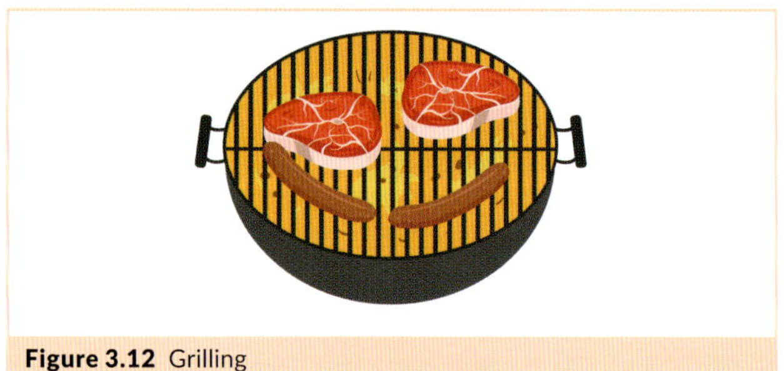

Figure 3.12 Grilling

Frying

Frying is a method of cooking food by total or part immersion in fat or oil, which is heated to between 150°C and 220°C. Convection currents in the fat or oil heat the outer surface of the food and the heat is then transferred to the food by conduction. Foods suitable for frying include meats, vegetables, fish and eggs. Foods suitable for deep-frying include battered or crumbed foods, such as fish, bananas and spring rolls.

Figure 3.13 Deep-frying

Wok cooking

Wok cooking is a method of cooking food in a deep, bowl-shaped frying pan with sloping sides, using a small amount of oil. Only a small section of the base comes in contact with the heat source, and food is tossed and

9780170495714

moved around the wok, over the very hot section. Less oil is used when cooking in a wok than when frying, and more nutrients are retained as the cooking process is fast.

Figure 3.14 Wok cooking

Moist methods of cooking food

Moist methods of cooking include boiling, poaching, steaming and stewing. When food is cooked by moist cooking methods, heat is transferred through water or steam to the food by conduction and convection.

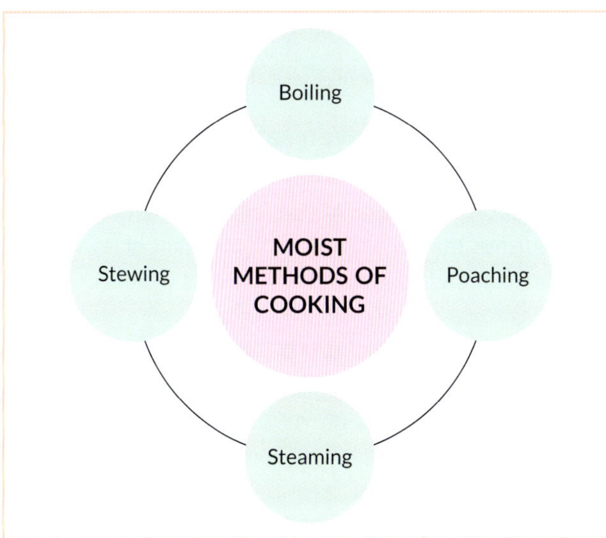

Figure 3.15 Moist methods of cooking

Boiling

Boiling is a method of cooking in water at 100°C. Rapid bubbling occurs over the surface of the saucepan and the whole surface of the food comes in contact with the water. Heat is transferred through convection currents. Foods suitable for boiling include soups, vegetables, pulses, cereals, pasta, red meat, poultry and jams.

Figure 3.16 Boiling

Poaching

Poaching is a method of cooking delicate foods in liquid at 85°C. The surface of the water does not bubble, but only trembles. Heat is transferred through convection currents. Foods cooked by poaching include fruit such as pears, meat such as chicken fillets, and eggs.

Figure 3.17 Poaching

Steaming

Steaming is a method of cooking food in the steam produced from boiling water. The food does not come into direct contact with the water. Convection currents carry the heat from the steam to the food's surface. This method is a healthier method of cooking as no fat or oil is used. Foods suitable for steaming include vegetables and fish, as well as food placed in a closed container such as puddings.

Figure 3.18 Steaming

Stewing

Stewing is a slow method of simmering food in a small amount of liquid. The extended length of time and the moist environment help to break down the connective tissue in tough cuts of meat. The liquid absorbs flavours and retains all the nutrients, and is served as part of the final dish. Heat is transferred by conduction and then convection. Foods suitable for stewing include tougher and cheaper cuts of meat and poultry, which are often stewed with vegetables, and fruit.

Figure 3.19 Stewing

Electromagnetic radiation

Electromagnetic radiation is when electric currents cause electrons within food to move rapidly and generate heat, which in turn cooks the food. Put simply, **radiation** is the transmission of energy though a medium such as food, and it is transferred by wave motion in the form of electromagnetic waves. The food does not come into contact with the heat source.

Induction cooking

An induction cooktop works by heating up a pot or pan, which then heats up the food. An electrical current is passed through a copper coil underneath the ceramic cooktop surface, which creates a magnetic field. If the pot is made of magnetic material – such as cast iron or some types of stainless steel – an electric current is produced in the pot. This causes the electrons in the pot to move rapidly and generate heat to cook food.

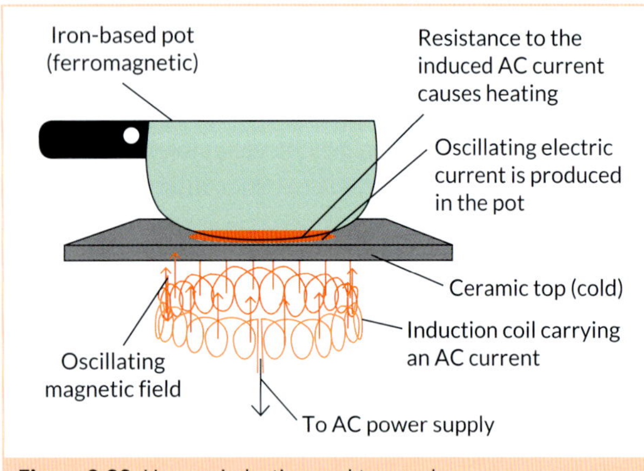

Figure 3.20 How an induction cooktop works

Microwave cooking

The microwave oven is an appliance that is found in most homes. Numerous families find a microwave very useful for cooking a wide variety of foods, thawing frozen foods and reheating pre-cooked foods.

While electric and gas ovens are based on radiant heat, microwave ovens use high-frequency radio waves that enter the food and cause its water molecules to vibrate. This creates heat, which in turn cooks the food. Food is not cooked according to a selected temperature, but by the number of watts the appliance has available; the power rating depends on the size of the microwave. The weight of the food, its water content and the power of the microwave all affect the cooking time required. Remember that the more food there is in a microwave oven, the longer the cooking time will be.

It is important to allow standing time after cooking in a microwave, because the food continues to cook after the microwave has been switched off. A good rule of thumb is that the standing time should equal the cooking time. Note that food cooked in a microwave will not brown as the surface temperature of the food does not reach a high enough temperature for the Maillard reaction to occur.

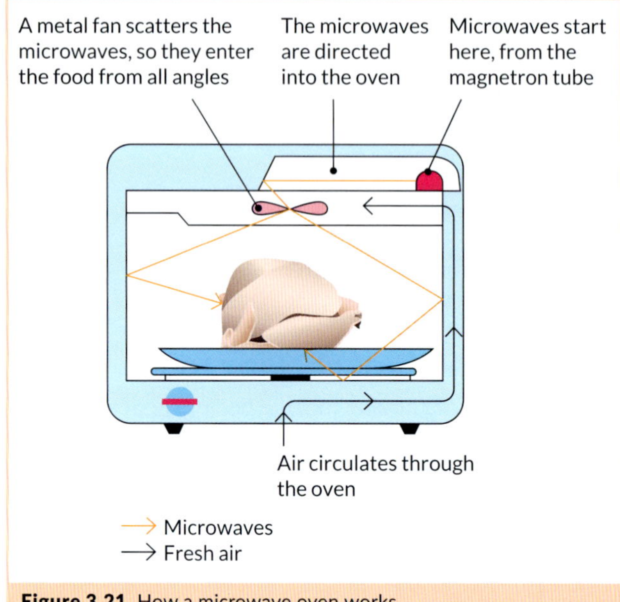

Figure 3.21 How a microwave oven works

Using a microwave oven safely

- Always carefully follow the manufacturer's instructions.
- Use dishes made of china, heat-resistant glass, ceramic, paper or appropriate plastics when cooking in a microwave.
- Do not use metal containers, as they may cause the microwave to arc and damage its magnetron. Metal containers also prevent the food from cooking because the microwaves are reflected away from the food.
- Do not turn a microwave on if it is empty; this can damage its magnetron.
- Use only microwave-safe plastic wrap or paper towel to cover food when cooking in a microwave. Never use aluminium foil, as this can cause the microwave to arc.
- Take care when removing plastic wrap from food after cooking – the escaping steam can cause a severe burn.
- Clean the microwave regularly by placing a small amount of detergent on a damp cloth. Heat the cloth in the microwave for one minute, then wipe the microwave clean.

9780170495714

PRACTICAL ACTIVITY 3.1

Cooking: Microwaved pappadams or prawn crackers

Aim

To be able to operate a microwave oven safely when cooking food.

Method

1. Place one pappadam or prawn cracker in a microwave.
2. Cook on high for 20 seconds. Observe the process, but stand well clear of the microwave.
3. If more cooking is required, increase the cooking time by a few seconds at a time.

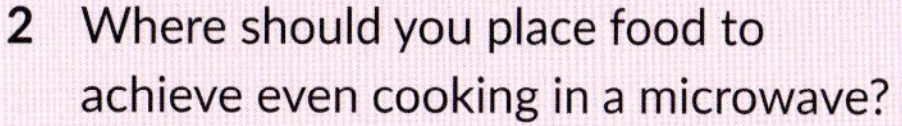

Analysis

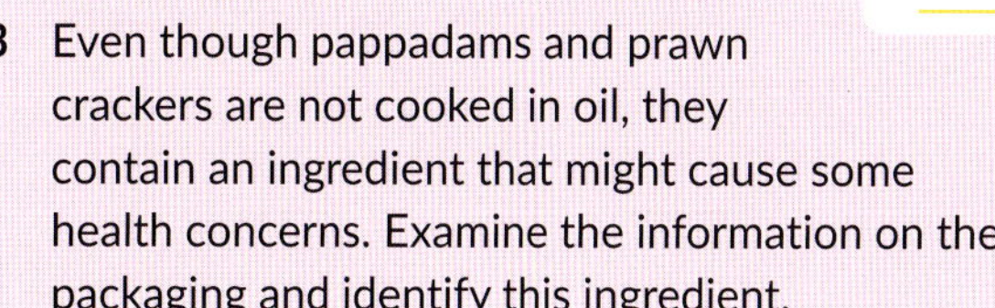

1. Describe your observations of the cooking process in the microwave.
2. Where should you place food to achieve even cooking in a microwave?
3. Even though pappadams and prawn crackers are not cooked in oil, they contain an ingredient that might cause some health concerns. Examine the information on the packaging and identify this ingredient.

Worksheet Practical activity 3.1

Conclusion

Identify three safety considerations you could now teach other people when cooking food in a microwave oven.

PRACTICAL ACTIVITY 3.2

Comparative food test and sensory analysis: How different cooking methods used to cook carrots produce different sensory properties

Aim

To determine how different cooking methods produce distinctive sensory properties in carrots.

Equipment

- 3 medium-sized carrots
- steamer
- container suitable for use in the microwave
- tray suitable for use in a conventional oven
- spray oil

Method

1. Preheat oven to 220°C and set up the steamer and saucepan with boiling water.
2. Peel each carrot and cut into 2-centimetre cubes.
3. Spray one-third of the carrot cubes with the oil and roast them until they are tender and golden brown.
4. Steam one-third of the carrot cubes until they are tender.
5. Place the remaining one-third of the carrot cubes in the microwave container and microwave until they are tender.
6. When all carrot cubes are cooked, place them on separate plates for a sensory evaluation.
7. Record your observations of the sensory properties of each carrot group in a table like the one below.

Worksheet Practical activity 3.2

Analysis

1. Compare the appearance of the carrots cooked by each cooking method.
2. Describe the aroma of each group of carrots. Why do you think they were different?
3. Taste each of the carrot groups. Describe the difference in flavour between each group.
4. Compare the texture of each group and explain why there are differences.
5. Outline the advantages of microwaving carrots.
6. Which carrots would you consider were cooked by the healthiest method? Justify your answer.

Conclusion

Identify the method of cooking that produced carrots with the most appealing sensory properties. Justify your answer.

Cooking method	Appearance	Aroma	Flavour	Texture	Sound
Roasting					
Steaming					
Microwaving					

Testing knowledge 3.2

11 Describe the difference between chopping and dicing carrots.

12 Explain the meaning of the term 'sauté' and why sautéing is considered to be an important cooking process.

13 Explain why it is important to sift flour when making a cake.

14 'Whisking' is a term found in many recipes. What does this term mean?

15 Create diagrams to demonstrate how cooking food by conduction differs from cooking food using convection heat.

16 Explain the benefits of cooking food in a wok.

17 Describe how heat is transferred during boiling. Explain the difference between boiling and poaching.

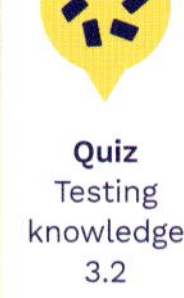

Quiz
Testing knowledge 3.2

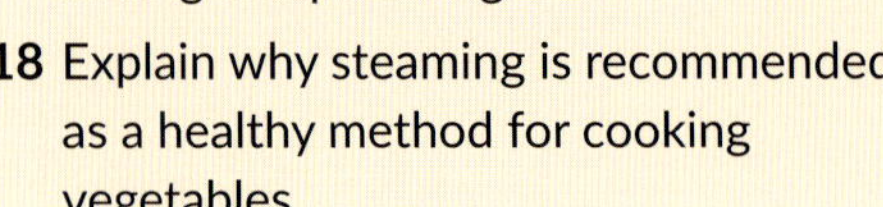

18 Explain why steaming is recommended as a healthy method for cooking vegetables.

19 Why is stewing a suitable method for cooking tough, cheaper cuts of meat?

20 Explain how a microwave oven cooks food.

Critical and creative thinking 3.1

Compare tools that can be used to carry out specific food preparation processes

Either:
Use Table 3.6 on pages 53–4 to select a food preparation process and compare suitable tools that could be used as part of that process.
Or:
Draw up a table like the one below, complete the example shown in it and answer the questions that follow.

1 Describe the characteristics of each tool.

2 Explain how the wooden spoon and plastic spatula are similar and different with respect to the characteristics you have observed.

3 Summarise your findings, including examples of food products for which you would use each tool.

Food preparation process: bind		
Description of process:		
Suitable tool or tools to carry out process	Wooden spoon	Plastic spatula
Sketch of tool		
Properties of materials used in tool		
Ease of use when carrying out the process		
How well the tool completes the process		
Ease of cleaning the tool		

9780170495714

DESIGN ACTIVITY 3.1

Designer muffins

Design brief

The Parents' Association at your school is organising a 'bake-off' for junior school students. The aim is to encourage students to bring healthy snacks to school, and to increase the consumption of vegetables and fruit consumed by Junior School students. Students are asked join in the fun of the school bake-off by designing and preparing a muffin that is suitable to bring to school as a healthy snack.

Investigating and defining

1 Based on the solution requirements and constraints in the design brief, develop four design criteria to evaluate the success of your designed solution.

2 Research a variety of flours, sweeteners and flavouring ingredients – such as fruit, vegetables, cheese, seeds and nuts – that would be suitable to include in a healthy muffin.

3 Develop a concept sketch with as many flavouring ingredients as possible that could be used in a muffin recipe.

4 Classify the flavours into 'sweet' and 'savoury'.

5 Copy and complete the table below by describing the role of each ingredient and listing examples of each ingredient.

Table 3.7 Functional ingredients in muffins

Functional ingredient	Functional role of the ingredients in muffin recipes	Examples of types of ingredient
Flour		
Sweetener		
Eggs		Chicken eggs
Flavourings		
Liquid		
Shortening		
Raising agent		

6 Prepare a recipe such as the Blueberry and Oat Breakfast Muffins on page 65 or the Spiced Apple Muffins on page 66 to explore the characteristics of the ingredients required in a muffin recipe.

Generating and designing

1 Use your knowledge of the functional ingredients of muffins to design two distinctly different designed solution options that meet the needs outlined in the design brief. If you are making a savoury muffin, remember to exclude the sweetener.

2 The Blueberry and Oat Breakfast Muffins recipe on page 65 or the Spiced Apple Muffins recipe on page 66, together with the quantities given in Table 3.8 on page 62, can be used as bases for your design.

3 After copying and completing Table 3.7 opposite, write out the recipe for your designed solution. Remember to give your recipe a creative name.

Worksheet
Design activity 3.1

Templates
Food order

Sensory wheel

Australian Guide to Healthy Eating

Planning and managing

1 Complete a food order for all the ingredients required for your recipe. Refer to page 12 for more information on food orders.

2 Make a list of the aspects of the production task that rely on you and your bench partner sharing and working collaboratively.

Producing and implementing

1 Prepare your designed solution, recording any changes you make to the ingredients or method during production.

Evaluating

1 Respond to your four design criteria in detail.

2 Describe the sensory properties – appearance, aroma, flavour, texture and sound – of your muffins.

3 Share your muffins with two other people and record their comments about them.

4 Taking into account your tasters' comments and your own experience, suggest improvements you could make to the ingredients and/or method if you were to make the muffins again.

5 Discuss your level of organisation during production. Identify areas for improvement.

6 Plot the ingredients for your muffins on a diagram of the *Australian Guide to Healthy Eating* and decide whether they are healthy. Justify your answer based on the recommendations of this food selection model.

iStockphoto.com /brebca

Figure 3.22 Vegetable muffins are a healthy snack.

Table 3.8 A recipe map table for muffins

Functional ingredient	Quantity required	Design option 1	Design option 2
Flour	2 cups		
Sweetener	¾ cup		
Eggs	1		
Flavourings	1 cup		
Liquid (usually milk)	¾ cup		
Shortening (usually oil)	½ cup or 125 millilitres or 125 grams		
Raising agent	1 teaspoon		
Designed solution:			
Justification:			

DESIGN ACTIVITY 3.2

Swirly scones

Scones are enjoyed in many parts of the world and are usually cooked in an oven. They are thought to have originated near Scone, a village in central Scotland, in the early 16th century. Scones were originally made in a triangular shape to represent the Stone of Scone or the Stone of Destiny, a red sandstone block that was traditionally used when Scottish kings were crowned. The first scones were made from oats and cooked on a griddle (flat pan) over a fire.

Today, scones are usually made from flour, and can be either plain or flavoured with a wide range of ingredients to make a sweet or savoury snack. They are a popular treat for morning or afternoon tea because they are quick to prepare and can be made and baked in the oven within 20 minutes. They are the main component of the well-known English Devonshire tea, which consists of warm scones, whipped or clotted cream, and berry jam served with a cup of tea.

iStockphoto.com/travellinglight

Figure 3.23 Scones with cream and jam are the main component of Devonshire tea.

Design brief

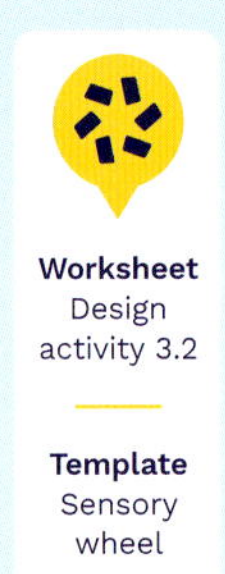

Start with a Basic Scone recipe (see page 40), and then design a new swirly scone that has a sweet or savoury filling as the swirl. The scone could also have a healthy focus.

Write a design brief, which should include:

- a description of the occasion at which the scones will be served
- who will be eating the scones
- where the scones will be served
- whether a sweet or savoury filling is required, and whether it needs to be healthy.

Investigating and defining

1 Use the solution requirements and constraints in the design brief to develop four design criteria to evaluate the success of your designed solution.

2 Undertake an online search of recipes for scones and make a list of ingredients that could be used to flavour a scone dough. Organise the information on a mind map, separating sweet and savoury filling ideas.

Generating and designing

1 Create two designed solution options for your scone fillings, using a recipe map table like Table 3.9 on the following page. Remember that each ingredient in the swirly filling has a specific function so you will need at least one ingredient in each functional role for the recipe to be successful.
   - Ingredients that melt will help to separate the layers in the swirl and provide moisture to the filling.

- Ingredients that have contrasting colour to the dough will ensure that the filling will be visible in the swirl pattern.
- Flavouring ingredients will add to the sensory appeal of the swirly scone.

In addition, a glaze will assist the scone dough to become a golden colour during baking.

2 Mix and match ingredients from your recipe map table to create two designed solution options.

3 Sketch how you intend to arrange or shape the scones for baking to make sure the swirl is visible.

4 Select your preferred designed solution and justify how you made your decision for the final design.

Table 3.9 A recipe map table for swirly scones

Functional role	Quantity required	Design option 1	Design option 2
Ingredient that melts (usually butter)	40 grams		
Ingredients that provide contrasting colour to the dough	⅓–½ cup		
Flavouring ingredients	• 4 tablespoons sugar or sweetening or • 1 teaspoon dried spice or • 4 tablespoons fresh herbs		
Glaze or topping (usually milk)	1 tablespoon		
Designed solution:			
Justification:			

Planning and managing

1 After selecting the filling and glaze, write out the recipe for your designed solution. It should include the ingredients for the swirly filling and the basic scone dough, and the method you will use to prepare the filling ingredients as well as the basic scone dough.

2 Complete a food order for all the ingredients required for your recipe.

Producing and implementing

1 Preheat oven to 210°C.

2 Prepare the filling ingredients for your scones.

3 Make up the Basic Scone recipe on page 40, up to the end of step 5.

4 Instead of patting out the dough, roll it out to a rectangle 30 × 20 centimetres in size, and then use the ingredients from your designed solution for the filling.

5 Gently spread the ingredient that melts over the dough, leaving a 2-centimetre strip on one long side uncovered.

6 Evenly sprinkle over the ingredients that will provide contrast in colour and flavour.

7 Brush the 2-centimetre strip with milk, then roll the dough from the opposite side.

8 Cut the roll into equal portions and arrange on a greased oven tray with the swirl visible.

9 Glaze before baking.

10 Bake for 12–15 minutes or until golden brown.

11 Take a photo of yourself with your swirly scones before taste testing.

Evaluating

1 Respond to the four design criteria you developed earlier to determine the success of your swirly scones.

2 Describe the sensory properties – appearance, aroma, flavour, texture and sound – of your swirly scones. Share your scones with two other people and record their comments.

3 Explain why it is important to have an ingredient in the filling of the scone that melts during baking.

4 List three health and safety issues you followed while preparing or baking the scones.

5 Discuss your organisation during production. Identify things that went well and areas for improvement.

6 If you were to make your scones again, what changes would you make – to either the ingredients or the method?

7 Write a paragraph to explain why a swirly scone is a healthier option than a cupcake for afternoon tea.

Figure 3.24 Swirly scones

Toasted Muesli

Muesli is great served with milk or yoghurt or nibbled on as a snack, since it is a very nutritious food. This recipe makes use of all of the types of measuring equipment utilised in food preparation.

INGREDIENTS

1 cup rolled oats
45 grams shredded coconut
1 teaspoon sunflower seeds
2 teaspoons sesame seeds
¼ cup skim milk powder
¼ cup unprocessed oat bran
⅓ cup All-Bran
1 tablespoon peanuts (optional)
1 tablespoon vegetable oil
1 tablespoon golden syrup
2 teaspoons honey
2 dried apricots, chopped
½ cup sultanas

SERVES ONE

METHOD

1. Preheat oven to 180°C.
2. Place the rolled oats, shredded coconut, sunflower seeds, sesame seeds, skim milk powder, oat bran and All-Bran in a large bowl.
3. Chop the peanuts, if using, and add to the other ingredients in the bowl. Mix well.
4. Place the vegetable oil, golden syrup and honey in a small saucepan. (Hint: measure the oil first, then dip the tablespoon you used for the oil into the golden syrup; it will then come off the spoon easily.)
5. Over a medium heat, bring the liquid ingredients to the boil. Remove from the heat immediately and pour over dry ingredients. Mix well.
6. Spread muesli in a thin layer on a baking tray. Bake for 5 minutes.
7. Remove from oven, stir carefully, then return to oven for another 5 minutes or until golden brown.
8. Remove from oven and cool.
9. Mix in chopped apricots and sultanas.
10. When completely cool, package in an airtight container.

EVALUATION

1. Explain how you accurately measured the dry ingredients.
2. Which tool did you use to stir the muesli? Explain why it was the most suitable tool.
3. Why was it important to remove the liquid ingredients from the cooktop as soon as they boiled?
4. Describe the safest way to remove the hot tray of muesli from the oven.
5. Classify the ingredients for the Toasted Muesli on a diagram of the *Australian Guide to Healthy Eating* and comment on how well this recipe meets the recommendations of this food selection model.

Mark Fergus Photography

9780170495714

Blueberry and Oat Breakfast Muffins

These muffins are ideal to serve for breakfast, as they contain oats, yoghurt and fruit, and are an alternative to a traditional breakfast of a bowl of ultra-processed cereal and milk. The recipe can be varied to include other fruits – such as raspberries, stewed apple or sultanas – or nuts.

INGREDIENTS

½ cup self-raising flour
½ cup wholemeal self-raising flour
½ teaspoon baking powder
1 teaspoon cinnamon
½ cup rolled oats
⅔ cup soft brown sugar
½ cup unsweetened Greek yoghurt
1 egg, lightly beaten
2 tablespoons vegetable oil
½ cup milk
½ teaspoon vanilla extract
1 cup frozen or fresh blueberries

METHOD

1. Preheat oven to 200°C.
2. Use patty papers to line a muffin tray (9 × ⅓ cup capacity).
3. Sift the flours, baking powder and cinnamon into a bowl. Add the husks to the bowl.
4. Add the oats and sugar to the sifted dry ingredients and stir through.
5. In a second bowl, combine the yoghurt, beaten egg, oil, milk and vanilla extract.
6. Make a well in the centre of the dry ingredients and add the yoghurt mixture. Gently fold the yoghurt mixture into the dry ingredients and then fold in the blueberries. Be careful not to over-mix, otherwise the muffins will be tough.
7. Spoon the mixture evenly into the patty papers in the muffin tray.
8. Bake for 15–20 minutes.
9. Test the muffins to see if they are ready; leave longer if necessary. The muffins are ready if they spring back when lightly touched with your finger, or if a fine skewer comes out clean and dry from the centre of a muffin.
10. Cool in tray for 10 minutes. Remove from pan and place on a cooling rack. It is best to allow the muffins to cool completely before taste testing.

Mark Fergus Photography

EVALUATION

1. Describe the sensory properties – appearance, aroma, flavour, texture and sound – of your Blueberry and Oat Breakfast Muffins.
2. Why is it important to sift the flours, baking powder and cinnamon in step 3?
3. Explain why it is important to fold the wet ingredients into the dry ingredients gently.
4. Identify the ingredients in the muffins that contain fibre.
5. Outline two rules for using the oven safely when cooking the muffins.

Spiced Apple Muffins

INGREDIENTS

1 cup canned apple slices
2 cups self-raising flour
½ teaspoon bicarbonate of soda
½ teaspoon cinnamon
1 teaspoon ground ginger
¼ teaspoon ground nutmeg
pinch ground cloves
¾ cup soft brown sugar
1 egg, lightly beaten
¾ cup milk
½ cup vegetable oil

Topping

1 tablespoon caster sugar
½ teaspoon cinnamon
½ small red apple, cored and very thinly sliced

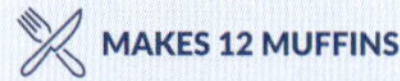
MAKES 12 MUFFINS

METHOD

1. Preheat oven to 200°C. Use patty papers to line a muffin tray (12 × ⅓ cup capacity).
2. Cut each of the canned apple slices into 3 pieces.
3. Sift the dry ingredients into a large bowl.
4. Combine the egg, milk and oil and whisk together.
5. Make a well in the centre of the dry ingredients and add the sliced canned apple. Mix in the wet ingredients and stir to combine. Do not over-mix, otherwise the muffins will become tough.
6. Spoon mixture evenly into the patty papers in the muffin tray.
7. Prepare the topping ingredients by combining the caster sugar and cinnamon in a small bowl.
8. Top each muffin with a thin slice of apple. Sprinkle with the sugar and cinnamon mixture.
9. Bake for 15–20 minutes.
10. Test the muffins to see if they are ready; leave longer if necessary. The muffins are ready if they spring back when lightly touched with your finger, or if a fine skewer comes out clean and dry from the centre of a muffin.
11. Cool in tray for 5 minutes. Remove from pan and place on a cooling rack.

Mark Fergus Photography

EVALUATION

1. Describe the sensory properties – appearance, aroma, flavour, texture and sound – of your Spiced Apple Muffins.
2. Explain why it is important to sift the dry ingredients in step 3 before mixing in the wet ingredients.
3. Identify another type of fruit you could use in the muffins if canned apple slices were not available.
4. Identify two safety rules you followed when baking your Spiced Apple Muffins in the oven.
5. Classify the ingredients of your Spiced Apple Muffins on a diagram of the *Australian Guide to Healthy Eating* and comment on the health rating you would give the muffins with reference to the amount of sugar and oil they each contain.

9780170495714

Mini Quiches

Pastry is traditionally used to encase egg mixture in a quiche. In this recipe, bread replaces the pastry, because it has a much lower fat content and is quick and easy to prepare. During baking, the heat in the oven coagulates the egg and sets the filling.

INGREDIENTS

8 slices wholemeal bread

30 grams butter, melted

1 rasher bacon, finely diced

2 spring onions, finely sliced

30 grams tasty cheese, grated

2 eggs

⅓ cup milk

20 grams butter extra for greasing the tray

METHOD

1 Preheat oven to 200°C.
2 Grease a muffin tray with the melted butter.
3 Trim crusts from bread. Roll each slice flat with a rolling pin to compress the slice; ensure there are no holes in the slices of bread.
4 Brush one side of each slice of bread with melted butter.
5 Carefully place the bread butter side down into small, greased muffin tins. Each slice will form a small cup with pleats in it.
6 Add bacon, spring onion and cheese to each bread case.
7 Lightly beat the eggs and milk with a fork, then pour equal quantities of this mixture into the bread cases.
8 Bake for approximately 15–20 minutes or until they are golden brown and the filling has puffed.
9 Remove from muffin tray and serve.

EVALUATION

1 Identify the process that the eggs undergo during baking.
2 Outline the main role that egg plays in the structure of a quiche.
3 Suggest some other ingredients that could be used to flavour the Mini Quiches.
4 Outline two rules for using the oven safely.
5 What are the nutritional benefits of using bread instead of pastry for the base of the Mini Quiches?

Mark Fergus Photography

Cheese and Ham Toastie with Mango Chutney and Pickled Cucumber

Many people consider a toastie to be a comfort food, as it is very satisfying and is linked to food they enjoyed as a child. Toasties are quick and easy to prepare as a snack and, when served with a side of coleslaw, make a great brunch or lunch dish.

INGREDIENTS

Pickled Cucumber
1 small Lebanese cucumber
1 teaspoon salt
3 tablespoons caster sugar
3 tablespoons rice wine vinegar

Cheese and Ham Toastie
2 thick slices sourdough bread
20 grams butter
1 tablespoon mango chutney
1 slice ham
30 grams mozzarella cheese, grated
30 grams mature cheddar cheese, grated

To serve
extra mango chutney (if desired)
coleslaw (see recipe on the following page, if desired)

METHOD

Pickled Cucumber

1. Slice the Lebanese cucumber into thin slices. Sprinkle with the salt and set aside for 15 minutes. Rinse and pat dry on paper towel. Place in a small bowl or lidded jar.
2. Place the caster sugar and rice wine vinegar in a small saucepan and stir over a low heat until the sugar is dissolved. Remove from the heat and allow to cool. Pour over the prepared cucumber.

Cheese and Ham Toastie

1. Place the bread on a clean board and spread one side of each slice with butter. Turn both slices over so the buttered side is down.
2. Spread the mango chutney evenly over the unbuttered side of each slice of bread.
3. Place the slice of ham on top of the chutney on one piece of bread.
4. Remove some of the pickled cucumber from the pickling liquid and place on top of the slice of ham.
5. Combine the grated cheeses in a bowl and mix together well.
6. Cover the ham and pickled cucumber with the cheese mixture. Cover with the other slice of bread and press down gently.
7. Heat a frying pan over a medium heat. Place the toastie into the frying pan, buttered side down, and press down on the toastie with a spatula.
8. Cook for 4–5 minutes until the toastie is golden brown and crispy.
9. Turn the toastie over carefully and cook the second side for a further 4–5 minutes until golden brown.
10. Remove to a board and cut in half. Serve while hot with extra chutney and coleslaw, if desired.

9780170495714

Coleslaw

INGREDIENTS

⅛ cabbage
½ carrot
¼ red capsicum
½ stick celery
1 spring onion
1 tablespoon coleslaw dressing
1 tablespoon French dressing

METHOD

1 Finely shred the cabbage and place in a large bowl.
2 Grate the carrot and finely dice the capsicum, celery and spring onion. Add to the bowl.
3 Toss the coleslaw dressing and French dressing through the prepared vegetables.

EVALUATION

1 Describe the sensory properties – appearance, aroma, flavour, texture and sound – of your Cheese and Ham Toastie with Mango Chutney and Pickled Cucumber.
2 Compare the two cheeses before and after cooking for flavour and melting properties.
3 Suggest other fillings that could be added to the cheeses for a new variety of toastie.
4 Create a diagram to demonstrate how to dice capsicum safely in step 2 of the recipe for Coleslaw.
5 Classify the ingredients for the Cheese and Ham Toastie with Mango Chutney and Pickled Cucumber served with a side of Coleslaw on a diagram of the *Australian Guide to Healthy Eating*. Write a paragraph to explain how well this meal meets the recommendations of this food selection model.

Mark Fergus Photography

Tacos with Spicy Meat Filling

INGREDIENTS

Spicy Meat Filling

2 teaspoons oil

¼ onion, finely diced

½ clove garlic, crushed

100 grams minced meat

1 tablespoon tomato paste

⅛–¼ teaspoon chilli powder

pinch salt, pepper and sugar

¼ teaspoon ground cumin

¼ teaspoon ground coriander

½ teaspoon Worcestershire sauce

¼ cup water

1 tablespoon canned kidney beans, rinsed and roughly chopped

Tacos

3 king-sized taco shells

½ tomato, diced

lettuce leaf, shredded

30 grams cheese, grated

chilli sauce

SERVES ONE

METHOD

Spicy Meat Filling

1 In a small saucepan, heat the oil and lightly brown the onion and garlic.

2 Add minced meat and stir until brown. Mash with a fork or potato masher to break the meat into small pieces.

3 Add the tomato paste, spices, Worcestershire sauce, water and kidney beans.

4 Simmer for 5–10 minutes until the water has evaporated and the mixture has thickened and is almost dry.

To assemble the dish

1 Preheat oven to 180°C.

2 Place the taco shells on a baking tray with the openings facing down so they remain open during cooking. Heat shells in oven for 4 minutes.

3 Fill with the Spicy Meat Filling, tomato, lettuce and cheese. Top with chilli sauce.

Additional serving suggestion

If you run out of taco shells, serve the Spicy Meat Filling on a bed of steamed rice in a serving bowl. Top with lettuce, tomato and grated cheese and finish with some chilli sauce.

EVALUATION

1 Why are the onion and garlic lightly browned before the minced meat is added to the Spicy Meat Filling?

2 Identify and describe the changes that occurred to the minced meat when it was cooked.

3 Why are the taco shells heated upside down?

4 Describe two safe work practices you should consider when removing food from the oven.

5 Classify the ingredients for the Tacos on a diagram of the *Australian Guide to Healthy Eating*. Comment on their nutritional rating.

Mark Fergus Photography

9780170495714

Cupcakes

INGREDIENTS

2 eggs
½ cup thickened cream
½ cup caster sugar
½ teaspoon vanilla extract
1 cup self-raising flour, sifted

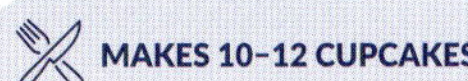

METHOD

1. Preheat oven to 200°C.
2. Use cupcake papers to line a patty cake tray (12 × ⅓ cup capacity).
3. Place the eggs and cream in a medium bowl and beat with electric hand beaters until combined.
4. Add caster sugar and vanilla extract and beat on high speed for 4 minutes or until the mixture is light and fluffy.
5. Sift in the flour and gently fold in using a metal spoon until there are no lumps.
6. Spoon equal quantities of cake batter into the cupcake papers.
7. Bake for 15–20 minutes or until pale golden and just firm to the touch.
8. Remove from oven and place on a cooling rack.
9. Decorate with glacé or butter icing.

EVALUATION

1. Describe the sensory properties – appearance, aroma, flavour, texture and sound – of your Cupcakes.
2. Explain how beating the eggs, cream, caster sugar and vanilla for 4 minutes contributes to the aeration of the Cupcakes.
3. Why is the flour folded into the recipe in step 5 instead of being beaten in?
4. Identify the processes that are responsible for the Cupcakes becoming pale golden during baking.
5. Identify the ingredients in the cupcakes that places them in the 'only sometimes and in small amounts' component of the *Australian Guide to Healthy Eating*. Predict the outcome for your long-term health if you were to eat cupcakes on a regular basis.

Mark Fergus Photography

CHANGES TO INGREDIENTS DURING COOKING

AERATION

Air is incorporated, increasing volume and creating light texture

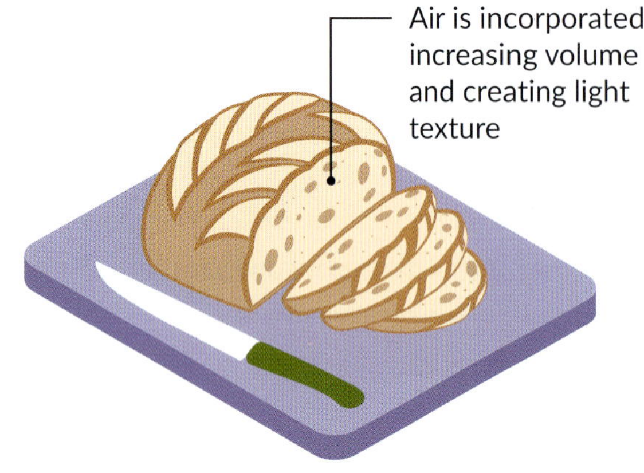

GELATINISATION

Starch absorbs liquid and swells and thickens with heat

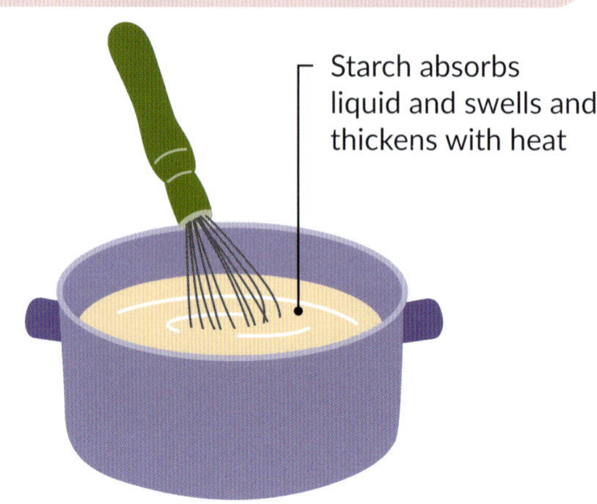

COAGULATION

Proteins thicken or set with heat

DEXTRINISATION

Starch turns brown and develops toasted flavour

CARAMELISATION

Sugar turns brown and develops rich flavour

MAILLARD REACTION

Protein and reducing sugar brown with heat

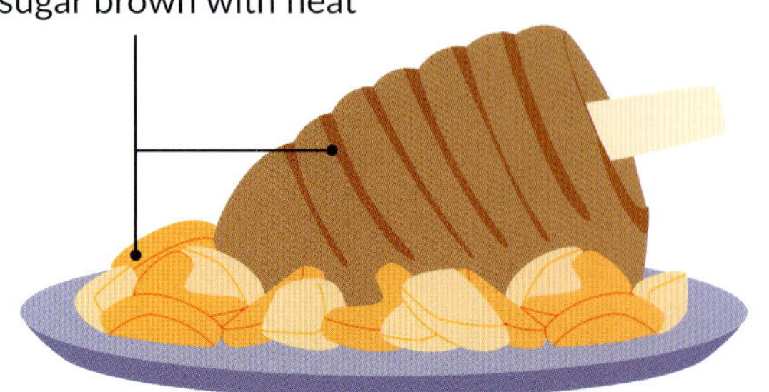

DENATURATION

Mechanical action of beating egg white increases its volume

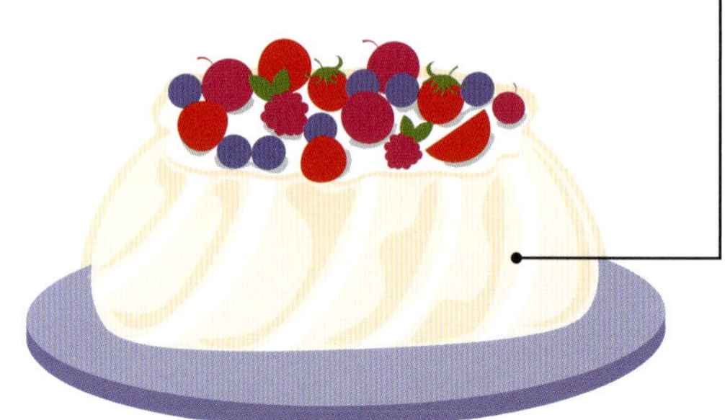

9780170495714

AERATION OF BAKED PRODUCTS

CHEMICAL

Bicarbonate of soda

Baking powder

MECHANICAL

Sifting

Creaming

Whisking

Rubbing in

BIOLOGICAL

Yeast

4 THE FOOD LAB

KEY TERMS

aeration the process of incorporating air into food products to increase their volume and create a light, airy texture

caramelisation the process that occurs when sugar (sucrose) begins to decompose when it is exposed to high temperatures (190°C) of dry heat

coagulation a form of denaturation that occurs when there is a permanent change in the protein from a liquid into a thick mass as a result of heat or the addition of acids

denaturation the permanent structural change of the protein molecules in food

dextrinisation the process that occurs when a starch is exposed to dry heat; the starch is broken down to dextrin, resulting in a change in colour to golden brown

fermentation the conversion of some of the sugar and starch molecules in a yeast dough to alcohol and carbon dioxide as the yeast ferments

gelatinisation the process that occurs when starch granules absorb liquid in the presence of heat, and thicken the liquid, forming a gel

kneading the process that makes gluten – the protein in wheat flour – stronger and more elastic so that it can begin to stretch and capture the carbon dioxide bubbles in a dough

Maillard reaction a reaction that turns food brown and creates pleasant, volatile, aromatic compounds when food is exposed to dry heat; it is a reaction between amino acids in protein and reducing sugars

proving a process in which a yeast dough is rested to allow time for fermentation to take place

reducing sugars include all monosaccharides and some polysaccharides; they react with amino acids to create the Maillard reaction

Worksheet
- Practical activity 4.1

Quizzes
- Testing knowledge 4.1
- Testing knowledge 4.2
- Testing knowledge 4.3
- Testing knowledge 4.4

To access resources above, visit **cengage.com.au/nelsonmindtap**

How ingredients change during preparation and cooking

A recipe is like a roadmap for creating food that you and others enjoy eating, while ingredients are the building blocks of a recipe.

As you become an experienced cook and understand how the physical and chemical properties of ingredients change during preparation and cooking, you will be confident in making decisions when preparing food. Knowing about the scientific reactions that occur when ingredients are combined at different stages in a recipe allows you to have greater control of the sensory properties in the final product.

Table 4.1 How the physical and chemical properties of ingredients change during preparation and cooking

Reaction	Description of reaction	Food examples
Producing volume and a light, airy texture		
Aeration	**Aeration** is the process of incorporating air into food products to increase the volume and create a light, airy texture. Food can be aerated by: • chemical ingredients – baking powder, or bicarbonate of soda plus an acid (tartaric acid or cream of tartar) • mechanical actions – sifting, creaming, whisking, beating, folding in or rubbing in • biological ingredients – yeast	Cakes (including sponge cakes), pavlova, souffle, bread, pastries
Creating structure and texture		
Denaturation	**Denaturation** describes the permanent structural change of the protein molecules in food. This can occur by the application of heat, mechanical action or the addition of acids.	Beaten egg whites, poached egg, marinated meat
Coagulation	**Coagulation** is a form of denaturation that occurs when there is a permanent change in the protein from a liquid into a thick mass as a result of heat or the addition of acids. The denatured proteins form a new structure that retains water; a semi-liquid becomes firm and maintains its shape.	Grilled meat, quiche, cooked egg, baked custard
Developing colour during cooking – browning		
Maillard reaction	The **Maillard reaction** turns food brown and creates pleasant, volatile, aromatic compounds when food is exposed to dry heat. It is a reaction between amino acids in protein and **reducing sugars**. It begins at about 154°C, and can be seen when sautéing on the cooktop or baking in the oven.	Cakes, roast meat, roast/baked vegetables, grilled seafood, biscuits, muffins
Dextrinisation	**Dextrinisation** is the process that occurs when a starch is exposed to dry heat; the starch is broken down to dextrin, resulting in a change in colour to golden brown.	Bread, scones, toast
Caramelisation	**Caramelisation** is the process that occurs when sugar (sucrose) begins to decompose when it is exposed to high temperatures (190°C) of dry heat. Hundreds of compounds are generated as the sugar decomposes, resulting in browning and enjoyable aromas. Sweet and acidic flavours are produced.	Gingersnaps, caramel sauce, toffee, caramelised onions
Thickening		
Gelatinsation	**Gelatinisation** is the process that occurs when starch granules absorb liquid in the presence of heat, and thicken the liquid, forming a gel.	Cheese sauce on cauliflower, white sauce in lasagna, cooked pasta, cooked rice

9780170495714

The ingredients in baked products

Baking is the process used to create food items including cakes, biscuits, pastries and yeast goods. They are baked in an oven and, during the cooking process, dry heat is circulated around the food. The high temperature activates chemical and physical changes to the ingredients resulting in delicious aromas, flavours and textures.

Flour

Flour is an essential ingredient in baked products because it provides volume and structure to the end product. Flour can absorb milk, eggs, water or butter during the mixing process to form batters and doughs. When flour is exposed to heat in an oven, the protein or gluten in it sets, preventing the final product from collapsing when it is cooled.

Two other processes that involve flour in baked products are the Maillard reaction and dextrinisation. The Maillard reaction assists in the colour development of baked products. It occurs when reducing sugars and the proteins in flour and eggs are exposed to heat in the oven, creating a golden brown colour. Dextrinisation occurs when the dry heat in an oven causes the starch in flour to dextrinise, contributing to the browned surface and delicious aroma that develops during baking.

Flours made from soft wheats – which are low in gluten and sometimes known as cake flours – are most suitable for baked products because they produce a soft texture. White, wholemeal or self-raising flour can be used to make cakes, biscuits, slices and pastries.

Sugar

Different types of sugars have been designed for specific purposes and to create different sensory properties.
In baked food products, sugar can:

- contribute to the golden brown colour
- provide a sweet flavour
- improve texture by creating a tender crumb
- aerate or increase volume when creamed with butter
- act as a preservative
- help yeast ferment in bread making.

There are several different types of sugar:

- A1 sugar – also known as white, granulated or table sugar – is the most common form of sugar. It has medium-sized crystals.
- Caster sugar has finer, smaller crystals than A1 sugar, and is often used in cakes, biscuits and desserts because it dissolves more effectively into the mixture and gives a more even appearance and texture.
- Icing sugar looks and feels like a white powder and is usually used to ice cakes and biscuits. Soft icing sugar has an anti-caking agent added to prevent lumps.

Figure 4.1 The characteristics of ingredients in baked products

- Brown sugar is a soft, moist sugar with fine crystals. Its distinctive flavour and colour are usually made by adding molasses to white sugar.
- Raw sugar is the sugar crystals before the refining process takes place.
- Demerara sugar has a pale golden colour and slight toffee flavour.
- Golden syrup is a golden-coloured, thick, sweet syrup made by processing molasses.

iStockphoto.com/kitamin

Figure 4.2 Types of sugar

Eggs

Eggs are a very versatile ingredient, with characteristics that enable them to perform many functions in food products.

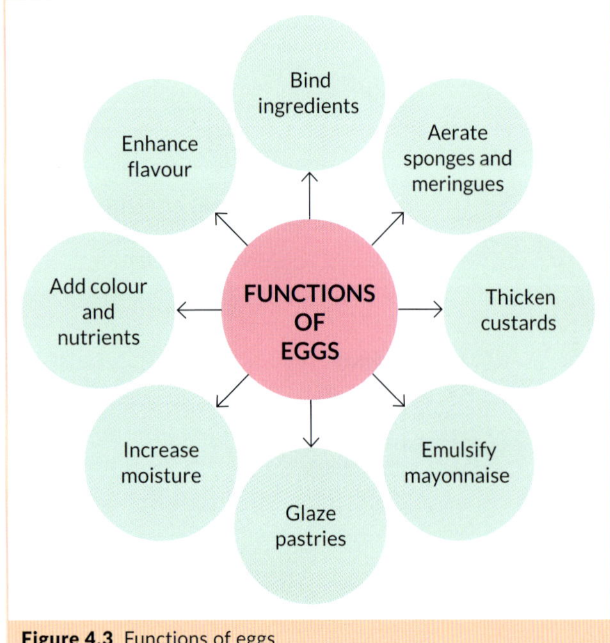

Figure 4.3 Functions of eggs

In baked products, eggs perform a range of complex functions when they are combined with other ingredients.

- During the mixing process, eggs trap bubbles of air, which then expand in the heat of the oven and aerate the mixture to produce a lighter texture. Separating, and then beating or whisking egg whites enables a large volume of air to be incorporated into the mixture, creating the light texture that is the feature of products such as sponge cakes and meringues.
- Eggs assist flour in building the structure of a product when they coagulate (become firm) during baking. In addition, during baking the Maillard reaction occurs and the colour of the cake, biscuits or pastry turns golden brown – this is the proteins from eggs and flour combining with the reducing sugars.
- The yolks of the eggs give a richness to baked products.

Most recipes featuring eggs are based on eggs weighing 55–60 grams.

Butter/shortening

Butter is a type of shortening that is made from the fat component of milk when it is separated and churned. Salted butter is used in most recipes, because the salt helps to balance the overall flavour of the product. Unsalted (or cultured) butter has a slightly softer flavour.

- In cakes and muffins, butter creates a moist, tender texture.
- In biscuits and pastry, butter is responsible for the characteristic short, crumbly texture and delicate flavour.
- In baked products made by the creaming method, butter helps to aerate the mixture and lighten the texture. During the creaming process, beating causes fat from the butter to trap small bubbles of air around the individual crystals of sugar, and so increase the volume of the mixture.

Butter is a perishable food and should be stored in the refrigerator to prevent the fats becoming rancid or going 'off'.

Margarine or blends of butter and margarine can be used as an alternative to butter in most recipes, but may slightly alter the flavour and texture.

Other ingredients

Milk

Milk is a liquid (moist) ingredient that is used in cake batters to help combine dry ingredients. The small fat component of milk contributes to the tender texture of cakes. In cakes made by the melt-and-mix method, milk keeps the batter soft and moist, allowing it to be beaten to incorporate air.

Raising agents

Raising agents are the ingredients that are responsible for aerating or leavening in many doughs, cakes and biscuits. During baking, they give lift to a dough or cake batter to produce a light, airy texture in the final product.

- Baking powder is a white powder that releases carbon dioxide (CO_2) when it comes in contact with moisture and heat.
- Bicarbonate of soda looks similar to baking powder. Because bicarbonate of soda is alkaline, it must be combined with an acidic ingredient (tartaric acid or cream of tartar) – as well as moisture and heat – before it produces CO_2.

Food manufacturers add baking powder to plain flour to produce self-raising flour for the convenience of consumers. Usually a low-protein flour is used to create light, soft-textured products.

As the bubbles of CO_2 are produced in a mixture, it begins to rise and its volume increases.

Flavouring/spices

Spices are flavouring ingredients made from the buds, bark, roots, berries or aromatic seeds of certain plants. In baking, they are usually used in their dry form and ground into a powder.

Vanilla is a spice with a warm, floral aroma that is often used to flavour baked products. It is available in liquid form as vanilla extract, or as a paste. Some other examples of spices often used in baked products are cinnamon, nutmeg and ginger.

Andris Tkachenko/Adobe Stock Photos

Figure 4.4 Spices can be ground into a powder to flavour baked products.

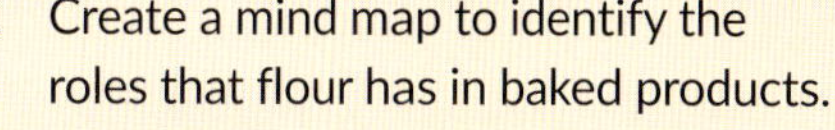

Testing knowledge 4.1

Quiz
Testing knowledge 4.1

1. Create a mind map to identify the roles that flour has in baked products.
2. Describe the chemical reaction involving starch that contributes to browning during the baking process.
3. Outline the different roles that sugar may have in a product that is baked in the oven.
4. Copy and annotate the diagram below to highlight the characteristics of different types of sugar.

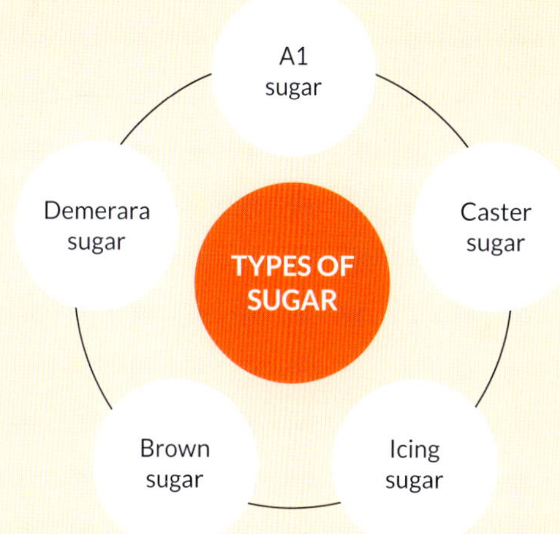

5. Explain how eggs can contribute to the aeration of a cake.
6. Describe the way butter helps to aerate cakes made by the creaming method.
7. List three important roles of milk in a cake batter.
8. Why are chemical raising agents used in many baked products?
9. Explain how chemical raising agents aerate cakes.
10. Why are spices included in recipes for baked products? Give two examples of spices that are often used in cake and biscuit recipes.

How baking works

During baking, exposure to dry heat in the oven causes physical and chemical changes to the ingredients in the mixtures of cakes, biscuits, pastries and yeast products. As the temperature rises, leavening occurs within the raw mixture. As the liquid evaporates to a gaseous steam, expansion of the molecules occurs, thus increasing the volume of the product and creating a light, porous texture.

The firm structure of baked products is formed when the protein in the egg becomes solid as it begins to coagulate, and the gluten – the protein in flour – sets to form a framework around the air bubbles.

Figure 4.5 A freshly baked butter cake

Rising

- Cake batter warms
- Chemical raising agents generate bubbles
- Bubbles within beaten eggs expand
- Water produces steam that pushes batter up

↓

Setting

- Egg protein unwinds, and reforms and combines with other ingredients
- Coagulation occurs
- Starch in flour absorbs remaining liquids
- Delicate crumb structure sets

↓

Colour formation

- Outside of cake browns
- Aroma and flavours develop
- Maillard reaction occurs
- Egg protein shrinks, so cake pulls away a little from cake tin

↓

Cooling

- Molecules are very active in cakes that are just out of the oven
- Texture is very soft and fragile
- Cake should be rested in tin for short time
- Cake should then be turned onto rack until completely cool

Figure 4.6 Stages in baking a cake

There is significant colour development on the outside of products during baking. The proteins combine with the reducing sugars to assist with the development of the golden brown colour on the crust. This process is the Maillard reaction. In addition, the starch in the flour undergoes the process of dextrinisation. The dry heat in the oven breaks down the starch molecules, which contributes to a brown crust on the outside. These chemical changes also result in the development of the delicious aromas of cakes and bread baking. Sugars undergo the process of caramelisation when heated to high temperatures and, as a result, also add to the golden brown colour and flavour of baked products.

Top tips for making cakes

1. Accurately measure ingredients, as the ratio between butter, sugar, flour and eggs has been carefully calculated for success.
2. Butter and line cake tins before starting to mix ingredients to minimise loss of volume from the mixture.

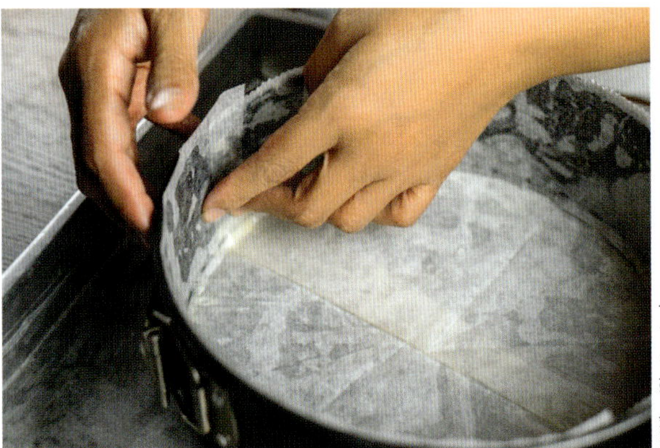

Figure 4.7 Butter and line cake tins before starting to mix ingredients.

3. Sift flour and other dry ingredients, such as cocoa and spices, before adding them to the mixture. This removes lumps that may not break up during mixing, incorporates air into the dry ingredients to help create a light texture, and combines all the dry ingredients.
4. Do not over-mix a cake batter; this causes the gluten to form long strands that will make the cake dense and heavy. Just mix until all the ingredients are combined and the batter is smooth.
5. In quick-mix recipes, melt the butter over low heat to prevent it from burning.
6. In recipes where the butter and sugar are creamed together, remove the butter from the refrigerator a few hours before creaming so that it will be at room

temperature; alternatively, soften it in a microwave for 10–15 seconds. When the butter is soft, it is easier to beat with sugar to aerate the mixture.

7 When you cook a sponge, it is important to be well organised, and to have all ingredients measured and tins prepared before you start to mix the batter. The mixture can lose volume if the procedure takes too long before baking.
8 Check cakes at least five minutes before the end of the specified cooking time to prevent overcooking.

Top tips for making biscuits

1. Allow space for biscuits to spread when you place them on the baking trays.
2. Bake biscuits at a low temperature, so that they can cook through without burning on the bottom.
3. If baking more than one tray of biscuits, swap the trays halfway through baking and turn the trays around, front to back. This helps to achieve even cooking if the oven being used has 'hot spots'.
4. After removing biscuits from the oven, loosen them with a metal spatula and allow them to cool on the tray.
5. Store biscuits in an airtight container so that they retain their crisp texture.

Figure 4.8 Freshly baked biscuits

Producing volume and a light airy texture in baked products

The ingredients and the processes used to combine them directly affect the aeration or leavening of cakes, biscuits, pastries and bread. This means that the size of the air bubbles, the tenderness of the crumb and the final sensory properties are influenced by one or more chemical, mechanical and/or biological actions.

Aeration of cakes, biscuits and pastries

Chemical aeration

Including food-safe chemicals – such as by using bicarbonate of soda, baking powder or self-raising flour – is one way to aerate a batter or dough to achieve a good-quality product. Sometimes, both chemical and mechanical raising agents are combined in the production, so that tiny air bubbles are created and then trapped and held by specific ingredients within the structure of a cake.

When using a chemical raising agent, it is important to work quickly and bake the product as soon as possible, because the raising agent is activated as soon as the mixture becomes moist. Working efficiently maximises the agent's aeration capacity, and the mixture is pushed up and out to increase the product's volume.

When using bicarbonate of soda, it is very important to accurately measure the raising agent, since too much will cause the baked product to taste bitter. To overcome the bitter after-flavour, bicarbonate of soda is used in combination with sweet ingredients, such as brown sugar or golden syrup, which have a strong, distinctive flavour that masks the bitterness. In contrast, baking powder, which has a mild flavour, is used in scones and muffins, as it does not affect their delicate flavour.

Mechanical aeration

Sifting, creaming butter and sugar together, whisking, beating and rubbing in are all processes that mechanically trap bubbles of air within a batter, pastry or dough to create a light, airy texture in the finished product.

When creaming butter and sugar together, the sharp edges of the sugar crystals help form small pockets of air in the butter. The mixture becomes lighter in colour and fluffier in texture, hence the descriptor – creamed butter and sugar should look like lightly whipped cream. During the cooking process, the tiny air pockets expand and contribute to the aerated texture and the increased volume of the end product.

The processes used to combine the key ingredients directly affect the size of the air bubbles, the tenderness of the crumb and the final sensory properties.

Mark Fergus Photography

Figure 4.9 Rubbing butter into flour

iStockphoto.com/mrcmos

Figure 4.10 Adding eggs to creamed butter and sugar

Table 4.2 Chemical and mechanical aeration

	Method	Aerating ingredients	Aerating action	Food examples
Chemical	Using bicarbonate of soda	Bicarbonate of soda – an alkaline white powder with a bitter flavour Must be combined with an acidic ingredient such as tartaric acid or cream of tartar Note: Is often combined with golden syrup or brown sugar to balance flavours	Becomes effervescent when mixed with a liquid and heat to produce CO_2	Anzac biscuits, gingerbread, steamed puddings
	Using baking powder	Baking powder – a combination of: • bicarbonate of soda • an acid such as tartaric acid or cream of tartar • starch filler	When mixed with moist ingredients, produces CO_2 to create a leavening action in the heat of baking	Scones, muffins
Mechanical	Sifting	Dry ingredients	Air is incorporated by passing dry ingredients through a sieve	All cakes and biscuits
	Creaming	Butter or margarine combined with sugar	Air bubbles are trapped by the fat around the sugar crystals Bubbles expand during cooking, causing the cake to rise	Butter cakes, fruit cakes, patty cakes, puddings
	Whisking	• Eggs and sugar, or • egg whites, or • egg whites and sugar	The protein in egg white stretches and traps tiny air bubbles	Sponge cakes, meringues, pavlovas
	Beating	Dry and wet ingredients	Air bubbles are introduced into the mixture using a circular motion	Quick-mix cakes
	Rubbing in	Flour and butter	The action of rubbing the butter into the flour with fingertips to resemble fresh breadcrumb coats the starch granules with fat and introduces air into dry ingredients	Scones, pastry

9780170495714

Aeration of bread and other yeast products

Biological aeration

Yeast is used as a leavening agent in many products, such as bread, pizza, crumpets and buns. Similar to baking powder, bicarbonate of soda or beaten egg white, yeast can aerate a dough to make it rise so that the final product has a light, cellular texture.

Yeast is neither plant nor animal, but a single-celled microscopic fungus. As yeast grows, it reproduces by budding; during this **fermentation**, the yeast converts the carbohydrates in the dough to CO_2, plus alcohol and water.

Natallia Harahliad/Shutterstock.com

Figure 4.11 Yeast fermenting

Fermentation is utilised in bread making to produce CO_2, which raises bread.

$$C_6H_{12}O_6 + \text{yeasts} \longrightarrow 2CO_2 + 2C_2H_5OH + \text{energy}$$

(glucose) → (carbon dioxide) (alcohol)

Budding of yeast

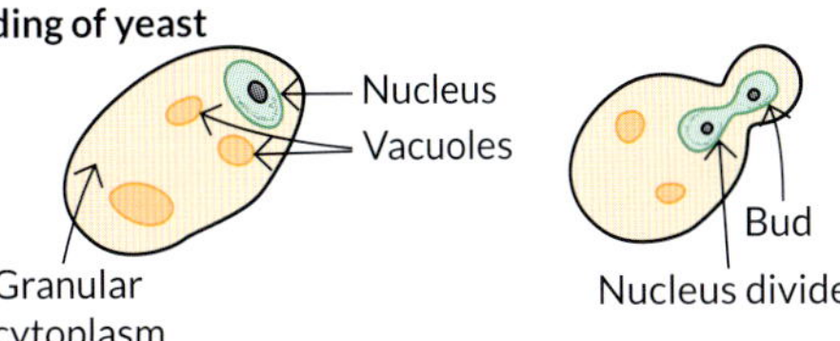

Figure 4.12 Reproduction of yeast cells

New Africa/Shutterstock.com

Figure 4.13 A pizza with a base made from yeast

Table 4.3 Biological aeration

	Method	Aerating ingredients	Aerating action	Products
Biological	Fermentation	• Yeast (fresh or dry), and • flour with high gluten content, and • warm water and/or milk	Fermentation – sugars and some of the starch are converted to alcohol and CO_2 as the yeast ferments	Bread, bread rolls, pizza bases, crumpets, buns

Table 4.4 Types of yeast

Active dry yeast	Compressed fresh yeast
Anton Starikov/Shutterstock.com	Jiri Hera/Shutterstock.com
• Looks like tiny, dehydrated granules • Available in sachets or large packets • Stored in the refrigerator or freezer • The yeast cells are alive but dormant, because of the lack of moisture • When mixed with a warm liquid, the cells are activated and begin to grow and produce alcohol and CO_2	• Has a creamy colour, is firm but with a moist texture, and has a sweet aroma • Purchased in blocks • Must be stored in an airtight container in the refrigerator because it becomes stale quickly • Loses its ability to leaven a dough as it becomes stale

PRACTICAL ACTIVITY 4.1

Scientific experiment: The growth of yeast

Try these tests to determine the ingredients and the environment that yeast requires to grow and produce good bread. You will need:

- 8 test tubes
- 8 balloons
- 8 × 1 teaspoon fresh or dried yeast
- 6 × ½ teaspoon sugar
- 4 × 1 teaspoon flour
- 2 × 1 teaspoon bread improver
- 4 × 50 millilitres warm water (36.5°C)
- 4 × 50 millilitres iced water.

Method

1. Put 1 teaspoon of yeast, then 50 millilitres of iced water into four of the test tubes. Put 1 teaspoon of yeast, then 50 millilitres of warm water (36.5°C) into the other four test tubes.
2. Following Figure 4.14, add other ingredients to the test tubes, as shown. Stir each test tube well.
3. Stretch a balloon over the neck of each test tube and watch what happens after 5, 10, 15 and 30 minutes.

Results

Record your results.

Conclusion

1. After observing each test, determine the ingredients and environments in which yeast grows best.
2. What implications can you draw from the results about the temperature of the environment required for making bread?

Worksheet
Practical activity 4.1

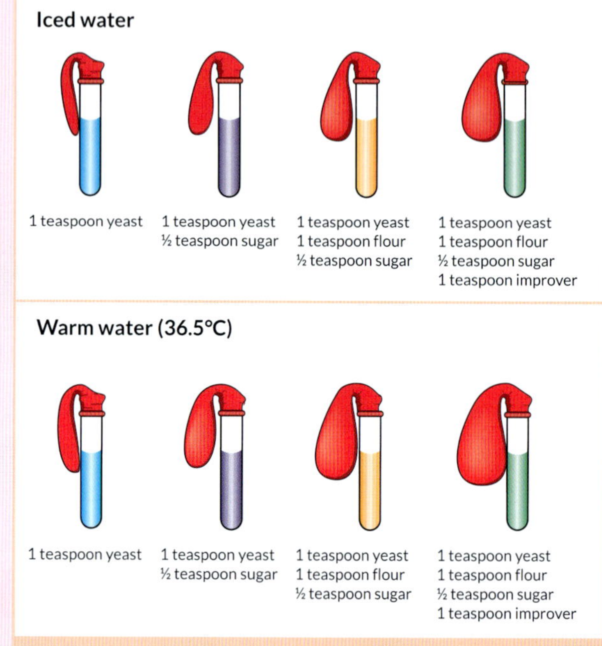

Figure 4.14 Add other ingredients as shown.

Figure 4.15 Freshly baked sourdough baguette

9780170495714

Processes in yeast baking

Fermentation
- Yeast is mixed with moisture and food in a warm environment
- As the yeast ferments, sugar and starch are converted to alcohol and CO_2

↓

Kneading
- Kneading strengthens the gluten in the flour so it is more elastic
- The dough stretches and captures the CO_2 bubbles

↓

Proving
- The CO_2 bubbles are trapped within the gluten structure of the dough in a warm environment
- Fermentation continues and the size of the dough increases

↓

Shaping and second proving
- The dough is knocked back and kneaded again to evenly distribute the air cells for an even texture
- The dough is shaped before proving for a second time in a warm environment

↓

Baking
- The high temperature in the oven kills the yeast and evaporates the alcohol
- A strong aroma of freshly baked yeast product is created
- The structure of the bread is set and the outside becomes golden brown

↓

Cooling
- The bread is done if it sounds hollow when tapped on the bottom
- The bread is removed from the tin to retain the texture of the crust
- The molecules are very active, so the bread is cooled for 30 minutes before slicing

Figure 4.16 Stages in producing a yeast product

Top tips for making and baking yeast doughs

1. Use flour with a high gluten content so that the dough has the strength to capture all the bubbles of CO_2 produced by the yeast during fermentation. Adding bread improver to the flour speeds up the fermentation of yeast and improves the quality of the dough.
2. **Knead** the dough, using the heel of your hand and a firm action. This will develop the gluten and ensure a good structure in the finished product.
3. **Prove** the dough in a warm environment that is free from draughts. Cover with oiled plastic wrap or a damp tea towel to prevent the surface of the dough drying out and forming a crust as it proves.
4. Prove shaped rolls close to each other on the tray so that they can support one another and not collapse during proving.
5. Spray the dough with water just before it goes into the oven – this helps the dough to stay moist so it can expand at the start of baking and create a good crust.
6. Cook yeast doughs in a hot oven.
7. The dough is cooked if it is golden brown and sounds hollow when tapped. The browning on the outside of the loaf occurs as a result of the Maillard reaction and dextrinisation.

Testing knowledge 4.2

11. Design a flow chart that demonstrates the changes that occur to a raw cake batter when it bakes in the oven.
12. Create an annotated diagram to demonstrate the chemical changes that occur to products when they are baked.
13. Sifting, creaming, whisking, beating and rubbing in are processes that incorporate air into a cake. Describe how each process achieves aeration.
14. Explain why it is important for butter to be at room temperature before creaming.

Quiz Testing knowledge 4.2

15 When making biscuits, explain why it is recommended to:
   a swap trays around in the oven halfway through baking
   b store biscuits in an airtight container.

16 Describe how creaming butter and sugar together contributes to the aeration of a cake batter.

17 Explain why brown sugar or golden syrup is included in recipes that use bicarbonate of soda as a raising agent.

18 Outline the three aspects of the environment that need to be considered when fermenting yeast.

19 Create a mind map that provides some tips for making a successful yeast dough.

20 Identify one observable change that occurs in the following stages to produce a loaf of bread.

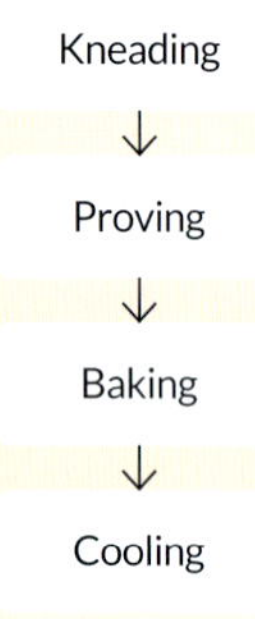

Creating structure in baked products

The function of proteins in flour, eggs, milk and butter is to contribute to the structure of baked products. This means that they set through the heating that occurs to the batters or doughs during baking. The way that proteins interact with other ingredients varies, and results in the different sensory properties in recipes.

Denaturation

Denaturation describes the permanent structural change of the protein molecules in food. This can occur by the application of heat, mechanical action or the addition of acids.

Imagine you are using a microscope, and you can see the individual strands of protein in raw meat, eggs or fish. The protein will look like a loose coil with water molecules attached to its outside surface. When the protein is heated – that is, cooked – it is 'denatured' and the coils tighten. If meat is overcooked, the coils continue to tighten and squeeze the water molecules out, so the meat is dry and tough to eat.

Coagulation

Coagulation is a form of denaturation and occurs when there is a permanent change in the protein from a liquid into a thick mass as a result of heat or the additions of acids.

For example, when eggs are poached, the physical structure alters – the white changes from a clear, thick liquid to a white, firm texture and the yolk changes from a yellow liquid to having a firm texture. The addition of a small amount of acid such as vinegar decreases the pH, so the proteins become less stable and more likely to denature and coagulate. By combining heat and an acid during poaching, the egg coagulates and holds its shape at a lower temperature, making the end product more appealing.

iStock.com/JoeGough

Figure 4.17 Poached eggs

When beating egg whites to make meringue, a small amount of cream of tartar – which is an acid – can be added. This addition of acid in the early stages of beating stabilises the egg white foam.

Using mechanical action to aerate eggs

Eggs – either whole, yolks or whites – trap air as they are beaten to create a foam made from millions of bubbles that will increase the volume and create a light texture in a mixture.

MaraZe/Shutterstock.com

Figure 4.18 Beaten egg whites aerate food products.

9780170495714

Egg whites

Egg whites can aerate products such as meringues, souffles and sponge cakes. When mechanical action is applied to egg whites – that is, they are whisked or beaten – the protein, ovalbumin, is denatured, creating a foam that will aerate a mixture. As the egg whites are beaten, the protein is stretched, forming a thin, elastic membrane that traps air bubbles. This mesh-like structure stabilises the air bubbles in the foam, and this trapped air expands when heated in the oven. The expanding bubbles lift the meringue, souffle or cake until the temperature of the mixture is hot enough to set the egg protein.

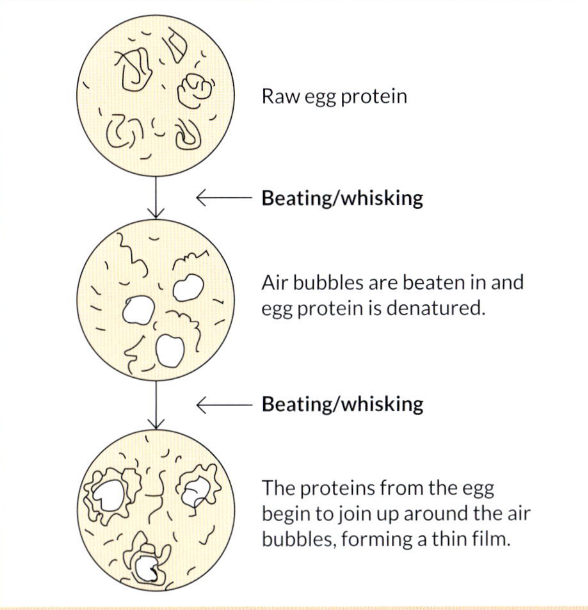

Figure 4.19 Stages in beating an egg white foam

Overbeating egg whites overstretches the protein molecules, causing the foam to collapse. Adding sugar to a beaten egg foam will stabilise the mixture and maximise its volume when baked.

Chatham172/Shutterstock.com

Figure 4.20 Egg whites aerate a cheese souffle.

Top tips to aerate egg whites

To maximise the foaming properties of egg whites when preparing them in the kitchen, it is important to:

- ensure that the whites are at room temperature
- ensure that the whites contain no trace of egg yolk and therefore no fat
- make sure all utensils are clean and dry
- avoid using plastic utensils, because they tend to retain fatty substances and inhibit the foam development.

Note: If egg whites have to be measured, 1 egg white is equivalent to 30 millilitres or 1½ tablespoons.

Elena Demyanko/Shutterstock.com

Figure 4.21 Beaten egg whites give this sponge cake its light and airy texture.

Addition of acids

Acidic ingredients such as vinegar and lemon juice are often combined with flavourings to make a marinade in which to soak meat, poultry or fish before cooking.

The primary function of an acidic ingredient in a marinade is to tenderise. This means the acid begins to denature or break down the protein bonds within the structure of the muscle fibres in the food, so that it becomes softer and easier to chew.

The length of time the ingredients are marinated varies. Meat and poultry can be marinated for several hours or overnight, whereas fish only requires a few minutes of marinating, because it has very little connective tissue. Meat, poultry or fish should be kept in the refrigerator while it is marinating to prevent the growth of spoilage microorganisms.

The benefits of the marinating process are two-fold: the texture of the food is improved, and the flavours have time to develop before cooking.

stockcreations/Shutterstock.com

Figure 4.22 Marinades enhance barbequed meat.

When selecting additional ingredients for a marinade – for example, sauces – avoid those with a high sugar or thickener content, as they will burn during cooking. These ingredients can be added later, just after cooking.

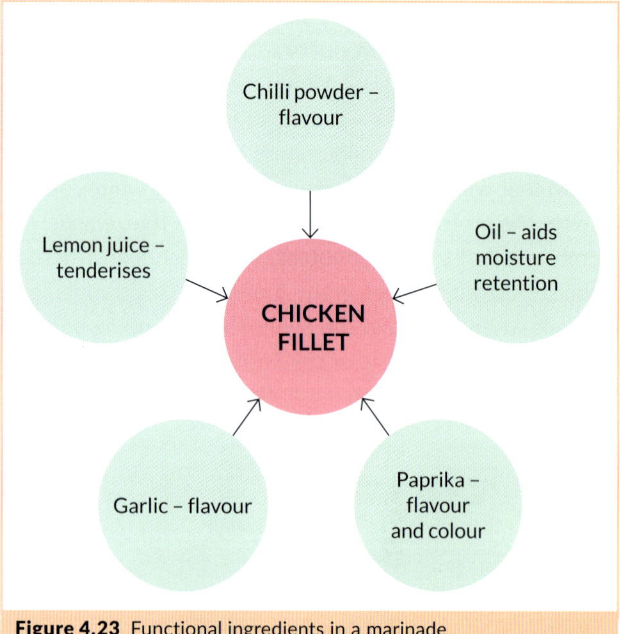

Figure 4.23 Functional ingredients in a marinade

Activity 4.1

Cooking around the world with acids

1 Research recipes that incorporate acids and are typical of the cultural food in specific parts of the world. The acid ingredient may be used to enhance the texture and/or flavour of the main ingredient.

   a Select one country per region, and then select one recipe that reflects its cuisine and includes an acid.

   b Record the functional ingredients for each product in a table similar to the one below.

Region	Country	Recipe (including reference)	Ingredient being marinated	Other main ingredients	Acidic ingredients	Other flavourings
Europe						
The Americas						
Asia						
North Africa						

2 Annotate your table as follows:

   a Identify whether each of your selected recipes was a main meal, side dish or snack food.

   b Identify whether the acidic ingredients were included to improve the texture and/or flavour of other ingredients.

   c Classify the other flavourings as spices, herbs or sauces.

3 After considering the anticipated sensory properties of your four selected recipes, which one appeals to you most? Explain your answer.

9780170495714

Testing knowledge 4.3

Quiz
Testing knowledge 4.3

21 Identify what can cause proteins to denature during the preparation and cooking of food.
22 Describe how heat alters protein molecules during cooking.
23 Explain how the structure of protein changes during coagulation.
24 Poached eggs are a popular food for breakfast. Describe how poaching alters the structure of eggs.
25 The mechanical action of beating or whisking egg whites is used to produce meringues. Draw a flow chart or diagram to demonstrate how egg whites change to aerate and increase the volume of a meringue.
26 Select an image – photo or diagram – that demonstrates each 'top tip' to aerate egg whites. Remember to record the source of each image.
27 List two ingredients that are classified as acids and are used in food preparation.
28 Explain why meat and poultry are marinated before cooking.
29 Why must marinading food be stored in the refrigerator?
30 Why is fish marinated for a short time only?

Colour development using dry heat

When protein, starch and sugar are exposed to the high temperatures of dry cooking methods, chemical reactions such as the Maillard reaction, dextrinisation and caramelisation occur. These reactions brown food products, and also aid the development of delicious aromas and flavours.

Maillard reaction

The Maillard reaction turns food brown and creates pleasant volatile aromatic compounds when exposed to dry heat. It is a reaction between amino acids in proteins and reducing sugars. It begins at about 154°C, so the browning effect can be seen when sautéing on the cooktop or baking in the oven.

A simple way of understanding the Maillard reaction is:

amino acid + reducing sugar + heat → browning + flavours

Dry cooking methods – such as baking, roasting, grilling and frying – heat the proteins that are combined with reducing sugars, and contribute to flavour development as well as browning. Also, as the surface of food browns, some dehydration occurs and the surface starts to crisp up, which in turn creates a contrast in textures when you bite into the food.

Baking a cake in the high temperature of the oven allows you to observe the Maillard reaction through the oven window. The reducing sugars and proteins in the cake batter begin to interact, and the chemical reaction browns the top and outside edges of the cake. The sugar caramelises, and delicious flavours and aromas develop.

In baked products

When baking cakes, biscuits, muffins and yeast doughs, the Maillard reaction occurs as reducing sugars and proteins from flour, egg or milk in the mixture are exposed to high temperatures in the oven. This reaction produces a golden crunchy crust on the baked food and a desirable flavour, often likened to roasted coffee or cocoa. The higher the sugar content, the darker brown the surface appears.

Green Creator/Adobe Stock Photos

Figure 4.24 The Maillard reaction produces a golden crunchy crust on baked food and a desirable flavour.

Frying crumbed food

Crumbing is a process that can be used with meat, poultry and fish to create a crisp, golden brown coating on the outside of the food. It is best suited to small, tender pieces of food that can be cooked quickly.

The food is first coated in seasoned flour to create a dry surface, and is then dipped into a beaten egg mixture, or the beaten egg mixture is brushed over the food; this makes a sticky surface for the crumbs to adhere to. Crumbs are then pressed onto the surface of the food. The crumbed food should be rested in the refrigerator for 30 minutes – this allows the moisture to be evenly distributed in the crumb coating, and helps to prevent the crumb coating from falling off the food during cooking.

stockcreations/Shutterstock.com

Figure 4.25 Crumbed fish fillet

Finally, the crumbed food is exposed to a dry cooking method – frying. The high temperature of the hot oil activates the Maillard reaction on the outside of the food, creating a golden brown crust. The heat inside the crumb coating steams the food.

Some examples of protein foods that can be crumbed and fried are lamb cutlets, beef schnitzels, chicken fillets and fillets of fish. Crumbed slices of vegetables such as eggplant and zucchini are delicious too.

Crumbing fish patties

Step 1: Place seasoned flour on paper towel. Coat the fish patties in the seasoned flour.

Mark Fergus Photography

Step 2: Beat together egg and milk. Place the patties on a flat plate. Brush the egg mixture over the patties.

Mark Fergus Photography

Step 3: Place breadcrumbs on a paper towel. Firmly press the breadcrumbs onto the patties with a metal spatula.

Mark Fergus Photography

Dextrinisation

Dextrinisation is the process that occurs when starch is exposed to dry heat – the starch is broken down to dextrin, resulting in a significant change in colour on the outside of food. The formation of dextrin alters the sensory properties of food. It changes the colour of starch from white to brown, and develops a slightly sweet, nutty flavour. A desirable 'cooked' or earthy aroma is produced. The texture of the food is altered too, because the dry heat dehydrates and crisps up the outside layer, so it becomes crunchy.

Examples of dextrinisation in cooking are:

- toasting a slice of bread, which turns brown
- baking a loaf of bread – the crust of outside edge turns brown
- browning a roux for gravy
- baking starchy vegetables, such as potatoes, to develop a brown outer layer.

Toasted Pictures/Shutterstock.com

Figure 4.26 The dextrinisation of starch develops the colour and flavour of toast.

9780170495714

Caramelisation

Caramelisation is the process that sugars undergo where they are heated to high temperatures and their molecules begin to break apart.

Mark Fergus Photography

Figure 4.27 Caramelising sugar

Consider table sugar crystals, which have no colour or aroma but are sweet to taste in their original form. When they are heated, a high temperature creates lots of new and different compounds that can be bitter or very aromatic, and create a deep brown colour. The more sugar is cooked, the less sweet it becomes and the darker its colour.

When making caramel, the sugar is usually dissolved in a little water – this enables the sugar to be heated over high heat immediately and reduces the danger of burning, as it allows the process of caramelisation to take place over a longer period of time. Only refined white caster sugar or table sugar should be used to make caramel, because these sugars have the fewest impurities that can cause crystallisation during cooking.

To achieve the specific colour and flavour required for a product, a sugar thermometer will help to take the guesswork out of making caramel.

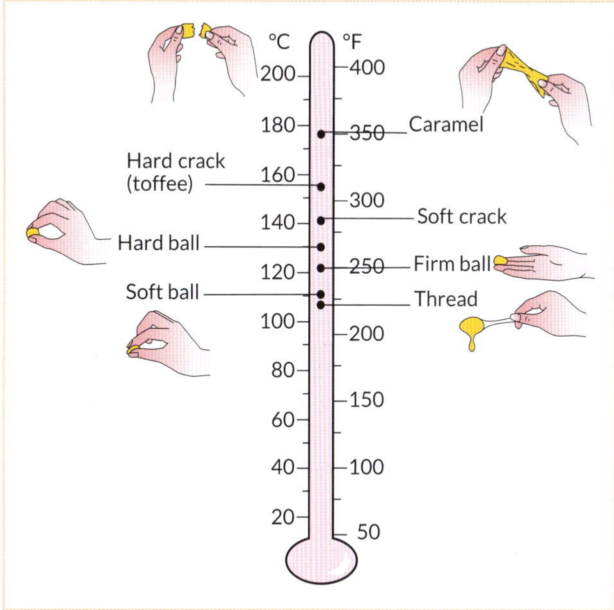

Figure 4.28 Stages in caramelisation

Thickening a mixture

Gelatinisation

Gelatinisation is the process that occurs when starch granules absorb liquid in the presence of heat and thicken the liquid, forming a gel.

Flour, which is used as a key ingredient in many food products, is made from grinding the endosperm – that is, the starch part – of cereal grains, which have been processed to remove the outer layer or husk.

Starch is not soluble in water. When starch is blended with water and allowed to stand, it will settle to the bottom of the bowl. If the starch solution is heated, the starch granules absorb the water and begin to swell. The swollen granules then take up more space, and the solution becomes more viscous and thickens to form a gel. During heating, the mixture – for example, a cheese sauce for lasagna – must be stirred to keep the starch evenly dispersed so that the end product is smooth and lump-free.

Boiling or steaming vegetables that are high in starch, such as potatoes, allows the starch granules to absorb moisture and swell or break down. For example, when making mashed potatoes, the raw potatoes have a firm texture to begin with, but after boiling they become soft and fluffy as the starch gelatinises.

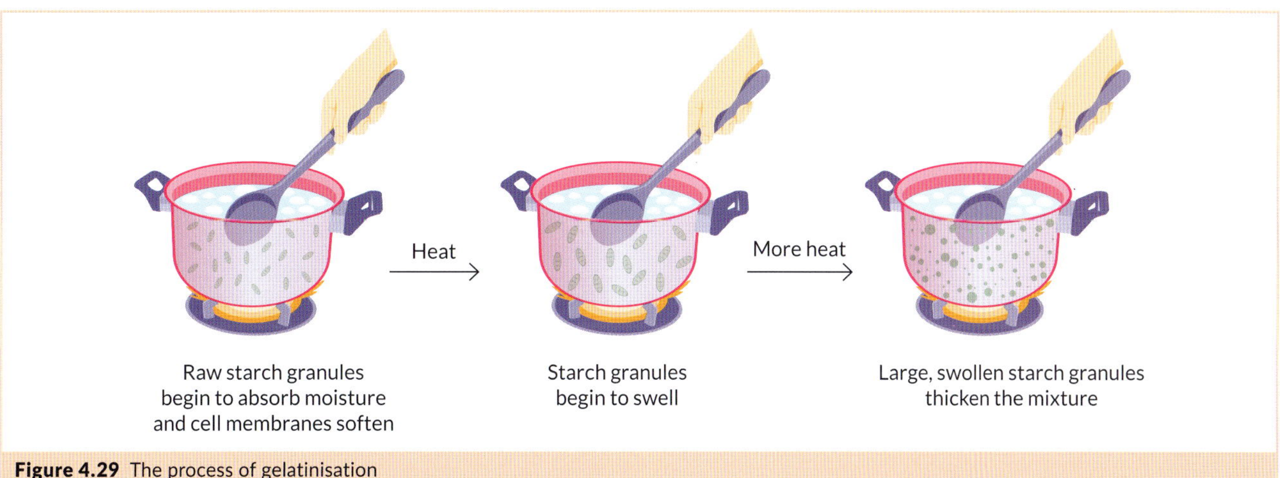

Figure 4.29 The process of gelatinisation

9780170495714

Ingredients such as rice and pasta are boiled or simmered in water or stock. Exposing these starchy foods to moisture and heat at the same time allows the starch granules to absorb the liquid, become tender and gelatinise. As pasta cooks and the starch expands, the protein network of the gluten sets or coagulates, and the pasta will hold its shape.

skaman306/Moment/Getty Images

Figure 4.30 As pasta cooks, the starch gelatinises.

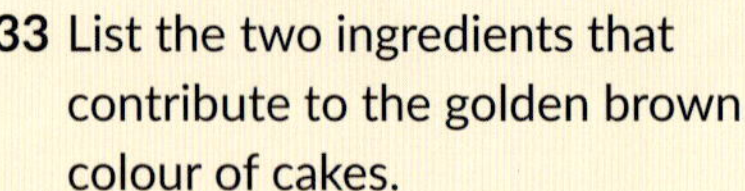

Testing knowledge 4.4

Quiz
Testing knowledge 4.4

31 Define the term 'Maillard reaction'.

32 Identify the dry methods of cooking that can activate the Maillard reaction.

33 List the two ingredients that contribute to the golden brown colour of cakes.

34 Explain the processes that occur when crumbed foods are fried.

35 Explain how dextrinisation changes the sensory properties of some foods.

36 What is caramelisation?

37 Why is refined sugar the best type of sugar to caramelise?

38 Which part of a cereal grain is used to thicken mixtures?

39 Describe the process of gelatinisation.

40 Explain how gelatinisation occurs when rice and pasta are cooked.

Critical and creative thinking 4.1

Comparing sensory properties of food

Copy and complete the following table to compare the sensory properties of food cooked using different methods, and identify the processes responsible for the differences.

Food items	Description of the similarities and differences of the food items	Processes responsible for the differences in sensory properties
Fresh bread Toasted bread		
Raw tuna Seared tuna		
Boiled meat (casserole) Grilled meat		
Steamed potatoes Roast potatoes		

9780170495714

Anzac Biscuits

Chemical aeration occurs in the production of Anzac Biscuits. Bicarbonate of soda is the alkaline ingredient which combines with an acidic ingredient – golden syrup – to produce the light, crisp texture. The intense flavour of golden syrup balances out the bitter flavour of bicarbonate of soda.

INGREDIENTS

1 cup flour

¾ cup caster sugar

1 cup rolled oats

1 cup coconut

125 grams butter

2 tablespoons golden syrup

1 teaspoon bicarbonate of soda

3 tablespoons boiling water

METHOD

1. Preheat oven to 180°C.
2. Sift the flour into a basin. Add the sugar, rolled oats and coconut.
3. Melt the butter and golden syrup in a small saucepan.
4. Dissolve the bicarbonate of soda in boiling water and add to melted butter and syrup.
5. Add the melted mixture to the dry ingredients and mix well.
6. Roll into 3-centimetre balls and place on an oven tray. Flatten with a fork or spatula.
7. Bake for 15–20 minutes.
8. Allow to cool for 5 minutes on the tray and then lift onto a cooling rack.

EVALUATION

1. Describe the sensory properties – appearance, aroma, flavour, texture and sound – of your Anzac Biscuits.
2. Why is it important to heat the golden syrup with the butter in this recipe?
3. What is the purpose of the rolled oats in these biscuits?
4. Predict the effect of cooking the biscuits at 220°C rather than 180°C.
5. Identify the ingredients in this recipe that are classified in the 'only sometimes and small amounts' section of the *Australian Guide to Healthy Eating*. Explain why these ingredients should not be included as a regular part of the daily diet.

iStockphoto.com /Kateryna Kravchuk-Rudomotkina

Chocolate Sponge

Beating egg whites and sifting the dry ingredients are the mechanical actions that create the light, fluffy texture of a sponge cake. Chemical raising agents also contribute to the high volume of the cake.

INGREDIENTS

spray oil and extra flour, for preparing the tins

½ cup cornflour

1 teaspoon plain flour

½ teaspoon bicarbonate of soda

½ teaspoon cream of tartar

1 tablespoon cocoa

1 tablespoon golden syrup

3 eggs (60–70 grams), at room temperature

½ cup caster sugar

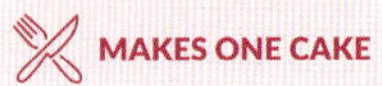

EVALUATION

1. Why were the dry ingredients sifted twice before being folded into the egg mixture?
2. Explain the role of the bicarbonate of soda, the cream of tartar and the golden syrup in the mixture.
3. Identify the processes you followed when separating and beating the egg whites to maximise the volume of the sponge cakes.
4. Describe the tests you used to determine when the sponge cakes were cooked.
5. Classify the ingredients for the Chocolate Sponge on a diagram of the *Australian Guide to Healthy Eating*. Explain why foods such as the Chocolate Sponge are included in the 'only sometimes and in small amounts' section of the *Australian Guide to Healthy Eating*.

METHOD

1. Cut two circles of greased paper and line two 20-centimetre round cake tins. Lightly grease with spray oil and dust with plain flour, tapping to remove extra flour.
2. Preheat oven to 180°C. Check the oven racks to make sure both tins can sit side by side on shelves near the centre of the oven.
3. Sift the cornflour, plain flour, bicarbonate of soda, cream of tartar and cocoa twice.
4. Slightly warm the golden syrup in a microwave before measuring. This makes it easier to fold through the mixture.
5. Separate the egg whites from the yolks, placing the whites in a large, clean, dry bowl and retaining the yolks.
6. Beat the egg whites until stiff, then beat in the egg yolks one at a time.
7. Gradually add the sugar one tablespoon at a time until a stiff meringue is formed. Add the golden syrup and beat until combined.
8. Add the sifted dry ingredients and lightly fold through with a metal spoon.
9. Divide the mixture between the two greased and lined tins. Tap lightly to remove any large air bubbles.
10. Bake for 15–20 minutes. When cooked, the cakes will begin to leave the sides of the tins and will spring back when lightly touched with the fingers.
11. When cooked, tip out each cake onto a sheet of baking paper and remove the paper lining. Quickly turn the cakes over and allow to cool on a cake rack.
12. The cakes can now be sandwiched with whipped cream. The tops can be spread with cream and topped with strawberries or iced with chocolate icing.

Mark Fergus Photography

9780170495714

Basic Bread

Yeast is the biological raising agent in bread. While the bread dough is proving, the fermentation of yeast takes place and the bubbles of carbon dioxide that are produced increase the volume of the dough.

INGREDIENTS

2½ cups bread flour

1½ teaspoons freeze-dried yeast

1¼ teaspoons bread improver

½ teaspoon salt

½ teaspoon sugar

½ teaspoon oil

200–250 millilitres warm to hot water

water spray

canola oil spray for greasing tray

METHOD

1. Sift all dry ingredients together. Mix in the oil and approximately two-thirds of the hot water to make a moist dough. Gradually add the remaining water, if required.
2. Cover the surface with plastic wrap and leave in a warm place to prove and double in size.
3. Turn out onto a floured board and lightly knead into a smooth dough.
4. Make into the desired shape and place on a greased tray. Lightly cover with oiled plastic wrap. Leave shaped loaf or rolls in a warm place to double in size.
5. Remove the plastic wrap and spray lightly with water before placing in the oven.
6. Bake a loaf in a preheated oven at 230°C for 20–25 minutes or bake rolls for 15–20 minutes or until golden brown. Bread will sound hollow when tapped on the base if it is cooked.

EVALUATION

1. Describe the sensory properties – appearance, aroma, flavour, texture and sound – of your Basic Bread.
2. List the three environmental factors that are important to ensure that a yeast dough rises.
3. Why is it necessary to prove a yeast dough?
4. Explain why a yeast dough is kneaded before shaping.
5. Discuss why bread is classified into the 'grains (cereal) foods' section of the *Australian Guide to Healthy Eating*.

Mark Fergus Photography

Poached Egg

During the process of poaching an egg, the acid in the water and the temperature of the cooking solution cause the proteins in the egg to denature and coagulate. This results in the egg changing from a liquid ingredient to a firm mass that retains its shape.

For a delicious breakfast, mix half a smashed avocado with a squeeze of lemon, salt and black pepper. Spoon the avocado mixture on a slice of hot, buttered toast and top with the poached egg.

INGREDIENTS

1 egg

1 teaspoon vinegar

To serve (optional)

½ avocado, smashed

squeeze of lemon juice

hot, buttered toast

salt and black pepper

SERVES ONE

METHOD

1. Crack the egg into a jug or cup.
2. Fill a frying pan with water to about 2 centimetres deep. Add a teaspoon of vinegar. The vinegar will lower the temperature at which the protein in the egg coagulates and help to set the egg.
3. Bring the water to the boil. Once boiling, stir briskly with a wooden spoon to create a whirlpool.
4. Gently pour the egg into the centre of the swirling water. The swirling water helps to keep the egg in a round shape.
5. Turn down the heat so that the water is just simmering.
6. Gently cook the egg for 2–3 minutes or until just set.
7. Carefully lift the egg from the water using a slotted spoon.

EVALUATION

1. What is the purpose of adding vinegar to the water when poaching the egg?
2. What does the swirling water do to the egg during poaching?
3. Describe how you can tell if water is simmering.
4. List one safety rule to consider when cooking a poached egg.
5. Explain why, from a nutritional perspective, poaching is a better method of cooking eggs for breakfast than frying.

Mark Fergus Photography

9780170495714

Cheese Omelette with Bacon

When the eggs in the omelette are poured into a hot frying pan, they begin to coagulate – the liquid begins to set, becomes firm and holds its shape due to coagulation. A golden brown colour forms on the surface of the omelette touching the pan as a result of the Maillard reaction.

INGREDIENTS

Cheese Omelette

2 eggs

1 tablespoon water

2 shakes pepper

1 teaspoon butter

¼ cup cheddar cheese, grated

1 tablespoon parsley, finely chopped

Bacon

2 rashers bacon, diced

SERVES ONE

METHOD

Cheese Omelette

1. Mix the eggs with the water and gently beat with a fork until well combined.
2. Season with the pepper.
3. Melt the butter in a non-stick frying pan over medium heat until it is foaming, but not brown.
4. Pour the egg mixture into the frying pan. Gently shake to ensure the omelette mixture covers the base of the pan.
5. Carefully lift the edges of the omelette with a spatula so that the mixture can run to the edges and set.
6. When the mixture is almost set, sprinkle the grated cheese and chopped parsley over the front half of the omelette.
7. Carefully lift the back half of the omelette over the top of the cheese and parsley.
8. Turn onto a warmed plate and serve immediately. Serve with microwaved or fried bacon.

Microwaved Bacon

Place a sheet of absorbent paper on a plate. Spread the diced bacon over the paper and cover with another sheet of absorbent paper. Microwave on high for 1–2 minutes or until crisp.

Fried Bacon

Place bacon in a frying pan and fry over medium heat until the bacon is crisp and lightly browned. Drain on absorbent paper.

Mark Fergus Photography

EVALUATION

1. Why is it important to beat the eggs and water until they are well combined?
2. Describe the changes that occur to the texture of the egg mixture as it cooks in the hot pan.
3. Identify the reactions that occur during the cooking of an omelette.
4. Discuss why it is important to lift the edges of the omelette.
5. Classify the ingredients for this recipe on a diagram of the *Australian Guide to Healthy Eating*. Explain why the Cheese Omelette with Bacon should only be eaten for breakfast occasionally.

Sweet Potato, Cauliflower and Red Lentil Curry

In this recipe, the high temperature and dry heat of roasting changes the sensory properties of the sweet potato and cauliflower. The Maillard reaction occurs and creates a golden colour, baked aroma and a slight bitter-sweet flavour in the vegetables. The red lentils thicken the curry and add a protein element to the dish.

INGREDIENTS

Sweet Potato and Cauliflower

¼ cauliflower, approximately 200 grams

350–400 grams orange sweet potato

2 tablespoons olive oil

pinch salt and pepper

Red Lentil Curry

½ onion, chopped

2 cloves garlic, chopped

3 centimetres fresh ginger, chopped

1 tablespoon olive oil

¼ cup Thai yellow curry paste

2 teaspoons fish sauce

1 cup vegetable stock

200 millilitres coconut milk

1 makrut lime leaf

½ cup split red lentils

50 grams baby spinach leaves

To serve

flatbread or steamed rice

lime or lemon wedges

METHOD

Sweet Potato and Cauliflower

1. Preheat oven to 200°C. Line a baking tray with baking paper.
2. Cut the cauliflower into florets, keeping some of the stem attached.
3. Peel the sweet potato and cut into 3-centimetre chunks.
4. Add two tablespoons olive oil and a pinch of salt and pepper into a clean plastic bag. Add all the cut cauliflower and sweet potato. Twist the bag closed, then massage the vegetables to coat them with oil. Turn onto the lined baking tray and arrange the vegetables in single layer.
5. Roast the vegetables in the oven for 20–25 minutes or until they have begun to turn golden brown. Remove from the oven and set aside.

Red Lentil Curry

1. Using a food mill, blend the onion, garlic and ginger into a paste.
2. In a medium to large saucepan, heat one tablespoon oil on medium-low heat. Add the onion paste and sauté gently for 1–2 minutes so the paste is just beginning to turn very pale golden colour.
3. Add the curry paste and fish sauce and stir through the onion paste.
4. Add the stock, coconut milk and lime leaf and bring to simmer. Add the red lentils, stir to combine the ingredients and then cover with a lid. Cook gently for 15–20 minutes or until the lentils are almost soft. Add an extra half cup of water if the mixture is becoming dry.
5. Add the roasted vegetables to the sauce and cook for 2 minutes, then stir in the baby spinach leaves until they begin to wilt. Remove the lime leaf.
6. Spoon into two serving bowls and add a squeeze of lime juice. Accompany with steamed rice or pieces of flatbread.

EVALUATION

1. Describe the sensory properties – appearance, aroma, flavour, texture and sound – of your Sweet Potato, Cauliflower and Red Lentil Curry
2. Identify the steps in the recipe where the Maillard reaction contributes changes in the properties of ingredients.
3. Explain why it is essential to the final sensory properties of the recipe to sauté the curry paste before adding in liquids.
4. Recommend three other dishes for a dinner that would complete the curry, if it was served as part of a shared table. Explain why you have selected each item to be part of the meal.
5. Classify the ingredients for this recipe on a diagram of the *Australian Guide to Healthy Eating*. Discuss how well the Sweet Potato, Cauliflower and Red Lentil Curry meets the guidelines of this food selection model.

9780170495714

Mark Fergus Photography

Bruschetta with Caramelised Onions

Onions are a naturally sweet vegetable and, when they are cooked slowly with dry heat, their flavour and colour deepen. The onions first soften, then change to a blond colour, and with more heating they become golden brown. The Maillard reaction and caramelisation change the sensory properties of the raw onions.

INGREDIENTS

Caramelised Onions

3 brown onions

2 tablespoons olive oil

¼ teaspoon salt

2 tablespoon brown sugar

1 tablespoon balsamic vinegar

Bruschetta

4 fresh asparagus spears

100 grams fresh ricotta

40 grams Greek-style feta cheese

pinch black pepper

4 thick slices sourdough bread, 1–2 days old

parsley or dill leaves to garnish

METHOD

Caramelised Onions

1. Peel the onions and cut them in half vertically. Lay the flat side on the chopping board and slice evenly into thin, half-moon slices. The onions must be sliced evenly, otherwise they will caramelise unevenly.
2. Heat the oil in a large frying pan over medium to low heat. Add the onion slices, sprinkle on the salt and sauté slowly for 15–20 minutes. A sprinkle of salt during cooking helps to draw out the moisture.
3. Stir occasionally to prevent the onion from catching on the bottom of the frying pan. Do not turn up the heat, as this will cause the onions to burn.
4. When the onions are softened and tinged golden, add sugar and balsamic vinegar.
5. Stir over low heat for another 5 minutes until the onions are slightly sticky.

Note: the Caramelised Onions can be used in a burger or on top of blanched green beans or broccoli. They can be stored in the refrigerator for 4–5 days.

Bruschetta

1. Cut the asparagus spears in half and blanch in boiling water for 1–2 minutes or until they are just tender when tested. Drain, refresh in iced water and drain again.
2. In a small bowl, combine the ricotta, feta and pepper.
3. Preheat the grill or a grill frying pan and cook each slice of bread on each side until toasted. Remove from the heat.
4. Divide the cheese mixture evenly and spread it on the toasted bread.
5. Dollop a spoonful of Caramelised Onions onto each slice of toasted bread.
6. Arrange the asparagus pieces on top of the onions and garnish with parsley or dill leaves.

9780170495714

EVALUATION

1. Identify the process and describe the changes that occurred to the sliced onion during cooking.
2. Why was salt added to the onion slices at the beginning of the cooking process?
3. Identify the reaction that occurred when the bread was toasted.
4. How did blanching alter the sensory properties of the asparagus?
5. Suggest three other combination of ingredients – protein and vegetables – that could be used to top bruschetta.

Mark Fergus Photography

Cauliflower, Cheese and Leek Gnocchi

Gelatinisation is responsible for the change in the texture of the gnocchi when it is cooked in boiling water. It is also evident during the production of the cheese sauce in steps 3 and 4. When heat is applied, the mixture thickens as the starch in the flour absorbs the liquid and changes its physical structure.

INGREDIENTS

300 grams cauliflower

1 leek, white part only

20 grams butter

½ cup chicken or vegetable stock

salt and pepper

250 grams store-bought potato gnocchi

Cheese Sauce

20 grams butter

1 ½ tablespoons plain flour

1 cup milk

pinch nutmeg

salt and pepper

30 grams cheddar cheese, grated

Topping

2 sprigs fresh thyme

60 grams fresh breadcrumbs (made from stale sourdough or ciabatta, if possible)

30 grams cheddar cheese, grated

METHOD

Vegetables

1 Cut the cauliflower into bite-size florets. Cut the leek in half lengthwise, then slice into 1-centimetre half-moon shaped pieces.

2 In a heavy-based saucepan with a tight-fitting lid, melt the butter over medium to low heat. As soon as the butter starts to bubble, add the cauliflower and leek. Stir gently so that the vegetables are coated with butter. Seal with a lid and cook gently for 5 minutes. Stir occasionally and do not allow the mixture to brown.

3 Add the stock and season with salt and pepper. Cook for 3 minutes uncovered so that the liquid evaporates and the cauliflower is just tender – it should still have some bite.

4 Remove from the heat and carefully spoon onto two foil baking trays, approximately 12 x 16 centimetres in size.

Cheese Sauce

1 In a small saucepan melt the butter over medium heat. Do not brown.

2 Add the flour and stir over the heat to form a thick paste. Remove from the heat.

3 Gradually add the milk, a quarter of a cup at a time, stirring to remove lumps before each addition. Add the nutmeg and season with salt and pepper.

4 Return the saucepan to the heat and bring to the boil, stirring consistently. Simmer for 1 minute and remove from the heat. Stir through the grated cheese.

5 Pour over the cooked cauliflower mixture, dividing the sauce equally between the foil baking trays.

Gnocchi

1 In a saucepan of boiling water, cook the gnocchi according to the instructions on the packet.

2 Drain the gnocchi, divide it into two portions and gently fold it into the cauliflower mixture on the foil baking trays.

Completing the dish

1 Preheat oven to 200°C.

2 Remove the leaves from the sprigs of thyme.

3 In a small bowl, mix the fresh breadcrumbs, grated cheese and thyme leaves.

4 Sprinkle over the cauliflower and gnocchi mixture and bake for 15–20 minutes until golden brown.

9780170495714

EVALUATION

1 Describe the sensory properties – appearance, aroma, flavour, texture and sound – of your Cauliflower, Cheese and Leek Gnocchi.

2 Sometimes this type of baked recipe is described as an 'au gratin' dish. Which cultural cuisine does 'au gratin' originate from and what are the characteristics of this style of recipe?

3 Which ingredients in the cheese sauce allow it to thicken? Identify the process that thickens the sauce and describe how it occurs.

4 Explain why gnocchi is classified as a type of pasta.

5 Suggest other dishes that could be served with the Cauliflower, Cheese and Leek Gnocchi to ensure that a main meal would meet the recommendations of the *Australian Guide to Healthy Eating*.

Mark Fergus Photography

WHERE AUSTRALIAN NATIVE FOODS ARE SOURCED

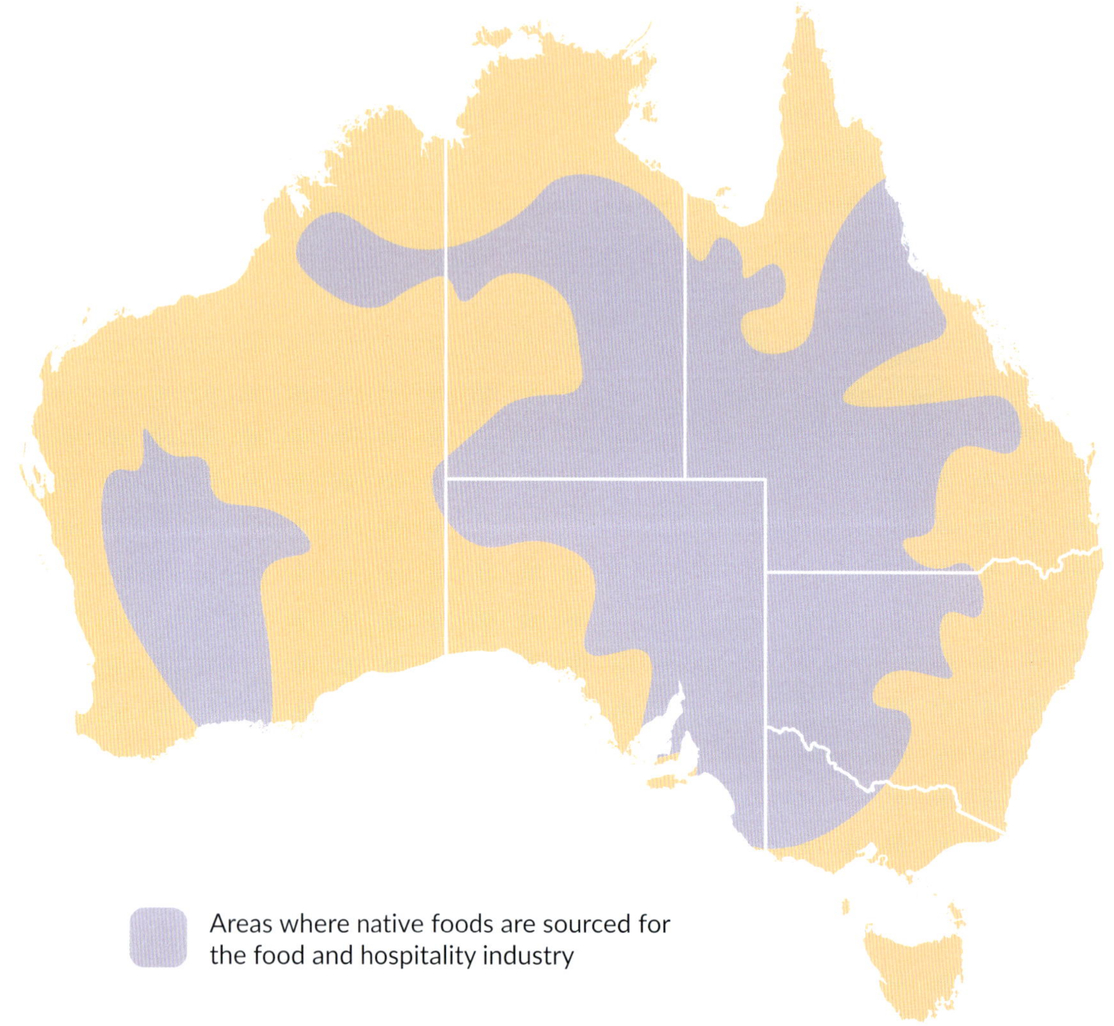

Source: https://gowrievictoria.org.au/changing-of-the-wurundjeri-seasons/

SOURCES OF NATIVE FOODS

PLANTS

- Yams
- Bush tomatoes
- Quandong fruit
- Kakadu plums
- Macadamia nuts

ANIMALS

- Ants
- Witchetty grubs
- Fish
- Kangaroos
- Emus
- Turtles
- Crocodile

9780170495714

5 INDIGENOUS FOODS

KEY TERMS

fire-stick farming the practice, undertaken by Australia's First Nations peoples, of using fire to burn vegetation so as to make animal hunting easier and to reorganise the composition of the plants and animals in the area

wild native foods the huge variety of edible native Australian herbs, spices, mushrooms, fruits, flowers, vegetables, animals, birds, reptiles and insects

Templates
- Decision table
- Sensory wheel
- Aboriginal and Torres Strait Islander Guide to Healthy Eating

Worksheets
- Practical activity 5.1
- Design activity 5.1

Quizzes
- Testing knowledge 5.1
- Testing knowledge 5.2

To access resources above, visit **cengage.com.au/nelsonmindtap**

NATIVE TOOLS & TECHNIQUES FOR FINDING FOOD

HUNTING & FISHING

- Spears
- Boomerangs
- Clubs
- Nets
- Harpoons
- Canoes

GATHERING

- Digging sticks
- Baskets
- Coolamons
- Grinding stones
- Stone tools
- Scrapers

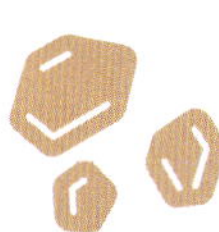

9780170495714

5

The foods of Australia's First Nations peoples

Before the arrival of Europeans and the beginning of colonisation, the traditional diet of Australia's First Nations peoples was diverse and nutrient-rich, and varied depending on their geographical location and the seasons. Those groups living near the sea or rivers had access to a diet abundant in fish, yabbies, mussels and other shellfish. In desert areas, protein was sourced from lizards or goannas. In temperate regions, kangaroos, witchetty grubs, possums, emus and other birds and their eggs were consumed. First Nations peoples across the country supplemented their diet with roots, yams, nuts, seeds, native greens, vegetables, nectars and bush fruits. However, the availability and quantity of food fluctuated based on climate, water supply and movement through different areas.

Australia's First Nations farmers

For many years, historians believed that before European colonisation, Australia's First Nations peoples were nomads who relied solely on hunting and gathering to collect their food supplies. However, this view is now disputed.

In his book *Dark Emu* (Magabala Books, 2014), Bruce Pascoe argues that Australia's First Nations peoples practised complex land management, and sustainable agriculture and aquaculture. He states that for over 50 000 years, they cared for their land using traditional methods that were sustainable and supplied them with all the food they needed. They considered the growth and regeneration cycles of plants, animals and birds, allowing them to sustain their food supply for thousands of years. Pascoe's research shows evidence that First Nations Australians used sophisticated systems; for example, they built dams and wells, and planted, irrigated and harvested crops. They also used sheds and secure vessels to store any surplus supplies for future use.

Archaeological evidence indicates that Australia's First Nations farmers used hoe-like implements to break up and loosen the soil, and that they bundled the cut grasses in heaps or stooks. Further evidence shows that they grew various crops, such as yams, native millet, macadamia nuts, fruits and berries. When yams were dug, a new 'crown' was always planted in the same place to ensure that the new crop would enable the food supply to be sustained. First Nations farmers also reared animals, such as dingoes, possums and emus until they were strong enough to survive on their own. They moved caterpillars to new breeding areas where there was a better food supply, and carried fish stock across the country to new lakes and streams where they could thrive.

Pascoe maintains the evidence reveals that, far from being hunter-gatherers, Australia's First Nations peoples were 'farmers without fences'.

Fire-stick farming

Fire was another tool used by Australia's First Nations peoples to help them secure and sustain their food supply. They used **fire-stick farming** to open up clearings of land so they could cultivate crops such as native millet, forming a patchwork of burnt and unburnt areas. This practice also helped to bring on new green grass that provided food for grazing animals, such as kangaroos, and lured them closer for hunting. Furthermore, Australia's First Nations peoples understood that many native plants require fire to stimulate flowering or seed germination.

Most importantly, Australia's First Nations peoples recognised that managing fire and making it work for them prevented uncontrolled fires that would wipe out their food supplies.

Ben Nottidge/Alamy Stock Photo

Figure 5.1 First Nations land management practices – including using fire to decrease the risk of bushfire and to regenerate flora – are being employed throughout Australia. This image shows an example in Cairns, North Queensland.

Aquaculture in Australia's First Nations communities

Aquaculture is a type of farming related to food sourced from waterways. For First Nations Australians living close to water, aquaculture was an important way in which communities farmed and caught their food. They designed and developed sophisticated tools and techniques.

In the Budj Bim National Park, near Lake Condah in western Victoria, there is evidence of early aquaculture, including stone traps and the remains of a large eel and fish farming system. This was built about 6600 years ago, making it even older than the Egyptian pyramids.

The Lake Condah area experienced volcanic activity when Budj Bim (previously known as Mount Eccles) erupted, tens of thousands of years ago.

9780170495714

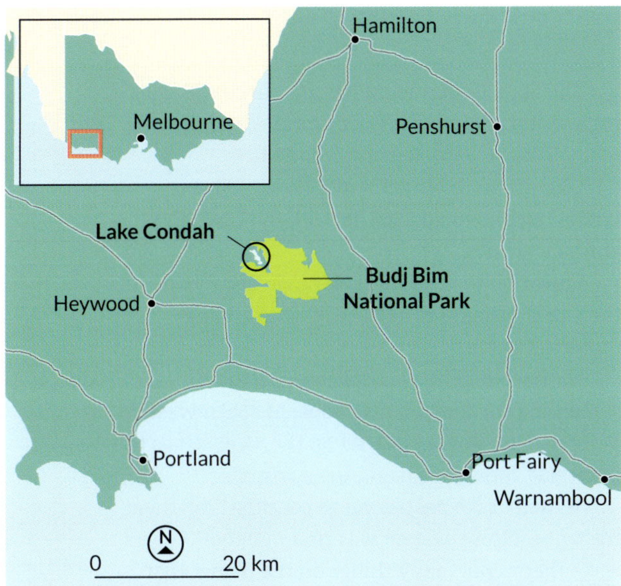

Figure 5.2 The location of the Budj Bim ancient Gunditjmara aquaculture system

The resulting lava flow changed the course of creeks and rivers and created a rocky landscape of basalt rocks, forming a network of pools, canals and channels in the surrounding area. The Gunditjmara people used engineering methods to create banks, form walls and channel the water flow into the pools and lake. Eels and fish were trapped in these natural depressions or holding ponds, which were formed by funnel-shaped rock structures. Hand-made nets of very fine sticks that were woven together or secured with native grasses were designed with one or more narrow entrances, which prevented eels and fish from swimming out once caught.

The eels provided a valuable and reliable source of food for the people in this area and oil for warmth. The Gunditjmara people smoked eels in the tree hollows and traded the smoked eels with nearby groups. There is also evidence of the wooden domed houses where people once gathered.

Weblink Lake Condah aquaculture

To learn more about Lake Condah, watch the video 'Lake Condah aquaculture' on the ABC Education website.

Andrew Bain/Alamy Stock Photo

Figure 5.3 Budj Bim eel traps

CASE STUDY 5.1

World Heritage status for Budj Bim

Read the newspaper article below and the previous information about Budj Bim, and then answer the questions that follow.

Ancient Indigenous aquaculture site Budj Bim added to UNESCO World Heritage list

Matt Neal, *ABC News*, 6 July 2019

A south-west Victorian Indigenous site that is older than the pyramids has been added to the UNESCO World Heritage List.

After more than a decade of hard work and lobbying, the Budj Bim Cultural Landscape near Portland was accepted onto the list at a meeting in Baku, Azerbaijan on the weekend.

The site was created about 6600 years ago by the Gunditjmara people, who used stones to build an elaborate series of channels and pools to harvest eels from Lake Condah.

There is also evidence at the site of stone dwellings, and trees that may have been used to smoke or preserve the eels that were caught.

Not only does Budj Bim bust the myth that all Indigenous people were nomadic and not agriculturally inclined, it is also considered one of the oldest aquaculture sites in the world.

Denis Rose, project manager for Gunditj Mirring Traditional Owners Aboriginal Corporation, said it had been a long journey to UNESCO recognition, but a valuable one.

'We first talked about this in 2002,' Mr Rose said.

'It's a very exhaustive process.

'We based it on a lot of evidence, and now that it's been decided, I'm extremely happy.'

He said the listing had three main benefits – recognition of Gunditjmara achievements on a global scale, increased protection for the site, and the potential tourism boost.

'There are a number of reports that say that once a place is declared as a world heritage site, tourism increases dramatically,' Mr Rose said.

The State Government has announced $8 million for a visitor centre and major works at the site to ready it for an expected visitor influx.

Glenelg Shire Mayor Anita Rank said the whole region would benefit from the UNESCO announcement.

Elevation to the World Heritage List means the site is recognised as having 'outstanding universal value'.

Mr Rose said he was delighted to think something the Indigenous people of south-west Victoria built now appeared on the same list as the pyramids, Stonehenge and the Acropolis.

'When I take people out to country I tell them this aquaculture system was first built 6600 years ago – there's not many things on the planet that still exist today that are older than that.'

Source: https://www.abc.net.au/news/2019-07-06/indigenous-site-joins-pyramids-stonehenge-world-heritage-list/11271804

Analysis

1 Describe the way the Gunditjmara people were able to engineer the landscape to develop a sustainable aquaculture system at Budj Bim.
2 What techniques did the Gunditjmara people use to trap eels at Budj Bim?
3 Develop a logical argument to explain how this form of aquaculture management challenges the perception that Australia's First Nations peoples were only hunter-gatherers.
4 Explain the importance to the Gunditjmara people and to the local community of gaining World Heritage status for Budj Bim.

The seasons

The seasonal calendar used in most Western societies is based on the Gregorian calendar, which has specific dates to mark the beginning of each season – summer, autumn, winter and spring. By contrast, Australia's First Nations peoples have traditionally used their acute sense of observation and understanding of the natural world to tell them when the seasons change and how this affects the food supply. Depending on whether they live in Northern, Western or South-Eastern Australia, the seasonal calendar may have five or more seasons. Australia's First Nations peoples take into account the water supply, when plants that grow food are ready for harvesting, and the breeding cycles of animals. They also consider the weather, fire patterns and the position of the stars in the sky. The interaction between the natural phenomena is complex, highly localised and specific to each language group across Australia.

Knowledge of every single food source has a story dating back thousands and thousands of years. Food is not just about supplying fuel for the body. The traditional ways and stories about food are passed on to the next generation when people are sitting together and eating on Country. Traditionally, this seasonal knowledge of food systems is shared aurally. In recent years, CSIRO has partnered with some First Nations groups to record information about food in seasonal calendars as a tool to demonstrate and communicate First Nations peoples' connection to – and use and management of – Country.

In the seasonal calendar, each season has a purpose. Australia's First Nations peoples look at the weather conditions and the types of plants that are growing well at that specific time of the year. These markers will indicate whether animals will feast on these plants and be a source of nutritious food for the community.

Most native fruits have a short season, so when they are available, they are eaten as a treat or for their health benefit. For example, the Kakadu plum is a nutritious fruit that has a high vitamin C content and is considered desirable despite its sour/bitter flavour profile.

Weblink
Indigenous calendars

9780170495714

Sources of food

The Australian environment supplies a wide range of foods, providing a nutritious diet for First Nations peoples. These foods are known as **wild native foods** – a term that refers to the huge variety of edible native Australian herbs, spices, mushrooms, fruits, flowers, vegetables, animals, birds, reptiles and insects.

All First Nations Australians were vitally involved in the business of food, which required the ability to plan, an eye for detail and well-developed bush skills. Women learnt the locations of every yam patch and fruit tree in their territory, and the times at which the trees bore fruit. They followed bees to their hives and stalked small animals. Men learnt from their Elders to become clever hinterland hunters, and to trap birds and other animals, such as kangaroos and wallabies. They learnt the art of camouflage and were able to swim underwater to catch fish and other sea creatures.

Detailed knowledge and careful planning were essential to ensure the sustainability of the environment, particularly as different species of animals and plants were found in different climate zones and rainfall areas. Native flora and fauna were used as food in every area, and food collection, preparation, cooking and distribution were major daily activities.

Generally, First Nations Australians collected sufficient foods for their needs at the time. They were always concerned to ensure that the animals and plants in an area could regenerate and replenish themselves naturally, and so continue to provide a wide range of foods into the future.

Table 5.1 Methods of obtaining food

Examples of methods of obtaining food	Examples of food types
Spearing	Animal foods – kangaroo, fish
Digging	Plant foods – grains, roots, edible grubs
Collecting in a dilly bag	Plant foods – fruit, vegetables, seeds, kernels/nuts
	Sweet foods – nectar, honey ants
	Eggs

Table 5.2 Wild bush foods obtained from plant sources

Fruits	Vegetables	Seeds	Roots	Nuts	Flowers
Bush tomato	Bulrushes	Wattleseed (acacia)	Yam	Bunya	Native fuchsia
Lilly pilly	Pigweed	Wild rice	Bush potato	Macadamia	Honey grevillea
Native passionfruit	Waterlily	Millet and grass seeds	Waterlily root	Kurrajong	Flowering gum
Bush banana	Mangrove			Moreton	
Illawarra plum				Bay chestnut	
Quandong					
Nonda plum					

Bush tomato

Wattleseed

Lilly pilly

Bush banana

Quandongs

Yams

Bunya nut

Moreton Bay chestnuts

Table 5.3 Wild bush foods obtained from animal sources

Mammals	Birds	Insects	Reptiles and amphibians	Seafood
Kangaroo	Duck	Ant	Crocodile	Barramundi
Wallaby	Emu	Locust	Turtle	Crab
Koala	Pigeon	Witchetty grub	Snake	Eel
Wombat	Cockatoo	Honey bee	Frog	Freshwater bream
Bandicoot	Parrot	Moth	Goanna	Clam
Water rat	Mallee fowl			Oyster
				Pippie
				Yabby

Wombat

Mallee fowl

Witchetty grub

Bandicoot

Water rat

Crocodile

Barramundi

Goanna

iStock.com/danicachang, iStock.com/Albert Wright, Auscape/Universal Images Group/Getty Images, Auscape/Universal Images Group/Getty Images, SOPA Images/LightRocket/Getty Images, iStock.com/ozgurcankaya, iStock.com/warat42, iStock.com/MoMorad

The diet of Australia's First Nations peoples

A traditional First Nations Australian diet consists mainly of plant foods that have had little or no processing, and that therefore supply high levels of fibre. These plant foods are very rich in nutrients, particularly vitamins and minerals. The seeds of many native grasses, in particular, contain much higher levels of protein and fat than cultivated cereal crops. Animal foods – especially kangaroo and freshwater bream – contribute approximately 50 per cent of the total energy intake in a traditional First Nations Australian diet.

After European colonisation, many First Nations Australians were displaced from their traditional lands and, as a result, were prevented from collecting, hunting, cultivating and eating their traditional foods. This led to the loss of traditional knowledge and skills in food choice, food preparation and cooking techniques. For First Nations Australians who lived on cattle stations, on missions or in government settlements, the only way to obtain food was through the communal dining rooms. They became increasingly dependent on British food rations, such as flour, sugar, tea, jam and tinned or salted meat.

These changes in the diet of First Nations Australians had a devastating impact on their health and led to the development of high rates of chronic disease, including diabetes, stroke, and heart and kidney disease.

The *Aboriginal and Torres Strait Islander Guide to Healthy Eating*

The *Aboriginal and Torres Strait Islander Guide to Healthy Eating* provides information for First Nations Australians on the amount and types of foods that they should be eating for better health and wellbeing. This guide is intended for use by health professionals and educators throughout Australia in a range of First Nations communities.

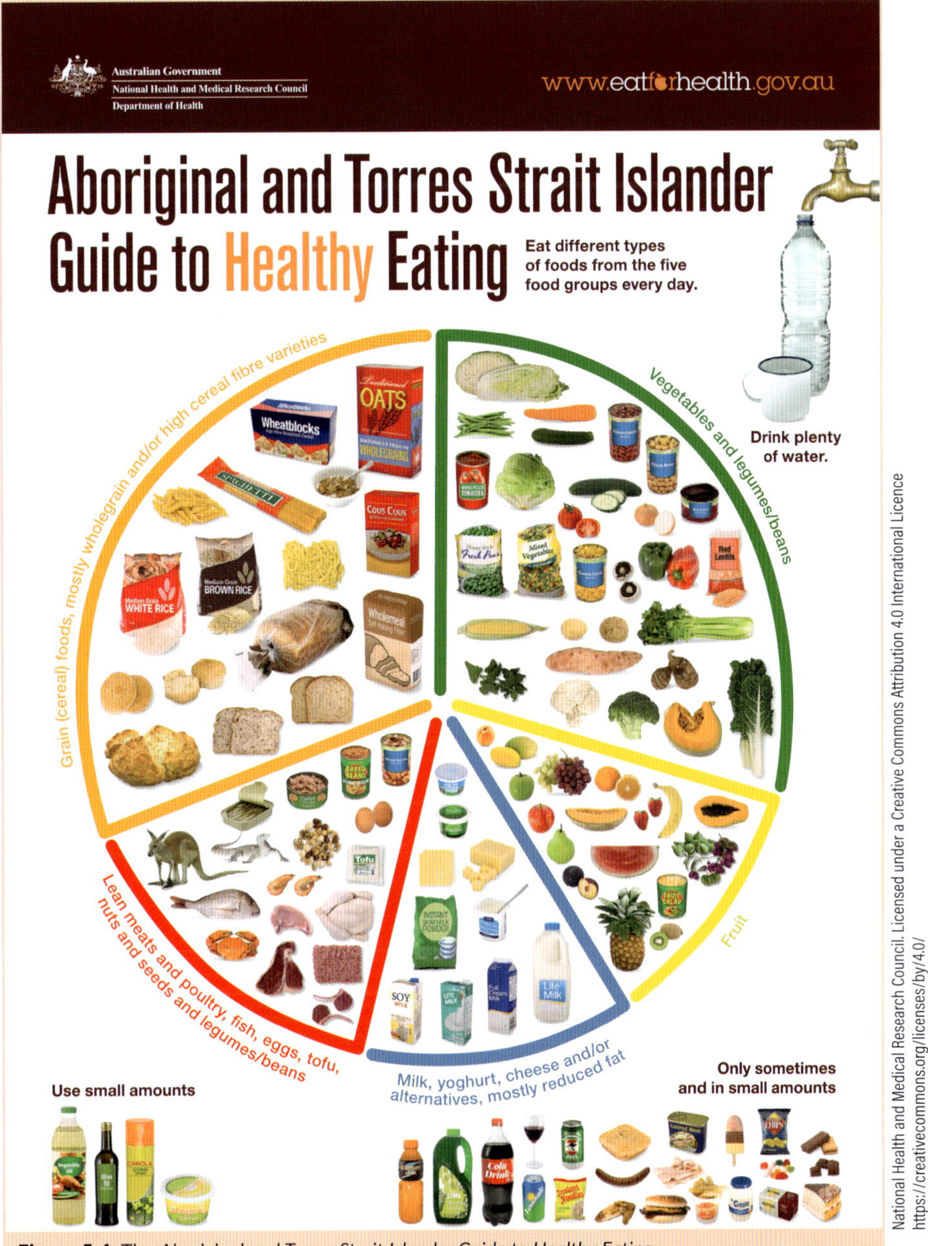

Figure 5.4 The *Aboriginal and Torres Strait Islander Guide to Healthy Eating*

Activity 5.1

Comparing food models

1. Describe the similarities between the *Australian Guide to Healthy Eating* on page 162 and the *Aboriginal and Torres Strait Islander Guide to Healthy Eating*. Consider the shape of the guide, food groupings/sections, and the proportion of each group/section.
2. Compare the main differences between the foods contained in the vegetables section of both guides. Hypothesise as to why there are differences in the types of vegetables included in these food selection models.
3. Analyse the differences between the meat sections of each guide.
4. Use a Venn diagram to compare the fruit sections of both guides.
5. Draw conclusions about why tinned, dried and long-life foods are included in most sections of the *Aboriginal and Torres Strait Islander Guide to Healthy Eating*.
6. Develop a logical argument to explain why it was necessary for the Australian Government to develop a separate guide to healthy eating for Aboriginal and Torres Strait Islander peoples.

Testing knowledge 5.1

Quiz
Testing knowledge 5.1

1 Create a mind map to demonstrate the complex agricultural and land management systems widely used by Australia's First Nations peoples.
2 Describe how fire-stick farming is a valuable tool used by Australia's First Nations peoples.
3 What do the remains of Aboriginal aquaculture at Budj Bim illustrate about First Nations Australian farm management systems?
4 Explain why knowledge of the seasons was essential for Australia's First Nations peoples in their search for food.
5 How is knowledge about traditional foods, their growing patterns and nutritional value shared between generations of Australia's First Nations peoples?
6 Make a list of the foods traditionally supplied by Australia's First Nations women and those supplied by Australia's First Nations men.
7 List the nutritional advantages of the traditional First Nations Australian diet.
8 What percentage of animal foods contribute to the traditional First Nations Australian diet?
9 Describe how European colonisation had an influence on the diet and health of Australia's First Nations peoples.
10 Explain the impact that a Westernised diet has had on the health of Australia's First Nations peoples.

Native foods in today's menus

Not long ago, it was difficult to find native Australian ingredients in our fresh food markets, on the supermarket shelves or on restaurant menus. However, they are now becoming more mainstream, and many restaurants across Australia are experimenting with native ingredients and showcasing their unique flavours. In the early years, the wild bush foods supplied to restaurants and food manufacturers were all gathered in their natural habitats, but it soon became obvious that if the industry was to be more than a novelty, wild harvesting was unsustainable and the crops would have to be farmed. In recent years, companies run by First Nations Australians have used traditional knowledge – handed down from generation to generation – to produce sustainable, high-quality, native foods for the consumer market and restaurant trade. These foods include salt bush, finger limes, warrigal greens and lemon myrtle.

Kangaroo meat is an example of a native food that is appearing in today's menus.

Kangaroo meat

The kangaroo is a marsupial that is native to Australia, and was one of the principal animals hunted and eaten by First Nations Australians and the early European colonisers. As more land was opened up for grazing, beef and lamb replaced kangaroo meat.

Meat from wild animals such as kangaroo are lower in fat than the meat of domesticated animals. Wild animals must scavenge for food and move around the countryside in search of food. Their energy intake and expenditure are closely balanced, meaning fewer fat stores are accumulated.

Kangaroo meat is cooked like any other game meat; that is, seared on a high heat to seal in meat juices, cooked lightly, and then rested to tenderise the meat. It is important not to overcook kangaroo meat as it is lean, and quickly dries out and becomes tough. Kangaroo meat can be purchased as rump steak or loin fillets, or it may be diced, minced or made into sausages.

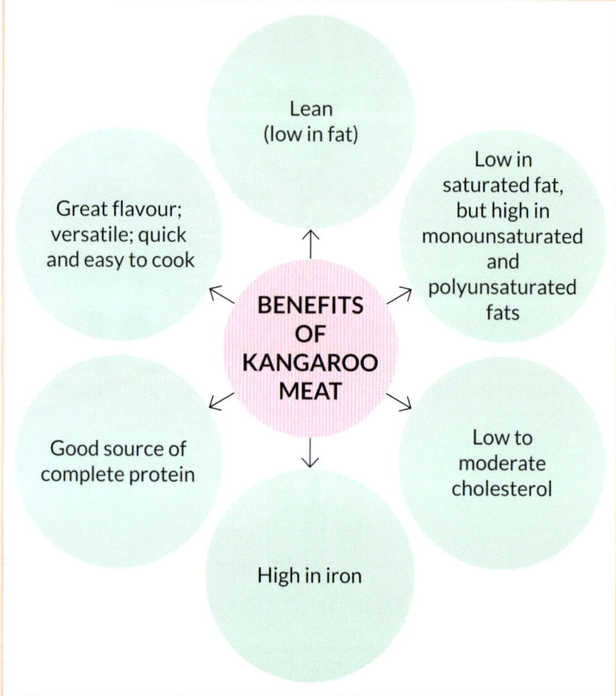

Figure 5.5 The benefits of kangaroo meat

Figure 5.6 Kangaroo fillet served as a meal.

seeshooteatrepeat/Shutterstock.com

9780170495714

Activity 5.2

Food analysis: Comparing kangaroo and other meats

Examine Table 5.4 and visit the Australian Wild Game Industry Council website. Find information about different cuts of kangaroo meat and kangaroo recipes, and then answer the following questions.

1 Why does kangaroo meat have one of the lowest fat contents of any meat?

2 Compared with beef, what are the nutritional advantages of kangaroo meat?

3 Compared with chicken, what are the nutritional advantages of kangaroo meat?

4 Identify a cut of kangaroo meat that would be suitable to use in a stir-fry recipe.

5 What is the recommended cooking time for stir-fried kangaroo?

6 With other members of your class, discuss the following questions:

a Do you believe that kangaroo meat is a viable alternative to other meats? Remember to think about cost, flavour, time of cooking, nutrient value, ethics and environment.

b Why do you think kangaroo meat is not commonly seen on many menus?

c Why are many consumers reluctant to eat kangaroo meat?

d What strategies can you suggest to increase consumer awareness of the benefits of eating kangaroo meat?

Weblink Australian Wild Game Industry Council

Table 5.4 A comparison of kangaroo and other meats

Meat	Protein (%)	Fat (%)	Iron (milligrams per 100 grams)	Cholesterol (milligrams per 100 grams)	Kilojoules (per 100 grams)
Kangaroo	24	1–3	2.6	56	500
Lean lamb	22	2–7	1.8	66	530
Lean beef	22	2–5	3.5	67	500
Lean pork	23	1–3	1.0	50	440
Lean chicken breast	23	2	0.6	50	470
Rabbit	22	2–4	1.0	70	520

Cooking with native ingredients

Australia produces a wide variety of native ingredients – including bush tomatoes, finger limes and macadamia nuts – that are very versatile and can be used in a variety of sweet and savoury dishes.

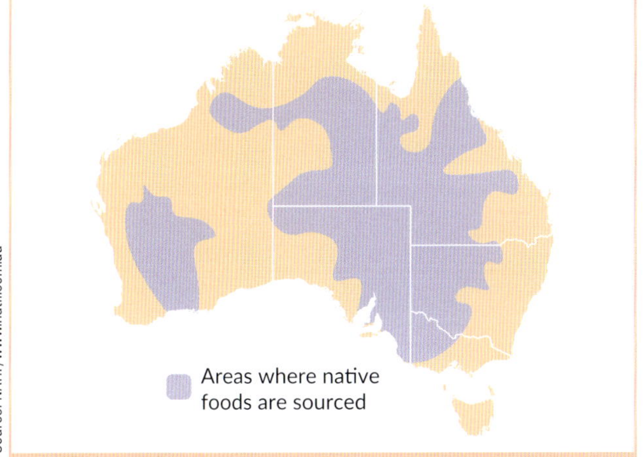

Figure 5.7 Areas where native foods are commercially sourced in Australia today

Bush tomato

One popular native ingredient is the fruit of the small shrub known as the bush tomato, or desert raisin. The fruit, which is about the size of a blueberry, must be allowed to ripen on the bush; otherwise, it will be toxic. The bush tomato is a good source of carbohydrates and vitamin C. It is best used after being ground to a powder in a food processor. The ground bush tomato is called akudjura, and has an intense flavour.

Figure 5.8 Bush tomatoes

The bush tomato can be:

- used to enhance the flavour of savoury dishes and give them a spicy, piquant taste
- added to soups, casseroles, pasta sauces, chutneys, relishes, pizzas, risotto or sauces – in fact, any dish that traditionally uses cultivated tomatoes.

Finger lime

Finger limes, commonly known as citrus caviar, are an exquisite fruit grown in the rainforests of Queensland and New South Wales. Varieties include different skin and flesh colours. Each fruit contains lime green or pink pearl-like crystals that explode in the mouth, giving a lime taste sensation.

Finger limes can be:

- added to salads
- used with seafood
- used in desserts
- added to cocktails.

Tommy Atthi/Shutterstock.com

Figure 5.9 Finger limes

Lemon myrtle

Lemon myrtle is derived from the aromatic leaf of a large tree found in Queensland. The leaf contains essential oils, which give it a wonderful perfume and spicy lemon flavours.

Stephanie Jackson - Aust wildflower collection/Alamy Stock Photo

Figure 5.10 Lemon myrtle

Lemon myrtle can be:

- infused in hot liquids to release its oils and then used in recipes in which dried herbs are not suitable
- used as a fresh herb; for example, draped over fish or chicken fillets before baking, or as a substitute for makrut lime leaves
- dried, crumbled and sprinkled over meat or fish before baking
- used to flavour teas, vinegars, oils, dressings or desserts
- added to breads, muffins, biscuits, sauces or mayonnaise
- used as an oil to flavour cream or yoghurt.

Native pepper

Native pepper is also known as mountain pepper. The native pepper plant has two products that can be used in food preparation – the berry and the leaf. The berries are dark blue to black. The leaves and berries have an aromatic peppery flavour, with a hotness rating between that of a peppercorn and a chilli.

blickwinkel/Alamy Stock Photo

Figure 5.11 Native pepper

Native pepper berries and leaves can be:

- dried, and then crushed or ground and used to make curries or other dishes containing chilli; breads and pastries; or chutneys for use on meat or fish
- used as an infusion to make a refreshing herbal tea, vinegars, oils or dressings.

Wattleseed

Wattleseed is one of the most popular and best known of all native Australian foods. The small, black seeds of the wattle are crushed after dry-roasting to produce grounds that contain coffee, chocolate and hazelnut flavours.

A small quantity of wattleseed is brought to the boil to soften the grounds. The liquid is then drained off, and both the liquid and the grounds are used as flavouring.

Figure 5.12 Wattleseed

Wattleseed can be used:

- as a substitute for coffee to make a 'wattlechino' (wattle cappuccino), using 1 teaspoon of ground seeds per cup; wattleseed is caffeine-free

9780170495714

- to flavour syrup to be poured over pancakes and puddings
- to flavour cream, ice-cream, mousses and meringues
- in biscuits and cakes
- in bread, muffins, pastry, pancakes and pasta. Always add the wattleseed towards the end of the mixing, because it affects the gluten and toughens the flour.

Macadamia nuts

The macadamia nut is an ingredient that is considered to be typically Australian. It is native to south-east Queensland and northern New South Wales, where it grows in the rainforests, close to streams. Macadamia nuts are mainly harvested between May and June, when they naturally fall to the ground. The nut has a high oil content and is consequently high in fat and energy.

Macadamia nuts can be purchased raw or oven-roasted. The raw nuts are best used in cakes, biscuits and puddings, while the oven-roasted nuts are best used in salads, ice-cream and mousses. Macadamia nuts can be used in recipes such as tarts, cakes and biscuits to replace walnuts, almonds and hazelnuts. They can also be combined with breadcrumbs to cover fish or chicken.

Macadamia nut oil can be extracted from the nuts. This can be used in place of other oils – for example, in salad dressings or for frying – or in cakes to replace butter.

Figure 5.13 Macadamia nuts

JIANG HONGYAN/Shutterstock.com

Activity 5.3

Research task: Australian native foods

Native Australian Traditional Indigenous Foods (NATIF) is a non-Aboriginal owned social enterprise business working on Boonwurrung Country, sourcing ethical and sustainable premium native foods. It sells native foods to the wholesale and retail markets.

Visit the NATIF website to broaden your knowledge of native foods in Australia.

1 Go to the 'About' page to read about the organisation and then answer these questions:
   a Explain why many native foods are not always available fresh and need to be preserved.
   b Discuss how the systems in the native food industry are beneficial for some Australian First Nations communities.

Weblink
NATIF

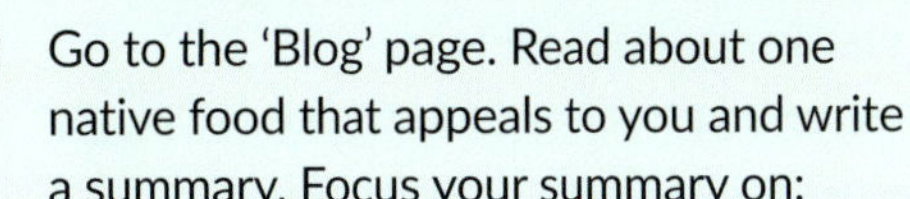

2 Go to the 'Blog' page. Read about one native food that appeals to you and write a summary. Focus your summary on:
   - growing habits of the plant
   - sensory properties of the food
   - nutritional value of the food
   - preservation techniques used
   - two recipes that you would like to trial and taste test. Justify your choice.

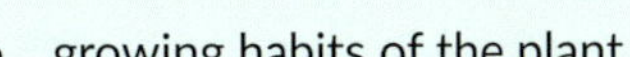

PRACTICAL ACTIVITY 5.1

Sensory analysis: Understanding native flavours: Wild bush crisps

Aim

To understand the sensory properties of native food flavourings.

Equipment

- 4 different wild bush flavourings, ground; for example, aniseed myrtle, akudjura (ground bush tomato), lemon myrtle and native pepper
- 2 small pitas, cut open to make 4 pieces
- macadamia oil (or spray)
- oven tray

Method

Worksheet
Practical activity 5.1

1 Preheat oven to 180°C.
2 Lay out the four pieces of pita, and lightly brush or spray one side of each piece with macadamia oil.
3 Lightly sprinkle with ground native flavouring – use one of the selected flavourings per piece of pita.

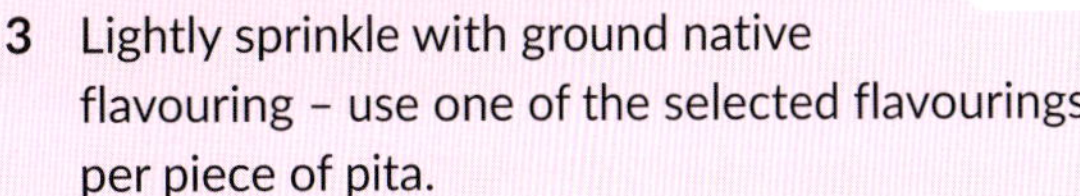

4 Place on an oven tray and bake for approximately 5–10 minutes or until crisp and lightly browned.
5 Complete a sensory test on the four different wild bush crisps. Remember to observe the appearance and aroma before undertaking the taste test.

Wild bush flavouring	Appearance	Aroma	Flavour
1			
2			
3			
4			

Results

Record your results in a table similar to the one above. Refer to the sensory wheel on page 4 to assist you in describing the sensory properties.

Analysis

1 Describe the similarities in and/or differences between the four types of wild bush crisps.
2 Which wild bush crisp was the most visually appealing? Why?
3 Which wild bush crisp had the most appealing aroma? Why?
4 Out of the four tests, which flavour was the most appealing to you?
5 Recommend which native flavouring would be best to serve with each of the following foods: pasta, chicken, meat, vegetables, cheese.

Conclusion

Which native flavour did you enjoy the most? In your opinion, which native flavour would have the greatest consumer appeal? Why?

Note: The wild bush crisps can be used as tasty alternatives to bread or crackers to serve with pâté or dips, or to accompany soup.

Testing knowledge 5.2

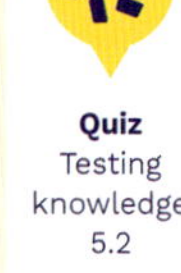

Quiz
Testing knowledge 5.2

11 Explain why kangaroo meat is considered to be a healthy food choice.
12 List three native food flavourings that complement savoury foods such as soups and fish.
13 Identify the traditional name for the bush tomato. List three main uses of this native food.
14 Explain why bush tomatoes would have been a good nutritional supplement in a traditional First Nations Australian diet.
15 Describe the flavour profile of wattleseed. List the steps you would need to take to prepare wattleseed for use in a recipe.
16 Why is it important to add wattleseed towards the end of mixing when adding it to bread or muffins?
17 Which native foods can be used as an oil?
18 When are macadamia nuts harvested, and in what form can they be purchased?
19 Identify the main nutrient content of macadamia nuts.
20 Suggest five ways in which macadamia nuts can be used in cooking.

Critical and creative thinking 5.1

Promoting native foods

1 Develop a promotional video clip or television advertisement to promote kangaroo meat as a viable ingredient in everyday meals. The promotional material should include the following information:
- the nutrient value of kangaroo meat
- why kangaroo meat is a versatile ingredient to use
- recipe ideas
- other relevant information.

2 Write a paragraph for a food website that discusses:
- ethical issues that would support kangaroo meat production as a food source
- arguments against the use of kangaroo meat as an ingredient in meals.

Suggest an appropriate platform on which to post your arguments.

9780170495714

DESIGN ACTIVITY 5.1

New café menu

Design brief

'Going Wild' is a First Nations Australian nursery in the Grampians that grows a range of native ingredients. The 'Going Wild' café, which is attached to the nursery, is about to update its spring lunch menu to several new warm savoury dishes, each with a suitable accompaniment to make a complete meal. The new menu items should reflect the First Nations Australian nature of the nursery by including at least one native ingredient.

Investigating and defining

1 Develop five design criteria to judge the success of your solution to the problem outlined in the design brief.
2 Prepare a recipe containing native flavours such as Calzones with Silverbeet and Native Flavours (page 117) or Kangaroo Meatballs (page 118) to explore the variety of flavours and textures you could incorporate into your designed solution.
3 Use online recipes, recipe books and food magazines to research warm savoury recipes and an accompaniment suitable for a light lunch that could incorporate at least one native ingredient.
4 Following your research, identify four recipes that could be designed solution options for the warm lunch menu items.
5 Compete a table similar to the one below to record the types of native ingredients used, a suitable accompaniment and an appealing feature of each recipe.

Recipe idea (including reference)		Native ingredient included	Accompaniment	Appealing features of the recipe
1				
2				
3				
4				

Generating and designing

Construct a decision table (see the example on page 12). Select the designed solution option you consider would be the best designed solution. Justify why your choice is the best solution and explain why you did not select the other designed solution options.

Planning and managing

1 Prepare a food order.
2 Before producing your warm savoury lunch dish and accompaniment, write up a production plan, noting any safe work practices to be followed, and identify the major processes to be used.
3 Make a list of the aspects of the production task that will rely on you and your bench partner sharing and working collaboratively.

Producing and implementing

1 Prepare the product.
2 Note any modifications made to the recipe during production.

Evaluating

1 Respond in detail to your five criteria to evaluate the success of your warm savoury lunch dish and accompaniment.
2 Describe the sensory properties – appearance, aroma, flavour, texture and sound – of the warm savoury lunch dish and accompaniment.
3 What, in your opinion, was the most difficult aspect of the production? Explain.
4 Discuss any improvements you would make if you were to produce this dish again.
5 Classify the ingredients for your warm savoury lunch dish and accompaniment on a diagram of the *Australian Guide to Healthy Eating* and discuss how well this meal meets the recommendations of this food selection model.

Scones with Lemon Myrtle

Scones are made from a bland flour mixture, and so are ideal to showcase native ingredients in recipes that are quick to produce.

INGREDIENTS

3 cups self-raising flour

1 tablespoon lemon myrtle powder

2 teaspoons caster sugar

¼ teaspoon salt

1 cup thickened cream

1 cup chilled soda water

2 tablespoons milk, for glaze

butter or lemon curd, to serve

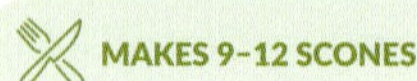

Mark Fergus Photography

METHOD

1 Preheat oven to 220°C. Grease a baking tray.
2 Sift the flour, lemon myrtle powder, caster sugar and salt into a large bowl. Return any lemon myrtle that remains in the sieve to the sifted ingredients.
3 Make a well in the centre of the dry ingredients and pour in all the cream and soda water at once. Mix gently with a spatula until the ingredients are just combined. The mixture will be soft and slightly sticky.
4 Lightly flour the surface of the bench and turn out the dough. Flour your hands and lightly knead the dough until it is just smooth.
5 Use your hands to gently flatten the dough into a disc that is 3 centimetres thick.
6 Flour a scone cutter or a glass that is 6 centimetres in diameter, and cut out 9–12 scones. Remember to re-flour the cutter or glass before cutting out each scone.
7 Transfer the scones to the baking tray and arrange them close together. This assists in the aerating process during baking. Glaze the surface of the scones with milk.
8 Bake at 220°C for 12–15 minutes or until the scones are well risen, golden brown on top and firm to touch.
9 Transfer to the scones to a cooling rack and cover with a clean tea towel.
10 Allow the scones to cool slightly before serving them fresh with butter or lemon curd.

Additional serving suggestion

The Scones with Lemon Myrtle are delicious served with a slice of Jarlsberg cheese.

EVALUATION

1 Describe the sensory properties – appearance, aroma, flavour, texture and sound – of your Scones with Lemon Myrtle.
2 List the ingredients that contribute to the aeration of the scones.
3 Explain why there is minimal kneading when making a scone dough compared to intense kneading when making bread.
4 How does covering the scones with a clean tea towel as they cool influence the sensory properties of the final product?
5 Suggest other native ingredients that could be used to flavour a scone dough, and accompaniments for serving.

9780170495714

Calzones with Silverbeet and Native Flavours

INGREDIENTS

Dough

1 cup bread flour

1 teaspoon dried yeast

¼ teaspoon sugar

½ teaspoon salt

1 tablespoon olive oil

⅓ cup warm water

Filling

6 stalks of silverbeet

30 millilitres olive oil, plus extra for drizzling

1 garlic clove, chopped

pinch dried chilli flakes, plus extra for sprinkling

½ teaspoon wild bush flavoured salt

freshly ground black pepper

semolina, for dusting

½ lemon, to serve

Mark Fergus Photography

METHOD

Dough

1 Sift the flour, dried yeast, sugar and salt into a large bowl. Stir in the oil and warm water and mix to a soft dough. Knead the dough on a lightly floured board until smooth and elastic.

2 Place dough in an oiled bowl, cover with plastic wrap and stand in a warm place for 15–20 minutes until doubled in size.

Filling

1 Preheat oven to 200°C and line a baking tray with baking paper.

2 Wash the silverbeet and shred into 5-millimetre pieces.

3 Fill a large saucepan with water and bring to the boil. Blanch the silverbeet in the boiling water for 2 minutes or until wilted. Drain, then immediately plunge the silverbeet into ice-cold water. Drain well and spread onto absorbent paper or a clean tea towel to dry.

4 Heat the olive oil and garlic over medium heat. Add the chilli flakes and the blanched silverbeet and season with wild bush flavoured salt and freshly ground black pepper. Cool for 10 minutes, stirring occasionally, until the leaves are dark green and the stalks are tender. Set aside.

To make the Calzones

1 Knead the dough lightly. Divide into two and form two balls.

2 Lightly dust a clean surface with semolina, then roll out each dough ball into a 15-centimetre round that is about 3 millimetres thick. Transfer the discs of dough onto the lined baking tray before filling.

3 Divide the silverbeet filling between the two discs of dough and spread it over one side of each disc, leaving a 2-centimetre border. Fold each of the other sides over the filling, pinch the edges together and pleat to prevent the mixture from escaping.

4 Drizzle with the extra olive oil and sprinkle with the extra chilli flakes. Bake in the oven for 5–10 minutes or until golden brown.

5 Transfer the calzones to a serving board or plate. Serve with a squeeze of lemon.

EVALUATION

1 Describe the sensory properties – appearance, aroma, flavour, texture and sound – of your Calzones with Silverbeet and Native Flavours.

2 What is the purpose of covering the dough with plastic wrap and standing it in a warm place for 15–20 minutes in step 2?

3 Why is it important to plunge the silverbeet into ice-cold water in step 5?

4 Explain why the recipe suggests you should place the dough on a baking tray before filling and folding the calzone.

5 Classify the ingredients for the Calzones with Silverbeet and Native Flavours on a diagram of the *Australian Guide to Healthy Eating*. Discuss how well this recipe meets the recommendations of this food selection model.

Kangaroo Meatballs

These meatballs can be served as an entrée or be eaten as finger food. They could be served with an accompaniment – such as a sauce or chutney – that features native ingredients.

INGREDIENTS

1 small onion, grated

1–2 tablespoons oil

¼ cup fresh breadcrumbs

1 tablespoon commercial bush tomato sauce or relish

1 small egg, beaten

4 shakes pepper

250 grams kangaroo meat, finely minced

1 tablespoon cream

¼ cup cornflour

fresh chives or parsley, very finely chopped, to serve

METHOD

1. In a medium frying pan, gently fry the onion in 2 teaspoons of oil until soft. Cool.
2. In a medium bowl, mix the breadcrumbs with the bush tomato sauce or relish.
3. Stir in the egg, pepper and cooked onion.
4. Add the kangaroo meat, breadcrumb mixture and cream and mix well.
5. Roll mixture into balls about 3 centimetres in diameter and then roll the balls in the cornflour.
6. Heat the remaining oil in the frying pan and cook the meatballs in batches, shaking the pan until all sides are brown.
7. Drain on absorbent paper. Transfer to a serving plate and garnish with finely chopped chives or parsley. If desired, pierce each meatball with a toothpick to serve.

EVALUATION

1. Why does this recipe use fresh breadcrumbs instead of dry breadcrumbs?
2. Suggest some other native flavours that could be used in the Kangaroo Meatballs.
3. Identify two ingredients in the Kangaroo Meatballs that undergo physical and chemical changes during frying. Identify the processes and explain how they affect the sensory properties of the final product.
4. What side dishes could you serve with the Kangaroo Meatballs to build a healthy meal?
5. Classify the ingredients for the Kangaroo Meatballs and the side dishes you have suggested on a diagram of the *Australian Guide to Healthy Eating*. Discuss how well this meal meets the recommendations of this food selection model.

Mark Fergus Photography

9780170495714

Tea Cake with Native Spice

INGREDIENTS

1 cup self-raising flour

1 egg

60 grams butter, room temperature

½ cup caster sugar

½ teaspoon vanilla extract

⅓ cup milk

Topping

1 tablespoon caster sugar

1 teaspoon Davidson's plum powder

1 tablespoon butter, melted

METHOD

1 Grease an 18-centimetre round cake tin and line the base with baking paper.
2 Preheat oven to 180°C.
3 Sift the flour into a small bowl.
4 Break the egg into a small bowl and beat with a fork to combine the white and yolk.
5 Place the butter and sugar in a large bowl and use an electric hand-held beater to cream them together until the mixture is pale in colour and resembles whipped cream.
6 Add the vanilla extract and half the beaten egg to the creamed mixture and mix until just combined. Add the remainder of the egg and beat to combine.
7 Use a large metal spoon or spatula to fold in the milk and flour alternately, one-third at a time. Mix until the ingredients are just combined.
8 Spoon the mixture into the lined cake tin and bake for 20 minutes or until the cake is pale golden in colour, firm to touch and beginning to shrink from the sides of the tin.
9 Allow to stand for 5 minutes and turn the cake onto a cake rack to cool.
10 Combine the first two topping ingredients in a small bowl.
11 While the cake is hot, brush it with the melted butter and sprinkle over the sugar mixture, using a metal teaspoon.

Mark Fergus Photography

EVALUATION

1 Describe the sensory properties – appearance, aroma, flavour, texture and sound – of your Tea Cake with Native Spice.
2 What impact does sifting the flour have on the final quality of the cake?
3 Explain how you knew when the butter and sugar were sufficiently creamed to progress to the next step in the recipe.
4 Why was it important to fold rather than beat the milk and flour into the butter and sugar mixture?
5 From your research of native ingredients, suggest some alternative native spices that would be suitable for the topping of this cake.

Wattleseed Pavlovas with Macadamia Cream and Sugar Bark

INGREDIENTS

Wattleseed Pavlovas

1 teaspoon wattleseed

2 egg whites

½ cup caster sugar

½ teaspoon vanilla extract

½ teaspoon vinegar

Sugar Bark

⅔ cup caster sugar

⅓ cup water

Macadamia Cream

150 millilitres cream

40 grams macadamia nuts

1 tablespoon Nutella

METHOD

Wattleseed Pavlovas

1. Preheat oven to 160°C. Cover a baking tray with baking paper and draw 6 circles approximately 6 centimetres in diameter on the paper.
2. Pour 1 tablespoon of boiling water over the wattleseed and soak for 10 minutes. Strain through a fine strainer. Retain the grounds.
3. In a medium bowl, beat the egg whites into a stiff foam. Gradually add the caster sugar, one tablespoon at a time. Beat until the mixture is glossy and stiff.
4. Fold through the vanilla extract, vinegar and wattleseed grounds. Divide the mixture into 6 equal portions and shape into nests within each circle.
5. Place in oven at 160°C, then immediately reduce heat to 130°C and cook for 20–30 minutes or until the crust has dried. Cool.

Sugar Bark

1. Combine the caster sugar and water in a saucepan and stir over a low heat until the sugar has dissolved. Wash down any sugar crystals from the sides of the saucepan with cold water and a pastry brush.
2. Bring the syrup to the boil, then reduce the heat. Cook until it is pale gold in colour. Do not stir, as this will cause the syrup to crystallise.
3. Remove from the heat and allow the syrup to rest until the bubbles have subsided. Pour onto a foil-lined tray. Cool.
4. Break the toffee into shards that resemble bark.

Macadamia Cream

Beat the cream to soft peaks. Finely chop the macadamia nuts. Place the Nutella in a small bowl and add 1 tablespoon of the whipped cream. Stir the whipped cream through the Nutella until the Nutella is softened. Fold the softened Nutella and the chopped macadamia nuts through the rest of the whipped cream.

To assemble the dish

Divide the Macadamia Cream between the pavlova nests. Place the Sugar Bark vertically in the Macadamia Cream to decorate.

9780170495714

EVALUATION

1. When making the pavlova mixture, what steps should be taken during preparation to ensure that the egg whites beat into a stiff foam?
2. Identify the process that occurs when the egg whites are beaten. Describe how this contributes to the sensory properties of the final product.
3. Why are the vanilla extract, vinegar and wattleseed grounds folded through the pavlova mixture instead of being beaten into it?
4. Why is it important to dissolve the sugar before bringing the syrup to the boil?
5. Identify the ingredients in the Wattleseed Pavlovas with Macadamia Cream and Sugar Bark that place them in the 'only sometimes' component of the *Australian Guide to Healthy Eating*. Predict a possible effect on long-term health if these pavlovas were consumed on a regular basis.

Mark Fergus Photography

Chocolate and Macadamia Biscuits

INGREDIENTS

⅓ cup macadamia nuts, finely chopped

1¼ cups plain flour

½ teaspoon baking powder

125 grams butter

¼ cup soft brown sugar

⅓ cup caster sugar

1 teaspoon vanilla

1 egg, lightly beaten

½ cup white chocolate buttons, roughly chopped

METHOD

1 Preheat oven to 180°C.
2 Line two biscuit trays with baking paper.
3 Spread the chopped nuts onto one of the lined biscuit trays and cook for 6–8 minutes or until a light gold in colour. Set aside in a bowl to cool. Re-line the biscuit tray.
4 Sift the flour and baking powder into a small bowl.
5 In a medium bowl, cream the butter and sugars until light and fluffy. Beat in the vanilla.
6 Gradually beat in the egg.
7 Stir in the sifted flour and baking powder.
8 Add the cooled toasted nuts and chocolate and mix well.
9 Place dessert-spoonfuls of the mixture onto the lined biscuit trays, leaving enough room for spreading.
10 Bake at 180°C for 12–15 minutes or until golden and firm to touch.
11 Remove from the oven and allow to stand on the tray for 5 minutes.
12 Carefully remove with a spatula and transfer to a wire rack. Allow to cool.

EVALUATION

1 What property of macadamia nuts causes them to brown quickly?
2 Explain how you knew when the butter and sugars were sufficiently creamed.
3 Why is it important to sift the flour and the baking powder together?
4 Why are the biscuits allowed to stand on the tray for 5 minutes before they are transferred to the wire rack?
5 Classify the ingredients for the Chocolate and Macadamia Biscuits on a diagram of the *Australian Guide to Healthy Eating*. Use the data to justify why these biscuits are included in the 'only sometimes' section of this food selection model.

 9780170495714

Mark Fergus Photography

FOOD SECURITY

- Availability
- Accessibility
- Acceptability
- Adequacy
- Stability

FOOD SOVEREIGNTY

Individuals and communities have control over their own food systems

FOOD CITIZENSHIP

Individuals make informed choices at all stages of the food system about sustainability, ethics and health

FOOD SYSTEMS

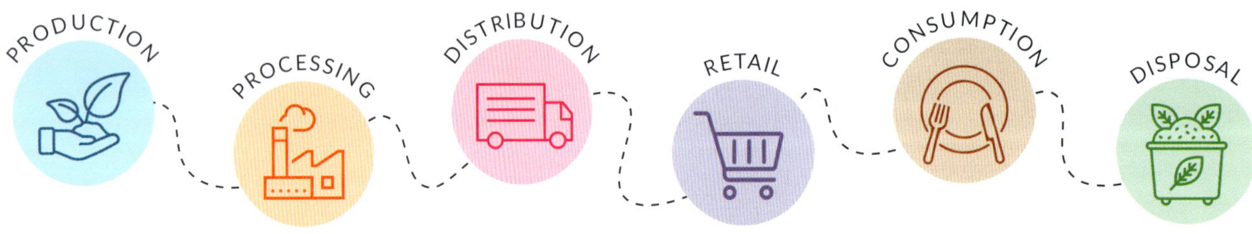

AUSTRALIAN SUPERMARKET SECTOR

37% Woolworths

28% Coles

18% Other (independent)

10% Aldi

7% Metcash (IGA)

SUSTAINABLE FOOD SYSTEMS

Economic sustainability

Social sustainability

Environmental sustainability

Cultural sustainability

CHALLENGES TO SUSTAINABLE FOOD SYSTEMS

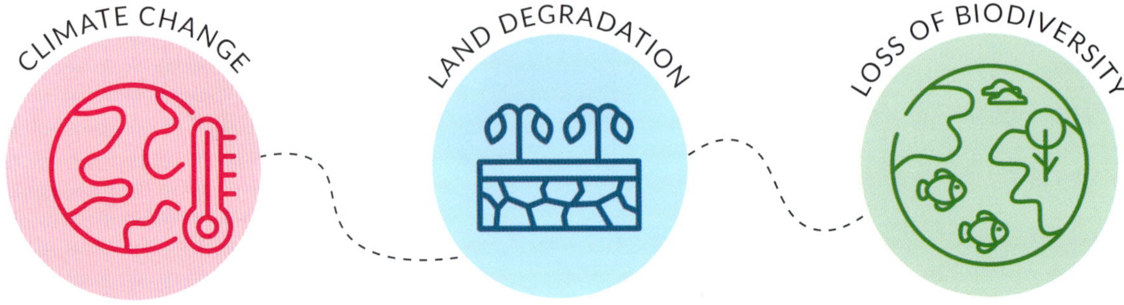

9780170495714

SUSTAINABLE FARMING PRACTICES

NO-TILL/MINIMUM TILLAGE FARMING

ORGANIC FARMING

PRECISION AGRICULTURE

PROTECTED CROPPING

VERTICAL FARMING

6 SUSTAINABLE FOOD SYSTEMS

KEY TERMS

biodiversity the vast array of living organisms that inhabit the planet and the interactions between them

food citizenship based on the idea that individuals are not just consumers at the end of the food chain, but active participants (or citizens) in the food system as a whole; food citizenship encourages individuals to support the environmental sustainability of the food system and the human rights of food producers through the food they buy and consume

food insecurity where an individual or family does not have access to sufficient food, or food of a nutritious quality, to meet their basic needs

food security the state that exists when all people, at all times, have physical, social and economic access to sufficient, safe and nutritious food that meets their dietary needs and food preferences for an active and healthy life

food sovereignty the right of individuals and communities to have control over their own food systems, challenging the dominance of multinational companies, with a particular focus on ensuring that food is produced ethically and sustainably, with a deep consideration for social justice, environmental sustainability and the wellbeing of local communities

food system a complex series of activities that enables food to move from farm to consumer, and includes the growing, harvesting, processing, transporting, manufacturing, consuming, disposing and recycling of food

Health Star Rating a front-of-pack labelling system that rates the overall nutritional profile of packaged food

no-till or minimum tillage farming a farming system that involves planting the new crop by directly drilling in between the rows of the previous crop without tilling the soil

organic farming an environmentally sustainable food production system; it produces crops and animals without the use of artificial chemicals, using natural systems instead

sustainable diets accessible, affordable, equitable and acceptable diets to society that have a low environmental impact

Templates

- Decision table
- Sensory wheel
- Australian Guide to Healthy Eating

Worksheets

- Practical activity 6.1
- Practical activity 6.2
- Practical activity 6.3
- Design activity 6.1

Quizzes

- Testing knowledge 6.1
- Testing knowledge 6.2
- Testing knowledge 6.3

Nelson MindTap

To access resources above, visit **cengage.com.au/nelsonmindtap**

9780170495714

Food security and sustainable diets

Everyone has a right to a nutritious food supply. However, according to the *Foodbank Hunger Report 2024*, almost half of low-income Australian families face **food insecurity**. To help overcome food insecurity, Australians are encouraged to develop food sovereignty and to become active food citizens.

Food security

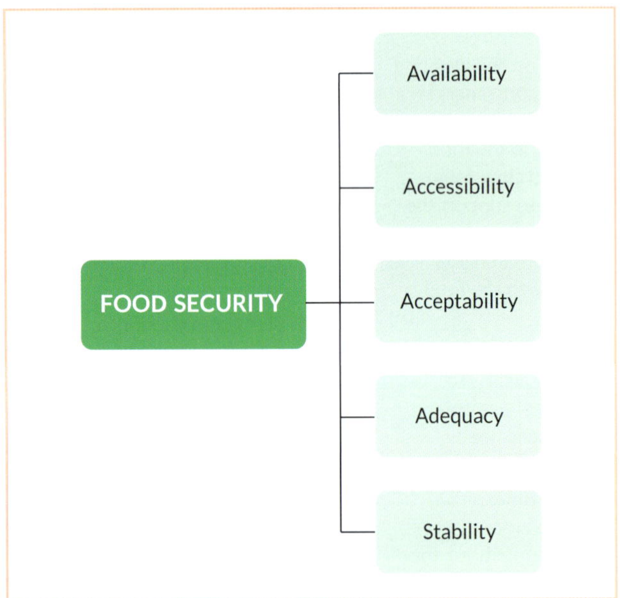

Figure 6.1 The five components of food security

Having a secure food supply is essential for a healthy life. **Food security** means having reliable access to sufficient safe, nutritious and culturally appropriate food for an active and healthy life. There are five components of food security:

- Availability – there must always be a sufficient supply of food for all people.
- Accessibility – people must have the physical and economic ability to obtain food.
- Acceptability – people must have access to food that is produced in a culturally acceptable manner.
- Adequacy – the food supply must be nutritious, safe, and produced so that it is environmentally sustainable.
- Stability – the food supply must be reliable and secure.

Food sovereignty

In Australia today, a few multinational companies have control over the production, processing, distribution, marketing and sale of a large majority our food supply. Many of the foods produced by these companies are ultra-processed, are high in fat, salt and sugar, and provide few of the nutrients we need for good health. **Food sovereignty** directly challenges the dominance of these companies, and promotes the right of individuals and communities to have control over their own food systems, with a particular focus on ensuring that food is produced ethically and sustainably, with a deep consideration for social justice, environmental sustainability and the wellbeing of local communities.

Some of the ways individuals can demonstrate their food sovereignty is to:

- purchase fresh produce directly from farmers through farmers' markets
- grow their own green leafy vegetables, carrots, tomatoes and herbs; for example, in their own vegetable garden or through participation in a community garden program
- subscribe to a 'paddock to plate' food delivery service, where fruit, vegetables, meat, dairy products and/or eggs are purchased directly from the producers, especially small-scale farmers.

iStock.com/JohnnyGreig

Figure 6.2 Australian families can purchase fresh vegetables directly from small-scale farmers.

Food citizenship

Every day, we as individuals make choices about the foods we eat. **Food citizenship** is a concept promoted by the Food Ethics Council, based on the idea that individuals are not merely consumers at the end of the food chain, but active participants (or citizens) in the food system as a whole. Food citizenship encourages individuals to recognise their role in supporting both the environmental sustainability of the food system and the human rights of food producers by making conscious choices about buying and consuming foods that prioritise sustainability and social justice.

Developing food citizenship also enables our community to become more food secure. Food citizenship encourages people to develop food-related behaviours that help to support an ethical and sustainable food

9780170495714

system. Food citizens take responsibility for their own health by consuming a healthy diet, and they also act to improve the food supply – and therefore the health outcomes – for other members of the community.

Figure 6.3 How to be a responsible food citizen

Sustainable diets

A **sustainable diet** is one that has a positive impact both on the health of individuals and on the environment. Such a diet has strong links to – and promotes – food security, as it is economically affordable and therefore accessible to everyone, regardless of their income. It is also equitable; that is, it is fair and reasonable and ensures that everyone is treated equally.

The concept of sustainable diets encompasses food being culturally acceptable to different ethnic and religious groups in our society. For example, some religious groups require animals to be killed and processed in a particular way. If this is done according to their religious regulations, such as Kosher or Halal, these communities will be able access food that is culturally acceptable to them.

Importantly, a sustainable diet must have a low environmental impact so that future generations have sufficient resources – such as soil and water – available to produce the food they need to feed themselves.

Australia's food system

A **food system** is the path that food travels from 'paddock to plate'. Australia's food system involves a complex series of stages that begin with growing or producing the food, its journey onto food processors and its distribution to food retailers. Consumers then purchase the food that they consume at home, at school, in the workplace, or at a café or restaurant. Disposal of food waste and recycling are critical at every stage in the food system, and involve encouraging food rescue, composting waste, and recycling packaging waste.

Weblink Food systems

The stages in a food system

Australia's main food system is complex and highly industrialised, and produces a large volume of affordable food for consumers.

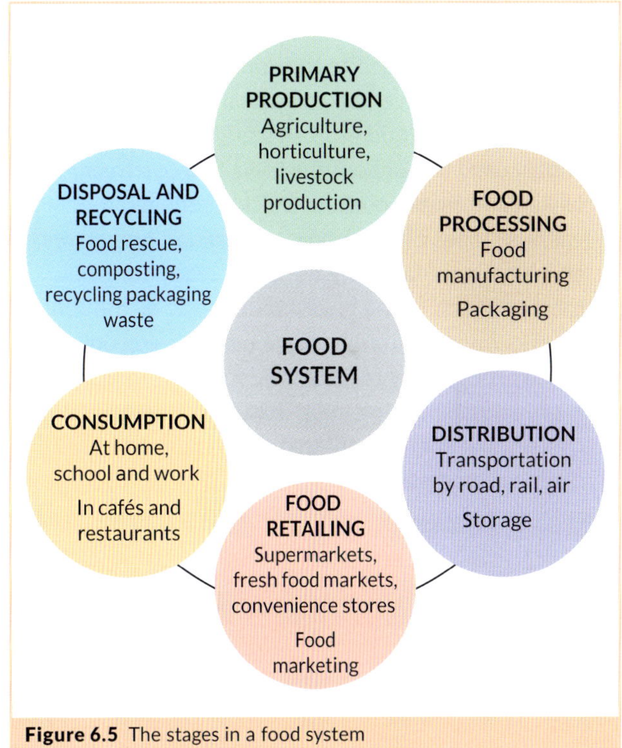

Figure 6.5 The stages in a food system

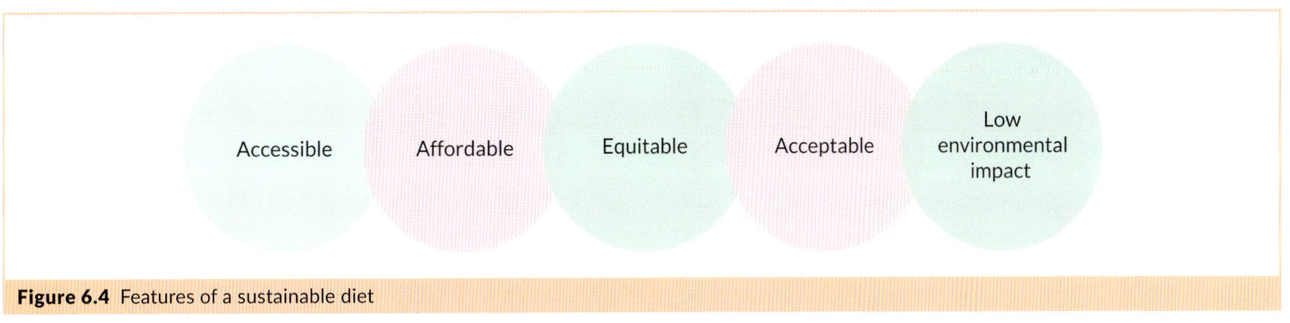

Figure 6.4 Features of a sustainable diet

- **Primary production:** Australian farmers grow and produce a wide range of primary produce – including cereals, fruit and vegetables, meat, fish, poultry, dairy products, legumes and nuts – that provide Australian consumers with safe and nutritious food. Farmers will either transport fresh food products such as vegetables and fruit to markets, or send the raw produce to food processors and food manufacturers.
- **Food processing:** Food processors turn the raw food products they receive from primary producers into other products; for example, they make milk into butter or cheese that is then ready for sale. Food manufacturers can then use products received from food processors to make 'value-added' or convenience products, such as chilled or frozen meals, biscuits and cakes, and pasta and pasta sauces. Many of these types of products include a wide variety of ingredients and are described as 'ultra-processed'. Food products are packaged to keep food safe and so that it can be transported and distributed safely.
- **Distribution:** Efficient transportation is essential to move food through each stage of the food system. Rail systems are used to transport unpacked bulk produce such as cereal grains from large grain storage silos to food processors. Road transport is widely used throughout Australia to move fresh, chilled and frozen produce quickly between the various stages of the food system. Food processors and food manufacturers transport ready-for-sale food products to the huge supermarket distribution centres for short-term storage before being delivered to supermarkets throughout Australia by road. High-value perishable food items, such as Australian oysters and crayfish, are transported by air to overseas markets.

KDS Photographics/Shutterstock.com

Figure 6.6 Wheat is transported by train from large grain storage silos in rural Australia.

- **Food retailing:** Consumers shop for food at supermarkets, fresh food markets, farmer's markets, greengrocers, butchers, convenience stores and small local stores. Food marketing is an essential component of the food retail sector, and food manufacturers use social media, television, magazines and instore displays to promote their products to consumers.
- **Consumption:** Food for family meals may be prepared at home using ingredients purchased from food retailers. Today, many consumers purchase meal boxes or meal kits that are delivered to their home and contain all the ingredients needed to make a healthy, delicious meal that minimises food waste. Purchasing takeaway food, using a meal delivery service, and eating at a café or restaurant are also popular with many consumers.

Daria Nipot/Shutterstock.com

Figure 6.7 Many families purchase meal kits that contain all the ingredients needed to make healthy, delicious meals.

- **Disposal and recycling:** Food waste is a major problem that occurs at every stage of the Australian food system. Farmers throw out perfectly edible food that does not meet supermarket specifications for size, shape or colour. Food manufacturers produce waste when they turn fresh vegetables and fruit into new products, such as pre-prepared meals. Supermarkets waste food that is not sold or is past its use-by date. The major foods wasted by Australian households are vegetables, fruit and bread.

  To address the problem of waste, many farmers, food processors, food manufacturers, supermarkets and restaurants donate excess food that is good to eat to food rescue organisations such as OzHarvest and Second Bite. These organisations use the donated food to prepare meals for people in need, thus preventing the food from going to landfill. In addition, households may compost fruit and vegetable scraps or add them to their green waste, so that the scraps can be composted in their local government facility. It is also important to recycle food packaging for reuse where possible to avoid it going to landfill.

9780170495714

Activity 6.1

'Field to Feast: Roma Tomatoes'

Weblink Field to Feast: Roma Tomatoes

Go to SBS OnDemand and access the video 'Field to Feast: Roma Tomatoes', featuring Thanh Truong – the 'Fruit Nerd'.

The video focuses on tomato growers Merc Bros & Co, which is based in Shepparton, Victoria. Merc Bros & Co grows a range of different varieties of tomatoes, including Roma tomatoes for the Italian community to use in making traditional passata, as well as Roma tomatoes and vine-ripened tomatoes for the supermarket sector.

After watching the video, answer the following questions.

1 Outline the stages in growing Roma tomatoes, as shown in the video.
2 Describe the physical properties of gourmet tomatoes grown in fields compared to those grown in a greenhouse.
3 Explain how the growers test the ripeness of the tomatoes before harvest.
4 How has technology improved the system of grading the tomatoes before they are packaged?
5 Create a flow chart of the stages in making a traditional Italian tomato passata, as shown in the video.
6 Read the recipe for Napoli Sauce on page 245 and explain how this recipe differs from the traditional Italian recipe shown on the video.
7 Why are some food products transported through the food system by road while others are moved by rail?
8 Sketch a map of your neighbourhood. List the places away from home and school that you and your friends are likely to consume food.
9 Make a list of retail outlets where members of the local community can purchase food in your neighbourhood.
10 Create a mind map of areas in our food system that create food waste.
11 Describe how farmers, food processors and food manufacturers can help minimise the amount of food waste sent to landfill.

Testing knowledge 6.1

Quiz Testing knowledge 6.1

1 What is meant by the term 'food security'?
2 Explain why it is important for individuals to demonstrate their food sovereignty.
3 Identify one example of how members of your household could become responsible food citizens.
4 How will producing food with a low environmental impact ensure that our diet is sustainable?
5 Create a flow chart of the stages of the food system that would turn wheat into a breakfast cereal such as Weet-Bix.

Food labelling

Food labels are an essential component of Australia's current food system. They are an important tool for consumers to use to help them make healthy food choices. Food Standards Australia New Zealand (FSANZ) sets the labelling requirements that apply to all packaged food sold in Australia and New Zealand, and these are contained in the Australia New Zealand Food Standards Code.

Figure 6.8 All packaged food sold in Australia must be labelled according to the Australia New Zealand Food Standards Code.

Food labels include information such as a use-by or best-before date, a list of ingredients the product contains, nutrition information, any ingredients that may cause an allergic reaction, any food additives, instructions on how to store and prepare the product, any advisory or warning statements, the country of origin, and the name and address of the manufacturer.

Nutrition information panel

Some of the most useful information on a food label is in the nutrition information panel. This panel must show the amount of nutrients per serve and per 100 grams contained in the product. This information helps consumers when shopping, as they can compare the nutritional content of similar products. It is better to compare the nutritional information per 100 grams, as this standard is set by FSANZ. The amount per serve, on the other hand, is determined by the manufacturers and can vary between brands of the same product.

Weblink FoodSwitch app

However, FSANZ has specified that some products do not need to provide a nutrition information panel, including:

- foods sold unpackaged
- foods made and packaged at the point of sale, such as bread made and sold in a bakery
- herbs, spices, packaged water, and tea and coffee because they have no significant nutritional value.

Source: https://www.foodstandards.gov.au/consumer/labelling/panels

Shoppers can download the FoodSwitch app onto their smartphone to take with them when they shop to help them make healthier food choices. This app allows consumers to use their phone camera to scan the barcode of food products to find out information about what is in the food they are buying, and suggestions for healthier alternatives.

The app also contains information about the Health Star Rating of a product. Each product is rated for its total fat, saturated fat, sugar and salt content. The traffic light ratings show if the product is low (green), medium (amber) or high (red) in these ingredients, based on widely accepted nutritional standards.

Ingredient list

All ingredients must be listed on a product's food label in descending order by weight.

The label must display a list of the most common food allergens, even if they are present in very small amounts. The foods likely to cause an allergic reaction that must be shown on a food label include peanuts, tree nuts, eggs, cow's milk, fish, crustacea, sesame seeds, soybeans, gluten and lupins.

Food additives must be shown in the ingredients list by their category – such as food colouring or flavour enhancer – name and number. Additives that may cause problems for some people include the flavour enhancer monosodium glutamate (MSG) (621), yellow food colouring (2G107) and benzoates preservatives (210, 211, 212 and 213).

Nutrition content claims and health claims

Nutrition content claims

A food label may make a claim about the nutrition content or substances in a food; for example, by stating that the product is 'low fat', 'reduced salt' or 'high fibre'. Before making these claims, a food manufacturing company must make sure the product meets the criteria set out by FSANZ. For example, in order to make a claim that a product is 'high fibre', the manufacturer will need to ensure that the product contains more than the amount of the fibre specified in the Australia New Zealand Food Standards Code.

Figure 6.9 To be labelled with the nutrition content claim '0 added sugar', YoPro yoghurt in a pouch must not contain any added sugar.

Health claims

Health claims are different to nutrition content claims, as they make a link between a food, or a nutrient or substance in a food, and an effect on health. There are two types of health claims:

- **General level health claims** refer to a nutrient or substance in a food and its effect on a health function. An example of a general level health claim is: 'Calcium is good for bones and teeth.'
- **High level health claims** refer to a nutrient or substance in a food and its relationship to a serious disease or to a biomarker of a serious disease. An example of a serious disease health claim is: 'Diets high in calcium may reduce the risk of osteoporosis in people 65 years and over.' An example of a permitted biomarker health claim is: 'Phytosterols may reduce blood cholesterol.'

9780170495714

PRACTICAL ACTIVITY 6.1

Worksheet
Practical activity 6.1

Comparing the labelling information on different types of rice crackers

Figure 6.10 Rice crackers are a popular snack food.

Aim

To develop an understanding of the importance of food labelling information to consumers.

Method

1 Select two different brands of rice crackers. Make sure the products have the same flavour profile; for example, both are 'original' crackers.

2 Examine the label on the packet of each type of rice cracker, and then copy and complete the following table.

Results

Food labelling information	Cracker 1	Cracker 2
Total fats per 100 grams		
Total energy per 100 grams		
Total sodium per 100 grams		
Additives		
Allergen information		
Net weight		
Country of origin		
Any nutrition content claims		

Analysis

1 Was there a difference in the amount of fat per 100 grams in each product? Explain your results.

2 Examine the food label and identify the ingredient or ingredients that contribute to the fat content of the cracker.

3 Was there a difference in the amount of energy per 100 grams in each product? Explain your results.

4 How would the results you observed in question 1 affect the amount of energy the crackers contain?

5 Which brand of rice cracker had the lowest sodium content?

6 Did either of the products contain any additives that you do not recognise? Why do you think these ingredients are included in the rice crackers?

7 Did the rice crackers contain similar allergen information? Explain your results. Why is this information required on a food label?

8 Were you surprised by the information about the country of origin? Why is this information important to consumers?

9 Describe any similarities or differences in any nutrition content claim made on each food label. What information would the manufacturer need to provide before making this claim?

Conclusion

Based on the information you have gained from examining the food labels of different brands of rice crackers, justify why food labels are an important source of information for consumers.

6

PRACTICAL ACTIVITY 6.2

Comparing the sensory properties of different types of rice crackers

Worksheet Practical activity 6.2

Template Sensory wheel

Aim

To determine which type of rice cracker is the most suitable as a snack for teenagers.

Method

1 Select two different brands of rice crackers. Make sure the products have the same flavour profile; for example, both are 'original' crackers.
2 Taste test each of the different crackers and record your description of the appearance, aroma, flavour, texture and sound in a table like the one following.
3 Write up your sensory analysis using technical terminology.
4 Rate each product on a scale of 1–5, where 1 = poor and 5 = great.

Results

Sensory property	Cracker 1	Cracker 2
Appearance		
Aroma		
Flavour		
Texture		
Sound		
Sensory rating (1 = poor; 5 = great)		

Analysis

1 Was there any observable difference in the appearance of the two types of rice crackers? Explain your answer.
2 Could you detect a difference in the aroma of the two types of rice crackers? Examine the food labels to determine which ingredients might contribute to the aroma of the crackers.
3 Which rice cracker had the sweetest flavour? Examine the food labels to determine which ingredients might contribute to the sweetness of the crackers.
4 Did the rice crackers have a similar texture and sound when eaten? Suggest a reason why the textures of each rice cracker are similar.

Conclusion

Based on the information from the taste test you have just completed and the information on food labelling in Practical activity 6.1, which rice cracker would you recommend as a healthy snack for teenagers? Justify your answer.

Health Star Rating system

The **Health Star Rating** system is a front-of-pack labelling system that rates the overall nutritional profile of packaged food and gives it a rating from half a star to five stars. It is a voluntary system that provides a quick and easy way to compare similar packaged foods, such as breakfast cereals. The more stars, the healthier the choice. However, it is important to remember that while the Health Star Rating system can help consumers select the best choice between different products of the same type, it is not designed to compare foods in different categories; for example, to compare breakfast cereal with yoghurt or baked beans.

Although most packaged food products in supermarkets have a nutrition information panel that provides details of the nutrient content, it takes time to compare the nutrient values in similar products before deciding which one to purchase. The benefit to consumers of the Health Star Rating system is that while shopping, they only need to glance at the front of the package to determine which item will help them make a healthy food choice.

The Health Star Rating system was developed by Australian, state and territory, and New Zealand governments in consultation with industry, public health and consumer groups.

Many health experts are now calling on the federal government to make the Health Star Rating system

9780170495714

mandatory. A report released in November 2024 by The George Institute for Global Health stated that 'just 36 per cent of intended supermarket items carried a Health Star Rating'. The audit of products by The George Institute also found that:

> *food companies continue to be highly selective about which products they chose to display the rating: recent peer-reviewed analysis of more than 21 000 products identified 61 per cent uptake among five-star rated products compared to 24 per cent of products with three or fewer stars.*

Source: https://www.georgeinstitute.org.au/media-releases/mandatory-health- star-ratings-a-must-after-10-years-of-food-industry-inaction

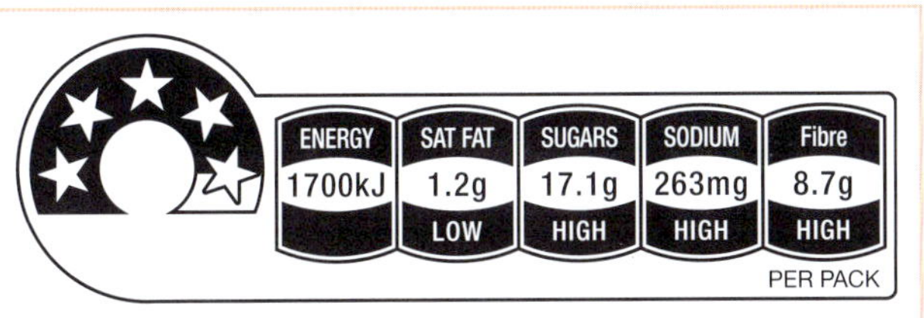

Figure 6.11 The Health Star Rating system provides key information about nutrients in packaged products.

Smart food labelling

As well as providing shoppers with information about the ingredients and nutrient content of food, food labels can include a wide variety of embedded smart technology to entice consumers or help them decide which products to purchase.

- A shopper can use their smartphone to scan a QR (quick-response) code on a food label to find out more information about the product. For example, a new variety of flat peaches called 'Ondine' was launched onto the Australian market in 2024. As part of its marketing campaign, the company has included a QR code on stickers on the fruit that allows the purchaser to explore information about the new peaches, including their sensory properties, why they are considered to be a healthy snack, suggestions about how to cook with them, and where to buy them locally. Other information can include where the key ingredient in the food is grown and details about the food's journey through the food supply chain.

Figure 6.12 A QR code on food packaging, such as the packaging of these eggs, allows customers to explore information about the eggs.

- New packaging materials that change colour are being developed to indicate the freshness of perishable food. Smart sensors in the packaging material can detect any changes, such as in the moisture or gases given off by vegetables. As vegetables pass their peak of freshness, the type and amount of gases they give off alter, and this triggers the packaging material to change colour, indicating that the vegetable may no longer be fresh but the food will still be safe to eat.
- Augmented reality is becoming a more common feature of food labelling. Augmented reality uses computer-generated digital information such as visual images and sound to provide a 'real-world' experience for users. Shoppers can use their smartphone to scan a food label and read or watch a video about the food product. This enables them to interact directly with the food label, and read the story behind the product and how and why it was produced. By scanning the label, they could also learn how to store, cook and serve the product to enhance its sensory properties and/or avoid food waste. For some consumers, watching a very short video on the nutritional benefits of including the product in their diet might be simpler and more appealing than trying to decode the nutrition information panel on the food label.

Australian supermarkets

Shopping for food is a regular weekly activity for most Australian families. Visiting a nearby fresh food market might be the preference for some people, while for others the most convenient way of purchasing their weekly food supplies is to shop instore or online at one of the major supermarkets.

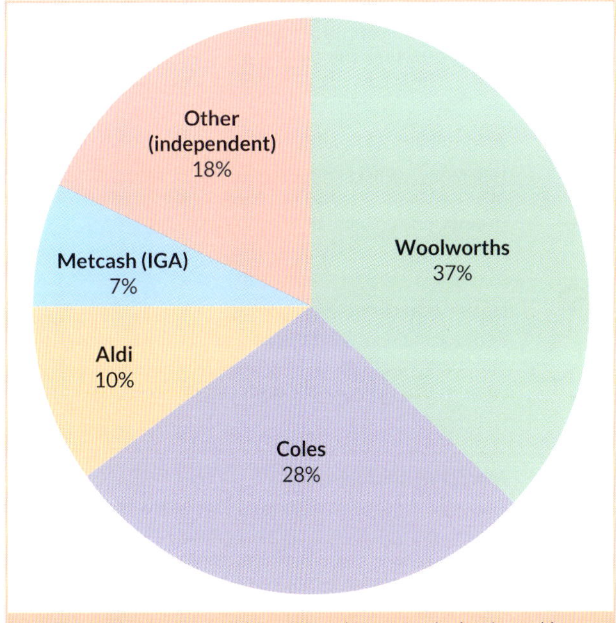

Figure 6.13 The Australian supermarket sector is dominated by Woolworths and Coles.

In Australia, the supermarket sector is dominated by Woolworths and Coles, which together account for 65 per cent of total grocery sales. The German company Aldi and Metcash, which includes IGA, together account for 17 per cent.

However, the cost-of-living crisis that many households have experienced over the past few years has had an impact on the food budgets of a large number of Australians and has led to significant changes in the supermarket sector.

- More Australians are now doing their weekly food shop at less expensive retailers, including Aldi and Costco.
- More shoppers are now choosing 'home-brand' products rather than more expensive brands.
- Families that have become vulnerable to food insecurity are now seeking food support, either on a short-term or longer-term basis, from organisations such as Foodbank and OzHarvest.

iStock.com/chameleonseye

Figure 6.14 Most Australians regularly buy their food at a supermarket.

The influence of the supermarket sector on the health of Australians

Research has shown that not many Australians select their food based on the *Australian Dietary Guidelines*; they are more likely to purchase products that are inexpensive and extensively marketed – such as ultra-processed foods and sugary drinks – when shopping at their local supermarket. An excessive consumption of foods high in saturated fat, salt and sugar by many Australians has resulted in an increasing incidence of overweight and obesity. Data from the National Health and Medical Research Council (NHMRC) released in June 2024 shows that 66 per cent of Australian adults aged over 18 and 26 per cent of children aged 2–17 are living with overweight or obesity. When we consider the number of aisles and shelving space dedicated to savoury snack foods, potato crisps, corn chips, sugary beverages, sweet treats, lollies and biscuits in supermarkets, it is unsurprising that these ultra-processed products are some of the most widely consumed by Australians.

A report titled *Inside our Supermarkets: Australia 2024* by Gary Sacks and Jasmine Chan explains that many of the products supermarkets sell are having a negative impact on the health of consumers, and are contributing to the unhealthy diets and excess weight of many Australians. The authors argue that the food industry, and particularly the supermarket sector, must be at the forefront of change by creating healthier food environments that will improve the diet and nutritional status of Australian consumers. The report identifies priority areas where it feels the supermarket sector must be more proactive.

PRIORITY RECOMMENDATIONS FOR THE SUPERMARKET SECTOR

1 Healthy food sales targets

Set company-wide targets to increase the proportion of sales from healthy products, and publicly report progress against this target each year.

2 Affordability of healthy food

Implement concrete actions to improve the affordability of healthy products, and limit price promotions on unhealthy products.

3 Healthy checkouts

Remove unhealthy products, such as confectionery and sugar-sweetened beverages, from displays near registers across all stores nationally.

4 Marketing to children

Reduce exposure of children to unhealthy food marketing, including eliminating use of promotional techniques (e.g. cartoon characters, interactive games, collectible campaigns) with strong appeal to children in relation to unhealthy products and brands.

5 Healthier products

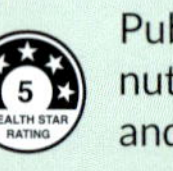

Publicise specific, time-bound targets for reducing nutrients of concern (sodium, sugar, saturated fat, and trans fat) and energy/portion sizes of own-brand products. Routinely report on progress towards commitments and targets.

6 Better nutrition labelling

Implement in-store and digital nutrition information strategies to guide consumers to purchase healthier products (e.g. displaying the Health Star Rating on instore shelf tags and online). Monitor and report the impact of these strategies on the healthiness of purchases.

7 Work with suppliers and branded food manufacturers

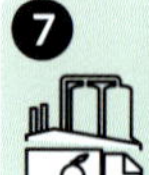

Work across the supply chain to improve nutrition-related practices, including with respect to product development, nutrition labelling and promotion practices.

Source: Inside our Supermarkets Australia 2024, https://www.insideourfoodcompanies.com.au/_files/ugd/2e3337_3a20bc9def2d4c49ba82518f92356ea1.pdf

Figure 6.15 The *Inside our Supermarkets: Australia 2024* report's key recommendations

9780170495714

CASE STUDY 6.1

Examining popular supermarket foods

Read the newspaper article below, and then answer the questions that follow.

These 10 popular supermarket foods are worse for you than you probably think

Susie Burrell, *The Age*, 27 May 2024

There are plenty of supermarket foods you likely know are not healthy. Soft drinks, potato chips and confectionery are just some of the foods that have very few essential nutrients and are so heavily processed they bear little resemblance to whole, natural foods.

Then there are the foods that look healthy, and sound healthy. In many cases, they are marketed as healthy and may even be found in health food aisles.

Indeed, our perception of them is so positive that we are more than happy to serve these foods regularly at home. Yet a closer look at the nutritional panels and ingredient lists may reveal they are not as good for you as commonly assumed.

Margarine

There is nothing healthy about refined vegetable oils being processed to make a spread, no matter what packaging or marketing campaigns may suggest. Certainly, there are spreads that are lighter, made with different oils or that have plant sterols added, but they are certainly not adding anything positive nutritionally to the diet. While we need good fats in the diet, these are always best from natural food sources – avocado, extra virgin olive oil, nuts and seeds; and if you do choose to have a spread with your favourite Vegemite on toast, a thin spread of butter is no worse than a margarine – in both cases, the less of both, the better. Or, even better, switch to avocado or a 100 per cent nut spread and avoid adding extra processed vegetable oil into your diet as much as possible.

Dips

It may be their bright colours and association with vegetables that lead us to assume that dips are healthy. While there are a couple of true vegetable-based dips, the reality is that most commercial dips have a base of cream or oil with small amounts of vegies added for colour and naming rights. This basically means that tucking into a tub of dip means you are actually tucking into a whole lot of fat and calories. If you truly love dip, take a magnifying glass to find the couple of vegetable-based dips in which the base is 60–70 per cent vegetables.

iStock.com/OlgaLepeshkina

Figure 6.16 Not all store-bought dips are as healthy as they might seem

Wraps

This is another food that is often perceived as a lighter, healthier option than bread. A quick scan of the ingredient list on a packet of wraps will reveal a long and convoluted list of processed ingredients that are required to turn a grain into a flat piece of bread that can be eaten for weeks, if not months, after it was made. Sure, you can find some wholegrain varieties, but, in general, wraps are much more processed than wholegrain breads, and certainly not a better option nutritionally.

Quick-cook oats

When it comes to breakfast options, wholegrain oats are one of the best options nutritionally. With a low glycaemic load and plenty of nutrients, plus offering a good amount of soluble fibre to help to reduce blood cholesterol levels, you can't go wrong with a bowl of oats to start the day. What's important to know, though, is that quick-cook oats are not the same as whole regular oats, especially when they come heavily processed with loads of extra sugars added. If quick-cook oats are your go-to, take a quick scan of the pack as in some cases this 'healthy' breakfast will contain up to 10g of added sugars, which somewhat negates the benefits of an oat-based breakfast. If oats are your thing, stick to whole oats when you can, or seek out the 'no added sugar' sachets if you need a quick-cook option.

>

Crackers

There are a handful of wholegrain minimally processed crackers in supermarkets but there is also a significant number of popular brands that have a base of white flour, palm oil, sugars and flavours that translate into a snack that is anything but healthy. Just because a cracker looks relatively plain does not mean that it is not ultra-processed, so make sure you grab crackers to team with cheese that have a wholegrain or seed base if you are keen to reduce your intake of ultra-processed, flavoured foods.

Fruit yoghurt

The yoghurt aisle can be a confusing place, even for nutrition professionals, with a wide range of sweetened and unsweetened Greek and natural yoghurts. While yoghurt is generally a nutritious food, the reality is that there is only a handful of 'no added sugar' flavoured and fruit-based options. Rather, most sweet yoghurts, both for children and adults, contain a decent amount of added sugar, which means that a tub of yoghurt can contain upwards of 15–20g of sugars per serve.

Chicken nuggets

For busy families, pre-crumbed chicken can be a convenient, cost-effective and child-friendly option for quick and easy midweek meals. While there are a small handful of products that stack up pretty well nutritionally, with up to 70 per cent chicken, there are plenty more that are pretty underwhelming, with as little as 40 per cent chicken along with plenty of fillers and ingredients that are not usually found in a piece of crumbed chicken. This means if pre-crumbed chicken is on your weekly shopping list, it's time to take a closer look at ingredient lists and look for options that contain at least 60 per cent chicken.

Processed meat snacks

Don't let the growing popularity of processed meat snacks for both children and adults fool you. Processed meat that is cured with nitrates, which may be listed as food additives 249–252 on ingredient lists, is known to cause changes in the digestive system. These changes can increase the risk of developing some types of gut cancers, which is why the World Health Organization suggests that the intake of processed meats be limited in the diet. Seeing an increasing number of snacks for both children and adults that encourage the consumption of processed meat as a snack is completely contrary to this public health advice. Processed meat may offer protein, but so do many other foods including dairy, fish, nuts and seeds minus the addition of nitrates into the diet.

Trail mix

How is it possible that a mix of fruit, nuts and seeds could be anything but healthy? While nuts and seeds are a rich natural source of essential nutrients, dried fruit and confectionery, which form the base of many trail mix blends found in supermarkets, do not offer much except extra calories, fats and sugars. As such, if you enjoy trail mix, you are best to make your own.

Protein bars

You could be forgiven for thinking that foods routinely found in the health food section are actually healthy and, while protein bars can offer a concentrated serve of protein, they can also have as many calories as a small meal alongside a long list of processed ingredients required to turn protein into a tasty bar. Sure, protein bars can be a convenient way to get protein on the run, but you will always be better to opt for fresh, whole foods for your protein hit and leave the bars as a back-up option.

Source: *The Age*, May 27, 2024, https://www.theage.com.au/goodfood/tips-and-advice/these-10-popular-supermarket-foods-are-worse-for-you-than-you-probably-think-20240524-p5jgcl.html

Analysis

1 Which of the food items listed in this article were you surprised to find were not considered to be a healthy food choice? Why?

2 Why does Susie Burrell, the author and nutritionist, recommend using avocado or a 100 per cent nut spread on your morning toast or on crackers rather than margarine?

3 Explain why wraps are considered to be a less nutritious food choice than wholegrain bread.

4 Outline the nutritional reasons why parents should avoid serving chicken nuggets as a meal to their children. Suggest an alternative meal suitable for children that would be more nutritionally sound. Justify your choice.

5 Why are processed meat snacks such as sliced ham or kabana considered to be an unhealthy food choice?

6 Wraps, margarine and dips are all commonly used as a lunch ingredient. Recommend alternative ingredients that are minimally processed that a family member could use to prepare a healthy lunch.

9780170495714

Testing knowledge 6.2

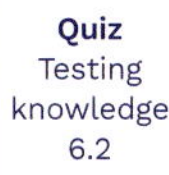

Quiz Testing knowledge 6.2

11 Explain why food labels are an important part of Australia's food system.

12 When analysing food products, why should the nutrients per 100 grams be compared rather than the amount per serve?

13 Outline how the ingredients are listed on a food label and the types of ingredients that must be included on the label.

14 Explain the difference between a nutrition content claim and health claim.

15 What is the Health Star Rating system and why is it a useful tool for consumers when shopping?

16 Justify why some food manufacturers include a QR code on their food labels.

17 Explain how new packaging materials will help consumers when shopping for vegetables in the supermarket.

18 Create a simple graph that demonstrates the total grocery sales of each of the major Australian supermarket companies.

19 Describe one way in which supermarkets have a negative influence on the health of Australians.

20 Select one of the priority recommendations for the supermarket sector from the *Inside our Supermarkets: Australia 2024* report and construct an argument about how implementing this recommendation would create a more health-focused shopping environment.

Sustainable food systems

The United Nations defines 'sustainability' as 'meeting the needs of the present without compromising the ability of future generations to meet their own needs'.

Sustainable development is based around four key areas or pillars: economic, social, environmental and cultural sustainability.

Economic sustainability

Aims to ensure that all resources are used efficiently, that food manufacturing companies remain economically viable, and that nutritious food is accessible and affordable for all Australians.

Social sustainability

Focuses on ensuring that communities are equitable and fair; for example, by supporting and promoting workers' rights through Fairtrade.

Environmental sustainability

Works towards ensuring that environmental resources are preserved and maintained, and not degraded or depleted at any stage within the food system.

Cultural sustainability

Ensures that a community's beliefs and practices are fostered and protected.

Figure 6.17 The four pillars of sustainable development

PRACTICAL ACTIVITY 6.3

Worksheet Practical activity 6.3

Sensory analysis: Taste testing alternative sources of protein

Aim

To determine whether there are cultural barriers to eating alternative sources of protein such as dehydrated crickets, mealworms or ants.

Method

1 Conduct a taste test of three different types of alternative protein snacks, such as snack crickets, mealworm snacks and ant candy.

2 Record your results in a table like the one on the following page.

Results

Alternative protein snack	Appearance	Aroma	Flavour	Texture	Sound	Sensory rating
Snack crickets						
Mealworm snacks						
Ant candy						

Analysis

1 Which alternative protein snack had the most appealing appearance, and which had the least appealing? Why?

2 Identify the alternative protein snack that had the sweetest flavour.

3 Did one of the alternative protein snacks have a more appealing aroma than the others?

4 Suggest a reason why the texture of each alternative protein snack is different.

5 Consider the sound each food made when you first chewed it. Did hearing the sound affect your enjoyment of the snack? Explain why or why not.

Conclusion

Based on the information from the taste test you have just completed, do you think alternative sources of protein, such as those derived from insects, would be culturally acceptable to most consumers even if the sensory properties were enjoyable? Justify your decision.

Challenges to sustainable food systems

The Australian farming community has a reputation for producing some of the best and most nutritious foods in the world – ensuring that, as a nation, we are food secure. However, there are a number of challenges for the nation to address if we are to continue to produce food sustainably into the future.

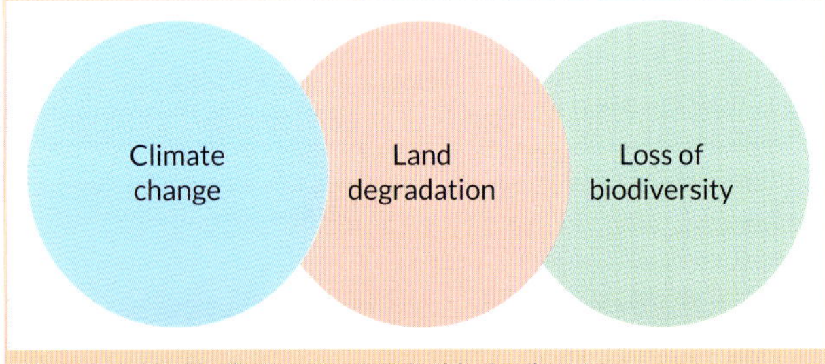

Figure 6.18 Challenges to sustainable food systems

Climate change

Food production, along with the production of many other things we rely on today, uses a wide range of resources. Some of these are renewable, such as sunlight and water, while others are non-renewable. The use of non-renewable resources – including oil, coal and gas – produces greenhouse gases and has been shown to have a direct impact on climate change.

Climate change has led to air, land and sea temperatures increasing more than 1.5°C in many parts of Australia since around 1910. Rainfall patterns have also changed, with some areas in the north of Australia being subjected to more frequent and severe flooding, while areas in the south and west of the continent are now more prone to drought. Increasing soil temperatures, floods and droughts make it more difficult for farmers to grow crops and raise livestock.

Staple food crops, such as wheat, are susceptible to the effects of climate change, as the increasing air and land temperatures and lack of rain have diminished crop yields and allowed pests and diseases to attack crops. Ocean temperatures too have started to rise. This will impact on fish colonies, as many cool-water species will not be able to survive in the warmer temperatures.

Another of the major impacts of climate change is the increased production of carbon dioxide (CO_2) in the atmosphere. While CO_2 is essential for plant growth, an excess of it reduces the nutrient content of some plants, especially the amount of protein and minerals present in wheat, soybeans and rice.

iStock.com/Ken Griffiths

Figure 6.19 Australia's production of wheat will be severely affected by climate change.

9780170495714

Land degradation

Degradation of the land limits the amount of carbon the soil can hold. It not only restricts plant growth, but also impacts on the variety and amount of biodiversity in the soil; that is, the numerous microbes, fungi and invertebrates that live in the soil. One of the main factors that can cause soil to degrade is climate change. Extreme weather conditions – especially drought or extreme fire events – deplete the soil moisture and leave it vulnerable to wind and water erosion.

Land degradation is also caused by human actions, such as deforestation, over-cultivation of crops, over-grazing on pastureland, and poor use of irrigation systems leading to salination of soils and a lowering of natural water tables.

As land is degraded, agricultural and livestock production become less sustainable.

iStock.com/kristypics

Figure 6.20 Salination can cause farming land to be unsustainable.

Loss of biodiversity

Biodiversity is a term used to describe the vast array of living organisms that inhabit the planet and the interactions between them. Biodiversity encompasses all the different species of plants, animals, bacteria, fungi, viruses and other microbial organisms that live on the Earth, as well as their genetic makeup and entire ecosystems.

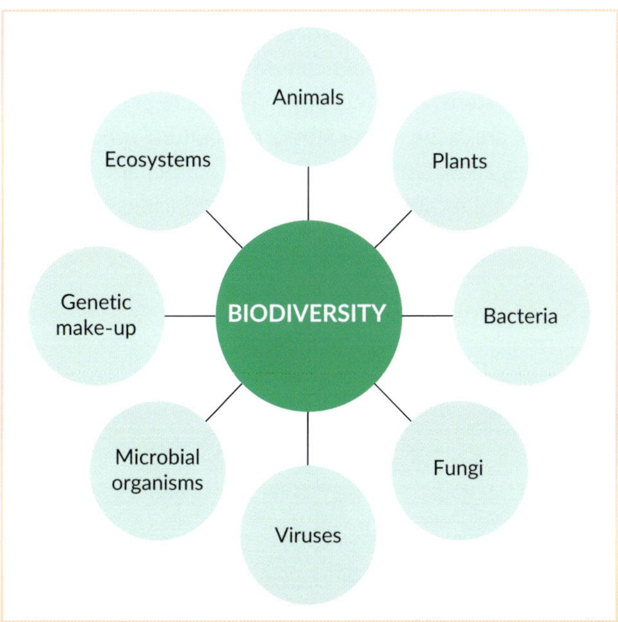

Figure 6.21 The variety of living things that make up biodiversity

Climate change is one of the biggest threats to the viability of our biodiversity. As the climate has warmed, the number and intensity of droughts and bushfires has increased. These events have had a critical impact on bee populations, as the increased temperatures and lower levels of rainfall affect the ability of trees to flower and therefore provide pollen and nectar for bees. Overuse of fertilisers and herbicides has caused waterways to become polluted and has led to a significant loss of diversity in the native plants and animals – including fish, birds and frogs – that live in or near rivers and streams.

Biodiversity is critical in sustaining food production. The billions of microbes that live in the soil cycle nutrients and break down organic matter to enable the soil to remain fertile and crops to grow. Bees and birds are essential for crop pollination; without them, fruit trees – including apple, pear, avocado and almond trees – and crops – including wheat and canola – would not produce fruit or grain. Other species, such as native frogs and bats, help reduce insect pests that could feed off crops.

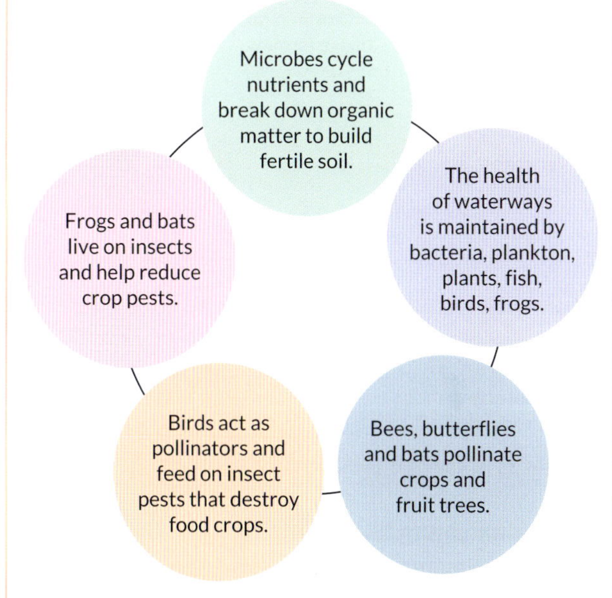

Figure 6.22 The importance of biodiversity in sustaining food production

M Rutherford/Shutterstock.com

Figure 6.23 Bees are essential for the pollination of fruit trees.

Making our food system sustainable

Figure 6.24 How to practise sustainable food habits

Implementing sustainable farming practices

Today, most Australian farmers understand the need to produce food crops sustainably so that the land remains viable and productive for future generations.

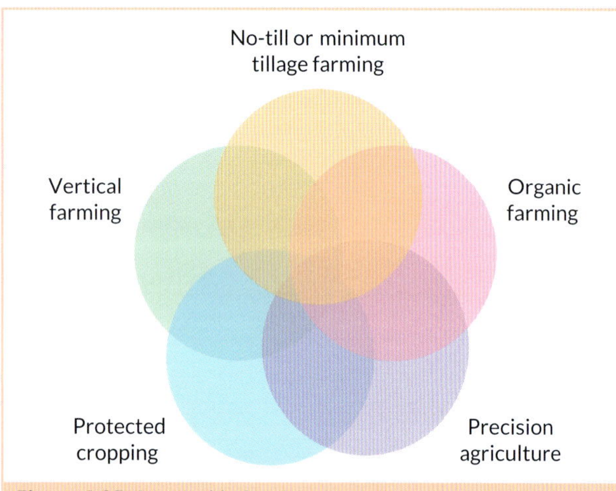

Figure 6.25 Sustainable farming practices

No-till or minimum tillage farming

No-till or minimum tillage farming is commonly used across Australia in broadacre farming, where crops such as wheat, canola, corn and chickpeas are grown on large tracts of land. This type of farming involves planting the new crop by directly drilling in between the rows of the previous crop without tilling the soil. This type of farming has been shown to be sustainable as the stubble from the previous crop is not removed, and so the nutrients it contains are returned to the soil as the stubble breaks down. In addition, the retained stubble helps to stabilise the soil and acts as a mulch, so that more moisture is retained. It also holds the soil in place, helping to prevent wind and water erosion.

Figure 6.26 Retaining stubble in minimum tillage farming

Organic farming

Organic farming is an environmentally sustainable food production system, as it produces crops and animals without the use of artificial chemicals, using natural systems instead. These natural systems can include the use of organic fertilisers, composting, crop rotation and biological pest control. Such systems minimise any environmental damage, and help to ensure the sustainability of the land.

Precision agriculture

Precision agriculture is a farming approach that uses data and technology to optimise resource use and improve crop yields by applying the right inputs at the right time, in the right place and in the right amount. It uses GPS and sensors to assist farmers to increase their profitability and minimise their environmental impact. Farmers can use data collected through satellite images to create accurate maps – for example, of their field boundaries and irrigation systems – as well as to monitor crop growth and predict yields.

Soil sensors can be used to detect the temperature, moisture and nutrient content of the soil so that farmers can water or fertilise their crops as required.

Agricultural drones are now widely used by farmers across Australia to evaluate the health of crops and detect whether there are any pests or diseases present on the plants. Farmers also use agricultural drones to survey their irrigation systems to detect any malfunction or identify crops that are facing water stress. Agricultural robots and drones allow farmers to apply the correct amount of agricultural fertilisers, pesticides or herbicides at the correct rate and to target specific plants when they need it.

Figure 6.27 Farmers can use agricultural drones to survey the health of their crops.

9780170495714

Precision agriculture is a far more sustainable form of agriculture, as it allows farmers to minimise the agricultural resources they use and to ensure that their crop yields are maximised.

Protected cropping

'Protected cropping' is term used to describe growing horticultural crops in a protected environment, such as a shade house, greenhouse, glasshouse or polytunnel; or placing a protective structure over the top of fruit trees in an orchard. This form of horticultural production is becoming more widely used across Australia as farmers grapple with the higher temperatures and drought or floods linked to climate change.

iStock.com/pierluigipalazzi

Figure 6.28 Many green leafy crops are now grown in protected environments to shield them from variations and extremes in the weather.

Protected cropping is considered to be a sustainable farming practice as it allows vegetables and fruit to be grown throughout the year in a controlled environment. Protecting plants in this way helps minimise crop loss as a result of extreme weather events – such as hailstorms or frost – bird attacks, pests and diseases. The closed environment also reduces the amount of water required as all water is recirculated, and little is lost through evaporation.

Greg Brave/Shutterstock.com

Figure 6.29 Victorian cherry farmers cover their trees in netting to prevent losing their valuable crop as a result of bird strikes, hail and fruit fly.

Vertical farming

Vertical farming is a form of protected cropping that uses a stacked indoor horticultural system to produce a range of vegetables and fruit, such as tomatoes, green leafy vegetables, herbs and strawberries. The produce is grown in an enclosed, protected, climate-controlled environment, usually in the absence of soil and natural light. In one of the main types of vertical farming, the plants receive their nutrients and water hydroponically, directly into their root systems, and can be grown all year round.

The use of hydroponic systems is considered to be sustainable as they are very water efficient and can reduce water consumption by 90 per cent in comparison with traditional horticultural practices. In addition, as vertical farming takes place in an enclosed environment, the plants are not affected by the extremes of climate, such as drought or floods.

However, there are some disadvantages to this type of food production in terms of its sustainability. Research has shown that vertical farming is very energy intensive, as it requires the use of artificial lighting. It also uses synthetic fertilisers to provide nutrients for the plants to grow.

YEINISM/Shutterstock.com

Figure 6.30 Vertical farming uses a stacked indoor horticultural system to grow crops such as green leafy vegetables.

Buying food in season

Buying food in season improves the sustainability of our food system. Fruit and vegetables are at their peak of flavour and nutrient content at specific times of the year. For example, cherries and berries – including blackberries, strawberries and raspberries – have their most intense colour and flavour in summer; while citrus fruit – especially navel and Cara Cara oranges – are at their best in winter to early spring; and asparagus, broccoli and sugar snap peas are at their peak in spring.

Buying fresh fruit and vegetables when they are in season means that they are abundant and therefore less expensive than at other times of the year.

Purchasing and eating food in season also improves the sustainability of our food system as it reduces the cost of transporting fruit and vegetables large distances across the continent; for example, bringing strawberries from Queensland to Victoria during winter, when they are out of season in Victoria. This in turn minimises the amount of greenhouse gases produced when food is stored and transported in refrigerated trucks.

In addition, eating seasonally supports local farmers and helps to reduce food waste.

iStock.com/AzmanJaka

Figure 6.31 Strawberries and cherries have their best flavour and colour when they are in season in summer.

Making ethical food choices

The food choices we make impact on the sustainability of our food system.

- Purchasing organic food improves the sustainability of the food system, as organic food is produced using natural fertilisers and pesticides rather than synthetic products. These natural systems minimise any damage to the environment, and help to ensure the viability and productivity of the land for future generations.
- Reducing the amount of meat and poultry we consume and following a flexitarian diet will reduce the impact our food choices have on the environment. A flexitarian diet is based mainly on plant foods – such as wholegrains, legumes, vegetables and nuts – but allows meat and other animal products in moderation. Livestock, particularly dairy and beef cattle, emit large amounts of methane, a greenhouse gas that contributes to global warming. Reducing our reliance on meat consumption will also help to improve soil health, reduce water pollution and increase biodiversity.
- Purchasing products that use only sustainably produced palm oil will help to ensure the survival of endangered species, such as Sumatran orangutans and tigers.
- When new palm oil plantations are developed, huge areas of rainforests are bulldozed and burnt – resulting in the destruction of biodiversity and the release of CO_2 into the atmosphere, which accelerates climate change. Rainforests are home to many rare and endangered species that use them as a source of food, or as corridors through which to move in search of food.

Vladislav T. Jirousek/Shutterstock.com

Figure 6.32 The expansion of palm oil plantations destroys rainforests that are home to endangered Sumatran orangutans.

The Roundtable on Sustainable Palm Oil (RSPO) was established in 2004 to minimise the impact of palm oil on the environment and on endangered species. When shopping, look for products that display the RSPO logo.

Wahavi/Alamy Stock Photo

Figure 6.33 The RSPO logo

Minimising food waste

Food waste is an ongoing problem in all sectors of the Australian food system. According to data released in January 2025 by the federal Department of Climate Change, Energy, the Environment and Water, approximately 7.6 million tonnes of food are wasted annually in all sectors of the food supply chain – that is, when food is grown, processed and transported, sold in supermarkets and fresh food markets, prepared and served in cafés and restaurants, and consumed at home: see Figure 6.34.

9780170495714

WHERE FOOD GETS WASTED
FROM PADDOCK TO PLATE
22% PRIMARY
17% MANUFACTURING
3% DISTRIBUTION
7% WHOLESALE / RETAIL
3% INSTITUTIONS
16% HOSPITALITY
32% HOUSEHOLDS

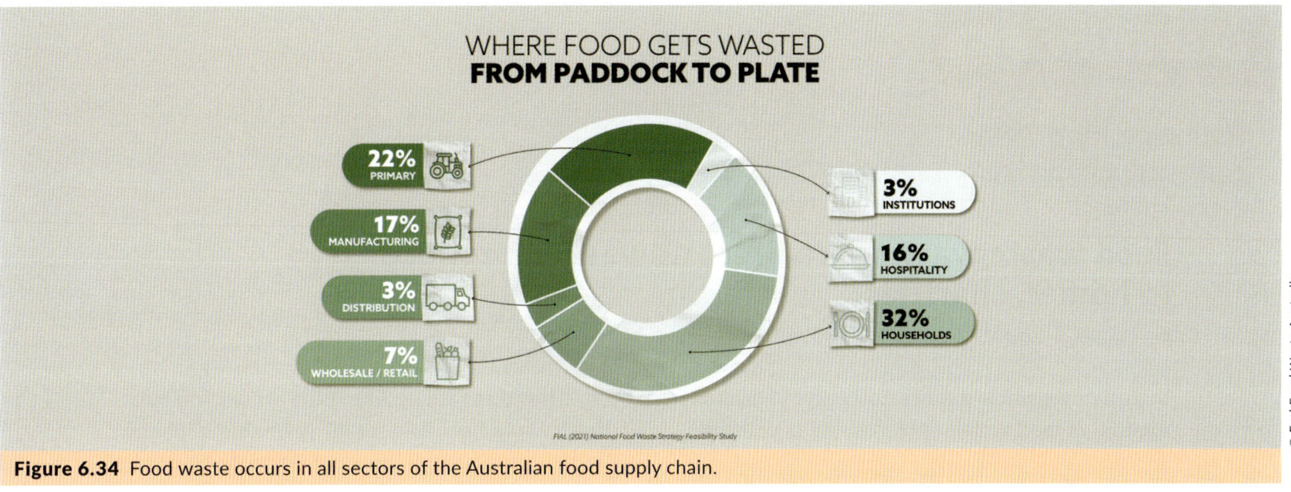

Figure 6.34 Food waste occurs in all sectors of the Australian food supply chain.

Why is food waste a problem?

According to the Australian National Food Waste Strategy Feasibility Study, there are a wide range of economic and environmental effects of food waste: see Figure 6.35.

Improving the sustainability of the food system by reducing food waste

Reducing food waste will help our food system become more sustainable by:

- reducing greenhouse gas emissions, especially methane given off by food waste as it breaks down in landfill
- preventing the loss of valuable water and soil nutrients used to grow food
- reducing the use of agricultural chemicals used to produce the grains, legumes, fruit, vegetables, nuts, seeds and animals we use as a source of food
- saving the energy used to grow, process, transport and sell food.

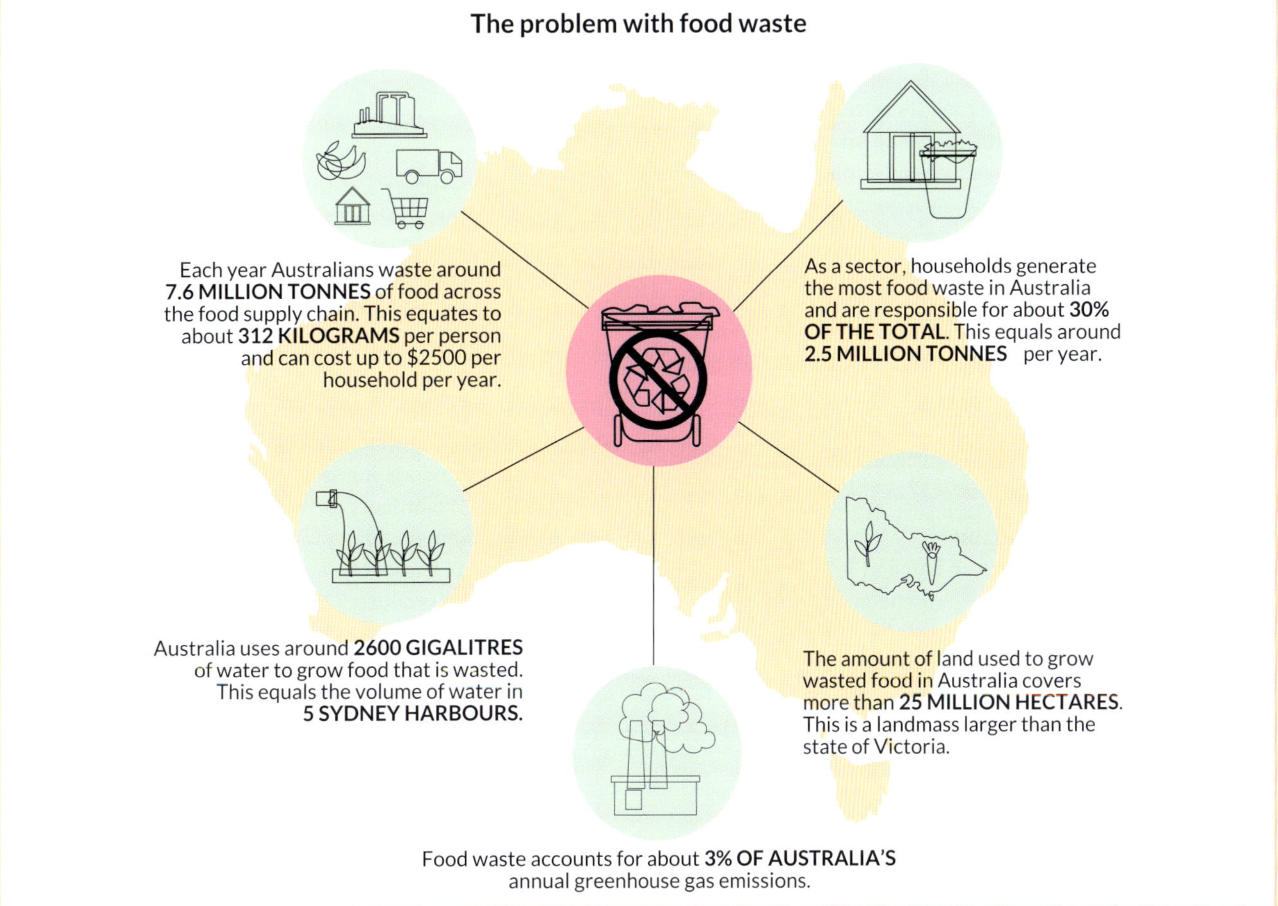

The problem with food waste
Each year Australians waste around 7.6 MILLION TONNES of food across the food supply chain. This equates to about 312 KILOGRAMS per person and can cost up to $2500 per household per year.
As a sector, households generate the most food waste in Australia and are responsible for about 30% OF THE TOTAL. This equals around 2.5 MILLION TONNES per year.
Australia uses around 2600 GIGALITRES of water to grow food that is wasted. This equals the volume of water in 5 SYDNEY HARBOURS.
The amount of land used to grow wasted food in Australia covers more than 25 MILLION HECTARES. This is a landmass larger than the state of Victoria.
Food waste accounts for about 3% OF AUSTRALIA'S annual greenhouse gas emissions.

Figure 6.35 Food waste creates significant problems for the sustainability of Australia's food system.

6

CASE STUDY 6.2

Reducing food waste

Read the newspaper article below and then answer the questions that follow.

Food waste warriors redirect unwanted produce to homes, charities

Pip Courtney, *ABC News*, 27 October 2024

Victorian vegetable grower Catherine Velisha says Australia's food system is broken but it isn't the supermarkets' fault.

Each season, supermarkets won't buy a portion of the cauliflowers she grows because they are too small, slightly yellow from sunburn or have tiny spots of leaf rub.

'There's not one player that's to blame, that's all of us, that's growers, supermarkets and shoppers,' Ms Velisha said.

'As individuals we feel like we're quite powerless but we're actually quite powerful and we can make decisions every day that can potentially change a whole food system along the line.'

She said consumers could influence change by not being so picky.

'Supermarkets go off what's left on the shelves, so if you're rummaging through cauliflowers or broccolis or apples and picking out certain ones, the data shows they sell those certain ones, and then that makes the spec,' she said.

Australians waste nearly 8 million tonnes of food every year, ranking it as the world's tenth most wasteful nation, according to End Food Waste Australia (EFWA).

Of the food wasted, 70 per cent of it is edible.

EFWA's Melissa Smith said throwing it out cost households $50 a week and the country $36 billion a year in wasted resources.

'Globally food waste has a bigger carbon footprint than flying, which in Australia it's about 3.5 per cent of our emissions,' Ms Smith said.

Finding a solution

University mates Josh Ball and Josh Brooks-Duncan started a food rescue business in Melbourne four years ago after being shocked at seeing edible produce being dumped.

Farmers Pick pays growers for produce which will be, or has been, rejected due to size, colour or shape.

The produce is then boxed and sold to consumers.

Mr Ball said it took farmers a while to realise their offers to buy unwanted produce were 'legit'.

'They'd say things like 'Why would you buy that? We've been told for 40 years this is rubbish, literally, or we should just put it in a paddock and let it break down',' he said.

The company distributes 120 000 kilograms of produce each week from warehouses in Melbourne and Brisbane.

Mr Ball said the company also bought unsold fruit and vegetables at central markets.

'We're always looking for new farmers to partner with and work together,' Mr Ball said.

'I think it's an easy pitch.

'It's like, "Hey instead of putting it in a paddock, put it in a box and we'll pay you".'

Working to a target

The federal government has set a target of halving food waste by 2030.

EFWA's Smith said the target was environmentally, socially and economically important.

'When we waste food we waste everything that went into making it, so that's the land, the water, the diesel, the fertiliser, and the staff hours,' she said.

'In dollars, this adds up to literally billions a year for horticulture with a landmass the size of Victoria used every year to grow food that gets wasted.'

To reduce food loss, EFWA offers an auditing service to help food businesses identify waste and redirect it instead of going to landfill.

Providores Simon George and Sons started donating fresh produce to charity Fare Share in Brisbane after undergoing an EFWA audit.

Fare Share volunteers turn out 1.5 million meals a year from donated produce.

Chef James Fien said he and his committed volunteers could produce 5 million meals a year if more food was delivered to him rather than the tip.

'The need is exponential, we've now got families that never in their life ever thought they'd reach out for some sort of food assistance, including those where someone has a job,' he said.

'We aren't even scratching the surface when it comes to food waste.'

Source: https://www.abc.net.au/news/2024-10-27/food-waste-warriors-redirect-unwanted-produce-homes-charities/104512870

9780170495714

Figure 6.36 About 1200 meals a week are made from the food donated by Simon George and Sons.

Analysis

1. Explain why vegetable grower Catherine Velisha says Australia's food system is broken.
2. Identify three key statistics related to food waste listed in the article.
3. Create a mind map to illustrate why and how Josh Ball and Josh Brooks-Duncan sought to find a solution to food waste.
4. Explain why End Food Waste Australia says that the federal government's target of halving food waste by 2030 is important.
5. Describe the role that Fare Share plays in reducing food waste in Australia.

Innovations in food production

Recent technological innovations – such as the use of artificial intelligence, the Internet of Things and 3D printing in the growing, manufacturing, transport and selling of food – have improved the safety and sustainability of the food system.

The use of artificial intelligence in grain storage facilities

Grain producers have introduced artificial intelligence (AI) sensors into their silos to measure real-time data, such as the temperature, moisture levels and air quality in the storage facility. This data helps them to detect environmental conditions that might lead to a loss of grain due to mould, bacteria or insect infestation. The technology improves the sustainability of the food system by reducing food waste.

The use of the Internet of Things in the food supply chain

The 'Internet of Things' (IoT) is a term used to describe physical objects or devices that connect and exchange data with other devices and systems. They may use communication networks, such as the internet or Bluetooth.

6

In relation to food, the IoT can use smart sensors to track food products with a short shelf life through the food supply chain to ensure that the product is transported at the optimal temperature and does not breach food safety requirements, thus minimising food waste.

The use of 3D-printed foods

While 3D printing of food is still in its infancy, it has the potential to help overcome some of the issues relating to food security and sustainability in the food system. A 3D printer uses edible 'inks' to build layers of ingredients that can be stacked, and sometimes cooked using lasers. It can be used to produce foods for individuals who have special nutritional needs or gastro-intestinal health issues; for example, those who require foods that require little chewing.

The use of 3D-printed foods may enable the food system to be more sustainable, as it will prevent the unnecessary use of natural resources, such as soil and water used to grow food. It will also minimise food waste by only producing the amount of a given food that is needed, and will extend the shelf life of food.

3D printing can be used to produce nutritious foods using non-traditional ingredients, such as algae. Trials are currently underway to produce some popular foods such as pasta, pizza and meat using 3D printing.

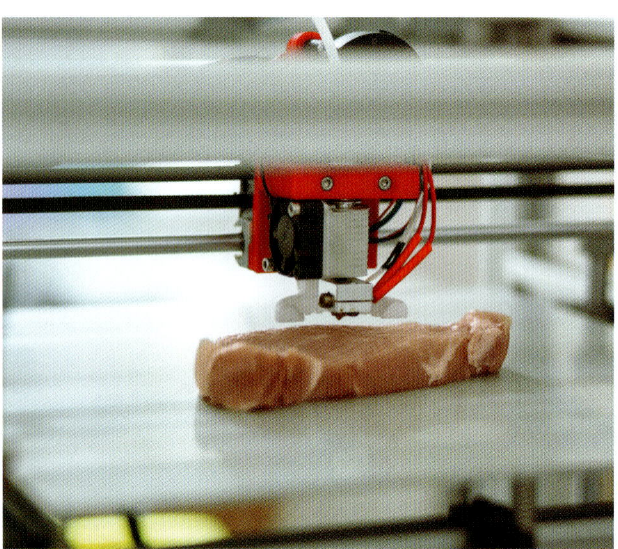

iStock.com/Zinkevych

Figure 6.37 A 3D printer recreating a piece of meat, imitating all the little details

Testing knowledge 6.3

Quiz
Testing knowledge 6.3

21 Explain why it is important for our food system to be economically sustainable.

22 Create a mind map to summarise the impact of climate change on food production.

23 Describe two significant threats to biodiversity in Australia.

24 Frogs, bats, plankton, plants and fish are all parts of the biodiversity. Discuss two ways in which they contribute to sustainable food production.

25 Create a diagram that demonstrates how precision agriculture assists farmers to produce food sustainably. Images and/or text can be used to communicate the information.

26 Explain why purchasing food in season improves the sustainability of our food system.

27 Provide one example that is not discussed in the text of how individuals can make an ethical food choice when shopping.

28 Create a mind map of the main problems associated with food waste in Australia.

29 Outline two ways in which reducing food waste will improve the sustainability of the food system.

30 Clarify how the 3D printing of food may improve the sustainability of the food system.

Critical and creative thinking 6.1

Gnocchi

Read the recipe for Gnocchi on page 152.

1 List the key ingredients used to make gnocchi.

2 Create a flow chart to demonstrate how the ingredients used in commercial fresh-chilled gnocchi would move through each stage in the Australian food system.

DESIGN ACTIVITY 6.1

Thai street food stall

The Summer Night Market has become an annual event in the calendar of the Queen Victoria Market. The market attracts people from across Victoria and tourists, who come to sample the cuisine of some of the many nationalities that now call Australia home.

One of the most popular dishes sold from the Thai Street Foods stall is a Thai caramelised chicken bowl served with an apple, carrot and ginger slaw.

9780170495714

Design brief

The owners of the Thai Street Foods stall would like you to suggest ideas for another recipe for a Thai-inspired food bowl that they could serve to give customers more choice. The food bowl must contain a protein ingredient, a marinade, a noodle or cereal product, fresh herbs and an edible garnish. It will be accompanied by the same apple, carrot and ginger slaw.

Worksheet Design activity 6.1

Templates Decision table

Sensory wheel

Australian Guide to Healthy Eating

Investigating and defining

1 Use the solution requirements and constraints in the design brief to develop four design criteria that will allow you to judge the success of your Thai-inspired food bowl.
2 Prepare the recipe for Thai Caramelised Chicken with Fragrant Rice Noodles served with Apple, Carrot and Ginger Slaw on pages 148–9 to develop an understanding of the ingredients and processes involved in preparing a Thai-inspired food bowl.
3 Use the internet, recipe books and food magazines to research the different types of ingredients used in Thai cuisine. Record these in a table like the one below.

Protein ingredient		Marinade ingredients	Noodle or cereal product	Fresh herbs	Edible garnish
1					
2					
3					
4					

Generating and designing

iStock.com/piyato

1 Using the information you have gathered from your research into suitable ingredients, develop two designed solution options for a Thai-inspired food bowl based on the solution requirements and constraints in the design brief.
2 Use a decision table (see the example on page 12) to help you select your designed solution.
3 Using the recipe for the Thai Caramelised Chicken with Fragrant Rice Noodles on page 148 as a guide to the quantities of ingredients you require, write up a new recipe for your Thai-inspired food bowl.

Planning and managing

1 Prepare a food order.
2 Before producing your Thai-inspired food bowl, write up a production plan, noting any safe work practices to be followed, and identifying the major processes to be used.

Producing and implementing

1 Prepare the product.
2 Note any modifications or changes you made during production.

Evaluating

1 Describe the sensory properties – appearance, aroma, flavour, texture and sound – of your Thai-inspired food bowl.
2 Evaluate your product according to the previously established criteria to evaluate the success of your Thai-inspired food bowl.
3 Share your Thai-inspired food bowl with two other people and record their comments about the sensory properties of the meal.
4 Suggest any modifications you would make to the recipe if you were to produce it again. Explain how you think these modifications will improve the dish.
5 Classify the ingredients of your Thai-inspired food bowl served with apple, carrot and ginger slaw on a diagram of the *Australian Guide to Healthy Eating*. Comment on how well this dish meets the recommendations of this food selection model.

Thai Caramelised Chicken with Fragrant Rice Noodles

INGREDIENTS

Fragrant Rice Noodles

1 makrut lime leaf

150 grams Thai rice noodles

½ cup mint leaves, coarsely chopped

½ cup coriander leaves, picked

½ long red chilli, seeds removed and thinly sliced

1 tablespoon lime juice

2 tablespoons roasted unsalted cashews or peanuts, roughly chopped (optional)

Thai Caramelised Chicken

2 skinless chicken thigh fillets (approximately 300 grams total weight)

2 tablespoons brown sugar

1 tablespoon soy sauce

2 tablespoons lime juice

1 tablespoon fish sauce

15 grams ginger, peeled and finely grated

2 makrut lime leaves

Garnish

1 tablespoon roasted unsalted cashews or peanuts, roughly chopped (optional)

1 tablespoon coriander leaves, picked

To serve

Apple, Carrot and Ginger Slaw (see recipe on the following page, if desired)

SERVES TWO

METHOD

Fragrant Rice Noodles

1. Finely slice the makrut lime leaf lengthwise, parallel to the vein of the leaf. Discard the vein – it is very tough so it is not used in this recipe.
2. Fill a large saucepan with water and bring to the boil. Add the rice noodles and cook for approximately 5–6 minutes or according to the instructions on the packet.
3. Drain, rinse well in cold water and drain thoroughly.
4. Place the noodles in a large bowl and add the mint, coriander, chilli, lime leaf, lime juice and nuts (if using). Stir to combine.

Thai Caramelised Chicken

1. Cut the chicken into 3-centimetre pieces.
2. Place the brown sugar, soy sauce, lime juice, fish sauce, ginger and lime leaves in a frying pan.
3. Stir the marinade mixture over a medium-high heat until the sugar has dissolved. Add the chicken pieces and simmer, stirring occasionally, for approximately 10–12 minutes or until the chicken is cooked.
4. Increase the heat slightly and cook for a further 2 minutes until the sauce has reduced and thickened. Remove the lime leaves.

To serve

Divide the Fragrant Rice Noodles between two serving bowls. Top with the Thai Caramelised Chicken and garnish with extra chopped nuts (if using) and coriander. Serve with Apple, Carrot and Ginger Slaw, if desired.

Mark Fergus Photography

9780170495714

Apple, Carrot and Ginger Slaw

INGREDIENTS

Slaw

⅛ green cabbage (approximately 150 grams)

½ medium carrot

½ red apple, such as Pink Lady or Jazz

1 tablespoon pickled ginger, drained and chopped

2 tablespoons coriander leaves, picked

Dressing

2 tablespoons rice wine vinegar

2 teaspoons caster sugar

1 tablespoon soy sauce

1 tablespoon fish sauce

1 tablespoon lime juice

METHOD

1 Shred the cabbage finely.
2 Use a sharp knife to julienne the carrot. Alternatively, use a julienne peeler if one is available.
3 Julienne the apple.
4 Combine the cabbage, carrot, apple, pickled ginger and coriander in a large bowl.
5 In a jar, combine the rice wine vinegar, sugar, soy sauce, fish sauce and lime juice. Place the lid on the jar and shake to combine.
6 Toss the slaw with the dressing until well coated.

EVALUATION

1 Describe the sensory properties – appearance, aroma, flavour, texture and sound – of your Thai Caramelised Chicken with Fragrant Rice Noodles served with Apple, Carrot and Ginger Slaw.
2 Identify the process that occurred when the rice noodles were cooked and describe the changes to the physical properties.
3 The ingredients used in the chicken marinade each perform a different role. Classify the marinade ingredients under the headings of sweetness, acidity and flavouring.
4 Identify the processes that occurred when the chicken was cooked.
5 Describe the physical changes to the properties of the chicken when it was cooked. Consider the changes in colour, texture and size.
6 Create a series of step-by-step diagrams to demonstrate to a new cook how to julienne a vegetable.
7 What is a temporary emulsion?
8 Explain why the dressing for the Apple, Carrot and Ginger Slaw is a temporary emulsion.
9 Classify the ingredients for the Thai Caramelised Chicken with Fragrant Rice Noodles served with Apple, Carrot and Ginger Slaw on a diagram of the *Australian Guide to Healthy Eating*. Explain how well this meal meets the recommendations of this food selection model.
10 Imagine you are a food influencer. Write a brief paragraph that you could post on your blog to encourage your followers to try these recipes.

Roasted Eggplant in a Spicy Sauce

INGREDIENTS

Roast vegetables

1 eggplant (approximately 400 grams)

2 tablespoons olive oil

¼ teaspoon salt

6 grinds black pepper

6 cherry tomatoes

Spicy Sauce

¼ teaspoon black mustard seeds

¼ teaspoon ground turmeric

½ teaspoon garam masala

½ teaspoon ground cumin

1 tablespoon olive oil

½ red onion, thinly sliced into half-moons

2 cloves garlic, crushed

2 centimetres fresh ginger, finely diced

½ long green chilli, deseeded and finely diced

½ cup diced canned tomatoes

100 millilitres water

¼ teaspoon salt

10 grams fresh coriander (2 tablespoons), chopped

2 teaspoons lemon juice

To serve

80 grams frozen peas, defrosted

1 spring onion, finely sliced diagonally

1 tablespoon coriander leaves, picked

1 teaspoon olive oil

6 grinds black pepper

1 tablespoon Greek-style yoghurt

2 teaspoons lemon juice

2 portions Plain Rice (see page 236)

SERVES TWO

METHOD

Roast vegetables

1. Preheat oven to 220°C. Line an oven tray with baking paper.
2. Cut the eggplant into 4-centimetre chunks. Toss in the olive oil, and sprinkle with salt and pepper. Arrange on the lined oven tray and roast for 20 minutes.
3. Add the cherry tomatoes to the oven tray and stir through the eggplant. Cook for about 6 minutes until the tomatoes have begun to blister and the eggplant is golden brown.
4. Remove from the oven and set aside.

Spicy Sauce

1. Collect the dry spices together – mustard seeds, turmeric, garam masala and cumin – and set aside.
2. Place the oil and sliced onion in a medium frying pan and cook over a medium heat for 5 minutes until the onion starts to brown. Stir occasionally.
3. Add the garlic, ginger and chilli and cook for 2 minutes. Add the dry spices and cook for a further 30 seconds until fragrant.
4. Add the diced tomatoes, water and salt, stir well and bring to a simmer. Reduce the heat to low and continue to cook for 15 minutes until the sauce thickens, stirring occasionally. Remove from the heat. Add the coriander and lemon juice.
5. Add the roast vegetables to the frying pan and roughly crush them, using a fork. The eggplant and tomatoes should still be recognisable.

To serve

1. Transfer the Roasted Eggplant in a Spicy Sauce to a serving bowl.
2. Place the peas, spring onion, coriander, oil and pepper in a small bowl and stir to combine. Spoon the pea mixture on top of the Roasted Eggplant in a Spicy Sauce. Leave a border so that the spicy sauce is visible.
3. Combine the yoghurt and lemon juice and drizzle over the pea mixture.
4. Serve with steamed rice.

9780170495714

EVALUATION

1 Describe the sensory properties – appearance, aroma, flavour, texture and sound – of the Roasted Eggplant in a Spicy Sauce.

2 List all the ingredients in this recipe that are classified as spices. Why are they fried in step 3 of making the Spicy Sauce?

3 Explain why roasting alters the physical properties of the eggplant.

4 Suggest three other dishes that would complement this recipe if it was served as part of a shared table when eating with family and friends. Justify your choices.

5 Classify the ingredients of the Roasted Eggplant in a Spicy Sauce on a diagram of the *Australian Guide to Healthy Eating*. Explain how this recipe will enable individuals to meet the recommended number of serves of vegetables they should eat each day.

Mark Fergus Photography

Gnocchi

INGREDIENTS

2 medium-to-large starchy potatoes

½ cup flour

½ teaspoon salt

1 teaspoon olive oil

½ egg, lightly beaten

To serve

Napoli Sauce (see page 245) or commercial pesto

parmesan cheese, grated

SERVES 2

METHOD

1. Boil the potatoes with their skins on until tender. Peel and put through a ricer (similar to a big Italian-style garlic press). Cool.
2. Mix together the potato, flour, salt, oil and egg into a soft dough.
3. Shape the dough into a fat loaf and set on a floured board. Cut off pieces and roll each one very lightly into a cord 1-centimetre thick. Cut each cord into 2-centimetre lengths.
4. Lightly roll each segment in the centre under your forefinger to give the piece a bow shape.
5. Set the gnocchi aside. The pieces should not touch.
6. Bring a large saucepan of water to the boil and drop in the gnocchi one at a time. They are cooked when they rise to the top. Remove immediately to a heat-proof dish using a slotted spoon. Cook 8 to 10 at a time.
7. Place on serving plate. Top with hot Napoli Sauce or commercial pesto. Garnish with parmesan cheese.

iStock.com/Ezumelmages

EVALUATION

1. Use a star diagram, such as the one below, to record the sensory properties of your Gnocchi.

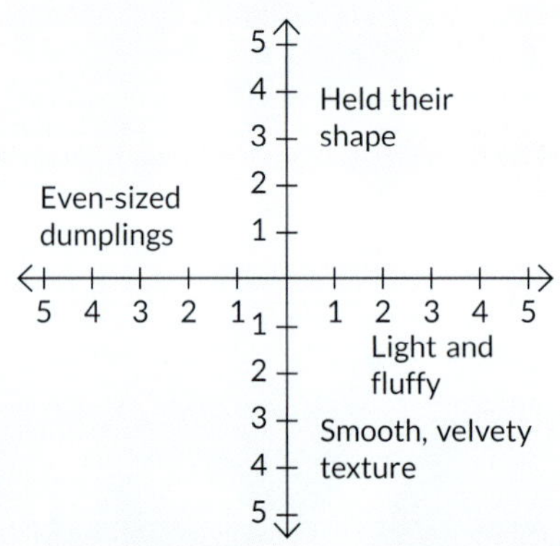

2. If you do not have a ricer, suggest another method you could use to process the potatoes.
3. Suggest an alternative method for shaping and rolling the Gnocchi other than that used in the recipe.
4. Why is it important to drop the Gnocchi into the boiling water one at a time and to remove each one when it rises to the top?
5. Write a paragraph that discusses the health benefits of the Gnocchi when served with Napoli Sauce or commercial pesto.

9780170495714

Basic Pasta Dough

INGREDIENTS

120 grams baker's flour or strong flour

¼–½ teaspoon salt

1–2 eggs

2 teaspoons olive oil

extra baker's flour or strong flour, for kneading

METHOD

1 Sift the flour and salt into a large bowl. Lightly beat the eggs in a small bowl.
2 Make a well in the centre of the flour and add the oil and some of the beaten egg.
3 Gradually mix the flour into the liquid ingredients, adding more beaten egg if necessary to form a firm dough.
4 Knead the dough for approximately 10 minutes or until very smooth and elastic, using extra flour as required. The dough should be able to stretch without cracking.
5 Cover the dough with plastic wrap. Allow to rest for 20 minutes.
6 Roll out by hand or use a pasta machine to cut into shapes.
7 When cut, hang over an oven rack or thin rod to dry.
8 Place in boiling water and boil for 2–3 minutes or until al dente ('tender to the bite'). You can test this by tasting a piece of the cooked pasta to check whether there is any uncooked starch remaining in the centre.
9 Drain and serve as desired.

EVALUATION

1 Why is it important to gradually mix the dry ingredients into the liquid ingredients?
2 What is the purpose of kneading the dough for 10 minutes?
3 Why is it desirable to rest the dough for 20 minutes before using it?
4 Explain why it is essential to cook pasta in plenty of boiling water.
5 Based on the principles of the *Australian Guide to Healthy Eating*, explain why pasta with a vegetable or simple meat sauce is considered to be a healthy meal.

iStock.com/simarik

Spaghetti Bolognese

INGREDIENTS

Bolognese Sauce

1 tablespoon oil

½ onion, finely diced

1 small clove garlic, crushed

125 grams minced beef

1 cup canned, diced tomatoes

2 tablespoons tomato paste

½ cup stock

¼ teaspoon dried basil

¼ teaspoon dried oregano

few shakes of pepper

1 tablespoon parmesan cheese, for garnish

Pasta

1/2 teaspoon salt

200 grams dried pasta

METHOD

Bolognese Sauce

1 Heat the oil in a medium-sized saucepan. Add the onion and garlic and gently fry for 2–3 minutes until just beginning to brown.

2 Add the minced beef and stir continuously until well browned.

3 Add the tomatoes, tomato paste, stock, herbs and pepper.

4 Bring the sauce to the boil. Reduce heat to simmer, cover with a lid and cook very gently for approximately 30 minutes or until the sauce has thickened. Stir occasionally during cooking to prevent sauce from sticking to the bottom of the saucepan.

Pasta

1 Fill two-thirds of a large saucepan with water. Place the lid on the saucepan and bring to the boil.

2 Add the salt to the water – this will add flavour and improve the texture of the pasta.

3 Gradually add the pasta to the boiling water. Stir once or twice only to separate the pasta.

4 Rapidly boil, uncovered, until the pasta is al dente. You can test this by tasting a piece of the cooked pasta to check whether there is any uncooked starch remaining in the centre. Cooking time will vary according to the type of pasta used. After the water has come to the boil, the pasta will take 12–15 minutes.

5 Drain the pasta in a colander.

6 Serve immediately with the Bolognese Sauce. Garnish with parmesan cheese.

Mark Fergus Photography

EVALUATION

1 Explain why the Bolognese Sauce is simmered for 30 minutes.

2 Why is salt added to the cooking water when boiling pasta?

3 Identify and describe the process that occurs to pasta when it is cooked.

4 Draw a flow chart to highlight the stages a packet of dried commercial pasta would follow in the Australian food system.

5 Classify the ingredients for the Spaghetti Bolognese on a diagram of the *Australian Guide to Healthy Eating* and explain whether it could be described as a healthy family meal.

9780170495714

Chicken Curry with Coconut Rice

INGREDIENTS

Chicken Curry

2 teaspoons oil

1 clove garlic, crushed

1 small piece fresh ginger, grated

1 green chilli, seeds removed and finely chopped

½ teaspoon lemongrass, finely chopped

2 skinless chicken fillets, sliced

1 small onion, quartered and separated

1 teaspoon coriander, roughly chopped

1 tablespoon light soy sauce

1 cup coconut cream

½ cup chicken stock

1–2 teaspoons green curry paste (according to taste)

1 extra chilli, to prepare as a chilli flower garnish

extra chopped coriander, to garnish

Coconut Rice

½ cup jasmine rice

½ cup full-cream coconut milk

½ cup water

Mark Fergus Photography

METHOD

Chicken Curry

1. Heat the oil in a wok over a medium flame and sauté the garlic, ginger, chilli and lemongrass for 1–2 minutes.
2. Add the chicken and allow to brown on both sides. Cook for 5–8 minutes. Do not turn the chicken pieces until each side has seared and the juices are sealed in.
3. Add the onion and quickly fry it, stirring.
4. Stir in the remaining ingredients and simmer for 15 minutes or until the chicken is tender.
5. Cut the tip off the extra chilli and slice the chilli lengthwise into thin strips, almost to the stem, to form petals. Soak in cold water for 15 minutes.
6. Serve with Coconut Rice and garnish with the chilli flower and more chopped coriander.

Coconut Rice

1. Place the rice in a strainer and rinse under cold water for approximately 1 minute or until the water runs clear.
2. Place the coconut milk and water in a small saucepan and bring to the boil.
3. Add the washed rice. Gently stir with a fork to separate the grains.
4. Cover with a tight-fitting lid. Lower the heat and simmer for 15 minutes. Do not remove the lid.
5. Leave to stand for a further 10 minutes for the rice to steam.
6. Fluff the rice with a fork. Serve with the Chicken Curry.

EVALUATION

1. Describe the sensory properties – appearance, aroma, flavour, texture and sound – of your Chicken Curry with Coconut Rice.
2. Why are the garlic, ginger, chilli and lemongrass sautéed before being combined with the chicken?
3. Identify and describe the process that causes the flesh of the chicken to become firm when it is cooked in step 2 of the Chicken Curry recipe.
4. Why is it important to rinse the jasmine rice well before cooking?
5. Classify the ingredients for the Chicken Curry with Coconut Rice on a diagram of the *Australian Guide to Healthy Eating* and explain whether it meets the recommendations of this food selection model.

NUTRIENTS FOR GOOD HEALTH

Fats are a concentrated source of energy; omega-3 is an essential fatty acid that supports the brain

The mineral calcium is essential for the formation of bones and teeth; the mineral iron helps form haemoglobin in blood and aids energy production

Vitamins support the development of new body cells and help the body absorb other essential nutrients

Carbohydrates provide the energy or fuel that each cell needs to function

Protein is used by the body for the growth and repair of all tissue

FOOD FOR A HEALTHY DIET

Eat a wide variety of nutritious foods for good health

THE BENEFITS OF EATING BREAKFAST

Eating breakfast boosts energy levels, improves concentration, and kickstarts the metabolism for the day

9780170495714

WATER

Drink at least six to eight glasses of water daily

Aids the digestion and absorption of other nutrients

Helps the removal of waste products from the body

Tap water contains fluoride, which helps to reduce tooth decay

viktorijareut/Adobe Stock Photos

7 EAT WELL, BE WELL

KEY TERMS

Australian Guide to Healthy Eating a pictorial representation of Guideline 2 of the *Australian Dietary Guidelines*: 'Enjoy a wide variety of foods from the five food groups each day'

breakfast the first meal eaten soon after waking up from a night's sleep

fast a period of time during which we eat nothing

food any substance that we eat or drink that provides the body with chemical substances called nutrients

food selection model a guide that visually represents the proportion of different food groups that should be consumed each day

glycogen energy stored in the muscle, tissue and liver

nutrients chemical substances in food that are broken down during digestion, including protein, carbohydrates, fats, vitamins and minerals

processed breakfast cereals grains such as corn, wheat and rice that have been softened by pre-cooking and then dried; most are fortified, meaning they have had vitamins and minerals added during processing

Templates
- PMI table
- Food order
- Production plan
- Sensory wheel
- Australian Guide to Healthy Eating

Worksheets
- Practical activity 7.1
- Practical activity 7.2
- Design activity 7.1

Quizzes
- Testing knowledge 7.1
- Testing knowledge 7.2
- Testing knowledge 7.3

Nelson MindTap

To access resources above, visit **cengage.com.au/nelsonmindtap**

9780170495714

Food and me

To maintain good health, people of all ages need to consume a diet of nutritious foods. Every day, we eat a variety of foods at various times, often in a range of different places. The reason we eat may be because we are hungry, or because we always have something to eat at a particular time of the day, or because we are at a social occasion at which food is served.

What is food?

Food is any substance that we eat or drink that provides the body with chemical substances called nutrients. Nutrients from the food we eat enable us to grow and repair body tissue. Nutrients also produce energy to fuel activity and body processes.

Most of the food we eat comes from plants or animals.

- A wide variety of plants are eaten by people around the world, including grains or cereals, vegetables, fruit, nuts and legumes.
- Animals that provide us with meat or dairy products include cattle, sheep, pigs, fish and seafood, goats, kangaroos, rabbits, crocodiles, deer and wild birds. Chickens, ducks and turkeys are a source of eggs, as well as meat.

Because plants and animals provide a range of different nutrients, health experts recommend that we include food from both sources as a part of a healthy diet.

Nutrients in food

The main **nutrients** found in food are protein, carbohydrates, including dietary fibre, and fats, vitamins and minerals. Figure 7.2 on the following page gives information about the different types of nutrients and their role in the body. During digestion, nutrients are broken down into tiny molecules that are absorbed into the bloodstream and then into the cells of the body.

Although it is not a nutrient, the water we drink or eat in food is essential to life. It is also absorbed into the bloodstream to provide the body with the fluid the cells need to process the nutrients.

Each type of food is made up of a different combination of nutrients, so it looks, tastes and feels different.

Figure 7.1 A range of foods from plant sources

9780170495714

Food sources of protein

Protein is found in:

- animal foods, including meat, milk, dairy foods, fish and eggs
- plant foods, including nuts, legumes and wholegrain cereals.

Food sources of minerals

- *Calcium* and *phosphorus* are found in dairy foods.
- *Iron* is found in red meat, lentils, green leafy vegetables and dried apricots.

Food sources of carbohydrates

- *Carbohydrates* are found in plant foods, such as cereals (bread, rice, pasta, noodles, dry biscuits) vegetables, fruit and legumes.
- *Sugars* are found in cakes, biscuits, confectionery, desserts and soft drinks.

PROTEIN

During digestion, the body breaks down protein into amino acids, which are the building blocks for the growth and repair of all body tissue.

MINERALS

Calcium, in conjunction with *phosphorus*, is essential for the formation of bones and teeth.

Iron helps form haemoglobin in the blood and assists in energy production.

FUNCTIONS AND SOURCES OF NUTRIENTS

CARBOHYDRATES

The main function of carbohydrates is to provide the energy or fuel that each cell needs to function.

Dietary fibre is not absorbed during digestion, but adds bulk to faeces.

VITAMINS

Vitamins enable body systems to function, including the development of new cells for growth and repair, energy production and the absorption of other nutrients.

FATS

Fats are a concentrated source of energy that provide twice as much energy per gram as carbohydrates and protein.

Omega-3 is an essential fatty acid that helps the brain and nervous system to develop and function.

Food sources of vitamins

- *Fat-soluble vitamins* are found in fish and vegetable oils, milk, cheese and nuts.
- *Vitamin A* is found in spinach, carrots, broccoli and salmon.
- *B group vitamins* are found in wholegrain cereals, green leafy vegetables and legumes.
- *Vitamin C* is found in oranges, kiwi fruit, capsicum, parsley and broccoli.

Food sources of fats

Fats are found in:

- animal foods, including meat, poultry, cheese, cream and butter
- plant foods, including margarine, olive oil, canola oil, peanut oil and avocado.

Omega-3 fatty acids are found in fish.

Figure 7.2 The functions and sources of nutrients

Testing knowledge 7.1

Quiz
Testing knowledge 7.1

1 Write a definition of 'food'.
2 Identify the two main sources of food. List three examples of food from each source.
3 Why is it valuable to eat a variety of foods from both plants and animals?
4 Explain the importance of protein in our diet. List two examples of protein from animal sources and two examples of protein from plant sources.
5 Outline the two main functions of carbohydrate in the body.
6 Explain why fats are a concentrated source of energy.
7 Why is it important to include omega-3 fatty acids in our diet?
8 Explain why it is recommended that we consume a range of foods each day that include vitamins.
9 Identify three types of food that are a good source of B group vitamins.
10 Why is calcium considered to be such an important mineral to include in the diet? Identify two main sources of calcium.

Selecting food wisely

Many factors influence our food choices, including our personal food preferences – those foods we like most and those we like least. Our cultural and religious background can also influence our food choices, as well as other factors such as ethical or environmental concerns, or the cost of food.

d3sign/Moment/Getty Images

Figure 7.3 Making healthy food choices

However, regardless of the factors that influence our food choices, it is important to eat a wide variety of foods from a range of food groups to ensure that we obtain all the nutrients we require for good health.

To help Australians make healthy food choices, the Australian Government, through the National Health and Medical Research Council (NHMRC), created the Eat for Health program. This includes the *Australian Dietary Guidelines* and the *Australian Guide to Healthy Eating*, both of which have been developed using the latest scientific evidence and expert advice.

The *Australian Dietary Guidelines*

The *Australian Dietary Guidelines* give up-to-date advice about the kinds and amounts of nutritious foods that we need to eat each day for good health and wellbeing. They provide a framework that can apply to all Australians, including those who are at risk of being overweight or developing type 2 diabetes. They also consider other factors, such as whether a woman is pregnant or breastfeeding.

The guidelines are based on age, gender, body size and the levels of activity people are involved in each day.

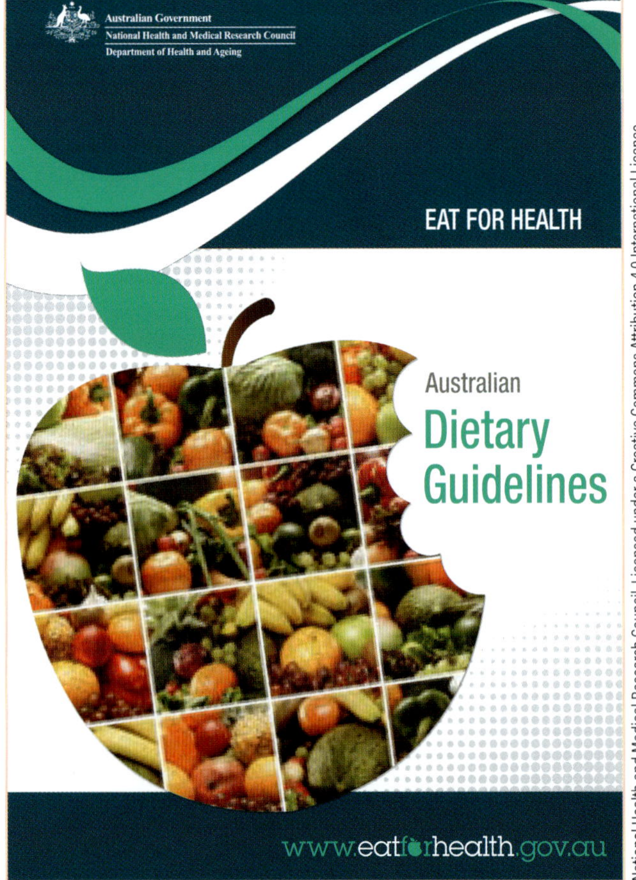

Figure 7.4 The *Australian Dietary Guidelines*

9780170495714

There are five guidelines. Each is considered to be equally important in terms of public health outcomes.

GUIDELINE 1

To achieve and maintain a healthy weight, be physically active and choose amounts of nutritious food and drinks to meet your energy needs

GUIDELINE 2

Enjoy a wide variety of nutritious foods from these five groups every day:

- Plenty of vegetables, including different types and colours, and legumes/beans
- Fruit
- Grain (cereal) foods, mostly wholegrain and/or high-fibre cereal varieties
- Lean meats and poultry, fish, eggs, tofu, nuts and seeds, and legumes/beans
- Milk, yoghurt, cheese and/or their alternatives, mostly reduced fat

And drink plenty of water

GUIDELINE 3

Limit intake of foods containing saturated fat, added salt, added sugars and alcohol

GUIDELINE 4

Encourage, support and promote breastfeeding

GUIDELINE 5

Care for your food; prepare and store it safely

Figure 7.5 The five *Australian Dietary Guidelines*

Review of the *Australian Dietary Guidelines*

The *Australian Dietary Guidelines* are currently under review and the revised guidelines will be published in 2026, after which they will be available on the Eat for Heath website. As a part of the review, the NHMRC will consider advice from the World Health Organization (WHO) that diets must not only be nutritious, but also sustainable. According to the WHO, 'there is an urgent need to promote diets that are healthy and have low environmental impacts. These diets also need to be socio-culturally acceptable and economically accessible for all.' (Source: https://www.who.int/publications/i/item/9789241516648)

Weblink
Australian Dietary Guidelines

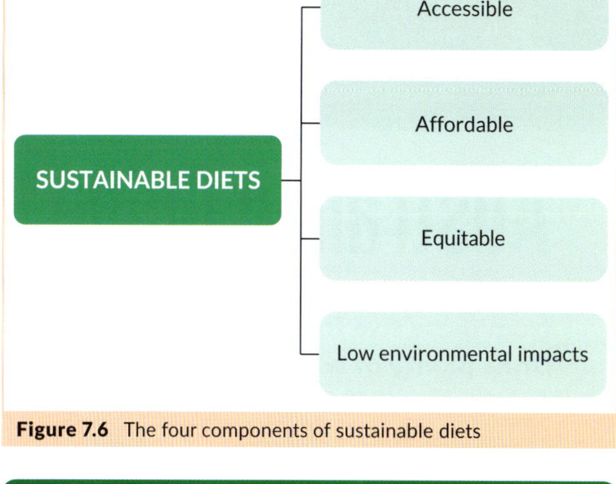

Figure 7.6 The four components of sustainable diets

Activity 7.1

The *Australian Dietary Guidelines* game or quiz

Create a game or quiz to help revise and test your knowledge of the key features of the *Australian Dietary Guidelines*. Your game or quiz should:

- be suitable to share with other class members.
- include eight questions and answers that will demonstrate an in-depth understanding of the guidelines.

Some examples of possible games or quizzes are:

- an online game or quiz program
- a series of flash cards
- a paper 'fortune teller'.

The *Australian Guide to Healthy Eating*

The ***Australian Guide to Healthy Eating*** is a **food selection model** that has been developed by the NHMRC, based on the information in the *Australian Dietary Guidelines*. It is a pictorial representation of Guideline 2: 'Enjoy a wide variety of nutritious foods from the five food groups every day'. The model is presented as a dinner plate containing the proportions of food from each group that should be eaten each day for good health: see Figure 7.7 on the following page.

The *Australian Guide to Healthy Eating* is based on eating a balanced diet that contains both plant and animal foods. However, it does contain alternatives for people who follow a vegetarian or vegan diet.

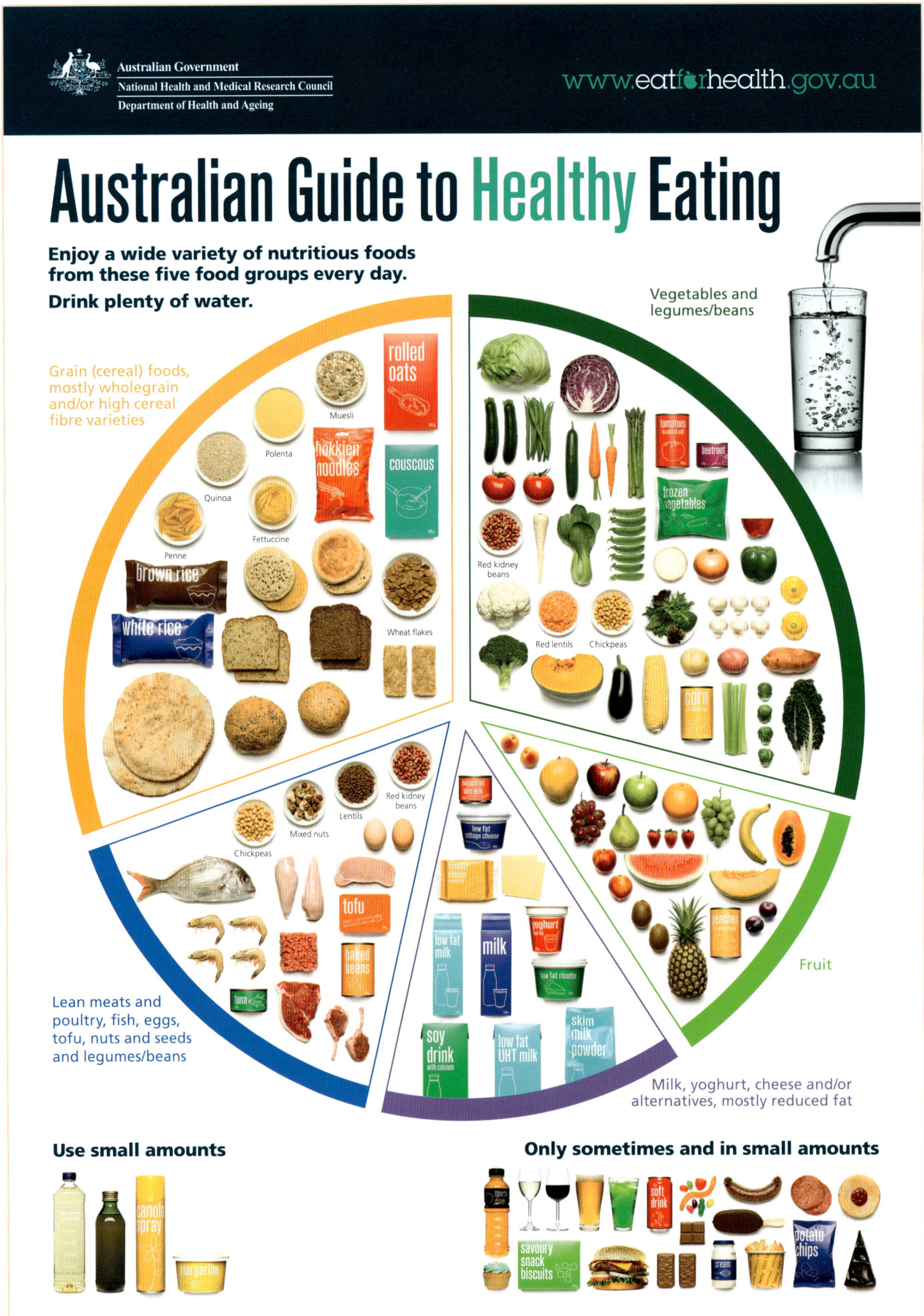

Australian Government
National Health and Medical Research Council
Department of Health and Ageing

www.eatforhealth.gov.au

Australian Guide to Healthy Eating

Enjoy a wide variety of nutritious foods from these five food groups every day.

Drink plenty of water.

Grain (cereal) foods, mostly wholegrain and/or high cereal fibre varieties

rolled oats
Muesli
Polenta
hokkien noodles
couscous
Quinoa
Fettuccine
Penne
brown rice
white rice
Wheat flakes

Vegetables and legumes/beans

tomatoes
beetroot
frozen vegetables
Red kidney beans
Red lentils
Chickpeas
corn

Lean meats and poultry, fish, eggs, tofu, nuts and seeds and legumes/beans

Red kidney beans
Lentils
Mixed nuts
Chickpeas
tofu
baked beans
tuna

Milk, yoghurt, cheese and/or alternatives, mostly reduced fat

evaporated skim milk
low fat cottage cheese
low fat milk
milk
yoghurt
low fat ricotta
soy drink
low fat UHT milk
skim milk powder

Fruit

peaches

Use small amounts

Only sometimes and in small amounts

Figure 7.7 The *Australian Guide to Healthy Eating*

9780170495714

Why is a food model useful?

The information contained in the *Australian Guide to Healthy Eating* helps people choose foods in the appropriate proportions each day for a healthy diet. By following this food model, the health and wellbeing of individual Australians – and therefore the general population – will be improved.

One clear advantage of following a food model is that people are more likely to maintain a healthy body weight. A healthy body weight will enable people to avoid many of the diet-related diseases associated with poor nutrition, including overweight and obesity, cardiovascular disease, type 2 diabetes and some forms of cancer.

The five food groups

To help ensure that we consume the amount and types of food and drinks we require each day to achieve optimal health and wellbeing, the *Australian Guide to Healthy Eating* is divided into five main food groups:

- vegetables, including different types and colours, and legumes/beans
- fruit
- grain (cereal) foods, mostly wholegrain and/or high-fibre cereal varieties
- lean meats and poultry, fish, eggs, tofu, nuts and seeds, and legumes/beans
- milk, yoghurt, cheese and/or their alternatives, mostly reduced fat.

The food within each of these groups provides a variety of nutrients we need for good health.

The *Australian Guide to Healthy Eating* also makes recommendations about the amount of food within each of these groups we should consume each day based on our age and gender.

Vegetables and legumes/beans

Eating a wide variety of vegetables of different types and colours, and legumes/beans in our diet each day is highly recommended as they provide us with many important vitamins and minerals, as well as dietary fibre. Potatoes, sweet potatoes and peas are a good source of carbohydrates. Other vegetables – including capsicum, broccoli and Asian greens – are a valuable source of vitamin C. Folate is found in many green vegetables, as well as in dried peas, beans and lentils. Legumes such as lentils, soybeans and chickpeas are an excellent source of protein and, when combined with grain foods, can be a good substitute for meat for people who adhere to a vegetarian or vegan diet. Legumes are also rich in carbohydrates and fibre, but low in fat and cholesterol.

In addition, including a range of vegetables in our diet is essential for maintaining good health as they are low in energy and can help us to maintain a healthy weight. This will therefore reduce the risk of becoming overweight and obese. It will also reduce the risk of some chronic diseases, including heart disease and some cancers.

It is wise to limit the consumption of fried vegetables, such as potato chips, as these foods are high in fat and are classified as a discretionary food.

Claudia Totir/Moment/Getty Images

Figure 7.8 It is important to select vegetables and legumes of different types and colours.

Fruit

Fruit contains a wide variety of nutrients including vitamin C and folate, along with potassium and dietary fibre. Fruit also contains natural sugars and therefore can be a source of carbohydrate. Like vegetables, fresh fruits are high in dietary fibre and low in kilojoules. Incorporating fruit in our diet each day may help to reduce the risk of heart disease and some forms of cancer. If selecting canned fruit, make sure you choose those that are canned in fruit juice rather than in a sugar syrup.

Limiting your consumption of fruit juice and fruit drinks is beneficial, as they are high in kilojoules and only contain small amounts of dietary fibre.

Tetra Images/Getty Images

Figure 7.9 Fruit contains a wide variety of nutrients.

Grain (cereal) foods, mostly wholegrain and/or high-fibre cereal varieties

It is important to include a wide range of wholegrain cereal foods in our diet as they provide our body with its main sources of carbohydrate for energy. Approximately 55 per cent of energy should come from foods in this group. Wholegrain cereals also provide a wide variety of other significant nutrients, including protein, dietary fibre, and vitamins and minerals, including folate, thiamin, riboflavin, niacin, vitamin E and iron. Consuming a variety of foods from this group may help to reduce our risk of developing heart disease, type 2 diabetes, excessive weight gain and some cancers. Because they are high in dietary fibre, wholegrain cereals are important in reducing the likelihood of developing colorectal cancer.

LauriPatterson/E+/Getty Images

Figure 7.10 Select mostly grain foods made from wholegrains.

Lean meat and poultry, fish, eggs, tofu, nuts and seeds, and legumes/beans

When selecting foods from this group, it is recommended that you choose lean cuts of meat and poultry. Lean red meat is a valuable source of iron, zinc and vitamin B12. All foods in this group are an excellent source of complete protein. However, foods from animal sources contain high levels of saturated fat and cholesterol, so you should select lean cuts of meat and serve poultry without the skin where possible.

It is recommended that you have two fish or seafood meals per week, as these foods are an excellent source of omega-3 fatty acids. Consuming fish and seafood on a regular basis can help reduce the risk of developing heart disease, stroke and dementia.

Processed meats such as sausages, salami, bacon and ham are not included in this group because their high fat and salt content means they are classified as discretionary choices.

SimpleImages/Moment/Getty Images

Figure 7.11 Fish is an excellent source of omega-3 fatty acids.

Milk, yoghurt, cheese and/or their alternatives, mostly reduced fat

A variety of foods from the 'milk, yoghurt, cheese and/or their alternatives' group should be eaten each day because they are a very valuable source of calcium, protein, iodine, riboflavin and vitamin B12. They also contain vitamin D and phosphorus, which help the body to utilise calcium, and therefore contribute to the development of strong bones and teeth. Health experts suggest that regular consumption of foods in this group can help reduce the risk of developing high blood pressure, heart disease, stroke, type 2 diabetes and some cancers. However, it is advised that you choose low-fat varieties of these foods to minimise your intake of saturated fat.

istetiana/iStock/Getty Images

Figure 7.12 Milk, cheese and yoghurt help build strong bones and teeth.

Foods outside the circle

Some foods sit outside the main circle of the *Australian Guide to Healthy Eating* – in the 'use small amounts' category and the 'only sometimes and in small amounts' category, which comprises foods classified as 'discretionary choices'.

9780170495714

Use small amounts

The 'use small amounts' category includes butter, table spreads and oils. It is important to include some fats in our diet because they provide us with essential fatty acids and the fat-soluble vitamins A, D, E and K. However, where possible, you should select unsaturated fats as they can help maintain blood cholesterol levels within the recommended range. Many foods within the main five food groups are good sources of essential unsaturated fatty acids. These include nuts, seeds, fish and avocado.

Monounsaturated and polyunsaturated oils and spreads used in cooking and dressings – such as sunflower oil, canola oil, peanut oil and olive oil – are all high in kilojoules and therefore should only be used in small amounts.

iStock.com/MarianVejcik

Figure 7.13 Oil should only be used in small amounts.

Discretionary choices

The foods and drinks that appear in the lower right-hand corner of the *Australian Guide to Healthy Eating* are described as 'discretionary choices'. This means they should be eaten 'only sometimes and in small amounts' because they are high in saturated fat, added sugar and/or added salt. These foods are high in kilojoules and low in essential nutrients, and are not necessary for a healthy diet.

An excessive consumption of discretionary foods can lead to a person becoming overweight or obese. In the future, this can result in the development of chronic diseases such as heart disease, stroke, type 2 diabetes and some forms of cancer. However, eating a small amount of discretionary foods occasionally as a treat can add variety and interest to your diet, and be part of celebrating special life events.

iStock.com/Anastasia_Stoma

Figure 7.14 Doughnuts and other discretionary foods should only be eaten sometimes.

Testing knowledge 7.2

Quiz
Testing knowledge 7.2

11 List four factors that can influence our food choices.

12 Outline two main reasons for the development of the *Australian Dietary Guidelines*.

13 What is one major advantage of following a food selection model such as the *Australian Guide to Healthy Eating*? Why is this important for our long-term health?

14 Draw a sketch of a carrot and annotate it to show the main nutrients found in vegetables.

15 Draw a knowledge map to highlight the nutritional advantages of consuming two pieces of fruit each day.

16 Explain how consuming foods from the 'grain (cereal) foods' category can help improve our long-term health.

17 Why is it recommended to try to eat fish or seafood meals at least twice a week?

18 List the nutrients found in a glass of cow's milk.

19 Explain why it is important to include some fats in our diet.

20 What are 'discretionary food choices' and why do they appear outside of the circle in the *Australian Guide to Healthy Eating*?

Activity 7.2

Test your skill

Answer the following 20 questions 'true' or 'false' to test your knowledge and understanding of nutrition.

1. Wholegrain cereals are a good source of dietary fibre.
2. Plant foods such as potato, sweet potato and corn are good sources of energy because they are high in fat.
3. It is important to include legumes in your diet as they are rich in omega-3 fatty acids.
4. We should gain most of our energy needs from plant foods.
5. Capsicum, broccoli, oranges, kiwi fruit and Asian greens are all very good sources of vitamin C.
6. Iron is important in the body as it assists in forming healthy blood and producing energy.
7. Calcium and phosphorus work together to build strong bones and teeth.
8. Fruit juice drinks are a great beverage to have for breakfast as they are high in dietary fibre.
9. Phosphorus, iron and calcium are all types of vitamins.
10. Fruit is low in sugar and dietary fibre.
11. Legumes are a good substitute for meat as they contain a high amount of protein.
12. It is best to select foods that contain saturated fats as these help to reduce blood cholesterol levels.
13. We should eat fish on a regular basis as it contains a large amount of omega-3, which is needed for the brain to function properly.

iStock.com/GMVozd

Figure 7.15 We should eat fish on a regular basis as it contains a large amount of omega-3: true or false?

14. It is important to include lean meat in our diet as it contains a high quantity of protein and iron.
15. Vitamin D and riboflavin help the body to utilise calcium.
16. Fat is a concentrated form of energy and provides four times as much energy per gram as carbohydrate.
17. You should have no more than one fish or seafood meal a week, as these foods are high in cholesterol.
18. Consuming a wide variety of cereal foods each day will help you to maintain a healthy weight and reduce the risk of developing type 2 diabetes.
19. Canned soft drinks are good substitutes for water as they only contain small amounts of sugar.
20. Processed meats such as sausages, bacon and ham are not included in the lean meat section of the *Australian Guide to Healthy Eating*, as they are high in fat and salt.

How did you score? Check your answers with your class teacher.

Results

20	Congratulations – what a star! You have learnt lots about the nutrients in food.
17–19	A fantastic effort! Check the answers you didn't get correct so that you get a perfect result next time.
14–16	A very good try. You have obviously learnt a lot about the nutrients needed for good health.
11–13	You have learnt quite a lot about nutrition but there is still a lot more to understand so that you can optimise your health.
8–10	You will need to work hard to learn more about the nutrients needed for a healthy diet and a healthy body.
5–7	You should recheck the information on the nutrients in food so that you are able to do better next time.
Less than 5	You have a lot of work to do to learn about the nutrients needed for a healthy diet and for good health.

9780170495714

Activity 7.3

Evaluating meals

Goal

To compare how two after-school snacks are similar or different using the *Australian Guide to Healthy Eating*

After-school snack 1	
Main snack	Cheeseburger
Sweet treat	KitKat
Beverage	Nudie orange juice

After-school snack 2	
Main snack	Cheese and ham toastie
Sweet treat	Doughnut
Beverage	Raspberry slurpee

Task

1 Classify the main ingredients of After-school snack 1 and After-school snack 2 in the table above on a diagram of the *Australian Guide to Healthy Eating*.

2 Identify some features of the main ingredients and cooking methods used in After-school snack 1 and After-school snack 2, and where they are classified on the *Australian Guide to Healthy Eating*.

Template Australian Guide to Healthy Eating

3 Describe how the main snack for After-school snack 1 and After-school snack 2 are similar or different based on these features.

4 Describe how the sweet treat and beverages for After-school snack 1 and After-school snack 2 are similar or different based on these features.

5 Explain what you know about the two after-school snacks and how well they reflect the recommendations of the *Australian Guide to Healthy Eating*.

6 What types of after-school snacks do you have? Do they meet the recommendations of the *Australian Guide to Healthy Eating*?

7 What changes can you make to your after-school snacks to meet the recommendations of the *Australian Guide to Healthy Eating*?

The Australian Guide to Healthy Eating template, https://www.eatforhealth.gov.au/sites/default/files/content/The%20Guidelines/N55_A4_DG_Food_Plate_Blank_HiRes.pdf

PRACTICAL ACTIVITY 7.1

Preparing and evaluating an after-school snack

Prepare the recipe for Cherry Tomato, Feta and Egg Surprise on page 44.

1. Complete a taste test as soon as you have cooked the dish and evaluate the sensory properties – appearance, aroma, flavour, texture and sound – of your Cherry Tomato, Feta and Egg Surprise.
2. How does this recipe rate, according to the *Australian Guide to Healthy Eating*?
3. Discuss whether this recipe would satisfy your hunger as an after-school snack.
4. If you had easy access to a kitchen after school, what other recipe would you prepare as a snack? Explain how well your chosen snack would meet the recommendations of the *Australian Guide to Healthy Eating*.

Worksheet Practical activity 7.1

Template Sensory wheel

CASE STUDY 7.1

Overhaul of nutritional guidelines

Read the article below, and then answer the questions that follow.

Experts urge overhaul of nutritional guidelines as kids skip fruit and vegies

Madeleine Heffernan, *The Age*, 11 February 2024

Natalie Stapleton, a dietitian and mother of two young children, says there's no one-size-fits-all approach to feeding kids. And she is one of many experts who say the country's nutritional guidelines are in desperate need of a review to reflect that.

'Every child will have unique nutrition needs. Nutrition intake for kids over the week varies,' says Stapleton, general manager of policy and advocacy at Dietitians Australia and mother to Riley, 3, and Liam, 18 months.

'At home, I focus on teaching my kids to recognise how eating food nourishes their bodies and minds, with the intention of helping them establish a positive relationship with food to guide them through life.'

'For breakfast this morning, we had oat porridge, baked beans and some fruit. It's not uncommon for us to serve up leftover family dinner for the kids' lunch the next day.'

Stapleton is among several child health experts calling for an overhaul of Australia's 32-year-old food and nutrition policy, as children's diets continue to fall short of national guidelines.

The latest data shows growing numbers of young people are skipping fruit and vegetables and falling outside normal weight ranges.

Dietitians Australia says poor diets are 'driven predominantly by food environments that promote and continually encourage unhealthy eating'. Excessive screen time, online fast-food orders, poverty and higher prices for healthy foods outside capital cities contribute to the challenge.

Stapleton says Australia's food and nutrition policy – now under review by the Department of Health – doesn't reflect the growth in highly marketed and ultra-processed foods, as well as the shift away from seasonal, locally grown food. 'Our food system has changed dramatically since 1992,' she says.

Dietitians Australia says an updated nutrition policy could include new taxes and marketing bans on unhealthy products, improved food labelling, help for schools to provide healthy foods, and the establishment of an independent statutory health promotion agency.

But change won't come fast. The health department says an updated food and nutrition policy won't be released until 2026. This includes changes to dietary guidelines, which currently advise Australians to be physically active; eat vegetables, fruit, grains, meat and dairy, and drink plenty of water; limit salt, sugars and alcohol; support breastfeeding; and store and prepare food safely.

How do children's diets fare?

Research by Dr Miaobing Zheng, a National Health Medical Research Council (NHMRC) early career research fellow, has found that children develop

9780170495714

unhealthy lifestyle patterns, such as consuming an energy-dense and nutrient-poor diet, as early as 18 months. She says there's also a strong link between parents and their children's health and behaviours – children show higher-risk of obesity and developing unhealthy lifestyle patterns when their parents have higher body weight and unhealthy lifestyle behaviours.

Similarly, children in families who enjoy a variety of nutritious food from the five food groups are more likely to make their own healthy choices as they get older, the NHMRC says.

The latest Australian Bureau of Statistics (ABS) figures show the proportion of children who do not usually eat fruit daily has doubled over the last decade (from 2.8 per cent in 2011–12 to 5.8 per cent in 2022). Similarly, the proportion of children who do not usually eat vegetables daily has tripled, from 1.5 per cent compared to 4.5 per cent.

Growing proportions of children are underweight or overweight, the ABS says. Just 64.6 per cent of young people aged 5 to 17 were in the normal weight range in 2022, with 7.6 per cent underweight and 27.7 per cent classified as overweight or obese. In 2011–12, 69.1 per cent of that age group were in the normal weight range.

A 2021 survey of primary-school-aged children's lunchboxes found discretionary food – food that is not needed to meet nutrient requirements, but contributes to the enjoyment of eating – accounted for 44 per cent of energy intake, and most children consumed less than one serve of vegetables, meat and alternatives, or milk and alternatives during school hours.

Teens eat the most discretionary food of any age group. At seven daily serves, discretionary food accounted for 41 per cent of the energy consumed by teens aged 14 to 18, compared with about a third for adults.

Source: Experts urge overhaul of nutritional guidelines as kids skip fruit and vegies' by Madeleine Heffernan – *The Age:* February 11, 2024. https://www.smh.com.au/national/experts-urge-overhaul-of-nutritional-guidelines-as-kids-skip-fruit-and-vegies-20231127-p5en13.html (Accessed February 11, 2024)

Analysis

1 In relation to her children and eating, what does dietitian Natalie Stapleton focus on?
2 Create a mind map of the factors that Dietitians Australia considers are the main reasons many Australians consume a poor diet.
3 List the areas that Dietitians Australia would like to see included in an updated nutrition policy.
4 Select one of these factors and explain why you think it could help improve the diet of many Australians.
5 Explain why the latest Australian Bureau of Statistics (ABS) figures on children's fruit and vegetable consumption is of concern to health professionals.
6 Outline two implications for children's long-term health of consuming less than the recommended number of serves of fruit and vegetables.
7 Propose two strategies you think could encourage children and teenagers to increase their consumption of vegetables.
8 Explain why health professionals would be concerned that 44 per cent of children's energy intake comes from discretionary foods.
9 Suggest reasons why teens eat the most discretionary food of any age group.
10 Do you think that food manufacturers and supermarkets have a responsibility to promote healthy food? Justify your response.

Water

Water is an essential dietary component that we need to keep our bodies healthy. It is a major component of every cell in the body – approximately 70 per cent of the body is composed of water. It also assists in the digestion and absorption of other nutrients, and helps with the removal of waste products from the body.

We lose water from our bodies every day through perspiration, tears, urine and faeces, as well as from our lungs when we breathe out. Therefore, we should drink at least six to eight glasses of water daily to replace the water lost.

RUNSTUDIO/DigitalVision/Getty Images

Figure 7.16 It is important to drink plenty of water.

Water is found in most foods, but is especially high in fruits and vegetables, as well as in milk and other liquids such as fruit juices.

If we do not drink enough water or other fluids, our body gives signals to let us know that we need to drink more to prevent dehydration. One of these signals is urine colour – a dark or yellow colour means that the urine is too concentrated and that we need to drink more fluid. Light-coloured or clear urine means that our body is hydrated and fluid intake is adequate.

Drinking a large amount of tap water is also important because in many countries, including Australia, tap water contains fluoride, which helps to reduce the incidence of dental caries. In recent years, there has been a big increase in the popularity of drinking bottled water, and dentists have seen evidence of an increase in the number of children who have tooth decay.

Tap water versus bottled water

Australia has some of the safest and best tap water in the world. However, as noted above, many consumers have turned away from drinking tap water and are choosing to purchase bottled water instead. Some reasons suggested for this increased consumption are that bottled water is more convenient than tap water, is very easily accessible, has greater health benefits and tastes better. It is surprising then that blind taste-testing experiments show that most people cannot tell the difference between tap water and bottled water.

Health professionals have shown that bottled water is no more 'pure' than tap water. It is widely thought that all bottled water is sourced from 'natural springs' and other natural sources; however, some bottled water is simply treated or filtered tap water.

There are substantial economic advantages for families in using tap water rather than bottled water. Bottled water is more expensive to purchase than milk or many soft drinks. Tap water costs a very small fraction of the cost of purchasing a bottle of water. Depending on the amount of bottled water a family purchases over a year, this can mean considerable savings.

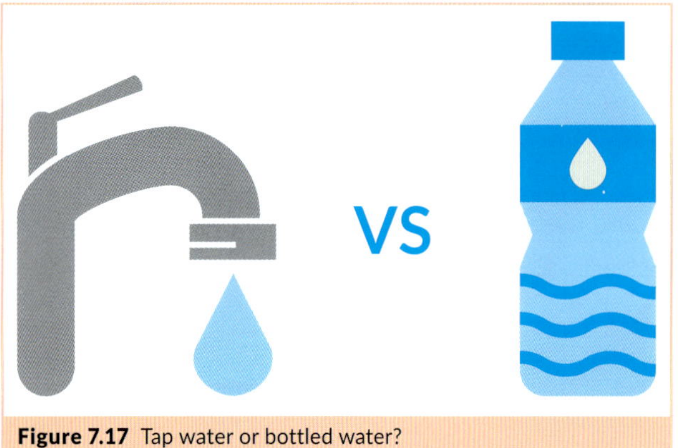

Figure 7.17 Tap water or bottled water?

Environmental impact of bottled water

There are also significant environmental costs associated with the production and purchase of bottled water. In Australia, much of the water that is bottled comes from underground aquifers. This can have a detrimental impact on farming communities by lowering the water table that farmers rely on for their stock and crops.

Another major problem is that the production of polyethylene terephthalate (PET) water bottles uses large amounts of energy and non-renewable resources, including oil and water. According to the Pacific Institute, it takes twice as much water to produce one PET water bottle than the amount of water the bottle can contain. Non-renewable resources in the form of energy and petroleum are used when the water bottles are filled in the factory and also when they are transported by road, rail, ship or air. Along the way, all of these methods of transportation release greenhouse gas emissions, particularly carbon dioxide (CO_2), into the atmosphere. In addition, much of the bottled water is refrigerated so that it is cold when purchased, and this too uses significant amounts of energy and adds to the production of greenhouse gases.

While many consumers attempt to recycle their plastic water bottles, most PET water bottles end up in landfill. Environmental pollution also occurs, as a large number of the water bottles are simply discarded and end up polluting our parks and waterways. They can take hundreds of years to break down. The 2023 report from Clean Up Australia Day shows that single-use plastic beverage bottles are one of three most commonly collected forms of rubbish.

Weblink Clean Up Australia: Rubbish Report

Reducing the consumption of bottled water

Figure 7.18 How to cut down on your use of bottled water

9780170495714

Breakfast helps you make better food choices

According to nutritionists, **breakfast** is the most important meal of the day. By eating a healthy breakfast, you provide your body with approximately one-third of the food you need for the day.

When you eat breakfast, you are 'breaking the fast' by supplying your body with its first food for the day after eight to 12 hours of sleep. A **fast** is a period of time during which we eat nothing. Eight to 12 hours is a long time to go without food – equivalent to at least a full day at school – and people of all ages and occupations need a healthy, adequate breakfast to refuel the body. Eating breakfast helps both children and adults to make better food choices during each day and throughout their lives.

The importance of breakfast

When you are asleep, your body is resting and does not require as much energy or kilojoules as when you are active during the day. The body uses the glucose produced from the carbohydrates you eat during the day and stores it in the muscle tissue and liver as **glycogen**. During the night, these glycogen stores are slowly released into the bloodstream to keep your blood sugar levels stable. Eating breakfast enables the body to replenish these stores of glycogen and provides a store of energy for the day's activities, as well as boosting metabolism. Eating a nutritious breakfast helps improve physical and mental performance and contributes important nutrients to the diet. Nutritionists recommend that breakfast be based on carbohydrates, preferably from fruits, wholegrain breads, cereals and grains, or vegetables.

iStock.com/lmgorthand

Figure 7.19 Eating a nutritious breakfast helps improve physical performance.

Breakfast improves brainpower

Research has shown that eating breakfast helps to improve mental performance. It boosts concentration, problem-solving ability, memory and mood, and enables us to think more quickly and clearly. Young people who eat breakfast have been shown to have better academic results than those who skip breakfast.

Breakfast helps maintain weight

Skipping breakfast does not help people lose weight since it often leads to overeating later in the day. If we eat a breakfast that is based on foods with a low glycaemic index, these foods will fill us up and stop us from feeling hungry. As a result, we are less likely to be tempted to reach for a snack food when we feel hungry in the middle of the morning or in the afternoon. Most commercial snack foods are high in fat and/or sugar, so they are energy-dense and can lead to weight gain.

International research undertaken in Brazil, the USA and Japan has concluded that eating breakfast is one of the main lifestyle factors linked to less body fat and better academic results in adolescent students. Results from these studies showed that skipping breakfast as little as once per week is associated with a higher level of body fat in teenagers – regardless of how active they are – and an increase in the risk of childhood overweight and obesity.

iStock.com/courtneyk

Figure 7.20 Eating breakfast will stop you from feeling hungry.

Breakfast eating habits

Many teenagers go to school in the morning without eating breakfast, or having drunk only fluids such as cordial, soft drink, tea or coffee, all of which have very little nutritional value. For those who do eat breakfast, research shows that during the school week, breakfast cereals, milk and bread are the most popular breakfast foods, and only a small number of people eat a cooked breakfast before going to school.

9780170495714

7

Activity 7.4

Breakfast eating habits

The following graphs summarise key findings from the 2011–12 National Nutrition and Physical Activity Survey. Examine Figures 7.21 and 7.22, and then answer these questions.

1 Which group consumed the most breakfast cereal?

2 Identify the percentage of boys and the percentage of girls who ate nothing for breakfast. Suggest two reasons why teenagers are likely to eat nothing for breakfast.

3 What was the most popular breakfast drink for boys and for girls? What are the health concerns for those who drink soft drink or cordial for breakfast?

4 Which group was most likely to eat toast or bread for breakfast?

5 What factors are likely to influence breakfast food choices for teenagers?

6 If carers laid out fruit, cereal, milk and juice, do you think teenagers who currently eat nothing would be encouraged to eat a healthy breakfast? Why or why not?

7 Suggest other strategies that you think could be used to encourage teenagers to eat a healthy breakfast.

8 Explain the effect that missing breakfast is likely to have on a student's performance at school.

9 List foods available at your school cafeteria that would be suitable for a healthy breakfast.

10 Recommend other breakfast foods suitable for teenagers that could be added to your school's cafeteria menu.

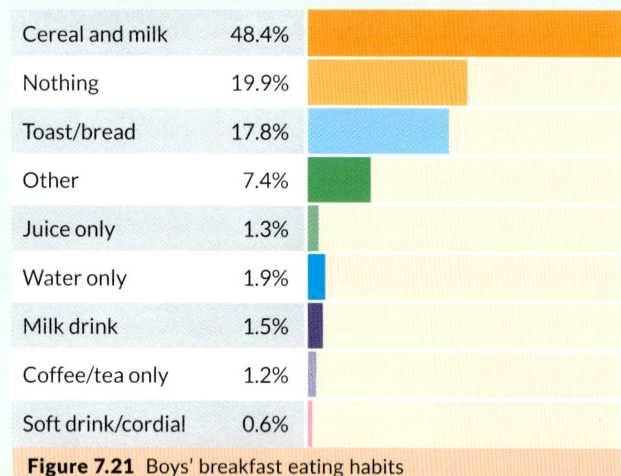

Figure 7.21 Boys' breakfast eating habits

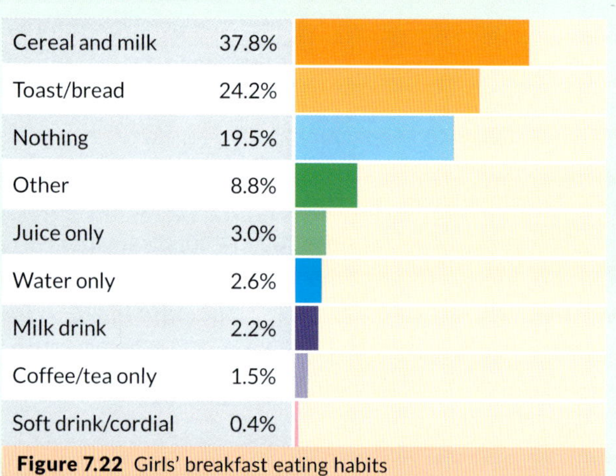

Figure 7.22 Girls' breakfast eating habits

Breakfast foods

The foods we eat for breakfast often depend on the cultural background of our family. Vegemite on toast is a breakfast staple that is loved (or loathed!) in many Australian households. A bowl of pho or sticky rice might be a breakfast food for families who migrated from Vietnam. Rice also features on the breakfast table of other South-East Asian nations, including Indonesia and Japan, while a stack of pancakes with maple syrup is popular in many parts of the USA.

Selecting a healthy food for breakfast can be challenging when we are rushing off to school or work.

Oats – a healthy breakfast cereal

Oats are often acknowledged as a healthy breakfast food because they are a complete wholegrain, meaning the bran, endosperm and germ are still intact; so they include valuable vitamins, minerals and dietary fibre. To make wholegrain cereals such as oats easier to digest, they are often cooked in water, milk or a combination of both.

Oats are a valuable breakfast food as they:

- are high in soluble fibre, which helps reduce blood cholesterol
- have a low glycaemic index, which means their carbohydrate is slowly absorbed into body systems, providing energy long after eating
- are a good source of B group vitamins, vitamin E, protein and minerals
- are associated with protective effects against heart disease in adults.

9780170495714

ArxOnt/Moment/Getty Images

Figure 7.23 Porridge made with rolled oats makes a healthy breakfast.

Types of oat cereals

Oat cereals come in different types:

- Rolled oats – which are used to make porridge, and are an ingredient in muesli and some biscuits – are processed by steaming and rolling, so the individual grains are flat and thin.
- Quick-cook oats are chopped up after steaming and rolling, so they are quicker to cook in the morning.
- Steel-cut oats are made by removing the outer husk, and then cutting the whole oat into several pieces using steel discs. They take longer to cook than the other varieties and have a nuttier flavour.

Regardless of the method of processing, all plain oat products have similar nutritional properties. However quick-cook and flavoured oat products often contain added sugar from extra ingredients such as dried fruit, and so are a less healthy option.

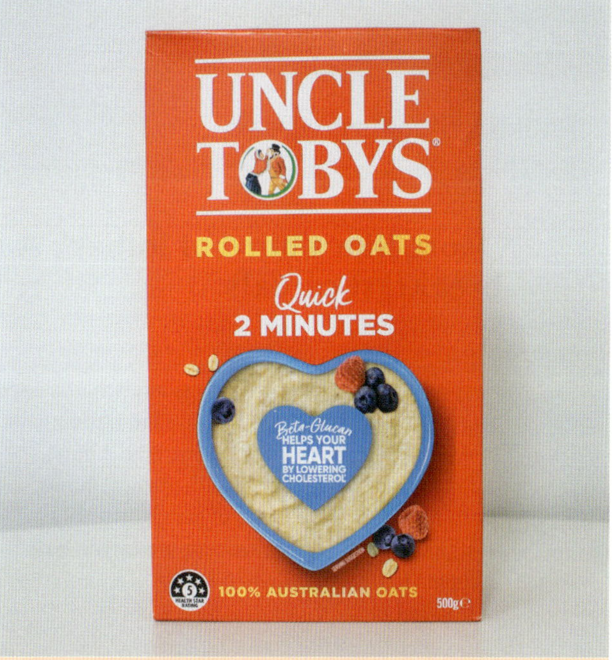

Figure 7.24 Quick-cook oats are a delicious start to the day.

Processed breakfast cereals

Processed breakfast cereals are grains such as corn, wheat and rice that have been softened by pre-cooking and then dried. This enables the grains to be eaten with cold milk. Most of these cereals are fortified, meaning they have had vitamins and minerals – for example, iron – added during processing.

Australia's most popular breakfast cereal is Sanitarium's Weet-Bix. It is said that if all the Weet-Bix breakfast biscuits eaten in Australia each year were laid end to end, they would circle the equator 2.8 times!

The best choice of processed breakfast cereal is one that is high in carbohydrates and fibre, and low in fat, salt and sugar.

iStockphoto.com/Teen00000

Figure 7.25 Weet-Bix is a popular supermarket cereal in Australia.

7

CASE STUDY 7.2

Ranking breakfast cereals

Read the article below, and then answer the questions that follow.

'More like a bowl of lollies': A dietitian ranks 10 popular breakfast cereals

Susie Burrell, *The Age*, 25 March 2024

Despite many cereals being considered a processed and high-sugar breakfast option, a large proportion of Australians start each morning with a bowl, data from the Australian Health Survey shows. With more than 450 types of cereal available in supermarkets, it is no wonder we can be left feeling a tad confused when trying to choose the best variety.

So, can breakfast cereal be a good choice nutritionally? How do you choose a good one, and what is the best way to enjoy it?

The importance of whole grains

Minimally processed wholegrain breakfast cereals, such as oats and bran-based options, are rich in essential nutrients, including slowly digested carbohydrates that support energy regulation.

They are also rich sources of dietary fibre, a nutrient essential for digestive health. When the diets of those who regularly consume a wholegrain breakfast cereal are compared with those who skip a daily bowl of grains, the latter group's overall intake of dietary fibre and B-group vitamins is significantly lower.

With the average Australian getting just two-thirds of the recommended daily intake of dietary fibre, and bowel cancer rates on the rise, it is safe to say that many of us could benefit by starting the day with a bowl of whole grains.

Considering blood glucose levels

One of the biggest issues nutritionally when considering breakfast cereals is the impact on blood glucose levels. Heavily processed breakfast cereals, including flakes, puffs, clusters and biscuits, are made using refined forms of whole grains.

When a whole grain is processed, it reduces the overall nutrient profile and creates a refined grain that is digested more quickly than whole grains, resulting in relatively high fluctuations in blood glucose levels.

High blood glucose levels are an issue for energy regulation and can damage cell health. For those with a genetic predisposition, this can increase the risk of developing blood glucose control issues including insulin resistance and type 2 diabetes. For this reason, when it comes to breakfast cereal, wholegrain, lower glycaemic index and lower carbohydrate choices are the best options nutritionally.

10 popular supermarket cereals ranked in order of health

1. Kellogg's All-Bran

With a massive 12g of dietary fibre per ½ cup serve, few cereals are as good for digestive health as the original All-Bran. While the second ingredient listed is sugar and a single serving contains 7g of sugar, nutritionally the digestive health benefits outweigh any negative associated with a small amount of sugar overall. This is thanks to the mixture of both soluble fibre, which plays a key role in appetite control, and insoluble fibre for a healthy gut.

2. Uncle Tobys Quick Oats

You can't go wrong with a 100 per cent oat-based product. While whole original oats are a better option for blood glucose control than quick-cook oats, oats of any variety are still a strong option nutritionally, offering a good amount of protein and dietary fibre. The key is to opt for the plain varieties and add some fruit, vanilla or cinnamon for sweetness, as the flavoured varieties can add 2–3 tsp of sugar to your wholegrain oat bowl.

3. Weet-Bix

It's hard to find fault with Australia's top-selling breakfast cereal. It does contain some added sugar and salt, but with a 97 per cent wholegrain base, 4g of dietary fibre per two breakfast biscuits, less than 1g of sugar per serve and a range of key nutrients including iron and B-group vitamins, Weet-Bix is a solid breakfast cereal choice.

9780170495714

iStock.com/SDI Productions

Figure 7.26 A large proportion of Australians start each morning with a bowl of cereal.

4. Kellogg's Special K

The more processed a cereal is, the more difficult it is to rate nutritionally as there are many factors to consider, including the percentage of whole grains, glycaemic load, added sugar and salt content and the overall nutritional profile. While Special K is highly processed, its relatively high protein content improves its ranking when compared to other varieties of flaked cereal.

5. Kellogg's Just Right

With 64 per cent whole grains and little added salt, Just Right offers a number of positive nutritional qualities. The downside is that it also contains several different added sugars, which bump its overall sugar content to almost 2 teaspoons per serve for just 4g of dietary fibre. This means it is not the best choice for those with blood glucose regulation issues.

6. Kellogg's Sultana Bran

While Sultana Bran is extremely high in dietary fibre, offering almost 7g per serve, and a mix of both insoluble and soluble forms to support gut health, the high proportion of dried fruit also significantly increases the overall sugar content. This means you are getting more than 12g of sugars in three-quarters of a cup of cereal.

7. Kellogg's Nutri-Grain

The second most popular breakfast cereal in Australia certainly tastes good, thanks to the 2 teaspoons of sugar per bowl. But nutritionally, the exceptionally low amount of dietary fibre in Nutri-Grain makes it a poor cereal choice. While it is heavily promoted for its high protein content and added minerals, including calcium and iron, it only suits those with extremely high energy demands. Think of it as a good source of energy rather than a healthy breakfast option.

8. Milo Cereal

Another cereal that is heavily marketed for the added nutrients it offers, Milo Cereal is heavily processed and does not stand out in any of the core areas as a nutritionally strong breakfast cereal will. With sugar the second ingredient and low amounts of fibre, whole grains and protein, there are much better breakfast cereal options for kids, in particular.

9. Kellogg's Corn Flakes

With a base of refined corn, sugar and salt, there are few nutritional positives in a bowl of Corn Flakes. Extremely low in dietary fibre and relatively high in salt, the high carbohydrate load coupled with very little dietary fibre means this cereal is not a good option for blood glucose control, satiety or digestive health.

10. Kellogg's Coco Pops

With a white rice and sugar base, and consequently, little protein or fibre, think of this cereal as more like a bowl of lollies with a few vitamins and minerals added rather than a nutritious breakfast option. While desperate parents may argue that it is better for children to eat something like this chocolatey mix for breakfast than nothing, as a dietitian I tend to disagree.

Source: Breakfast cereals ranked from healthiest to unhealthiest (smh.com.au) (Accessed 26.05.2024)

Analysis

1. From a nutrition perspective, explain why Susie Burrell (author and nutritionist) recommends we should select a breakfast cereal that is made up of wholegrains and is minimally processed.
2. Discuss how heavily processed breakfast cereals can impact on blood glucose levels.
3. Justify why having a bowl of Kellogg's All-Bran is a better choice for breakfast than a bowl of Kellogg's Nutri-Grain.
4. Explain why eating two Weet-Bix breakfast biscuits with milk could be considered a healthy breakfast option.
5. Identify the similarities between Milo Cereal and Kellogg's Corn Flakes that has led Susie Burrell to list these breakfast cereals as two of her least-preferred cereals.

Activity 7.5

Top-selling processed breakfast cereals

A survey by *Choice* magazine has identified the top-selling breakfast cereals in Australia. The nutritional information of each of the four top-selling processed breakfast cereals is shown in Table 7.1. Carefully analyse this information, and then answer the questions that follow.

Weblink
Uncle Tobys

Table 7.1 Nutrition information per 100 grams of major ingredients

	Sanitarium Weet-Bix	Kellogg's Nutri-Grain	Kellogg's Corn Flakes	Kellogg's Coco Pops
Energy	1480 kilojoules	1600 kilojoules	1620 kilojoules	1640 kilojoules
Dietary fibre	12.9 grams	2.7 grams	4.2 grams	1.7 grams
Total fat	1.3 grams	0.6 grams	1.8 grams	1.3 grams
Sugars	3.0 grams	32 grams	8.9 grams	32.3 grams
Salt/sodium	270 milligrams	480 milligrams	535 milligrams	330 milligrams

1 Which cereal has the most fibre?
2 Explain why this cereal is likely to contain the most fibre.
3 Why should we try to consume foods with a high fibre content?
4 Which cereal would people who wish to reduce the fat in their diet be advised to choose?
5 Which cereal is highest in sugar? What health concerns are associated with consuming cereals that are high in sugar?
6 Identify the cereal lowest in salt/sodium.
7 Why is it important to read the labels when making cereal choices?
8 Which cereal would nutritionists be most likely to recommend, and for what reasons?
9 Identify a television advertisement for a breakfast cereal. Describe the way that the advertisement promotes the cereal product. Which age group is it aimed at?
10 Visit the Uncle Tobys website and examine the nutritional information of its Traditional Oats. Discuss how Uncle Tobys Traditional Oats compare with Sanitarium Weet-Bix from a nutritional perspective, and whether they would be a nutritious food to eat for breakfast.

PRACTICAL ACTIVITY 7.2

Sensory analysis: Taste testing breakfast cereals

Worksheet
Practical activity 7.2

Aim

To determine which cereal is most suitable as a breakfast food for adolescents

Method

1 Conduct a taste test of four breakfast cereals. You may eat them either with milk or dry. Whichever method of tasting you choose, you must taste each cereal in the same way.
2 Wash and dry your bowl between samples.
3 Record your impressions of each cereal in a table like the one below.

Results

Cereal type	Appearance	Texture	Flavour
Sanitarium Weet-Bix			
Kellogg's Nutri-Grain			
Kellogg's Corn Flakes			
Kellogg's Coco Pops			

9780170495714

Analysis

1 Which cereal had the sweetest flavour?
2 Does your sensory test for sweetness agree with the information in Activity 7.5?
3 Which cereal took the longest to chew and swallow?
4 Suggest a reason why the textures of each cereal are different.
5 Was the cereal with the highest amount of dietary fibre the hardest to chew?
6 Which cereal had the most appealing appearance and which the least appealing? Why?

Conclusion

Based on the taste test you have just completed and the information in Activity 7.5, which cereal would you most recommend as a healthy breakfast food for a teenager? Justify your answer.

Eggs for breakfast

Eggs are a very affordable protein food, and contain a range of vitamins and minerals. They can form the basis of a cooked breakfast or a quick, easy family meal. A special cooked breakfast on a Saturday or Sunday morning can be a time when families relax and enjoy being together.

Eggs are a versatile ingredient that can be cooked in a variety of ways. One of the simpler methods of serving an egg is to have it either soft- or hard-boiled. Soft-boiled eggs are delicious when served with fingers – or 'soldiers' – of toast that can be used to dip into a runny yolk. Fried eggs (with or without bacon) are another popular breakfast or brunch food, and eggs can also be served scrambled or poached. Making an omelette is slightly more challenging, but allows you to add a range of other ingredients to the eggs to make a more substantial dish. Other delicious ways to serve eggs for breakfast or brunch include vegetable fritters, or baking them nestled within a spicy bean mix.

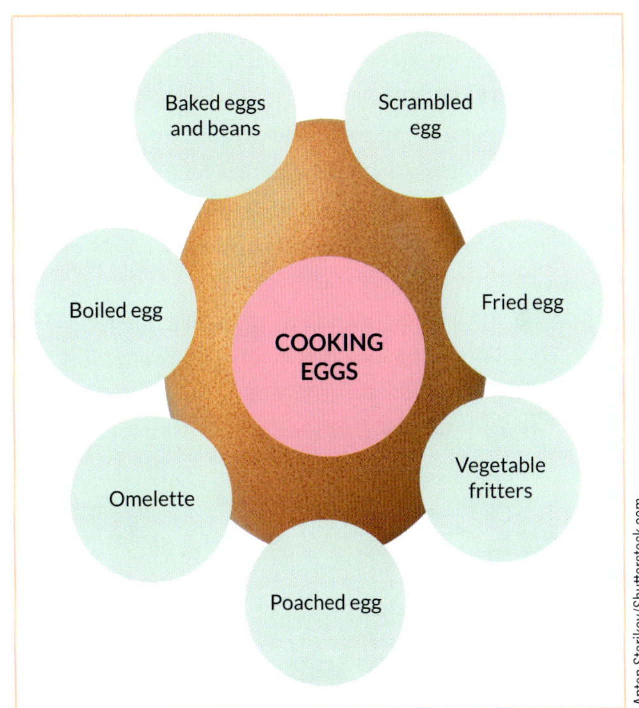

Anton Starikov/Shutterstock.com

Figure 7.27 Methods of cooking eggs for breakfast

Testing knowledge 7.3

21 What are the three main uses of water in the human body? List four of the best food sources of water.
22 Outline three environmental concerns associated with using bottled water.
23 Describe two ways in which you could reduce your use of bottled water.
24 What percentage of the day's food intake should breakfast provide?
25 Describe what happens to the body's energy store during sleep.
26 What are the benefits to students in eating breakfast?
27 Explain how eating breakfast will help individuals maintain a healthy weight.
28 List the main lifestyle benefits linked to eating breakfast identified in international research. Would you expect similar results for Australian adolescents? Why or why not?
29 Discuss why wholegrain cereals are recommended as a good breakfast food.
30 Create an annotated diagram of the benefits to young Australians of consuming oats for breakfast.

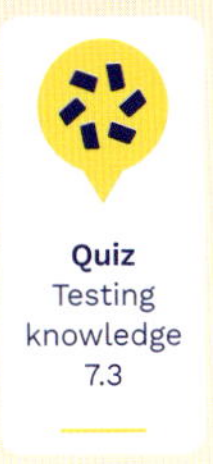

Quiz Testing knowledge 7.3

Critical and creative thinking 7.1

Healthy eating for teenagers

Work with a partner to prepare an animation or video clip to highlight the importance of one of the five food groups in the *Australian Guide to Healthy Eating*. Your animation or video clip should be appropriate to include in a health promotion campaign aimed at teenagers.

1 Make a list of the main points to be included in your animation or video clip.
2 Storyboard the key scenes of your animation or video clip.
3 Create your animation or video clip using an appropriate app.
4 Edit your animation or video clip to approximately 30 seconds of viewing time.
5 Present your animation or video clip to the class and obtain feedback on the positive aspects of it and any areas for improvement.
6 Evaluate your animation or video clip and its suitability to be included in a health promotion campaign aimed at teenagers.

iStock.com/Rodrigo Sepulveda

Critical and creative thinking 7.2

The benefits of eating breakfast

Part A: Prepare a breakfast summary foldable

Using an A4 sheet of paper, prepare a tri-fold summary foldable (see Figure 7.28) that summarises the key information about breakfast. Use the following headings:

- The importance of eating breakfast
- Breakfast eating habits: Boys compared to girls
- My perfect breakfast

List the menu, including drinks, of your perfect breakfast. Underneath your breakfast menu, draw a diagram of the *Australian Guide to Healthy Eating*. Classify the food items on your perfect breakfast menu on the diagram. Write a comment on the nutritional value of your perfect breakfast.

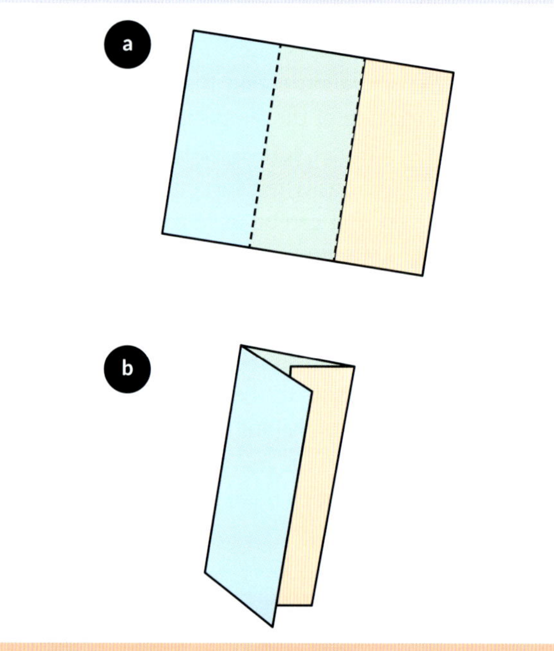

Figure 7.28 Preparing a summary foldable

Part B: Create a promotional poster

Create a promotional poster to advertise a new breakfast club that will be starting next term at your local primary school. Your goal is to promote the benefits of eating breakfast to students and to encourage them to come along. Use the information from your breakfast summary foldable about the importance of eating breakfast to assist you in preparing your poster.

9780170495714

DESIGN ACTIVITY 7.1

Wellness bowl

Worksheet
Design activity 7.1

Templates
PMI table

Food order

Production plan

Sensory wheel

Australian Guide to Healthy Eating

Wellness bowls – also known as poke bowls or Buddha bowls – are a trend that has the tick of approval from nutritionists and dietitians. Based on plant foods, these bowls are colourful, have a range of flavours and textures, and usually follow the recommendations of the *Australian Guide to Healthy Eating*. Importantly, wellness bowls are more interesting than a traditional green salad as they contain a mix of raw and cooked vegetables.

Design brief

Design and produce a wellness bowl that will be used as part of the promotion for Healthy Eating Week at your school to educate the community about the *Australian Guide to Healthy Eating*. The bowl will include cooked wholegrains or slow-release carbohydrates, raw and cooked vegetables, a protein ingredient, a dressing, and extra ingredients to add crunch and flavour. The photograph and recipe for the wellness bowl will be put on the school website to give families some new ideas for healthy meals.

Investigating and defining

1. Use the specifications in the design brief to develop three design criteria suitable for evaluating the success of the finished product.
2. Prepare the Wellness Bowl recipe on page 182 to develop an understanding of the ingredients and processes involved.
3. Research the history of wellness bowls.
4. Research some of the key ingredients often found in wellness bowls.
5. Explain why dietitians and nutritionist support the trend of wellness bowls.
6. List five ingredients that could be used in each component of a wellness bowl in a recipe map like the one above.

Figure 7.29 A recipe map for a wellness bowl

Generating and designing

1. Download three images demonstrating the presentation of wellness bowls that appeal to you.
2. Use a PMI (Plus, Minus, Interesting) table to record your thinking around the presentation of these three wellness bowls.
3. Explain why you would or would not use these design ideas in the presentation of your wellness bowl.

4 Use the ideas from your research and the information in your recipe map to generate two designed solution options to meet the need identified in the design brief. The design for your wellness bowl should include these five key components:
   - half a cup of cooked wholegrains or slow-release carbohydrates
   - a variety of raw and cooked vegetables with different flavours, colours and textures
   - a protein to help to keep you feeling full
   - a delicious dressing to bring everything together
   - extra ingredients that will make the bowl more interesting and delicious by adding crunch and flavour.
5 Design your final recipe. Use recipe conventions when writing it, so that it is ready for production.

Planning and managing

1 Prepare a food order.
2 Develop a production plan for the recipe. Identify the key processes to be used, and describe relevant safe work practices for each process.

Producing and implementing

1 Prepare and present your wellness bowl.
2 Photograph your wellness bowl.
3 Record any changes you made during production.

Evaluating

1 Evaluate the success of your wellness bowl using the design criteria you developed.
2 Describe the sensory properties of your wellness bowl. Consider the overall product rather than individual ingredients.
3 Classify the ingredients for your wellness bowl on a diagram of the *Australian Guide to Healthy Eating*. Explain how well this dish meets the recommendations of this food selection model.
4 Discuss any changes you would make to the ingredients or processes to improve your recipe.
5 Describe how you and your partner worked collaboratively during production. You could comment on setting up your work bench, sharing the cooktop and oven, washing the dishes and cleaning your work area.

Alexander Spatari/Moment/Getty Images

Figure 7.30 Wellness bowls are a very healthy meal.

9780170495714

Margherita Pizzas with a Cauliflower Crust

This recipe for pizza is a great alternative to traditional pizza for someone who has coeliac disease. Instead of using wheat flour to make the pizza base, it uses finely chopped cauliflower, with gluten-free pizza dough mix or ground almonds to give structure to the pizza base.

INGREDIENTS

½ small cauliflower (approximately 300 grams), cut into florets

⅔ cup (85 grams) gluten-free pizza dough mix

⅓ cup (30 grams) parmesan, finely grated

1 egg, lightly beaten

salt and pepper

4 tablespoons Napoli Sauce (see recipe on page 245)

½ cup mozzarella, grated

8 basil leaves

MAKES 2 × 16-CENTIMETRE PIZZAS

METHOD

1. Preheat oven to 200°C. Draw a 16-centimetre circle on two sheets of baking paper. Turn the paper upside down and place each sheet on a baking tray.
2. Whiz the cauliflower in a food processor until it resembles couscous. You will need approximately two cups of 'cauliflower couscous'.
3. Combine the cauliflower in a bowl with the gluten-free pizza dough mix, parmesan and egg. Season to taste with salt and pepper.
4. Place half the cauliflower mixture onto each sheet of paper and shape each into a 5-millimetre-thick pizza base.
5. Bake the pizza bases for 20 minutes or until dark, golden and firm.
6. Spread 2 tablespoons of Napoli Sauce over each pizza base. Top with the mozzarella.
7. Bake for a further 5 minutes or until the cheese is browned and melted.
8. Take care when transferring the pizzas to serving plates; the base will be more delicate than that of a traditional pizza.
9. Top with the torn basil leaves.

Note: you can substitute the gluten-free pizza dough mix with ⅔ cup of ground almonds to give a slightly nutty flavour to the dough.

Mark Fergus Photography

EVALUATION

1. Describe the sensory properties – appearance, aroma, flavour, texture and sound – of your Margherita Pizzas with a Cauliflower Crust.
2. Explain why this recipe would be ideal to serve to someone who suffers from coeliac disease.
3. Describe some other ingredients that would make delicious toppings for the cauliflower base.
4. What aspect of this production did you find most challenging? Explain why.
5. Classify the ingredients for the Margherita Pizzas with a Cauliflower Crust on a diagram of the *Australian Guide to Healthy Eating*. Comment on how well this dish meets the recommendations of this food selection model.

7 RECIPES

Wellness Bowl

Wellness bowls have become popular as they are very nutritious meals that are simple to make. Wellness bowls can be built from almost any grain, vegetable and protein to suit an individual's food and flavour preference. They are also very convenient to make, as each of the components can be made in advance and stored in the refrigerator.

INGREDIENTS

¼ cup brown rice

¼ teaspoon ground cumin

¼ teaspoon ground coriander

¼ teaspoon ground sweet paprika

½ cup canned chickpeas, drained and rinsed

olive oil spray

2 stems broccolini

1 egg

2 radishes

2 cherry tomatoes

½ cup baby spinach leaves

1 tablespoon flat leaf parsley, finely chopped

Aioli Dressing

1 tablespoon mayonnaise

1 tablespoon lemon juice

2 teaspoons extra virgin olive oil

¼ teaspoon crushed garlic

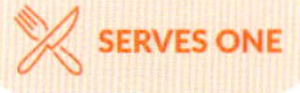

METHOD

1. Rinse the rice in cold water then place in medium saucepan. Add 2 cups of cold water and bring to the boil.
2. Stir once or twice and then reduce heat to simmer and cook for 25 minutes. Drain before serving.
3. Preheat oven to 200°C. Line a baking tray with paper. Combine the cumin, coriander and sweet paprika in a small bowl.
4. Pat the chickpeas dry with paper towel and spread onto the lined baking tray. Spray lightly with olive oil spray and sprinkle with the spice mix. Roast for 15 minutes or until the chickpeas are golden.
5. Cut the broccolini stems into four equal pieces. Place 4 centimetres of water in a small saucepan and bring to boil. Add broccolini and cook for 1 minute. Drain, rinse under cool water and drain again.
6. Place the egg in a small saucepan and cover with cold water. Bring to the boil, then reduce to a simmer and cook for 4–6 minutes, depending on how soft you like the yolk. Drain the egg, then refresh in cold water. Peel the egg and cut in half.
7. Slice the radishes thinly and cut the cherry tomatoes in half.

Aioli Dressing

Combine the mayonnaise, lemon juice, olive oil and crushed garlic in a small bowl and mix well.

To serve

Arrange the rice, chickpeas, broccolini pieces, spinach leaves, radish slices and tomato pieces in individual sections in a serving bowl. Place the egg halves on the top, with the yolk visible. Sprinkle with the chopped parsley and drizzle over the aioli dressing.

9780170495714

EVALUATION

1 Describe the sensory properties – appearance, aroma, flavour and texture – of your Wellness Bowl.
2 Describe one safety rule to follow when cooking the rice on the cooktop.
3 Explain why the broccolini is only cooked for 1 minute in step 5 of the recipe rather than for 5 minutes. Why is it important to rinse the broccolini under cool water after cooking?
4 Classify the ingredients for the Wellness Bowl on a diagram of the *Australian Guide to Healthy Eating*. How many serves of grains and vegetables does this recipe provide?
5 Explain whether you would rate this recipe as 'healthy' or 'very healthy', based on the recommendations of the *Australian Dietary Guidelines*. Justify your rating.

Mark Fergus Photography

RECIPES 7

Fried Rice

Fried rice is a staple dish in many Asian countries – including China, Japan, Vietnam and Indonesia – where it is often sold from street stalls. It is very adaptable recipe, as it can be made for one or two people or for a crowd. The ingredients can also be varied depending on what leftovers or vegetables are in the refrigerator. The secret to a preparing a successful fried rice is to use a pre-cooked, cold and dry long-grain rice, such as jasmine rice.

INGREDIENTS

4 cups water
⅔ cup jasmine rice
1 egg
1 tablespoon oil
1 rasher bacon, diced
1 clove garlic, crushed
½ onion, finely diced
¼ carrot, finely diced
¼ capsicum, finely diced
¼ stick celery, finely diced
¼ cup frozen peas
¼ cup corn kernels
1 tablespoon soy sauce

To garnish
2 spring onions, finely sliced
¼ cup bean shoots

METHOD

1. Bring the water to the boil in a saucepan.
2. Add the rice. Stir once or twice to separate the grains. Simmer 12–15 minutes until tender. Do not overcook.
3. Drain the rice and spread it in a thin layer on a tray covered with absorbent paper. Refrigerate uncovered until required. This process helps to dry out the rice so the grains remain separated.
4. Beat the egg with a fork in a small cup to combine the white and yolk.
5. Gently heat the oil in a wok. Add the beaten egg, tilt the wok to spread the egg into a thin layer and cook until set. Turn off the heat.
6. Remove the omelette from the wok and dice finely.
7. Reheat the wok. Add the bacon to the wok and stir-fry for 2–3 minutes. Drain on paper towel.
8. Add the garlic, onion, carrot, capsicum and celery to the wok and stir-fry for 2–3 minutes, until the onion is transparent. Add the frozen peas and cook for 1 minute.
9. Add the cooled rice, diced bacon, diced omelette, corn kernels and soy sauce. Toss over the heat until the ingredients are well combined and heated through.
10. Garnish with sliced spring onions and bean shoots and serve immediately.

Mark Fergus Photography

EVALUATION

1. Describe the sensory properties – appearance, aroma, flavour and texture – of your Fried Rice.
2. Why is it important not to overcook the rice in step 2?
3. Explain why the rice is refrigerated after boiling.
4. Identify the safe work practices you followed when boiling and stir-frying the rice.
5. Classify the ingredients for the Fried Rice on a diagram of the *Australian Guide to Healthy Eating* and comment on the nutritional value of the recipe.

9780170495714

Prawn and Vegetable Rice-paper Rolls

Rice-paper is a delicate, edible wrapper used to package delicious ingredients in a tasty low-fat snack. Roast pork, chicken or tofu could be used to replace the prawns for the protein component of this recipe.

INGREDIENTS

20 grams rice vermicelli

1 spring onion, finely julienned

¼ carrot, grated

1 tablespoon coriander, chopped

1 tablespoon mint, chopped

⅙ red capsicum, finely julienned

20 grams bean shoots

1 tablespoon cashews, finely chopped

3 teaspoons sweet chilli sauce

8 rice-paper sheets

8 prawns, cooked, peeled, de-veined and halved lengthwise

Dipping Sauce

½ lime, juiced

1 tablespoon sweet chilli sauce

1 centimetre lemongrass, finely chopped

2 teaspoons coriander, finely chopped

2 teaspoons fish sauce

METHOD

1. Soak vermicelli in warm water for 10–15 minutes or until softened. Drain well.
2. Combine vermicelli with spring onion, carrot, coriander, mint, capsicum, bean shoots, cashews and sweet chilli sauce. Mix well.
3. Assemble the rice-paper rolls one at a time. Soak a rice-paper sheet in warm water for about 30 seconds until just softened, then drain on a clean, dry tea towel.
4. Place a portion of the filling on one end of rice paper. Add prawn, then tuck in the ends and roll to enclose filling.
5. Cover with plastic wrap and repeat the process.
6. To make the dipping sauce, combine ingredients and mix well.
7. Serve rolls on a platter, accompanied by a bowl of dipping sauce.

EVALUATION

1. Describe the sensory properties – appearance, aroma, flavour and texture – of your Prawn and Vegetable Rice-paper Rolls.
2. What are some other products that could be substituted for the prawns?
3. What cereal product is used in the recipe and what is its function?
4. Discuss the nutritional advantages and disadvantages of including raw vegetables in snack foods.
5. Classify the ingredients for the Prawn and Vegetable Rice-paper Rolls on a diagram of the *Australian Guide to Healthy Eating*. Write a brief paragraph to explain whether the rolls would be suitable as a healthy snack.

Mark Fergus Photography

Baked Chicken Balls with Risoni and Peas

INGREDIENTS

Chicken Balls

½ medium zucchini, grated

250 grams minced chicken thighs

pinch salt

½ teaspoon lemon rind, finely grated

1 clove garlic, crushed

1 tablespoon parsley, finely chopped

2 sprigs thyme, leaves only, chopped

1 small egg

¼ cup grated parmesan

⅓ cup panko breadcrumbs

¼ red onion, grated

pinch black pepper

spray olive oil

Risoni and Peas

1 tablespoon olive oil

20 grams butter

½ brown onion, finely diced

1 clove garlic, crushed

2 sprigs thyme

¼ cup white wine

½ cup risoni (such as Barilla No. 26)

1 cup chicken stock

salt and pepper

½ cup frozen peas

Topping

2 teaspoons parsley, finely chopped

1 tablespoon grated parmesan

1 teaspoon lemon rind, finely grated

1 tablespoon pine nuts, toasted (optional)

METHOD

Chicken Balls

1. Preheat oven to 180°C. Line an oven tray with baking paper.
2. Sprinkle the grated zucchini with a pinch of salt and set aside for 5 minutes.
3. In a large mixing bowl, add the chicken mince, lemon rind, garlic, parsley, thyme, egg, parmesan, panko crumbs and pepper.
4. Squeeze the liquid from the zucchini and grated onion and add to the bowl. Mix well.
5. Roll a heaped tablespoon of chicken mixture into a ball and place on the oven tray. Repeat with the remaining mixture. Hint: wet your hand lightly before rolling the chicken balls, as the mixture is sticky.
6. Lightly spray the Chicken Balls with oil spray. Place in the oven and bake for 12–15 minutes until the Chicken Balls are just firm to touch – they will finish cooking when they are baked in the risoni.

Risoni and Peas

1. Heat the oil and butter in a frying pan over medium heat. Gently sauté the onion until it has softened and is just beginning to change colour.
2. Add the garlic and sprigs of thyme and fry for 1–2 minutes without browning. Add the wine and simmer for 2 minutes.
3. Add the risoni, chicken stock and salt and pepper to taste, and bring the mixture to a simmer.
4. Transfer the risoni mixture to an ovenproof dish and sit the Chicken Balls on top. Cover with a lid or aluminium foil and place in the oven for 15 minutes.
5. Remove from the oven and sprinkle the frozen peas over the top. Re-cover with the lid or aluminium foil and rest for 5 minutes.
6. Combine the topping ingredients and sprinkle over the Baked Chicken Balls with Risoni and Peas. Serve while hot.

9780170495714

EVALUATION

1 Describe the sensory properties – appearance, aroma, flavour, texture and sound – of your Baked Chicken Balls with Risoni and Peas.

2 Identify and describe the changes that occur to the chemical and physical properties of the chicken balls when they are baked in the oven.

3 Explain why it is essential to the final sensory properties of the recipe to sauté the onion and garlic before adding the risoni and stock.

4 Identify the reaction that occurred to the risoni when it was baked in the oven. Explain how this reaction affected the sensory properties of the risoni.

5 Classify the ingredients for the Baked Chicken Balls with Risoni and Peas on a diagram of the *Australian Guide to Healthy Eating*. Write a brief paragraph to explain how well this recipe meets the recommendations of this food selection model.

Mark Fergus Photography

Potato Salad with Egg and Tuna

This is a very colourful salad that makes a nutritious lunch dish or light meal on a hot summer's night. The variety of crisp vegetables gives a great textural contrast to the softer protein ingredients and, when combined with the tangy dressing, makes a delicious dish.

INGREDIENTS

6 chat potatoes

2 eggs

1 carrot, julienned

50 grams snow peas, with tops removed

¼ green capsicum, julienned

¼ yellow or red capsicum, finely diced

1 large ripe tomato, cut into wedges

6 black olives, halved

100 grams lettuce or salad mix or rocket

170 grams canned tuna, drained

1 tablespoon parsley, finely chopped

Dressing

1 small clove garlic, crushed into a paste (optional)

juice of ½ lemon

3 tablespoons olive oil

salt and pepper

½ teaspoon sugar

METHOD

1. Steam the chat potatoes for approximately 20 minutes or until tender.
2. Place the eggs into a saucepan filled with warm water. Bring the water to the boil. Reduce the heat to a simmer, and cook the eggs for exactly 8 minutes.
3. Remove the eggs from the heat and drop them into the sink to crack their shells. Place into a saucepan or bowl of cold water and allow to cool.
4. Shell the eggs, then slice into quarters.
5. Cut the potatoes into thick slices.
6. Bring a small saucepan of water to the boil and blanch the carrot for 2 minutes. Add the snow peas and cook for a further 30 seconds. Drain and refresh in cold water. Drain again.
7. Mix the garlic, lemon juice, olive oil, salt, pepper and sugar together.
8. Mix the chat potatoes, blanched carrots, snow peas, capsicum, tomato and olives together in a small bowl and toss with the dressing.
9. Place the lettuce or salad mix or rocket onto serving plates (or into two containers), and top with the tuna, boiled eggs and parsley.

Mark Fergus Photography

EVALUATION

1. Describe the sensory properties – appearance, aroma, flavour and texture – of your Potato Salad with Egg and Tuna.
2. Explain why it is important to crack the shells of the eggs immediately after they are cooked.
3. Describe one safety rule to follow when blanching the snow peas.
4. Why does the recipe suggest cutting the chat potatoes into thick slices, rather than fine slices?
5. Classify the ingredients for the Potato Salad with Egg and Tuna on a diagram of the *Australian Guide to Healthy Eating*. Discuss how well this recipe addresses the guidelines of this food selection model.

9780170495714

Chilli con Carne

INGREDIENTS

Spices

3 teaspoons ground cumin

2 teaspoons sweet paprika

½ teaspoon cayenne pepper

½ teaspoon chilli powder

1 teaspoon garlic powder

1 teaspoon onion powder

Con Carne

1 beef stock cube, Massel (gluten-free)

¾ cup hot water

2 cloves garlic

½ brown onion

½ red capsicum

1 tablespoon olive oil

200 grams minced beef

1 tablespoon tomato paste

400 grams canned diced tomatoes

400 grams canned red kidney beans, rinsed and drained

1 teaspoon brown sugar

¾ teaspoon salt

To serve

1 cup cooked rice

30 grams tasty cheese, grated

2 tablespoons sour cream

¼ avocado, cut into chunks

METHOD

1. Combine all the spices together and set aside.
2. Dissolve the beef stock cube into the hot water and set the beef stock aside.
3. Finely chop the garlic and dice the onion and capsicum.
4. Using a heavy-based medium saucepan, heat the oil over medium heat. Sauté the onion for 1 minute, then add the garlic and cook for 1 minute. Add the capsicum and sauté for approximately 2 minutes until the onion is translucent. Do not brown.
5. Increase the heat to high, add the minced beef and break it up with a wooden spoon while stirring. Cook for approximately 2 minutes until the meat changes colour.
6. Add the spices and cook for a further 30 seconds, stirring, until they become very aromatic.
7. Add the tomato paste, diced tomatoes, kidney beans, beef stock, brown sugar and salt.
8. Bring to the boil, then reduce to a simmer. Simmer gently, uncovered, for 20 minutes, stirring occasionally, unit the sauce has reduced and thickened.
9. Serve with rice, grated cheese, sour cream and avocado chunks.

EVALUATION

1. Describe the sensory properties – appearance, aroma, flavour, texture and sound – of your Chilli con Carne.
2. Explain why it is important to add the garlic after the onion has begun to cook.
3. Some of the vegetables were chopped or diced. Describe the differences between these two cutting processes.
4. What is simmering and how does it cook the ingredients in the Chilli con Carne?
5. Classify the ingredients for the Chilli con Carne on a diagram of the *Australian Guide to Healthy Eating* and explain how well this meal meets the recommendations of this food selection model.

Mark Fergus Photography

DIGESTION

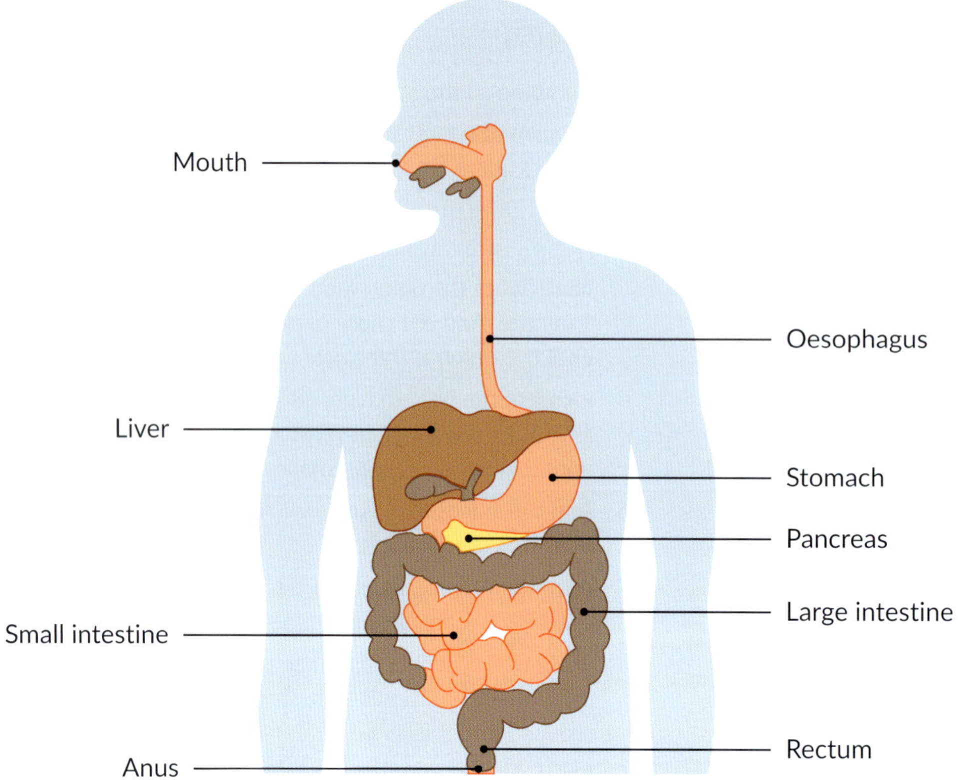

LIFESTYLE & DIET-RELATED ISSUES

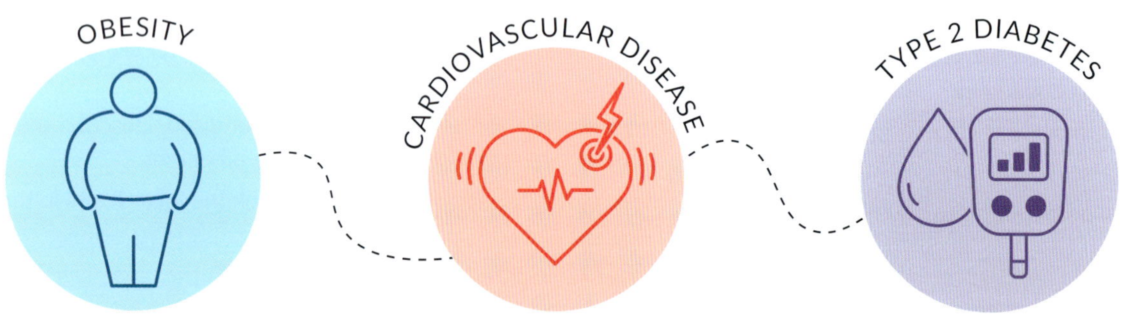

FOOD ALLERGIES

Abnormal immunological reactions to certain foods, usually proteins

FOOD INTOLERANCES

Chemical reactions to particular foods

9780170495714

HOW TO KEEP THE GUT MICROBIOME HEALTHY

Try to eat 30 different plant foods each week

Eat fruit and vegetables from all the colours of the rainbow

Include fermented foods like yoghurt and kimchi in the diet

Eat fruit, vegetables, olive oil and nuts that are rich in polyphenols

Cut down on ultra-processed foods

Drink plenty of water

8 EATING WELL FOR LIFELONG HEALTH

KEY TERMS

cardiovascular disease a general term used to describe a range of diseases, including heart disease, stroke and blood vessel disease

diabetes a disease where the pancreas is unable to produce sufficient insulin to enable the glucose produced during digestion to be absorbed into the bloodstream

digestion the breakdown of large pieces of food into smaller components that can be absorbed into the bloodstream

enzymatic hydrolysis a chemical digestive process that breaks down food by breaking the molecular bonds that hold the food together

FODMAP an acronym that stands for fermentable oligosaccharides, disaccharides, monosaccharides and polyols

food allergy an abnormal immunological reaction to food

gluten intolerance a condition where the gluten proteins present in wheat, oats, barley, triticale and rye cause a person to suffer intestinal discomfort after eating food that contains gluten

glycaemic index (GI) a ranking of carbohydrate foods based on the immediate effect they have on blood sugar levels

gut microbiome the complex ecosystem of microbiota that live in the gut

microbiota the microscopic living organisms such as bacteria, fungi and viruses that live in the small intestine

Templates
- Recipe map
- Decision table
- Food order
- Production plan
- Sensory wheel
- Australian Guide to Healthy Eating

Worksheets
- Practical activity 8.1
- Design activity 8.1
- Design activity 8.2

Quizzes
- Testing knowledge 8.1
- Testing knowledge 8.2
- Testing knowledge 8.3
- Testing knowledge 8.4
- Testing knowledge 8.5

Nelson MindTap — To access resources above, visit **cengage.com.au/nelsonmindtap**

9780170495714

Essentials for lifelong health and wellness

Ensuring optimal nutrition by adopting and maintaining healthy eating habits throughout life is the key to good health and overall wellbeing. A diet that follows the recommendations of the *Australian Dietary Guidelines* and the *Australian Guide to Healthy Eating* will provide us with a wide variety of nutrient-rich foods to support our growth and development, and our mental and emotional health. A nutrient-rich diet will enable our gut microbiota to thrive and our digestive system to function by breaking down the foods we eat and enabling essential nutrients to be absorbed. A well-balanced, nutrient-rich – rather than energy-dense – diet will also reduce the likelihood of becoming overweight or obese.

iStock.com/Fly View Productions

Figure 8.1 Eating a well-balanced diet is key to maintaining good health and overall wellbeing.

Digestion

To ensure that our body is absorbing all of the nutrients necessary for good heath, we need a healthy, well-functioning digestive system. **Digestion** is the breakdown of large pieces of food into smaller components that can be absorbed into the bloodstream.

The digestive system (see Figure 8.3) begins in the mouth and finishes at the anus. When you eat an apple or a ham and cheese sandwich, your teeth begin the process of digestion by grinding up the food. The saliva in the mouth contains the enzyme salivary amylase, which begins to break down the starch in the food into the simple sugar maltose, through a process called **enzymatic hydrolysis**. During this process, the salivary amylase breaks down the food by breaking the molecular bonds that hold the food together. When the food is swallowed, it is pushed down the oesophagus through a series of muscular contractions called peristalsis.

As the food enters the stomach, it passes through a sphincter, or valve, that stops it from going back up into the oesophagus. The muscles in the wall of the stomach churn and mix the food into a thick liquid called chyme.

The stomach is a highly acidic environment. The hydrochloric acid in the stomach begins the breakdown of protein into amino acids, and the enzyme gastric lipase – which is also found in the stomach – begins to break down the fat in the food.

Weblinks
How does the Human Digestive System work

How your digestive system works

From the stomach, the food passes through the small intestine, which is like a long tube about 6 metres long that is loosely coiled in your abdomen. The first section of the small intestine is called the duodenum. In the duodenum, more digestive juices are added to the food from the pancreas and liver. Bile produced in the liver and stored in the gall bladder breaks down (or emulsifies) the fat into even smaller particles. In the small intestine, a number of enzymes released from the pancreas break down fats, starch and proteins.

As the food moves through the rest of the small intestine, its nutrients are absorbed through the villi, or finger-like projections that line the wall of the intestine.

Any waste material that is not absorbed in the small intestine is moved into the large intestine or bowel. Here, any water is removed, and the waste material, called faeces, is then stored in the rectum before it is passed out of the body through the anus.

Gut health

A healthy gut is important to both our physical and mental health. There are over 100 trillion microscopic organisms that live in the gut (the small and large intestine) of all human beings. These microorganisms are called **microbiota**, and include over 400 species of individual bacteria, fungi and viruses. As a group, this complex ecosystem of microbiota that live in our gut is called the **gut microbiome**.

The microbiota help to break down food – turning it into nutrients our body can use, as well as supporting our immune system. The large intestine is densely populated with a ecosystem of microbiota; however, the small intestine is home to a smaller quantity and fewer species of microbiota. Each person has a microbiome that is unique to them – much the same way as we each have a unique fingerprint.

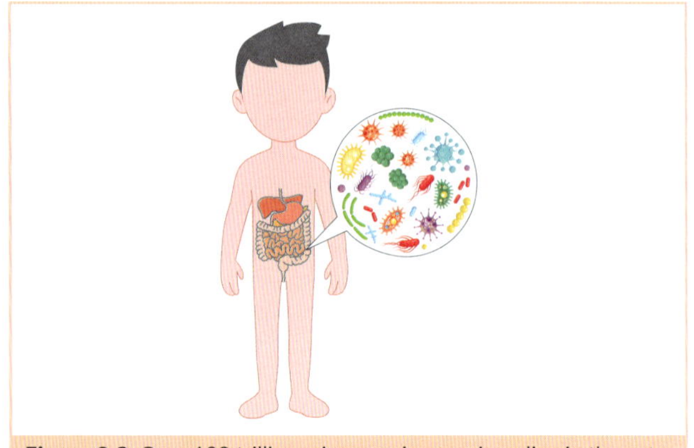

Figure 8.2 Over 100 trillion microscopic organisms live in the gut.

9780170495714

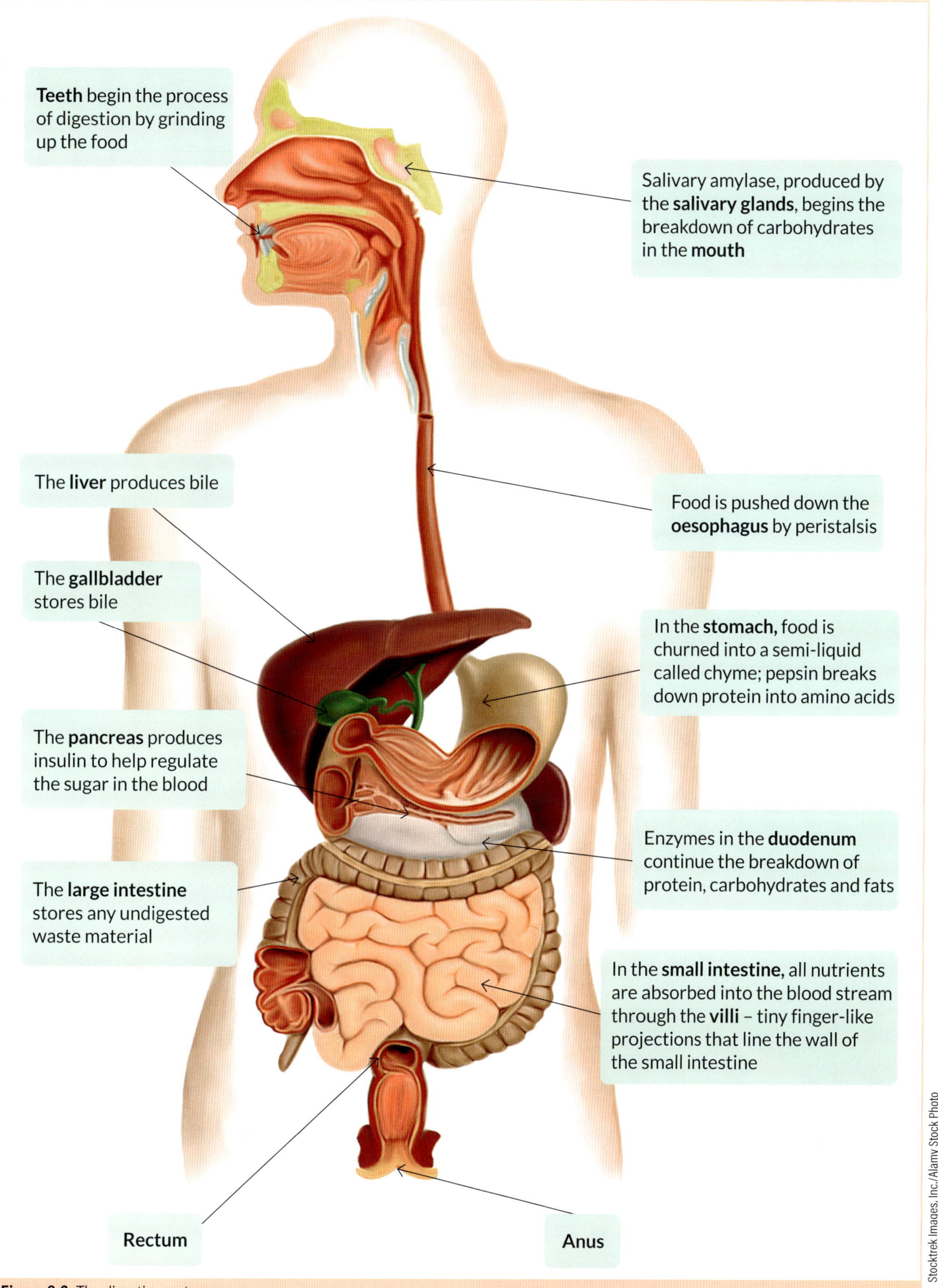

Figure 8.3 The digestive system

Tips for a healthy gut

The type and balance of bacteria found in our gut are affected by many lifestyle factors, including our level of physical activity, the amount of sleep we get and how much stress we are under. However, the most important factor to ensure a healthy gut is to eat a nutritious diet.

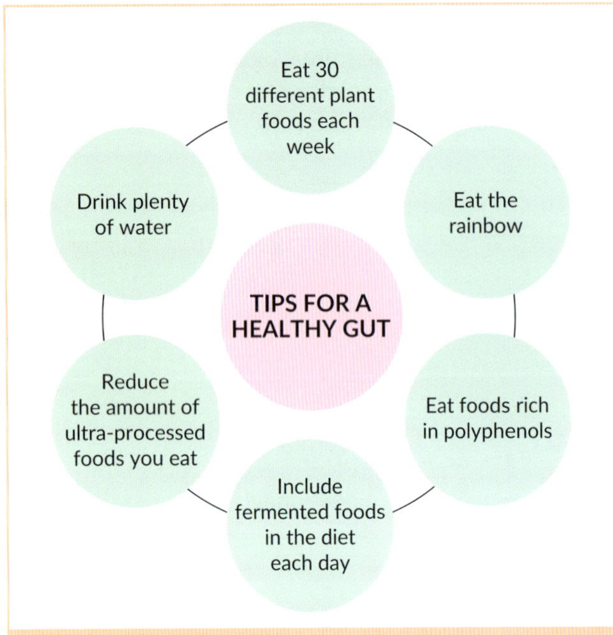

Figure 8.4 Tips for a healthy gut

1. Eat 30 different plant foods each week

Plant foods such as wholegrain cereals, vegetables, fruit, pulses, nuts and seeds are great for gut health, as they are high in dietary fibre and resistant starch. Resistant starch is a prebiotic compound that is not broken down or absorbed in the small intestine. Prebiotic substances are important to our health because they help to stimulate the growth of gut microbiota that live in the colon (the longest part of the large intestine), enabling them to thrive.

2. Eat the rainbow

It is important to include as many foods from the different colours of the rainbow as possible in your diet each week. Colourful fruit and vegetables – such as berries, apples, kiwi fruit, citrus fruit, bananas, grapes, spinach, red onion, carrots and capsicum – are nutrient-dense and they are also all high in dietary fibre, which microbiota love.

carlosgaw/E+/Getty Images

Figure 8.5 Eat the rainbow.

3. Eat foods rich in polyphenols

Many fruits and vegetables – as well as nuts, extra virgin olive oil, herbs, spices and dark chocolate – contain polyphenols. These are micronutrients, found in plants, that help our gut microbiome to thrive.

4. Include fermented foods in the diet each day

Yoghurt and fermented foods – such as kimchi, sauerkraut, miso and kefir – contain living microbes, which are known as probiotics. Probiotics are live bacteria, such as bifidobacteria and lactobacilli. Including a small amount of fermented foods in your diet each day helps support the microbiota that live in the small and large intestines.

Nina Firsova/Shutterstock.com

Figure 8.6 Fermented foods help the gut microbiota to thrive.

5. Reduce the amount of ultra-processed foods you eat

Eating whole foods that are minimally processed – rather than ultra-processed foods that are high in fat, salt and/or sugar – will allow the gut microbiota to thrive and not be outnumbered by harmful bacteria and other organisms that live inside the gut. Eating whole foods will also reduce the likelihood of gaining weight or suffering from gastric upsets.

Ultra-processed foods include biscuits, many breakfast cereals, ham, salami, meat pies, ready-made meals, cakes, doughnuts and packaged treats such as potato crips. These all contain significant amounts of added sugar, salt and saturated fats. Many of these foods also contain additives such as emulsifiers and artificial sweeteners, which can have a detrimental effect on the gut microbiota.

6. Drink plenty of water

Make water your drink of choice! Being well hydrated means that the gut microbiota can grow and there will be different types of microbiota populating the gut. Drinking plenty of water will also help remove toxins and other waste from the body, and assist in preventing constipation.

 9780170495714

CASE STUDY 8.1

Comparing juice to the whole food

Read the article below, and then answer the questions that follow.

Can you drink your fruit and vegetables? How does juice compare to the whole food?

Emma Beckett, Adjunct Senior Lecturer, Nutrition, Dietetics & Food Innovation – School of Health Sciences, UNSW Sydney, *The Conversation*, 2 July 2024

Do you struggle to eat your fruits and vegetables? You are not alone. Less than 5% of Australians eat the recommended serves of fresh produce each day (with 44% eating enough fruit but only 6% eating the recommended vegetables).

Adults should aim to eat at least five serves of vegetables (or roughly 375 grams) and two serves of fruit (about 300 grams) each day. Fruits and vegetables help keep us healthy because they have lots of nutrients (vitamins, minerals and fibre) and health-promoting bioactive compounds (substances not technically essential but which have health benefits) without having many calories.

So, if you are having trouble eating the rainbow, you might be wondering – is it OK to drink your fruits and vegetables instead in a juice or smoothie? Like everything in nutrition, the answer is all about context.

It might help overcome barriers

Common reasons for not eating enough fruits and vegetables are preferences, habits, perishability, cost, availability, time and poor cooking skills. Drinking your fruits and vegetables in juices or smoothies can help overcome some of these barriers.

Juicing or blending can help disguise tastes you don't like, like bitterness in vegetables. And it can blitz imperfections such as bruises or soft spots. Preparation doesn't take much skill or time, particularly if you just have to pour store-bought juice from the bottle. Treating for food safety and shipping time does change the make up of juices slightly, but unsweetened juices still remain significant sources of nutrients and beneficial bioactives.

Juicing can extend shelf life and reduce the cost of nutrients. In fact, when researchers looked at the density of nutrients relative to the costs of common foods, fruit juice was the top performer.

So, drinking my fruits and veggies counts as a serve, right?

How juice is positioned in healthy eating recommendations is a bit confusing. The *Australian Dietary Guidelines* include 100% fruit juice with fruit but vegetable juice isn't mentioned. This is likely because vegetable juices weren't as common in 2013 when the guidelines were last revised.

The guidelines also warn against having juice too often or in too high amounts. This appears to be based on the logic that juice is similar, but not quite as good as, whole fruit. Juice has lower levels of fibre compared to fruits, with fibre important for gut health, heart health and promoting feelings of fullness. Juice and smoothies also release the sugar from the fruit's other structures, making them 'free'. The World Health Organization recommends we limit free sugars for good health.

But fruit and vegetables are more than just the sum of their parts. When we take a 'reductionist' approach to nutrition, foods and drinks are judged based on assumptions made about limited features such as sugar content or specific vitamins.

But these features might not have the impact we logically assume because of the complexity of foods and people. When humans eat varied and complex diets, we don't necessarily need to be concerned that some foods are lower in fibre than others.

Juice can retain the nutrients and bioactive compounds of fruit and vegetables and even add more because parts of the fruit we don't normally eat, like the skin, can be included.

So, it is healthy then?

A recent umbrella review of meta-analyses (a type of research that combines data from multiple studies of multiple outcomes into one paper) looked at the relationship between 100% juice and a range of health outcomes.

Most of the evidence showed juice had a neutral impact on health (meaning no impact) or a positive one. Pure 100% juice was linked to improved heart health and inflammatory markers and wasn't clearly linked to weight gain, multiple cancer types or metabolic markers (such as blood sugar levels).

Some health risks linked to drinking juice were reported: death from heart disease, prostate cancer and diabetes risk. But the risks were all reported in observational studies, where researchers look at data from groups of people collected over time. These are not controlled and do not record consumption in the moment. So other drinks people think of as 100% fruit juice (such as sugar-sweetened juices or cordials) might accidentally be counted as 100% fruit juice. These types of studies are not good at showing the direct causes of illness or death.

What about my teeth?

The common belief juice damages teeth might not stack up. Studies that show juice damages teeth often lump 100% juice in with sweetened drinks. Or they use model systems like fake mouths that don't match how people drinks juice in real life. Some use extreme scenarios like sipping on large volumes of drink frequently over long periods of time.

Juice is acidic and does contain sugars, but it is possible proper oral hygiene, including rinsing, cleaning and using straws can mitigate these risks.

Again, reducing juice to its acid level misses the rest of the story, including the nutrients and bioactives contained in juice that are beneficial to oral health.

So, what should I do?

Comparing whole fruit (a food) to juice (a drink) can be problematic. They serve different culinary purposes, so aren't really interchangeable.

The *Australian Guide to Healthy Eating* recommends water as the preferred beverage but this assumes you are getting all your essential nutrients from eating.

Where juice fits in your diet depends on what you are eating and what other drinks it is replacing. Juice might replace water in the context of a 'perfect diet'. Or juice might replace alcohol or sugary soft drinks and make the relative benefits look very different.

On balance

Whether you want to eat your fruits and vegetables or drink them comes down to what works for you, how it fits into the context of your diet and your life.

Smoothies and juices aren't a silver bullet, and there is no evidence they work as a 'cleanse' or detox. But, with society's low levels of fruit and vegetable eating, having the option to access nutrients and bioactives in a cheap, easy and tasty way shouldn't be discouraged either.

Analysis

1. Evaluate the validity of the article. Consider the reliability of the sources of information used in the article, the context and the presentation of evidence.
2. Outline the main reasons that many people do not consume the recommended number of serves of fruit and vegetables.
3. Explain why some health professionals do not consider fruit juice to be as nutritionally valuable as whole fruit.
4. Draw conclusions about the effects of consuming fruit juice on the health of individuals.

Testing knowledge 8.1

1. Clarify why it is important for individuals to follow the recommendations of the *Australian Dietary Guidelines*.
2. Describe the digestive processes that take place in the mouth.
3. Explain the meaning of the term 'peristalsis'.
4. Outline what happens to food when it enters the stomach.
5. What is the duodenum? Explain the role of the duodenum in the digestive process.
6. Outline the role of bile in the digestive process.
7. Explain the role of the small intestine in the digestive process.
8. Describe the gut microbiota and the role they play in the digestive process.
9. What are prebiotic substances and how can they improve gut health?
10. Explain why it is important to include fermented foods in our diet every day.

Quiz
Testing knowledge 8.1

9780170495714

Lifestyle health issues

The way we live our lives has a major impact on our overall health and wellbeing. People who do not consume a healthy diet – and instead eat a diet that is high in saturated fat, salt and/or sugar – are more likely to gain weight and be at risk of developing one or more lifestyle health issues, or lifestyle diseases. These health issues are also linked to a lack of physical activity and other unhealthy practices, such as an excessive consumption of alcohol, or smoking or vaping. Lifestyle diseases include obesity, cardiovascular disease, type 2 diabetes and some forms of cancer.

Obesity

According to 2022 data from the National Health and Medical Research Council (NHMRC) released in June 2024, 66 per cent of Australian adults aged over 18 and 26 per cent of children aged two to 17 are living with overweight or obesity. A report prepared in 2023 from the World Obesity Atlas predicted that 47 per cent of all Australian adults will be living with obesity by 2035.

It is clear from the latest medical research that the spike in the number of Australian adults and children living with overweight or obesity is a direct result of our lifestyle, especially the foods we consume. Overweight and obesity are major health concerns, because they are linked to the development of a range of serious health problems including cardiovascular disease, type 2 diabetes and some cancers.

The level of overweight and obesity in Australian children has risen sharply and is now among the highest in the world, with more than 26 per cent of young people (or one in four) living with overweight or obesity.

As in other stages of life, obesity in childhood and adolescence can have a major impact on the health of young people. Type 2 diabetes and high blood pressure – once only seen in late adulthood – are now becoming more prevalent in young people who are overweight.

Much of the increase in the incidence of overweight and obesity in childhood is linked to the changing diets of Australian children. Although the amount of food young people eat has remained much the same as it was in the past, the energy density of the food being eaten has increased by approximately 11–15 per cent, resulting in a far higher kilojoule intake. This, combined with the fact that children today are far less physically active than children in past generations, has led to an increase in the number of children who are living with overweight or obesity.

Energy sources

To enable our bodies to function, we need energy. The body uses two main nutrients to generate energy: carbohydrate and fat. If the stores of these nutrients are depleted, the body uses its stores of protein to provide a secondary energy source. During digestion:

- carbohydrate is broken down – initially into glucose, and later into glycogen – and is stored mainly in the muscles of the body
- fat is broken down into fatty acids and is stored as adipose or fat tissue if it is not required for energy production
- protein is broken down into amino acids and is used for energy if all other sources have been exhausted.

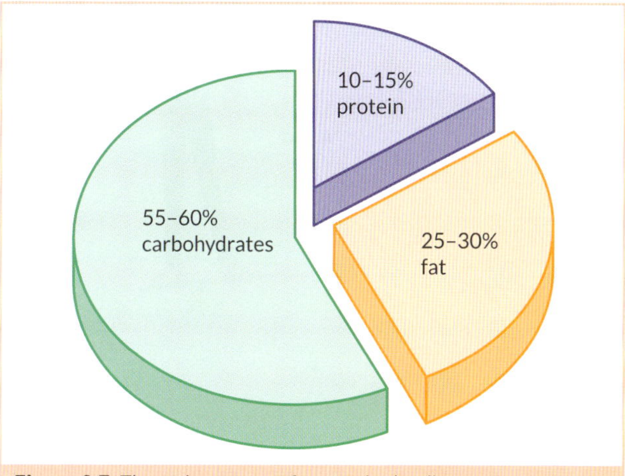

Figure 8.7 The main sources of energy in the diet

The body can produce 16 kilojoules of energy from every gram of carbohydrate, 37 kilojoules of energy from every gram of fat, and 17 kilojoules of energy from every gram of protein. Nutritionists recommend that we obtain 55–60 per cent of our energy from foods that are high in carbohydrates, such as cereals, grains, breads, fruits and vegetables. Only 30 per cent of our energy should come from fats, and 10 per cent from protein.

Detailed information on carbohydrates can be found in Chapter 9: 'Grain foods' and a discussion of fats can be found in Chapter 14: 'Only sometimes!'.

Energy balance

To maintain a healthy weight, it is necessary to achieve an energy balance, so that the energy we expend is equal to the energy we take in from food. However, sometimes our bodies require energy in addition to that supplied by the food we have just eaten. In these circumstances, our bodies need to call on their reserves, which are stored as adipose or fat tissue. If more energy is expended than is taken in, even over a fairly short period of time, weight loss may be the result. Alternatively, if we consume more food than our bodies are able to use in the form of energy, excess energy will be stored as fat tissue, leading to weight gain. See Figure 8.8 on the following page.

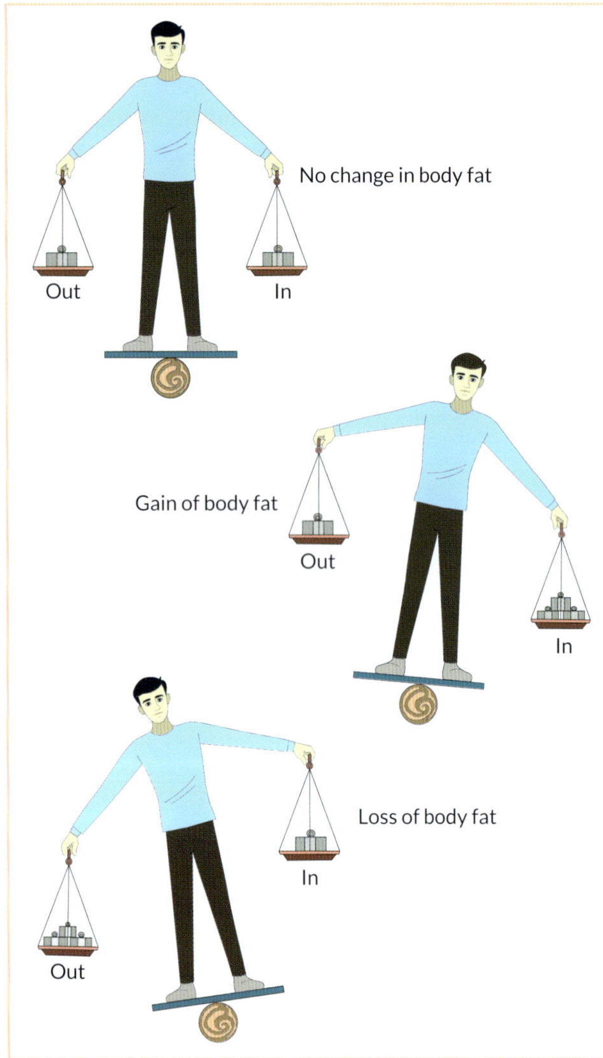

Figure 8.8 Energy balance

Although it would seem logical that two people who eat the same meals for breakfast, lunch and dinner would produce the same amounts of energy, this is in fact most unlikely. Each person's energy requirement is unique to them. However, there are several factors that influence the amount of energy each person requires.

- The basal metabolic rate (BMR), or the amount of energy the body requires to function, varies according to a person's build. The greater the person's size, and the more muscle than fat they have, the higher their BMR will be.
- Gender is a factor in determining energy needs – males have a higher BMR than females.
- Generally, the younger a person is, the more energy they require; in particular, children and adolescents who are rapidly growing have a higher BMR.
- The following people also have a higher BMR:
  - pregnant women
  - people involved in heavy manual labour
  - athletes with intense training schedules.

Testing knowledge  8.2

11 What percentage of Australian children are living with overweight or obesity?

12 Why is being overweight or obese considered to be a major health concern?

13 Outline two factors that have led to an increase in the incidence of overweight and obesity in childhood.

14 Identify the two main nutrients used in energy production.

15 What is glycogen, and where is it stored in the body?

16 Explain what happens to fat during digestion and how fat is stored in the body for future use.

17 How many kilojoules of energy are provided by one gram of fat and by one gram of carbohydrate?

18 Explain the change in energy balance that can lead you to lose weight and to put on weight.

19 What is the meaning of the term 'basal metabolic rate'?

20 List four factors that influence the amount of energy an individual requires.

Cardiovascular disease

Cardiovascular disease is one of the major causes of death in Australia. Cardiovascular disease is a general term that is used to describe a range of diseases, including heart disease, stroke and blood vessel disease. Approximately one in six – or 4.5 million – Australians suffer from some form of cardiovascular disease. According to the Heart Foundation, this equates to 10 per cent of the total population; and, of those suffering from cardiovascular disease, one person will die every 12 minutes; that is, 120 people each day. Men are far more at risk of dying from cardiovascular disease than women.

One of the main factors contributing to cardiovascular disease is being overweight or obese. An excess of body fat can impact on the cardiovascular system and contribute to coronary heart disease, high blood pressure and stroke. It is therefore clear that, to minimise the risk of developing cardiovascular disease, it is important to consume a healthy diet, exercise regularly and maintain a healthy weight.

9780170495714

Type 2 diabetes

Diabetes is a chronic health condition, and is the fastest-growing health condition affecting the Australian population. Diabetes occurs when the pancreas is unable to produce sufficient insulin to enable the glucose produced during digestion to be absorbed into the bloodstream.

There are two types of diabetes. Type 1 diabetes affects about 10–15 per cent of all people who suffer from diabetes. It is not a lifestyle disease, but occurs when the immune system damages the pancreas, meaning that the pancreas is unable to produce the hormone insulin. People who suffer from type 1 diabetes require a daily injection of insulin to break down the glucose in their bloodstream.

Type 2 diabetes is a far more common condition, accounting for 85–90 per cent of all diabetes cases. This type of diabetes is caused when the pancreas does not produce sufficient insulin to enable glucose to be absorbed into the bloodstream. Type 2 diabetes is considered to be a lifestyle disease, as people who live with this condition are often overweight or obese, are physically inactive, or have smoked. Research has shown that having another family member who has type 2 diabetes can increase the likelihood of developing this condition. Australia's First Nations peoples are much more likely to suffer from diabetes than other members of the Australian population.

The most recent data from the Australian Institute of Health and Welfare shows that in 2021, approximately 1.2 million Australians had developed type 2 diabetes. This is triple the number from the year 2000, when 400 000 Australians were diagnosed with this condition. This rise in the incidence of diabetes is mainly a result of the increasing number of people in the community who are overweight or obese. Another major concern is that type 2 diabetes is now beginning to be diagnosed in young people, rather than being confined to older adults, as was the case in the past.

Type 2 diabetes is often referred to as the 'silent killer', because some people who have it do not show any symptoms, leading to significant underreporting of the disease. It has been suggested that the real number of people suffering from type 2 diabetes is likely to be double the number of people who have been diagnosed with it.

Type 2 diabetes is a significant lifestyle disease as it can lead to cardiovascular disease, severe kidney damage and eye disease, and can even require the amputation of toes and limbs.

The most effective way for people with type 2 diabetes to control their condition is by consuming a healthy diet and participating in regular physical activity.

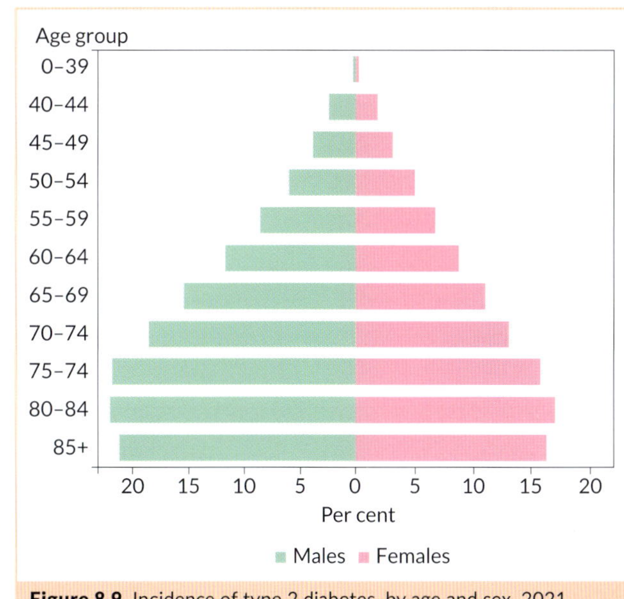

Figure 8.9 Incidence of type 2 diabetes, by age and sex, 2021

The glycaemic index

The **glycaemic index (GI)** was developed to help people with diabetes to better manage their blood sugar levels. The GI ranks carbohydrate foods based on the immediate effects they have on blood sugar levels. Carbohydrate foods that release energy into the bloodstream over a prolonged period have a low-GI rating. Carbohydrate foods that break down quickly during digestion – and therefore give an almost instant energy boost – have a high-GI rating. All foods are ranked from 0 to 100:

- 55 or less = low GI
- 56–69 inclusive = moderate GI
- 70 or more = high GI.

Today, the GI is utilised by many people within the community, including those who wish to lose weight or manage heart disease. Professional athletes also use the GI to enable them to develop their glycogen stores more effectively before competition, and also to recover quickly after an event.

Wischnewski, Jan/Foodcollection/Getty Images

Figure 8.10 Muesli made of oats and bran served with fruit is a low-GI breakfast.

8

Table 8.1 GI rating of foods

Food	GI rating
Prunes	29
Dried apricots	31
Fettuccini	32
Yoghurt (low-fat)	33
Chickpeas	33
Mixed-grain bread	34
Spaghetti	37
Apples	38
Sustagen	40
All-Bran	42
Porridge	42
Oranges	44
Baked beans	48
Peas	48
Carrots	49
Bananas	55
Basmati rice	58
White bread	70
French fries	75
Corn Flakes	84
Potato (baked)	85

Figure 8.11 Bürgen 85% Lower Carb bread has a low GI.

One of the most important benefits of a low-GI diet is that it makes us feel full for longer. This means that a diet based on low-GI foods will reduce the likelihood that we will become hungry between meals, meaning we will be less likely to feel tempted to indulge in snack foods.

Changing to a low-GI diet is not difficult – it simply involves swapping carbohydrate foods with a high GI for those with a low GI; for example, selecting wholegrain or sourdough bread instead of white bread; selecting a cereal such as muesli that is made of oats and bran and serving it with fruit for breakfast instead of a processed breakfast cereal; and selecting pasta instead of potatoes.

Activity 8.1

The glycaemic index

Weblinks
Better Health Channel
Diabetes Australia
Taste

Access the Victorian Government's Better Health Channel website. Search the site using the key words 'glycaemic index', and use the information you find to answer the following questions.

1 Explain why the GI is seen to be an important tool in selecting food, particularly for people who have been diagnosed with type 2 diabetes.
2 List three factors that influence the GI ratings of food.
3 What is the meaning of the GI symbol that is found on some foods?
4 Using the information on the GI ratings of foods in Table 8.1 or on the Diabetes Australia website, design a two-course meal that has a low GI rating. The Taste website may also be helpful.

Testing knowledge 8.3

Quiz
Testing knowledge 8.3

21 Define the term 'cardiovascular disease'.
22 Explain why cardiovascular disease is a major health concern in Australia.
23 Clarify how diabetes occurs and state how many people in the community are affected by this condition.
24 Copy and complete the diagram on the following page of the main risk factors for developing type 2 diabetes.

9780170495714

25 Explain why type 2 diabetes is often described as the 'silent killer'.

26 Describe the impact that type 2 diabetes can have on the health of individuals if it is not effectively managed.

27 What is the glycaemic index and why was it originally developed?

28 Why are some foods classified as having a low GI and others as having a high GI?

29 List three foods that have a low GI and three foods that have a high GI.

30 Outline the main benefits of a low-GI diet.

Strategies to enhance good health

One of the keys to good health is to maintain a healthy weight range throughout life. As mentioned earlier in this chapter, the best way to do this is to consume a diet that follows the recommendations of the *Australian Dietary Guidelines* and the *Australian Guide to Healthy Eating*. However, given the ready availability of a wide variety of ultra-processed foods that are high in fat and sugar, the extra-large portion sizes commonly presented to us by food manufacturers and food retailers, and the impact of food marketing, this may be easier said than done.

In addition to consuming a healthy diet, two of the most important strategies in maintaining a healthy weight are to make sure we keep physically active and minimise our intake of energy-dense foods (rather than nutrient-dense foods), such as snack foods. Reducing portion sizes is also important. These will mean that our energy intake and energy output are more likely to be in balance, and weight gain will be minimised.

Roy Morsch/The Image Bank/Getty Images

Figure 8.12 Staying physically active is important for our mental and physical health.

Stay physically active

Regardless of how old we are, it is important to stay active to optimise our physical and mental health. The Australian Department of Health and Aged Care has developed a set of physical activity guidelines to encourage all Australians to stay well.

AGE 5–17	AGE 18–64
Undertake at least 1 hour of moderate-to-vigorous aerobic activity each day Participate in vigorous activities at least 3 days per week	Be active on most days of the week Each week, aim for 2.5–5 hours of moderate activity or 1.25–2.5 hours of vigorous activity, or a combination

Figure 8.13 The physical activity guidelines

When combined with a healthy diet, regular exercise such as walking each day enables us to maintain a healthy weight and to minimise other lifestyle diseases, such as cardiovascular disease and type 2 diabetes. There are many types of wearable technology such as smart watches, as well as iPhone apps that allow individuals to track their exercise regime in real time.

The Heart Foundation Walking website encourages all Australians who want to improve their physical and heart health to join a local walking group. Over 300 000 people from across Australia have joined the Heart Foundation Walking community and linked up with groups of like-minded walkers in their local area.

Weblink Heart Foundation Walking

REGULAR WALKING CAN HELP ...

reduce the risk of heart disease and stroke

manage weight, blood pressure and cholesterol

prevent and control diabetes

reduce the risk of developing some cancers

maintain bone density

improve balance and coordination

improve mood and mental health

Figure 8.14 The benefits of walking

Consume a nutrient-dense diet

In order to maintain a healthy weight, it is important to be mindful that we must not only participate in regular exercise, but also consume a nutrient-dense rather than an energy-dense diet. While we all occasionally enjoy indulging in a delicious treat – such as a chocolate bar, a packet of crisps, a doughnut or muffin – eating them comes at a price. These snacks are energy-dense, and consuming them on a regular basis can lead to weight gain. For example, a 60-kilogram person would take nearly 75 minutes to walk off the 1430 kilojoules of energy provided by a chocolate doughnut decorated with M&M's.

Rawf8/Alamy Stock Photo

Figure 8.15 It would take almost 75 minutes to walk off the kilojoules in this chocolate doughnut.

Reduce portion sizes

Another reason why people's waistlines have expanded in recent years is because we are now eating far more than we used to – and far more than we really need to. Food portions have dramatically increased in the past few decades, and, as the amount of food we eat has risen, so too has the number of kilojoules we consume. In the 1980s, individual serves of soft drink were sold in 237-millilitre containers, whereas today, their usual size is between 500 and 600 millilitres. Similarly, cupcakes and scones have doubled in average size over the same period from 40 grams to 80 grams.

King-sized or 'Texas' muffins have become the norm. Cakes, slices and biscuits sold in bakeries, cafes and supermarkets are now so big that they are in fact large enough for two or three people to share. Similarly, a giant serve of popcorn at the movies is enough for the whole family. Many other treats, such as chocolate bars, are now served in king-sized packets or twin packs. 'Super-sized' meals available in fast-food outlets also provide far more food – and therefore more kilojoules – than we really need. This is sometimes called 'portion distortion'.

How has this happened?

- The food industry has been eager to increase serving sizes, which it dubs 'upsizing'. Food manufacturers know that customers like to feel that they are getting value for money, and so will be happier to pay a little more for a larger serve, than pay what seems to be a high price for a smaller portion.
- Another strategy manufacturers use is to bundle together food items to make a 'combo pack' or 'meal deal'. In such cases, the manufacturers win, because the profit margin they make from the additional items – such as drinks or fries – is usually high, so the actual cost to them is almost negligible.

An additional problem is that in Australia, serving sizes listed on packages of food are often inconsistent, and the serving size is determined by the food manufacturers rather than by regulation.

Evidence clearly suggests that making even a small reduction in the amount of food we eat can make a big difference in managing our weight. So, one of the best strategies is to resist king-sized treats and meal deals, and to only order small portions when eating out. Alternatively, share a sweet treat such as a muffin or cake with a friend. Eating more slowly will also enable your brain to register when you have had enough. Finally, only eat enough to satisfy your hunger – you can always leave some food on your plate!

9780170495714

Figure 8.16 Reducing portion sizes when cooking at home or eating out will help individuals manage their weight.

Testing knowledge 8.4

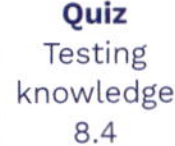
Quiz Testing knowledge 8.4

31 Outline two key strategies individuals can use to help them maintain a healthy weight.

32 What is the key message from the Australian Department of Health and Aged Care about physical activity, and why is this message considered to be so important?

33 Select two of the benefits of walking identified by the Heart Foundation and explain how they will improve the health of Australians over their lifetime.

34 Write a simple slogan that could be included on the Heart Foundation Walking app to promote two benefits of walking.

35 Explain why it is better to eat a piece of fruit or a few nuts after exercising rather than snacking on a chocolate bar.

36 Clarify how the portion sizes of food sold to consumers have changed in recent decades.

37 Explain the benefit to food manufacturers of 'upsizing'.

38 Outline another strategy other than upsizing that manufacturers have used to increase the sale of their products.

39 Explain why the serving sizes listed on a food label can cause confusion for Australian consumers.

40 List one strategy other than those suggested on page 202 that could assist individuals to reduce their portion size.

Food allergies and intolerances

Many people need to manage their diets to avoid particular foods to which they are allergic, such as peanuts, eggs or shellfish. Other people may have a hypersensitivity or intolerance to a food – for example, wheat or certain fruits – and so they must also be careful about the foods they select to eat.

Food allergies

A **food allergy** is an abnormal immunological reaction to food. A foreign substance, usually a protein, enters the bloodstream, and an antibody is produced to fight it. Each time the foreign substance enters the body, more antibodies are produced.

The reaction caused by a food allergy is usually physical and occurs within an hour of exposure to the food. Symptoms such as hives, rashes, hay fever, asthma, stomach pain or diarrhoea, headache or swelling of the face or eyelids may occur. In some cases, the physical symptoms of a food allergy can become more severe with each exposure, and can even be life-threatening; this is known as anaphylaxis. There is no cure for a food allergy, so the treatment is simply to avoid the problem food.

It is recognised that the risk of developing an allergy is much higher if another member of your family has an allergy. Foods that can cause allergic reactions in some people include milk, fish, shellfish, peanuts, eggs and legumes. Many children grow out of food allergies. However, peanuts are an exception – this allergy is often severe and lifelong.

Figure 8.17 Some foods that can cause allergic reactions: nuts, wheat flour, eggs, fish, shellfish and legumes

People with food allergies must read and understand food labels to ensure that they do not eat foods that are toxic to their bodies. Food Standards Australia New Zealand (FSANZ) aims to assist people with a food allergy by requiring food manufacturers to include information on labels if the food contains an ingredient that may cause a severe allergic reaction such as anaphylaxis, regardless of how small the amount of the ingredient is.

8

People involved in food preparation, either in the home or in the hospitality industry, should also take particular care when preparing and serving food to people with allergies to ensure that their food is not contaminated.

Food intolerances

Some people are born with a condition that makes it impossible for them to metabolise a particular food or nutrient. A food intolerance is not an immune response like a food allergy, but a chemical reaction to particular foods. People who are affected may lack a particular enzyme, or be unable to produce an enzyme in sufficient amounts to digest certain foods.

Food intolerance reactions are similar to food allergy symptoms, but are generally less severe. The reactions are delayed for 24–48 hours after exposure to the food, and the severity of the symptoms usually decreases the more the food is avoided. Foods to which some people may be intolerant include chocolate, wheat, cola, eggs, garlic, cucumbers and certain fruits; for example, oranges, strawberries, pineapples and tomatoes.

Food intolerances that are relatively common in our society include lactose intolerance, gluten intolerance, and intolerance to fermentable oligosaccharides, disaccharides, monosaccharides and polyols (FODMAPs).

Lactose intolerance

Lactose is the sugar (carbohydrate) found in milk and milk products. People who are lactose intolerant lack the enzyme lactase in their system, or have it in insufficient amounts to break down lactose in the small intestine. Instead, the lactose moves along the large intestine where the bacteria within the colon cause the lactose to ferment and produce carbon dioxide (CO_2). This may cause the bowel to retain water and produce symptoms of excessive wind, diarrhoea, bloating and abdominal pain.

The most effective way to manage lactose intolerance is to reduce the amount of lactose in the diet, rather than eliminating dairy foods entirely.

Figure 8.18 Food manufacturers produce a wide range of lactose-free dairy products.

Dairy foods are an excellent source of calcium and essential for the development of healthy bones. Today, food manufacturers have developed milk and a range of other dairy products with a reduced level of lactose for people who suffer from lactose intolerance.

Gluten intolerance

Gluten intolerance is a condition where the gluten proteins – which are present in wheat, oats, barley, triticale and rye – cause a person to suffer from significant intestinal discomfort after eating food that contains gluten.

The symptoms associated with gluten intolerance can vary widely, but include diarrhoea, constipation, cramping and flatulence. Some people may also feel bloated as a result of gas being produced as the body tries to break down the food. Others may feel extremely tired and weak, and have difficulty concentrating on tasks. Other symptoms include joint and muscle pains throughout the body. The onset of symptoms may not be immediate, and it may be many hours or days from the time the person initially consumed gluten to the onset of symptoms. Gluten intolerance does not cause damage to the small intestine.

The most effective way for people to manage a gluten intolerance is to follow a gluten-free diet. However, the total exclusion of gluten in not always necessary, as many people are able to tolerate a small amount of gluten if they consume it inadvertently. For some people, the inclusion of a probiotic in the diet can increase the proportion of healthy gut microbiota and reduce the symptoms of gluten intolerance.

As gluten occurs in a range of cereals, it is important to read food labels carefully. These cereals are widely used in food processing; therefore gluten is found in a range of commonly eaten foods, such as breads, pasta, pizza, cereals, cakes, biscuits, pies and soy sauce. It is also a common thickener in processed foods, making it difficult to avoid.

Although gluten intolerance and coeliac disease may have similar symptoms, they are very different conditions. **Coeliac disease** is an autoimmune disease in which the immune system reacts to the gluten in food and causes significant damage to the small intestine. It is not a food allergy or an intolerance to food. Coeliac disease is a lifelong health condition.

Cereal foods suitable for people with gluten intolerance

The following foods are made from cereals that do not contain gluten:

- rice and rice products, including rice pasta, rice cakes, rice crackers, puffed rice, baby rice cereal, rice noodles and rice bran
- buckwheat and sorghum
- maize (corn) and related products, including polenta, maize cornflour, pure corn chips, taco shells and popcorn

9780170495714

- sago, tapioca and arrowroot
- lentil flour and chickpea flour
- soy and soy products, including soy bran and potato flour
- gluten-free mixes for bread, pastry, pizza, cakes and biscuits.

FODMAP intolerance

FODMAP is an acronym that stands for fermentable oligosaccharides, disaccharides, monosaccharides and polyols. It describes a group of short-chain carbohydrates – including lactose, fructose, fructans and galactans – that may be poorly absorbed in the small intestine of some people.

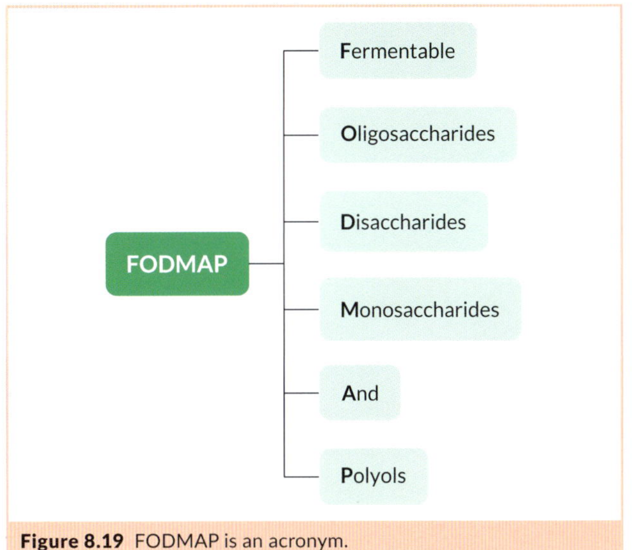

Figure 8.19 FODMAP is an acronym.

As FODMAP foods move slowly through the small intestine, they draw in water. Once they reach the large intestine, the gut bacteria uses short-chain carbohydrates (oligosaccharides, disaccharides, monosaccharides and polyols) as their source of food, and rapidly ferment them, producing gases. This causes the wall of the intestine to expand or become bloated, leading to cramping and severe pain. Other symptoms of FODMAP intolerance include flatulence, constipation and diarrhoea. FODMAP intolerance often occurs in people who suffer from irritable bowel syndrome (IBS).

Managing FODMAP intolerance involves replacing foods high in FODMAPs with those that have a lower FODMAP content. Experts say that a FODMAP diet is not a diet to be followed throughout life. Instead, it should initially be followed quite strictly for two to six weeks to allow the symptoms to subside. After that, individuals should start reintroducing some high FODMAP foods into their diet to learn the level at which they can tolerate these foods.

Weblink
FODMAP resources

Meat-free diets

Some people follow a meat-free diet. This is usually because their ethical or religious beliefs mean they are opposed to the killing of animals for food, or because of their views on health or the environment. Research has shown that following a plant-based diet can benefit health by reducing the risk of becoming overweight or obese, thereby also reducing the risk of developing lifestyle health issues, such as heart disease and type 2 diabetes.

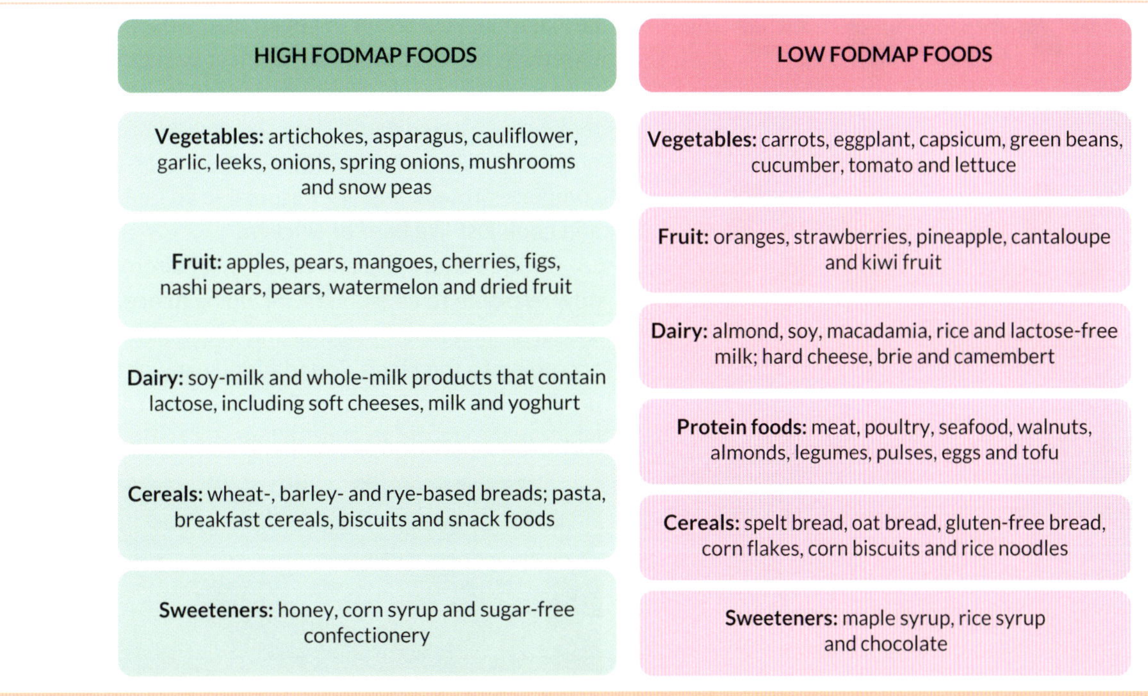

HIGH FODMAP FOODS	LOW FODMAP FOODS
Vegetables: artichokes, asparagus, cauliflower, garlic, leeks, onions, spring onions, mushrooms and snow peas	**Vegetables:** carrots, eggplant, capsicum, green beans, cucumber, tomato and lettuce
Fruit: apples, pears, mangoes, cherries, figs, nashi pears, pears, watermelon and dried fruit	**Fruit:** oranges, strawberries, pineapple, cantaloupe and kiwi fruit
Dairy: soy-milk and whole-milk products that contain lactose, including soft cheeses, milk and yoghurt	**Dairy:** almond, soy, macadamia, rice and lactose-free milk; hard cheese, brie and camembert
Cereals: wheat-, barley- and rye-based breads; pasta, breakfast cereals, biscuits and snack foods	**Protein foods:** meat, poultry, seafood, walnuts, almonds, legumes, pulses, eggs and tofu
Sweeteners: honey, corn syrup and sugar-free confectionery	**Cereals:** spelt bread, oat bread, gluten-free bread, corn flakes, corn biscuits and rice noodles
	Sweeteners: maple syrup, rice syrup and chocolate

Figure 8.20 High FODMAP foods and low FODMAP foods

People who follow meat-free diets include the following:

- A pescatarian follows a mainly plant-based diet and does not eat meat or poultry, but will consume some fish and seafood.
- A lacto-ovo vegetarian does not eat meat or fish, but will eat dairy products and eggs.
- A vegan does not eat meat or any other product that comes from animals; generally, a vegan does not use any product that comes from an animal.

Complementing proteins

Meat, poultry and fish provide our body with a range of essential nutrients, including complete protein, iron and omega-3 fatty acids. Therefore, it is essential that a person who follows a meat-free diet is able to obtain an adequate supply of complete protein. Soybeans and soybean products – such as soy milk and tofu – are good sources of complete protein. Quinoa also contains high levels of complete protein.

However, while some plant foods contain protein, they are generally lower in one or more of the nine essential amino acids that make up complete protein. To overcome this problem, it is important to complement proteins from plant sources to make up a complete protein. This is done by combining foods from cereals such as wheat, rice and pasta with pulses such as dried beans (soy, haricot or cannellini beans), lentils, chickpeas or nuts in the one meal.

Figure 8.21 Complementing proteins

Because people following a pescatarian, vegetarian or vegan diet do not eat meat, they may also have trouble gaining adequate amounts of iron in their diets. They need to make sure that they eat lots of green leafy vegetables, wholegrain cereals, dried fruit and legumes, which all contain non-haem iron. They should also eat foods high in vitamin C at the same time, because this vitamin helps the body to absorb iron from plant sources.

Figure 8.22 Some of the best non-meat sources of iron (per 100 grams)

Top row, left to right: Galayko Sergey/Shutterstock.com, iStock.com/Floortje; middle row, left to right: iStock.com/spafra, Happy Stock Photo/Shutterstock.com bottom row, left to right: New Africa/Shutterstock.com, bigacis/Shutterstock.com

Plant-based meat alternatives

Food manufacturers are now producing meat-free products to appeal to those consumers who follow a pescatarian, vegetarian or vegan diet, who are looking to reduce their consumption of meat based on the latest health advice, or who are trying to reduce the impact that a meat-based diet has on the environment. Many of these products are produced from plants, are high in complete protein, and are formulated to look, smell, taste and cook just like beef or chicken.

A wide variety of plant-based meat alternatives are now available to consumers, including mince, burgers, sausages, meatballs, nuggets and fillets. However, according to the Heart Foundation, many of these products are high in salt, so it is important to read the labels carefully and to choose low-salt options.

Figure 8.23 Plant-based meat alternatives can come in the form of burger patties.

iStock.com/MTStock Studio

9780170495714

PRACTICAL ACTIVITY 8.1

Worksheet Practical activity 8.1

Template Sensory wheel

Comparative food test: Comparing beef burger patties and plant-based burger patties

Aim

To compare the physical and sensory characteristics, and nutritional properties of beef burger patties and plant-based burger patties

Equipment

- commercially prepared fresh beef burger patties, such as Coles Finest
- commercially prepared plant-based burger patties, such as v2 plant-based burger patties
- frying pan

Method

1 Cook the fresh beef burger patties and the plant-based burger patties according to the manufacturers' instructions.

2 Read the label on each product, and then complete a comparison of the nutritional, physical and sensory properties of both types of burger patties. Use a table similar to the one below. Refer to the sensory wheel on page 4 for words to assist you.

Results

Physical and sensory properties	Fresh beef burger patties	Plant-based burger patties
Nutritional properties per 100 grams		
Energy (kilojoules)		
Total fat (grams)		
Sugar (grams)		
Sodium (milligrams)		
Protein (grams)		
Dietary fibre (grams)		
Quantitative measures (physical properties)		
Weight of one burger patty (grams)		
Height and length of one burger patty (centimetres)		
Colour of cooked product		
Qualitative measures (sensory properties)		
Appearance		
Aroma		
Flavour		
Texture		
Overall appeal 5 = like a lot 1 = dislike a lot		

Analysis

1 Describe the similarities and differences in the weight, height, length and colour of the fresh beef burger patties and the plant-based burger patties.

2 Identify the variety of burger patty that had the most appealing appearance when cooked.

3 Which variety of burger patty had the most appealing aroma and flavour?

4 Identify the variety of burger patty that had the most appealing texture or mouth feel.

5 Using the information in the nutrition panels of the two products, identify the product with the highest overall:
   a energy value; that is, fat and sugar
   b sodium content
   c protein content
   d dietary fibre content.
6 Which product would be preferable to include in a healthy diet? Why?

Conclusion

After analysing your results, which product would you prefer to use to make a burger? Justify your decision with reference to the physical and sensory characteristics and nutritional properties of each burger patty.

Activity 8.2

Selecting vegetarian meals

Access the Sanitarium website and search for information about vegetarian diets.

1 Explain three health benefits of following a vegetarian diet.
2 Record three food and nutrition tips for people who follow a vegetarian diet. Ensure that these tips are different from those listed in this chapter.
3 Examine two recipes on the Sanitarium website, and identify the ingredients in each of them that would provide a source of complete protein.
4 Using the information from the website, list four ways you could increase your consumption of wholegrain cereals.
5 Create a one-day eating plan for a vegetarian that includes breakfast, lunch, dinner and snacks based on the information you have discovered on the website.

Weblink
Sanitarium

Testing knowledge 8.5

41 Write a sentence to explain the meaning of the term 'food allergy'. List the foods most likely to cause food allergies.
42 Identify some of the physical symptoms that may occur if a person has a food allergy.
43 Explain how a food intolerance differs from a food allergy.
44 List some of the main foods that may cause a food intolerance.
45 What is a gluten intolerance and how does it affect the human body?
46 Copy and complete the diagram opposite to highlight some cereal foods that are suitable for people living with gluten intolerance.
47 What is FODMAP intolerance and how does it impact on the digestive system of a person with this type of intolerance?
48 Create a mind map to highlight the key features of a pescatarian diet, a lacto-ovo-vegetarian diet and a vegan diet.
49 Discuss how vegetarians can make sure they obtain an adequate supply of iron in their diet.
50 Why is it important to read the product label carefully before purchasing ultra-processed foods, such as a plant-based meat alternative?

Quiz
Testing knowledge 8.5

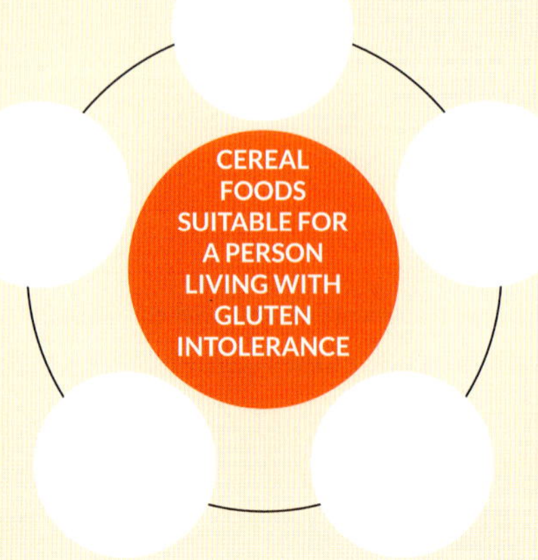

9780170495714

Critical and creative thinking 8.1

Debate

'The youth of today need to take greater responsibility for their health and future wellbeing.'

Work in small teams to develop arguments for the affirmative and negative in response to this statement.

DESIGN ACTIVITY 8.1

Risotto

A health magazine, *The A–Z of Gut Health*, is looking for new and exciting recipes for risotto that will assist readers to optimise their gut health.

Design brief

1 Write your own design brief for a risotto dish based on a classic risotto recipe. Develop your own specifications based on the '5 Ws':
- Who – for whom the risotto will be suitable
- What – a risotto that supports gut health by including a variety of different vegetables
- When – in what season the risotto will be served
- Why – why the risotto will be useful to include in the magazine
- Where – where the risotto will be served.

2 Format sentences or statements based on each of the above specifications into a paragraph that will become your design brief.

Investigating and defining

1 Based on the solution requirements and constraints in your design brief, develop four or five design criteria by which you can judge the success of your designed solution.

2 Prepare a recipe for risotto such as Mushroom and Pea Risotto on page 212 to develop

Worksheet Design activity 8.1

Templates Recipe map

Decision table

Food order

Production plan

Australian Guide to Healthy Eating

Figure 8.24 A recipe map for risotto

an understanding of the ingredients and processes involved in preparing a risotto.

3 Access the Taste website and answer the following questions.
   a List two important tips for making a perfect risotto.
   b Develop a list of popular vegetable, flavouring and complementary ingredients suitable to use in a risotto.

Generating and designing

1 Develop two designed solution options based on the solution requirements and constraints in your design brief. Use the recipe map on the previous page as a guide. Select the vegetable, flavouring and complementary ingredients to develop your own designed solution options.

2 Use a decision table (see the example on page 12) to help you select your designed solution.

Planning and managing

1 Prepare a food order.

2 Before producing your risotto, write up a production plan, noting any safe work practices to be followed, and identifying the major processes to be used.

Producing and implementing

1 Prepare the product.

2 Note any modifications or changes you made during production.

Evaluating

1 Evaluate the success of your risotto using the previously established design criteria.

2 Was the flavour of the product appetising? In your opinion, does the recipe require any further modification to enhance the product's flavour? Would you add or omit some ingredients if you were to make this product again?

3 What aspect of the production did you find most challenging? Outline how you managed this challenge.

4 Comment on your overall management of time for this task – discuss your designing and planning, as well as the production of the risotto.

5 Classify the ingredients of your risotto on a diagram of the *Australian Guide to Healthy Eating*. Comment on how well it meets Guideline 2 and to what extent it will support gut health.

DESIGN ACTIVITY 8.2

DinnerMagic

DinnerMagic is an online food ordering company that delivers boxed meal kits to the consumer's doorstep. Each food box contains a recipe card and all the fresh ingredients to prepare the chosen meal. The only ingredients the consumer may need to provide are pantry staples, such as olive oil, eggs, or salt and pepper.

Design brief

You have been asked to design a healthy main course meal that can be included in the new DinnerMagic range of spring recipes. The meal must support the guidelines of the *Australian Guide to Healthy Eating*, use seasonal ingredients and be suitable for a weeknight dinner. All DinnerMagic recipes must be able to be prepared in 30 minutes and serve two people.

The recipe card must include:

- a brief description of the meal
- a photograph of the finished meal
- a list of all of the pre-measured ingredients provided in the meal kit
- a list of any additional pantry items required
- a list of all the utensils needed to prepare the recipe
- a method with no more than six easy-to-follow steps.

Worksheet Design activity 8.2

Templates Decision table

Food order

Production plan

Sensory wheel

Australian Guide to Healthy Eating

Weblink Taste

Investigating and defining

1 Use the solution requirements and constraints in the design brief to develop five design criteria that will allow you to judge the success of your meal kit.

2 Undertake online research of the types of meals suitable to include in a boxed meal kit.

3 With other members of your class, create a short survey to give to teachers at your school to determine the popularity of home-delivered boxed meal kits. Focus your questions on:

9780170495714

- the number of days each week they use home-delivered boxed meal kits
- preferred recipes.

4 Based on the results of the survey, develop a list of the most popular boxed meal kit recipes.

Generating and designing

1 Use a range of recipe books, magazines or a website such as Taste to develop a list of four recipes or designed solution options that could be included in DinnerMagic's new spring range.

2 Complete a decision table similar to the one below.

Decision table: DinnerMagic home-delivered boxed meal kit		
Recipe name	Advantages	Disadvantages
Designed solution:		
Justification (use the information in your advantages and disadvantages in your discussion)		

Planning and managing

1 Prepare a food order.

2 Before producing your meal kit recipe, write up a production plan, noting any safe work practices to be followed, and identifying the major processes to be used.

Producing and implementing

1 Prepare the recipe for the meal kit you have designed.

2 Photograph the completed dish to add to the recipe card.

3 Note any modifications or changes you made during the production of the recipe.

4 Prepare the recipe card, including the photograph of the finished dish.

Evaluating

1 Evaluate the success of your meal kit using the previously established design criteria.

2 Share your completed meal kit with a friend and ask them to write a review of it that could be posted on the company website.

3 Describe the sensory properties – appearance, aroma, flavour, texture and sound – of your meal kit.

4 Give the recipe card to a friend. Ask them to read through the recipe, and, if possible, cook the dish at home. Ask them to comment on whether they find the recipe easy to follow and a helpful guide to the preparation and cooking of the dish. Considering this feedback from your friend, what improvements could you make to the recipe card?

5 Classify the ingredients of your meal kit on a diagram of the *Australian Guide to Healthy Eating*. Comment on how well it meets Guideline 2.

iStock.com/alvarez

Figure 8.25 A delicious mid-week family meal

Mushroom and Pea Risotto

INGREDIENTS

15 grams butter

2 spring onions, diced

¼ red capsicum, diced

⅓ cup arborio rice

1½ cups vegetable stock

4 mushrooms, diced

⅓ cup peas

black pepper

2 tablespoons parmesan cheese, grated

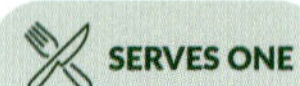

METHOD

1 Sauté the spring onion and capsicum in butter until soft and lightly coloured.
2 Add the rice and cook for a further 1 minute or until the rice becomes opaque.
3 Bring the vegetable stock to the boil in a saucepan, then reduce to a simmer. Stir half a cup of the hot stock into the rice.
4 Cover with a tight-fitting lid and simmer very gently until the rice has absorbed the stock.
5 Stir in a further half cup of hot stock and continue to simmer very gently until the rice has absorbed the stock.
6 Add the remaining hot stock and stir gently to loosen the rice from the bottom of the saucepan. Add the diced mushrooms and peas. Season with a little black pepper. Cover with the lid.
7 Continue to simmer very gently until all of the stock has been absorbed and the rice is plump and creamy. The risotto should take approximately 20–25 minutes to cook.
8 Add the parmesan cheese and serve immediately.

EVALUATION

1 Describe the sensory properties – appearance, aroma, flavour, texture and sound – of your Mushroom and Pea Risotto.
2 Identify and describe the changes that occur to the rice when it is simmered with the stock.
3 Explain why it is important to simmer the risotto very gently during the cooking process, rather than boil it quickly.
4 List the important health and safety steps to follow when preparing a risotto.
5 Classify the ingredients for the Mushroom and Pea Risotto on a diagram of the *Australian Guide to Healthy Eating*. Explain how well it meets the recommendations of this food selection model.

Mark Fergus Photography

9780170495714

Chicken Laksa

INGREDIENTS

1 tablespoon oil

2 tablespoons laksa paste

2 cups chicken stock

200 millilitres low-fat coconut milk

1 tablespoon fish sauce

1 makrut lime leaf

100g hokkien noodles

½ chicken fillet, finely sliced

4 baby corn, cut in half lengthwise

6 green beans, trimmed and chopped

¼ red capsicum, cut into strips

½ lime

½ cup bean shoots

2 tablespoons peanuts, chopped (optional)

1 tablespoon roughly chopped coriander leaves

SERVES TWO

METHOD

1 Heat the oil in a saucepan and add the laksa paste. Sauté for 2 minutes or until fragrant.
2 Add the chicken stock, coconut milk, fish sauce and makrut lime leaf and bring to the boil.
3 Add the noodles and simmer for 4 minutes, then add the sliced chicken and simmer for a further 3 minutes.
4 Add the corn, green beans and capsicum and simmer for a further 4 minutes.
5 Remove the makrut lime leaf and serve the laksa in deep bowls. Add a squeeze of lime juice to each bowl and top with bean shoots, chopped peanuts (if desired) and coriander.

EVALUATION

1 How does sautéing the laksa paste in oil in step 1 of the recipe add to the sensory properties of the finished dish?
2 Explain the difference between boiling and simmering.
3 Describe how the process of coagulation changes the form of the protein in chicken when it is simmered.
4 What are hokkien noodles? Explain how they add to the nutritional and sensory properties of the Chicken Laksa.
5 Classify the ingredients for the Chicken Laksa on a diagram of the *Australian Guide to Healthy Eating* and explain how well you think this recipe meets the guidelines of this food selection model.

Mark Fergus Photography

Gluten-free Patty Cakes

INGREDIENTS

30 grams butter

½ cup gluten-free self-raising flour

¼ cup caster sugar

1 egg

1 tablespoon milk

½ teaspoon vanilla extract

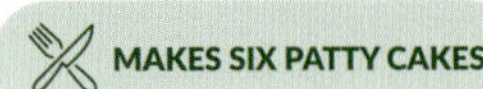
MAKES SIX PATTY CAKES

METHOD

1. Preheat oven to 200°C. Grease a patty cake tray (6 × ⅓ cup capacity) with butter or line it with paper cups.
2. Melt the butter without browning, either in the microwave for 30 seconds on high or in a saucepan over a gentle heat.
3. Sift all the dry ingredients into a bowl and add the egg, milk and vanilla extract.
4. Stir in the melted butter.
5. Mix well and spoon into the prepared patty cake tray.
6. Bake for 12 minutes. Do not overbake, as this can cause the patty cakes to become dry.

Variations

You can make the Gluten-free Patty Cakes into Gluten-free Butterfly Cakes for a perfect birthday party or afternoon tea treat. Alternately, you could decorate the Gluten-free Patty Cakes with piped swirls of butter icing.

Gluten-free Butterfly Cakes

1. Once the patty cakes are cooled, use a sharp knife to cut a circular cone shape from the centre of each cake. Slice the cone into half. These will become the 'wings' of the butterfly.
2. Place a small teaspoon of raspberry or strawberry jam in the hole in the centre of each cake.
3. Spoon a teaspoon of thickly whipped cream on the top of the jam. Gently press the cut wings into the cream to form the butterfly.
4. Dust with icing sugar to serve.

Mark Fergus Photography

EVALUATION

1. Describe the sensory properties – appearance, aroma, flavour, texture and sound – of your Gluten-free Patty Cakes.
2. Describe one safety rule you observed when using the oven to bake the patty cakes.
3. Identify and explain the process that causes the patty cakes to brown when baked in the oven.
4. How can you test if the patty cakes are cooked?
5. Identify the ingredients in this recipe that are classified in the 'only sometimes and in small amounts' section of the *Australian Guide to Healthy Eating*. Explain why these ingredients should not be included as a regular part of a daily diet.

9780170495714

Spicy Gochujang and Kimchi Noodles

INGREDIENTS

1 tablespoon sesame seeds

2 × 90-gram bundles dry udon noodles

2 teaspoons vegetable oil

⅓ red capsicum, finely sliced

½ cup kimchi, chopped

2 teaspoons gochujang paste (or more if you like it super spicy)

15 grams ginger, grated

1 clove garlic, crushed

2 teaspoons soy sauce

1 teaspoon honey

½ cup vegetable stock

2 spring onions, finely sliced

½ cup coriander, chopped

1 tablespoon roasted peanuts, finely chopped

salt and pepper

To serve

2 teaspoons extra coriander, chopped

1 egg, fried or boiled (if desired)

METHOD

1. Toast the sesame seeds in a small saucepan over medium heat for approximately 5 minutes or until lightly toasted. Remove the seeds from the saucepan and set aside to cool.
2. Bring a saucepan of water to the boil. Add the udon noodles and cook for 8 minutes. Drain, rinse well under cold water and drain again. Set aside.
3. Heat the oil in a large frying pan. Add the capsicum and stir-fry for 2 minutes. Add the kimchi and gochujang paste and stir-fry for a further 2 minutes.
4. Add the ginger and garlic and cook for 1 minute or until fragrant. Add the soy sauce, honey and stock, and simmer for a further 2–3 minutes or until the sauce has reduced a little and is slightly thick.
5. Add the cooked udon noodles, spring onions, coriander and peanuts. Season with salt and pepper and toss gently until well coated with the sauce.
6. Divide the noodles between two bowls and top with toasted sesame seeds and extra coriander, and a fried or boiled egg (if desired).

Mark Fergus Photography

EVALUATION

1. Describe the sensory properties – appearance, aroma, flavour, texture and sound – of your Spicy Gochujang and Kimchi Noodles.
2. Identify and describe the changes that occur to the chemical and physical properties of the udon noodles when they are cooked in step 2 of the recipe.
3. Udon noodles are made of wheat and therefore contain starch. Outline the digestive processes that occur in the mouth when you eat udon noodles.
4. Explain how eating foods such as kimchi can improve gut health.
5. Classify the ingredients for the Spicy Gochujang and Kimchi Noodles served with a fried egg on a diagram of the *Australian Guide to Healthy Eating*. Based on the recommendations of this food selection model, justify whether you would recommend this dish as a healthy meal for a family that included teenagers.

Koshari

Koshari is a traditional dish of Egypt. The tomato sauce in this recipe is flavoured with baharat, a Middle Eastern spice that has a smoky, paprika flavour. When served over a combination of rice, pasta and lentils, it makes a satisfying vegetarian meal.

INGREDIENTS

Baharat-spiced Tomato Sauce

2 teaspoons olive oil
½ onion, finely diced
1 garlic clove, crushed
1 teaspoon baharat
½ teaspoon smoked paprika
¼ teaspoon dried chilli flakes
¾ cup canned cherry tomatoes
½ cup canned chopped tomatoes
½ teaspoon caster sugar
⅛ teaspoon salt and pepper
2 teaspoons red wine vinegar

Rice, Pasta and Lentil Base

½ vegetable stock cube
½ cup hot water
½ cup canned brown lentils
⅓ cup basmati rice
⅔ cup small macaroni
salt and pepper

To serve

20 grams butter
½ brown onion, thinly sliced

SERVES TWO

METHOD

Baharat-spiced Tomato Sauce

1. Place the oil and onion in a medium saucepan and sauté over medium heat until tender. Add the garlic and cook for a further minute.
2. Stir in baharat, smoked paprika and chilli flakes and cook for 1 minute or until fragrant.
3. Add the cherry tomatoes, chopped tomatoes and sugar, and simmer over a low heat until the mixture has thickened slightly.
4. Stir in the vinegar and season with salt and pepper. Keep warm.

Rice, Pasta and Lentil Base

1. Combine the stock cube with the hot water.
2. Place the lentils in a sieve and wash thoroughly until the water runs clean. Drain well.
3. Rinse the rice under running water and place in a small saucepan with the vegetable stock, and bring to the boil. Immediately reduce the heat, cover with a lid and turn the heat to very low. Simmer for 10 minutes or until the rice has absorbed the stock. Do not uncover. Allow to stand for 10 minutes or until ready to serve.
4. Bring a large saucepan of water to the boil, add the macaroni and stir once or twice to prevent the macaroni from sticking to the bottom of the saucepan. Boil for approximately 8 minutes or until the macaroni is al dente. Drain well.
5. Return the macaroni to the saucepan and stir in the lentils and rice. Season with salt and pepper.

To serve

1. Heat the butter in a frying pan over medium-high heat, add the onion and fry for 8–10 minutes until golden brown and crisp.
2. Divide the Rice, Pasta and Lentil Base between two bowls and spoon the Baharat-spiced Tomato Sauce over the top. Scatter with the crisp onion.

9780170495714

EVALUATION

1 Describe the sensory properties – appearance, aroma, flavour, texture and sound – of your Koshari.
2 Explain why it is essential to the final sensory properties of the recipe to sauté the onion and garlic before adding the tomatoes when making the Baharat-spiced Tomato Sauce.
3 Identify and describe the changes that occur to the chemical and physical properties of the macaroni when it is boiled.
4 Explain the process that causes the onions to brown when fried and the impact this has on the sensory properties of the onion.
5 Classify the ingredients for the Koshari on a diagram of the *Australian Guide to Healthy Eating*. Write a paragraph to explain whether this recipe meets the recommendations of this food selection model and whether it would be suitable to serve as a vegetarian meal.

Mark Fergus Photography

EDIBLE GRAINS/SEEDS OF GRASSES

Corn

Barley

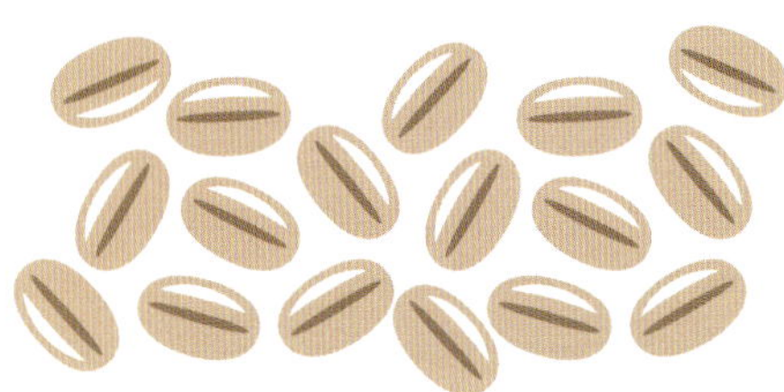

Oats

Rice

Wheat

Quinoa

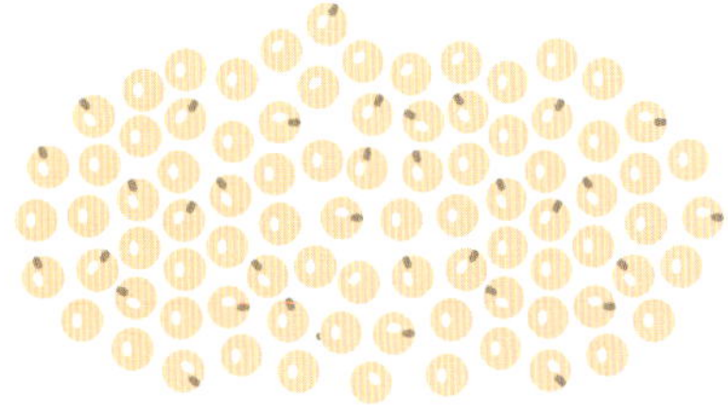

Rye

NUTRIENTS IN GRAINS

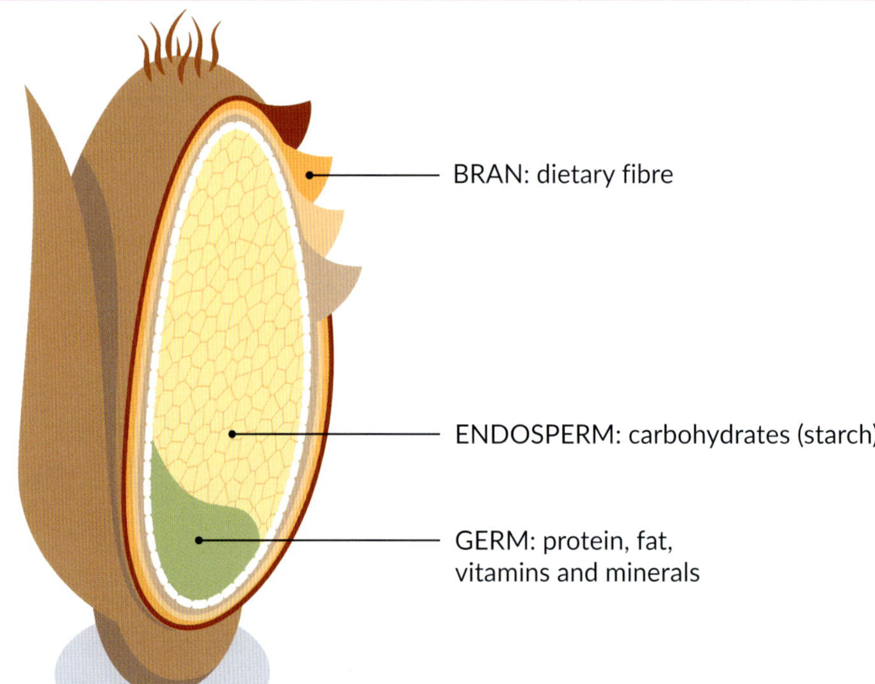

9780170495714

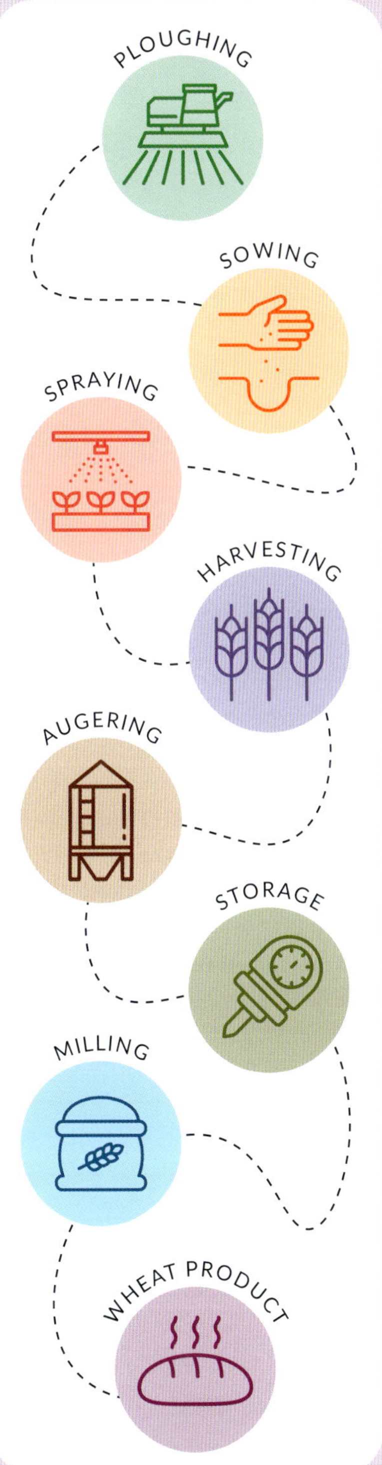

9 GRAIN FOODS

KEY TERMS

endosperm the main body of the grain, which is composed almost entirely of starch

gluten the main protein in wheat flour

grain (cereal) foods edible seeds of certain grasses, including wheat, oats, rice, rye, barley, millet, corn and quinoa

no-till or minimum tillage farming a farming practice that involves planting the new crop by directly drilling in between the rows of the previous crop without tilling the soil

stubble the stalks of the grain crop, which are left once the heads have been cut off during harvesting

Templates
- SWOT analysis table
- Food order
- Recipe map
- Sensory wheel
- Australian Guide to Healthy Eating

Worksheets
- Practical activity 9.1
- Practical activity 9.2
- Practical activity 9.3
- Design activity 9.1
- Design activity 9.2

Quizzes
- Testing knowledge 9.1
- Testing knowledge 9.2

To access resources above, visit **cengage.com.au/nelsonmindtap**

Grain foods in the *Australian Guide to Healthy Eating*

The message from nutritionists is to eat more **grain (cereal) foods**, with the stipulation for mostly wholegrain sources. Grains are plant foods and make up one of the largest segments of the *Australian Guide to Healthy Eating*. The grains in this group include wheat, oats, rice, rye, barley, millet, corn and quinoa.

Figure 9.1 The place of grain (cereal) foods in the *Australian Guide to Healthy Eating*

Grains for good health

All cereal grains provide a similar range of important nutrients. The nutrient value of the end product will largely depend on the amount of processing involved. For example, the removal of the bran and germ layer during milling to produce white flour or white rice reduces the end product's vitamin, mineral and fibre content.

The most nutritious grains are wholegrain cereals, which are an excellent source of carbohydrates and dietary fibre, yet are low in fat. Wholegrains also provide the body with small amounts of protein, B group vitamins and minerals. The bran of cereal grains is made up almost entirely of dietary fibre. Dietary fibre is important in the diet because it absorbs water, and therefore adds bulk and helps food to move through the digestive tract. Consuming a variety of mostly wholegrain or high-fibre grains may help to reduce the risk of developing diseases such as heart disease, type 2 diabetes and some cancers, and will help us avoid gaining excessive weight.

Figure 9.2 Wholegrain foods

Carbohydrates

According to health professionals, 50–60 per cent of our daily energy needs should come from carbohydrates. Carbohydrates are classified according to the number of molecules they contain:

- monosaccharides are the simplest form of carbohydrate
- disaccharides contain two monosaccharide molecules
- polysaccharides contain many monosaccharide molecules.

Monosaccharides and disaccharides are both forms of sugar, whereas polysaccharides are found in the form of starch and cellulose in fruit, vegetables and wholegrain cereals.

Health experts recommend that we select nutrient-dense carbohydrates such as pasta, wholegrain breads, fruit and vegetables as our sources of carbohydrate, rather than energy-dense foods containing added sugars.

Table 9.1 How much of the grain foods group should I eat?

Serves per day		
	12–13 years	14–18 years
Boys	6	7
Girls	5	7

Figure 9.3 What is a serve?

9780170495714

Types of grain foods

Grains or cereal foods are the edible grains or seeds of certain grasses. They have been important as a source of food throughout the world since the origins of humankind.

Wheat, oats, rice, rye, barley and maize (corn) are all grains that are processed in some way for us to eat. Specific varieties of grains are cultivated in each of the world's regions, as they are better suited to the climate and soil of particular areas. Due to their wide availability, grain foods are the staple foods of many regions and are often eaten several times a day.

Some 'ancient grains' – such as quinoa, millet, freekeh, farro, teff and buckwheat – are becoming more popular with some consumers as they are often marketed as being more nutritious than other grains. Quinoa, millet, teff and buckwheat, for example, are gluten-free, while teff and farro are higher in fibre than other grains.

The structure of a cereal grain

All cereal grains are made up of three main parts:

1 the bran, which is the cover or outer layer of the grain; this is made up mainly of dietary fibre
2 the **endosperm**, which is the main body of the grain; this is composed almost entirely of starch
3 the germ or embryo, which contains the nutrients needed for a new plant to grow – protein, fat, and some vitamins, minerals and carbohydrates.

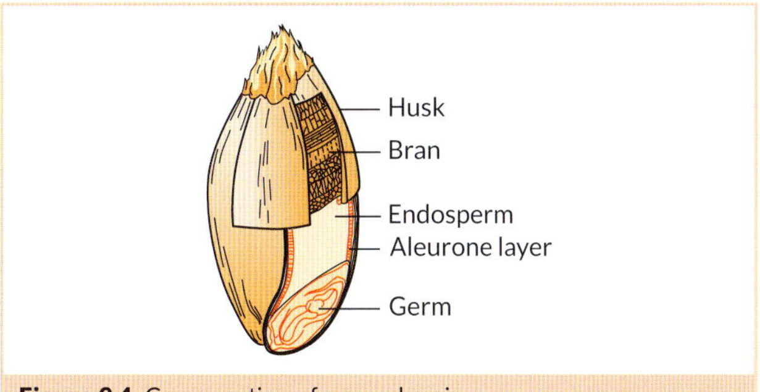

Figure 9.4 Cross-section of a cereal grain

Grain foods and their uses

iStock.com/OlekStock **Wheat** is used for breads, pasta, noodles, cakes, biscuits, extruded snack foods, bulgur and thickening agents.	iStock.com/Vasily Bokov **Oats** are used as porridge – as rolled oats/oatmeal – and in muesli and oatcakes.	iStock.com/Everyday better to do everything you love **Rice** is cooked and used in savoury dishes, puddings, breakfast cereals, biscuits, rice cakes and extruded snack foods.
Saravut Khusrisuwan/Shutterstock.com **Barley** is used in breakfast cereals and as a thickening agent in soups and casseroles.	totophotos/Shutterstock.com **Maize (corn)** is used as a fresh vegetable, and the kernels can also be canned, frozen or dried. It is used in breakfast cereals, polenta, tortillas, corn oil, corn syrup and popcorn.	iStock.com/Aleksandr Krotkov **Quinoa** is used as a whole or flaked cereal grain, as flour, or in breakfast cereals, biscuits, savoury dishes and snack foods. It is a very versatile ingredient and can be added to salads, soups, vegetable patties and muffins.

9

Rice

Rice is one of the most important staple foods eaten throughout the world. More than half of the world's population eats rice on a daily basis. It has been harvested for thousands of years, and recent archaeological finds indicate that it was present in China from around 3000 BCE.

Rice is particularly important in many Asian cultures – such as in South-East Asia, India and China – where it often forms the focal point of every meal, and is frequently served as an accompaniment to other, much spicier foods.

In Japan, numerous shrines to the rice god, Inari, dot the countryside; and in India and China, when young couples are married, they are showered with rice to ensure fertility. The custom of throwing confetti or rose petals at weddings stems from this tradition.

Approximately 40 000 varieties of rice are grown throughout the world. The main rice-growing areas are found in South-East Asia, Africa, the Middle East, Australia, North America, South America and Europe.

In Australia, the main rice-growing areas are in the Murrumbidgee and Riverina regions of southern New South Wales. These regions are most suitable for rice growing because there are extensive areas of flat land, suitable clay-based soils, and irrigated water supplies.

Rice is high in carbohydrates, and low in fat, sugar and salt. It is a good source of B group vitamins and is gluten-free.

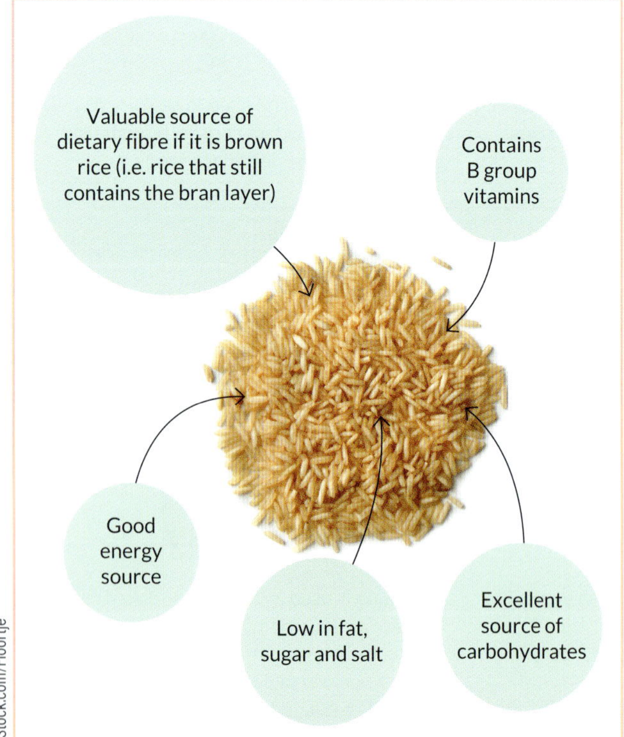

Figure 9.5 The nutrient value of rice

Types of rice

Rice is sometimes classified according to its colour. Brown rice has had its outer husk removed, but retains the bran; while white rice has undergone more processing stages, has been polished during milling, and has had its bran and germ removed.

There are thousands of different varieties of rice, but these are usually classified into three main groups.

1. **Long-grain rice:** The grains remain separate when cooked, and are light and fluffy in texture. Long-grain rices such as basmati and jasmine are generally used for savoury dishes, such as pilafs, and as accompaniments to curries. Long-grain rice may be either polished (white) or unpolished (brown or wholemeal).
2. **Medium-grain rice:** The grains are slightly rounder in shape than those of long-grain rice, stick together when cooked, and are moist in texture. The grains cling together more than those of long-grain rice, but are still separate when cooked. Medium-grain rice is widely used in Chinese cuisine. Arborio rice is a medium-grain rice that is soft in texture; it is ideal for preparing risotto because it can absorb a lot of liquid, such as stock, without becoming gluey.
3. **Short-grain rice:** The grains are much rounder in shape than those of long- and medium-grain rice. Short-grain rice is moist, tender and sticky when cooked. Traditionally, short-grain rice is used in Japanese cuisine for making sushi, and in Spanish cuisine for making paella. Classic rice puddings are also prepared using short-grain rice.

Figure 9.6 Three types of rice: arborio rice (medium-grain), basmati rice (long-grain) and Japanese sushi rice (short-grain)

Wild rice is not a true rice; it is the seed of an aquatic North American grass. It is not cultivated in a similar way to traditional rices, and is expensive to harvest. It ranges in colour from dark brown to black and has a very nutty flavour. Wild rice is often combined with other rice varieties to add colour and flavour to a dish.

Top tips for cooking rice

Rice is a versatile food and can be cooked in a variety of ways. When it cooks, it undergoes gelatinisation. This

9780170495714

is the process that occurs when starch granules in the endosperm (starch component) of rice grains absorb liquid in the presence of heat, thickening the liquid and softening and swelling the grain.

During cooking, rice increases in volume – one cup of uncooked rice will produce about three cups of cooked rice. You should allow approximately one-third of a cup or 70 grams of raw rice per person.

Care must be taken when using cooked rice, because bacteria quickly multiply in warm rice if it is not stored in the refrigerator. Sushi and rice salads must always be refrigerated after preparation and not left in a warm atmosphere. When using cooked rice in a recipe such as fried rice, always reheat the cooked rice until it is very hot to prevent bacterial growth.

eelnosiva/Shutterstock.com
Miracle Stock/Shutterstock.com

Figure 9.7 A comparison of uncooked and cooked rice

Rice is usually cooked in one of four ways: by rapid boil, absorption, rice cooker or microwave (see page 236).

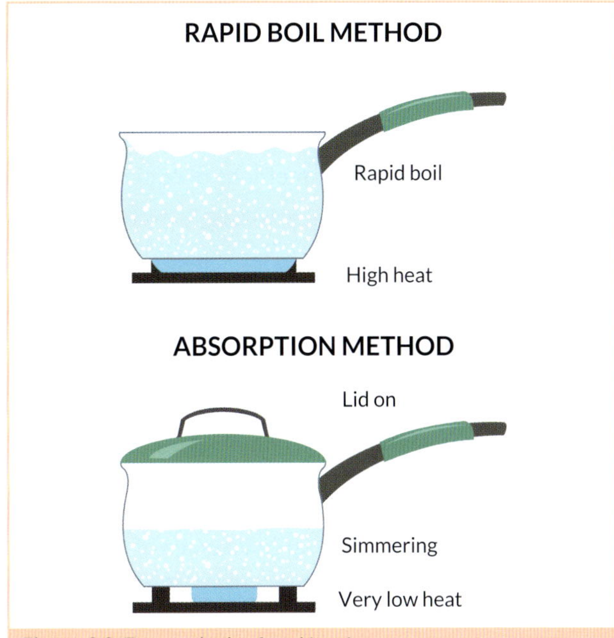

Figure 9.8 Two methods of cooking rice

PRACTICAL ACTIVITY 9.1

Worksheet Practical activity 9.1

Comparative food testing: Comparing methods of cooking rice

Aim

To compare the cooking time, volume, sensory properties and separation of rice grains cooked by different methods

Equipment

- 2 saucepans (one with a tight-fitting lid)
- rice cooker
- microwave-safe bowl
- 4 × ¼ cups Calrose rice

Method

1. Using the information on methods of cooking rice on page 236, cook a quarter of a cup of rice by each of the four methods.
2. After you have cooked each batch of rice (and drained it if necessary), compare the cooking time, volume, sensory properties and separation of grains.
3. Record your results in a table similar to the one below.

	Rapid boil	Absorption	Rice cooker	Microwave
Cooking time				
Volume				
Texture				
Flavour				
Separation of grains				

Analysis

1 Were there differences between the four methods in the times required to cook the rice until it was tender? Which method had the shortest cooking time? Which method took the longest?
2 Explain why rice increases in volume when cooked.
3 Was there a significant difference in the volumes of rice produced using each cooking method? Explain your answer.
4 Which cooking method produced the rice with the best texture and flavour?
5 Identify the method or methods of cooking that produced grains of rice that were clearly separated.

Conclusion

If you were going to cook rice for a meal, which method of cooking would you recommend? Why?

Sustainable rice production: Water-efficient farming

Today, Australian farmers are competing on the world market with small quantities of top-quality rice. One of the key issues facing sustainable rice production in Australia is that growers use large volumes of water to irrigate their rice paddocks. In recent years, Australian rice growers have become aware of the importance of developing environmentally sustainable systems to produce rice, and have developed a variety of strategies to improve water efficiency. The latest research shows that Australian-grown rice now requires 50 per cent less water than rice grown by other major rice-producing countries.

Some of the strategies rice producers have introduced to improve the sustainability of their farming practices are to:

- move towards irrigation systems that use less water than flood irrigation. The reduction in flood irrigation also has the effect of lowering water tables and minimising salinity, as less water is used. In addition, greenhouse gas emissions that are linked to flood irrigation are reduced with more efficient watering systems
- introduce laser levelling technology to level the ground where rice is to be grown, ensuring that water is evenly distributed
- use new technology to measure the amount of water that the rice crop requires, so that water is only released to the crop as it is needed. This gives farmers precise control of water both on and off the paddock, and has led to a 60 per cent reduction in the amount of water used in rice production
- develop closed rice production systems that recycle water and keep the water and nutrients on the property
- use high-yielding, shorter-season rice varieties that require less water to grow.

Fairfax Syndication/Nic Walker

Figure 9.9 Rice growing in New South Wales

Activity 9.1

Rice varieties

Visit the websites of the Rice Growers' Association of Australia, Sunrice and Tilda, and complete the following.

1 Draw a table similar to the one on the following page. Select an example of a rice variety in each classification, and then list its sensory properties when cooked and how it is used in food preparation. One example has been completed for you.
2 Explain why brown rice is considered to be a better nutritional choice than white rice.
3 Using information from the websites, list examples of rice that have been processed so that they cook or reheat quickly.

Weblinks
Rice Growers' Association of Australia
Sunrice
Tilda Rice

9780170495714

Rice variety	Properties of the grain	Uses in food preparation
Long-grain		
Jasmine	Tender texture; fragrant aroma	Asian dishes: stir-fries; fried rice
Medium-grain		
Short-grain		

4 Rice is used to produce a wide range of processed foods. Identify four rice-based processed food products that are widely available in supermarkets.

5 Environmental sustainability is an important issue for Australian rice growers.

  a Outline two strategies Australian rice growers are implementing to reduce chemical spray drift on their properties.

  b Describe two management practices Australian rice growers are using to improve habitat and biodiversity on their farms.

Testing knowledge 

1 Draw up a diagram like the one below to identify the main nutrients found in wholegrain cereals.

Quiz
Testing knowledge 9.1

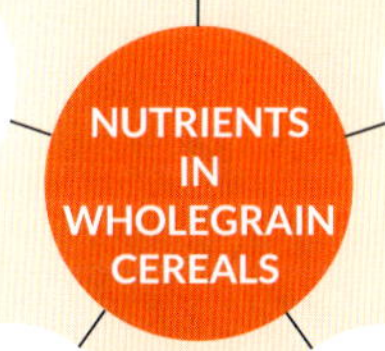

2 Outline the health benefits of including wholegrain or high-fibre cereals in your diet.

3 Explain the difference between monosaccharides, disaccharides and polysaccharides.

4 What are grain foods? Explain why different types of cereal grains are staple products of particular regions of the world.

5 Explain why, from a nutritional perspective, it is better to eat wholegrains rather than refined or 'polished' grains.

6 Develop a concept map that includes the different varieties of rice, as well as dishes made from each variety.

7 What is wild rice and how does it differ from other types of rice?

8 Identify the process that describes the changes in starch when rice is cooked. Explain how this process makes rice more palatable to eat.

9 List the different methods that can be used to cook rice. Which method would require the most accurate timing to prevent the rice from becoming overcooked?

10 Outline three strategies rice growers are using to improve the sustainability of their farms.

Wheat

Wheat covers more of the Earth's surface than any other grain crop, and is the staple grain food for much of the world's population.

Australian wheat farmers have an international reputation for producing top-quality wheat that is in high demand globally. The main wheat-growing areas in Australia are in Western Australia, New South Wales, South Australia, Victoria and Queensland.

Set out on the following page is the journey of wheat – from ploughing the paddock in preparation for sowing the crop, to the final product.

Wheat production: From paddock to plate

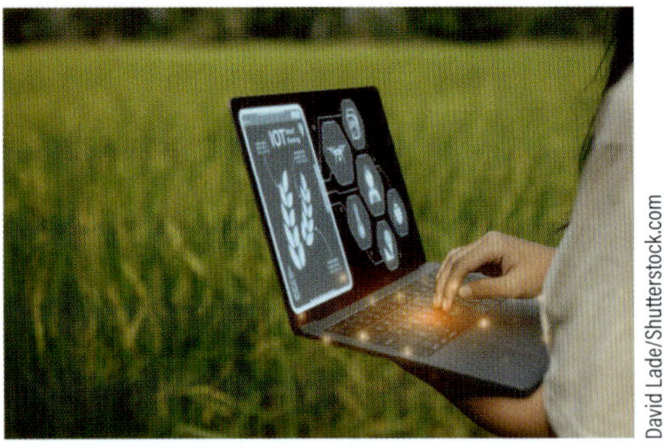
David Lade/Shutterstock.com

1. Technology in farming Farmers use smart farming technologies, The Internet of Things and precision agriculture to provide real-time data that will maximise their crop yield. For example, analysing the health of the soil, monitoring insect and pest populations, tracking local weather conditions and analysing their crop health and plant growth.

Bits And Splits/Shutterstock.com

2. Sowing: Seeds are planted by direct drilling in-between the rows of the previous crop.

Grindstone Media Group/Shutterstock.com

3. Spraying: The crop is sprayed with agricultural chemicals to reduce the growth of weeds, fungi and pests that diminish the final yield.

crbellette/Shutterstock.com

4. Harvesting: The growing season is completed by early summer, when farmers harvest the crop. The harvester strips the head of the wheat stalks and separates the grains from the chaff.

iStock.com/jodie777

5. Transporting: The grain is augered from the harvester into trucks to be transported to the grain receival centre or to be stored on the farm.

iStock.com/CUHRIG

6. Storage: Grain can be stored in silos, bunkers, grain bags or grain sheds. It is then transported for domestic use or to seaports for export.

GheorgheGarcu/Shutterstock.com

7. Milling: In flour mills, the grain is crushed and processed to different degrees to produce ingredients such as white flour, wholemeal flour and semolina.

iStock.com/MEDITERRANEAN

8. Products made from wheat flour: These include breads, cakes, biscuits, pastry, pasta, scones, noodles, breakfast cereals, pizza crusts, crumpets, and snack foods such as pretzels.

9780170495714

Sustainable wheat production: No-till or minimum tillage farming

Most wheat farmers work towards developing strategies to make their cereal production more sustainable and environmentally friendly. Sustainable farming involves farming practices such as **no-till or minimum tillage farming**, which maintain the land's productivity so that it will be available for future generations. This type of farming is a method of producing cereal crops such as wheat and canola that has significant environmental advantages. It also provides economic advantages for farmers, who state that when using it, their crops provide higher yields – especially during dry years.

No-till or minimum tillage farming involves leaving the stubble from the previous year's crop to enrich and stabilise the soil. **Stubble** refers to the stalks of the cereal crop, which are left once the heads have been cut off during harvesting. The new crop is planted by direct drilling in between the rows of the previous crop. This differs from conventional farming methods that involve ploughing or tilling the land as the first stages in preparing the soil before the crop is planted.

Jon Lewis/Alamy Stock Photo

Figure 9.10 Planting seeds by direct drilling

Leaving the stubble from the previous crop in the soil after harvest holds significant benefits for soil health. As the stubble breaks down, the nutrients it contains are returned to the soil. The stubble also provides a layer of mulch on the surface of the soil, ensuring that moisture is retained, which is especially important in dry years. A further advantage of retaining stubble on the land is that the roots of the stubble hold the soil in place, minimising wind erosion.

Along with leaving stubble in the soil, no-till or minimum tillage farmers use GPS technology to establish a system of controlled traffic lanes – or 'tramlines' – that their large machinery follow when sowing and harvesting crops. Using designated traffic lanes for machinery means that the soil is not crushed or compacted, but remains moist. GPS also enables farmers to use precision sowing; that is, to sow crops in between rows from the previous year's crop.

vidimamax/Shutterstock.com

Figure 9.11 Retaining stubble after the harvest

Gluten: The protein in wheat flour

Wheat flour has two major components: starch and protein. Its main protein is **gluten**. 'Gluten' was originally a Latin word meaning 'glue'. Gluten in wheat flour is itself made up of two component proteins: glutenin and gliadin.

While the protein gluten is primarily found in wheat, there is also some gluten in other cereal grains, such as rye, barley and oats.

Gluten is the component of flour that helps to form the structure of bread and other cereal products. In the bread-making process, the gluten in the dough stretches to form a skin that holds the bubbles of carbon dioxide (CO_2) – just like bubble gum – that are formed by the action of yeast. Without gluten, the CO_2 would escape, and the dough would not rise.

Hard or strong wheats contain the most gluten, and so are best for making bread. Gluten flour can be purchased at supermarkets and health food stores and added to other flours to make a strong dough. It is a creamy, greyish powder that has been separated from wheat endosperm, and is made up of strands formed by about 20 different amino acids. Each strand has a helix formation, which makes it look like a tiny spring, and, after water is added, a complex three-dimensional network is formed.

Figure 9.12 Strands of gluten in flour before hydration

When bakers mix and knead a bread dough, they are said to be 'developing the gluten', because the strands of gluten are being straightened and overlapped. This means the 'springs 'of gluten must be uncoiled, and the resulting straightened strands must then be recombined into a continuous, overlapping mesh. The new structure forms a three-dimensional network capable of retaining the CO_2 produced by the yeast during fermentation.

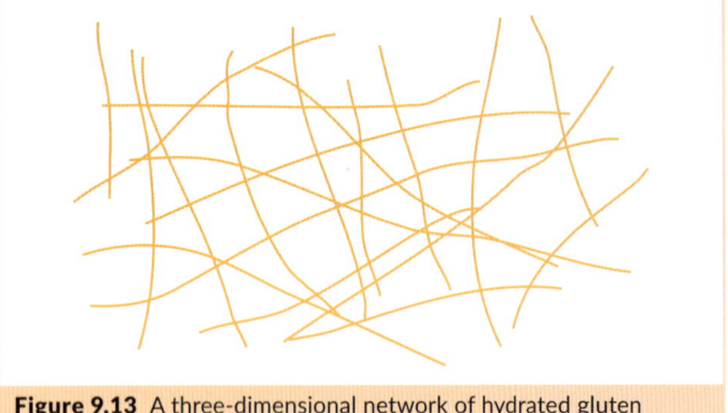

Figure 9.13 A three-dimensional network of hydrated gluten in dough

Australian wheat

Australia is one of the world's top producers of clean, white and food-safe wheat. It is the main grain crop grown here, and its high quality means that it is in demand both domestically and as an export commodity. Australian wheat makes excellent bread dough for large commercial bakery systems producing white loaves, because it is strong and has good oven lift and volume. The resulting loaves also have a bright white crumb and a golden crust. In addition, Australian strong wheat flour is suited to artisan-style bread and wholegrain products. In Asia, the demand for Australian soft wheat flours to make cakes and traditional cookies is increasing.

More than 30 per cent of Australia's wheat export is used to produce Asian-style noodles. This is because Australian wheat has the ideal ratio of starch to protein (gluten) for noodles, which produces an appealing mouth feel and texture. It is used to produce fresh noodles from the chill cabinet such as hokkien noodles, or dried noodles such as instant, ramen or udon noodles.

Australian wheat
Sliced bread
Artisan-style bread
Flatbreads
Cakes, biscuits and pastries
Pasta and couscous
Asian-style noodles
Steamed dumplings
Asian-style steamed bread

Figure 9.14 Food products made from Australian wheat

Clockwise from right to left: Itsanan/Shutterstock.com; Rawpixel.com/Shutterstock.com; Svetlana Monyakova/Shutterstock.com; Tatjana Baibakova/Shutterstock.com; Oqbas/Shutterstock.com; iStock.com/yipengge; iStock.com/LauriPatterson; iStock.com/vichie81; Center: caseyjadew/Shutterstock.com

9780170495714

Types of wheat flour in Australia

When preparing food products with flour, it is often useful to know the gluten content or the amount of protein in the flour to ensure success. In Australia:

- Plain white flour has a gluten content of around 10 per cent. It is sometimes referred to as 'all-purpose flour', and is suitable for baking, thickening, coating and binding.
- Self-raising flour is plain flour with an aerating or raising agent added. When combined with liquid and the application of heat, CO_2 is released and leavens or raises the batter or dough. It is low in gluten and so produces a soft texture, which is desirable in cakes.
- Bakers' flour (also called 'strong flour' or 'bread flour') has about 12 per cent gluten content. The gluten creates the framework for the yeast to grow in doughs that are used for bread and pizza. This flour is also used to make pasta.
- Wholemeal flour can be finely or coarsely milled, and contains all parts of the wheat grain. Products made with this type of flour are darker in colour and denser in texture than those containing white flour.
- Cake flour is milled from varieties of wheat with a low gluten content. It is ground into a very fine particle size and is chlorinated for extra whiteness. This type of flour is used in packet cake mixes and in the production of commercial cakes and biscuits.
- Cornstarch is a fine, white starch that can be made from wheat or maize (corn). It is used for thickening sauces and for baking light, delicate cakes such as sponges. Only cornstarch made from maize is gluten-free, so you should check the labels if you are making gluten-free products.

Bread

Bread is one of the oldest and most diverse foods in the world. Throughout the ages, bread has been an important staple food for many people, and has often been referred to as 'the staff of life'. Many types of grains have been used to make flour for bread throughout the centuries. However, only the flour from wheat and, to a lesser degree, from rye, can produce dough that is capable of holding the leavening gases produced by yeast sufficiently well to yield well-risen loaves with a fine, soft cellular structure.

The physical properties – that is, the shape, size and texture – of bread made from wheat flour are the result of the presence of gluten in the flour. All the processes of bread making – fermenting, kneading and proving – involve changing and improving the natural properties of gluten to make the dough strong enough to hold the bubbles of CO_2 that are produced during fermentation.

In traditional bread-making methods, this gluten modification occurs over several hours during fermentation of the dough. A 12-hour fermentation period may be required when yeast activity is low. Today's bread-making processes rely on other means of modifying the gluten, and enable good-quality bread to be made in two hours.

Dough development can be hastened by the presence of very small quantities of oxidising agents, known as bread improvers. Furthermore, some flours contain gluten that is too tough and strong; if this is the case, bread improvers are added to soften the flours in order to make a loaf with a soft texture.

FoodAndPhoto/Shutterstock.com

Figure 9.15 Various types of bread

Flatbreads

Some of the world's oldest and simplest breads are flatbreads. Flatbreads are usually quick to make, and are cooked on a cooktop, under a griller or in a very hot oven. They may or may not use yeast as a raising agent.

Many different baking techniques are used to make flatbreads, although the basic steps are the same as for loaf bread. The formula is very basic, and uses flour, water and salt.

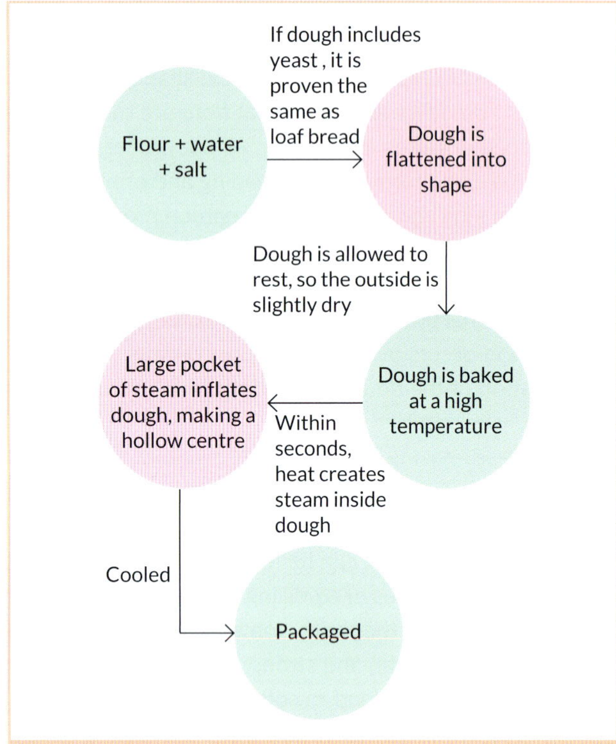

Figure 9.16 The production of flatbread

Flatbreads can be crisp or chewy, plain or rich, and come in many shapes, sizes and flavours. In some cultures, traditional flatbreads are used to wrap around food, and replace plates and cutlery. For example, in Greece, pita is used as a wrap for meat kebabs; and in India, naan and chapatti are designed so that people can scoop food from their plates into their mouths.

Table 9.2 Types of flatbreads

Name	Country/region	Type of flour
Pita	Middle East	Wheat
Lavash	Middle East	Wheat
Naan	India	Wheat
Chapatti	India	Wheat
Puri	India	Wheat
Pappadam	India	Lentil, rice, potato
Tortillas	Mexico	Corn
Branch bread	Scandinavia	Wheat, rye
Griddle oatcakes	Scotland	Oats
Roti	Malaysia	Wheat

Brent Hofacker/Shutterstock.com

Figure 9.17 Naan is an Indian flatbread.

PRACTICAL ACTIVITY 9.2

Sensory analysis: Taste testing different types of bread

Worksheet Practical activity 9.2

Template Sensory wheel

1. Taste test a range of different types of bread, including:
   - white
   - wholemeal
   - multigrain
   - rye blend
   - sourdough
   - flatbread.

   Focus on the sensory properties – appearance, aroma, flavour, texture and sound – of each type of bread. Record your results in a table.
2. Rate the breads from the one you liked the most to the one you liked the least. Justify your ratings.
3. Predict which bread has the highest fibre content. Explain the criteria you used to make your decision.
4. With a partner, brainstorm and record why there is such a large variety of breads available today.
5. If you were responsible for buying bread for your family, which type would you choose? When writing your response, consider the sensory properties, nutritional value and the way that bread is used in your home.

9780170495714

Labelling of bread

Food Standards Australia New Zealand (FSANZ) requires all food manufacturers to comply with food labelling laws. All packaged foods must be labelled with nutritional information about how much fat, protein, energy, carbohydrate and salt is in the food. The labels must also show the percentage of key ingredients and the main ingredients that may cause allergies.

In addition, there are two main methods of date marking that food manufacturers are required to use:

1 **use-by date:** for foods that should not be consumed after a certain date for health and safety reasons
2 **best-before date:** for foods that can be eaten for a while after the best-before date; they are safe, but may have lost some quality and/or nutrient value.

Because the freshness of bread is important to consumers, FSANZ has developed additional standards for marking the date on bread labels. As well a use-by or best-before date, bread and other baked products are also labelled with other dates if the bread has a shelf life of less than seven days:

1 **baked-for date:** for bread that has been baked up to 12 hours before the marked date
2 **baked-on date:** for bread that has been baked on the marked date.

The other strategy developed to assist consumers to know when packaged bread has been baked is the colours of the bag closure tags. The colours correspond to the day of the week the bread is baked: blue (Monday), green (Tuesday), red (Thursday), white (Friday) and yellow (Saturday). The colours are in alphabetical order, with blue corresponding to Monday because it starts with the letter 'B'. There is no colour for Wednesday or Sunday, because many bakeries traditionally do not operate on these days.

Figure 9.18 Baking day bag closure tags for packaged bread

Couscous

Couscous is a grain product made from semolina, the coarsely ground endosperm of wheat, with the addition of some wheat flour. It has been used as a staple food in North Africa and some Middle Eastern countries since the earliest times in recorded human history. Today, couscous is popular because it is a versatile food that makes a great alternative to pasta or rice.

Couscous can be purchased as either a fine grain or as a larger 'pearl' grain.

Fine-grain couscous is light and fluffy; it is ideal for serving with spicy foods or vegetable dishes, and is equally delicious served as a pilaf. The pre-cooked, 'instant' form of fine-grain couscous that is available in supermarkets is much simpler to prepare than the traditional method, which involves hours of preparation.

Svetlana Monyakova/Shutterstock.com

Figure 9.19 A dish made with fine-grain couscous

Pearl couscous is similar to fine-grain couscous, but is larger in size and the granules are slightly chewy in texture. Pearl couscous is often used as a substitute for pasta or rice and is perfect in salads and for adding to soups.

Elena Veselova/Shutterstock.com

Figure 9.20 A dish made with pearl couscous

Couscous for good health

As couscous is made from the endosperm of wheat, it is high in carbohydrates and is a good source of energy. Couscous is also high in vitamin B1 (thiamin) as well as vitamin B3 (niacin). Another advantage of incorporating couscous into the diet is that it is naturally low in fat.

PRACTICAL ACTIVITY 9.3

Product analysis: Getting to know couscous

Worksheet
Practical activity 9.3

1 Work with a partner to prepare one of the instant couscous packaged mixes available in supermarkets.
2 While the couscous is cooking, examine the packaging and record the ingredients that are included in the product.
   a What hints for serving have been included on the packaging?
   b In which country was the couscous made?
   c Why are the cooking instructions given in several languages?
3 Once your couscous is cooked, taste test the product. Rate the taste as 'excellent', 'very good', 'good', 'fair' or 'unacceptable'.
4 List the other ingredients that must be added to the couscous so that it cooks successfully.
5 Do you think the product would be improved by the addition of any other ingredients? What else could you add to improve the flavour of the couscous?

Testing knowledge 9.2

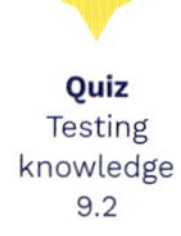
Quiz
Testing knowledge 9.2

11 Draw a flow chart that illustrates the stages in the growing and storage of wheat before it is processed into flour.
12 What is no-till or minimum tillage farming? Outline three reasons why farmers may use this method of producing grain crops.
13 What is gluten? Explain why gluten is an important component of flour that is used to make bread.
14 Why is wheat a valuable export for Australia?
15 Explain why there is an increasing demand for Australian wheat in Asian countries.
16 Describe the main differences between wholemeal flour and cake flour.
17 What is bread improver, and why do many commercial bread manufacturers use it in their production process?
18 Explain why flatbreads are a popular food throughout the world.
19 Outline the information that is included on the packaging of packaged bread that consumers can use to determine the freshness of the bread before purchase.
20 What is couscous? Identify the regions in the world where couscous is a staple food.

Critical and creative thinking 9.1

SWOT analysis of no-till farming

Complete a SWOT analysis (Strengths, Weaknesses, Opportunities, Threats) of no-till or minimum tillage farming.

Template
SWOT analysis table

Strengths:	Weaknesses:
Opportunities:	**Threats:**

Critical and creative thinking 9.2

Labelling of packaged bread

A food label is like the food's ID card. Outline the information that consumers can collect from the label on packaged bread and explain how this may affect their purchasing decisions.

9780170495714

DESIGN ACTIVITY 9.1

Speedy ramen bowl

One comfort food on a cold winter's afternoon is a hot bowl of soup. A packet of instant noodles that includes a dry noodle cake and a flavour sachet is the starting point for your warming bowl full of goodness – and it is just for you!

Design brief

The challenge is to design a ramen bowl for a quick snack or light meal for one person that can be prepared and served in approximately 15 minutes. Most of the ingredients for the speedy ramen bowl should come from the main segments of the *Australian Guide to Healthy Eating*, with only a minimum of ingredients from outside the circle. The ingredients in the most highly rated ramen bowls are not overcooked and are easy to eat with chopsticks and a spoon. Extra flavours are added as a garnish. Minimal kitchen equipment should be required to create the speedy ramen bowl.

iStock.com/ahirao_photo

Investigating and defining

1 Based on the solution requirements and constraints in the design brief, develop five design criteria to evaluate the success of your speedy ramen bowl.

2 Prepare the recipe for Instant Noodles in Soupy Goodness (page 240) if you have not made this type of product before.

3 Use the internet, commercial products and recipes from home cooks to research the types of ingredients that could be included in your speedy ramen bowl.

4 List four recipe ideas that appeal to you. Remember to include the source of the recipes. Make a quick note about why you selected each idea.

Worksheet Design activity 9.1

Templates Food order

Sensory wheel

Australian Guide to Healthy Eating

5 Using your recipes, classify the ingredients under the following headings in a table like the below.

	Recipe, with source and note about why it was selected	Protein	Vegetables	Flavourings to finish
1				
2				
3				
4				

Generating and designing

1 Create two designed solution options for your speedy ramen bowl, and annotate your ingredient and production ideas using diagrams like the ones below.

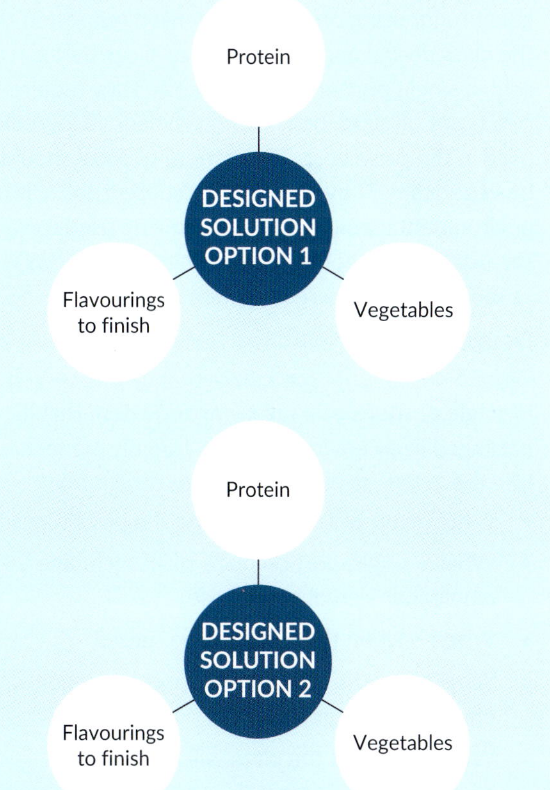

2 Select the designed solution option you prefer and justify your choice. Explain why you did not select the other option.
3 Write up your new recipe ready for production.

Hints for success

- Don't overcook the noodles.
- Cut the vegetables into pieces that are easy to pick up with chopsticks.
- Don't overcook green vegetables – they only need to be blanched, so they stay crisp and colourful.

Planning and managing

1 Prepare a food order.
2 Annotate a copy of your speedy ramen bowl recipe to indicate the tools and equipment needed in production and to highlight key safe work practices.

Producing and implementing

Produce your speedy ramen bowl and taste test it before evaluating the product.

Evaluating

1 Respond to the design criteria you developed.
2 Describe the sensory properties – appearance, aroma, flavour, texture and sound – of your speedy ramen bowl.
3 Discuss the effectiveness of your time management for this production.
4 What recommendations would you suggest to improve the quality of your speedy ramen bowl if you were to make the recipe again?
5 Classify the ingredients of your speedy ramen bowl on a diagram of the *Australian Guide to Healthy Eating*. Comment on how well this dish meets the recommendations of this food selection model.

DESIGN ACTIVITY 9.2

Pizza slices

Members of the school student council have approached your Food Studies class to support their latest fundraising program. They have requested that the class design and produce pizza slices that can be sold in single portions during recess. They require two types of pizza, both with a wholegrain ingredient used in the pizza base. One variety of pizza should be vegetarian. They also want to support the school's environmental policy, so the packaging used to serve the pizza slices should create minimal landfill, as well as meet food safety requirements.

Design brief

Write a design brief for a pizza that can be served in single portions as a takeaway food item that is packaged in an environmentally friendly manner. Use the '5 Ws' as the basis of your design brief:

- Who – who will be purchasing the pizza slices
- What – single-portion pizza slices including a wholegrain cereal in the base
- When – when the pizza will be served
- Why – why the pizza will support the school's environmental policy
- Where – where the pizza will be sold.

Worksheet
Design activity 9.2

Templates
Recipe map
Food order
Sensory wheel

Investigating and defining

1 Based on the solution requirements and constraints in your design brief, develop five design criteria to evaluate the success of your pizza slices.
2 Research recipe ideas for pizza bases that have some wholegrain ingredients; pizza toppings, including vegetarian options; and serving shapes and sizes.
3 Research possible designed solution options for suitable environmentally friendly packaging that creates minimal waste.

Generating and designing

1 Record the information from your research on a recipe map like the one on the following page to create two designed solution options.
2 Sketch and annotate your two designed solution options – the base, toppings and shape. Remember to note down why each of your designs is appropriate to serve in single portions at recess.
3 Select the designed solution option that best meets the design brief, and justify your choice. Explain why you did not select the other option.
4 Develop an interesting name for your new product, and write up your recipe so it is ready for production.

9780170495714

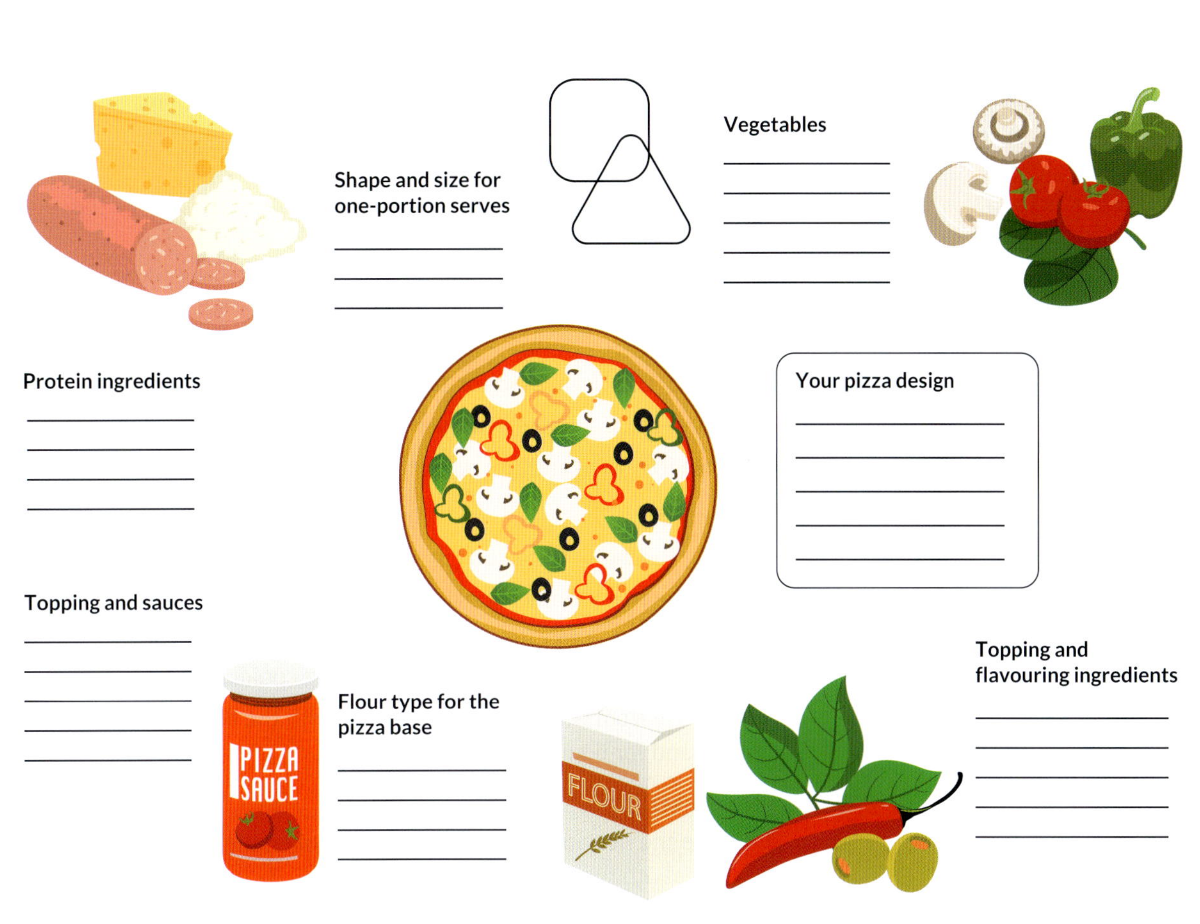

Figure 9.21 A recipe map for pizza

Planning and managing

1. Complete a food order.
2. Make a list of the aspects of the production task that rely on you and your bench partner sharing and working collaboratively.

Producing and implementing

1. Produce your new product.
2. Record any modifications you made during production.

Evaluating

1. Respond to the design criteria you developed, in detail.
2. Describe the sensory properties – appearance, aroma, flavour, texture and sound – of your pizza. Share your pizza with two other people and record their comments.
3. Comment on whether the dough base that included some wholegrain ingredients was firm enough to be eaten as a single serve of pizza without the topping spilling off it.
4. After the pizza was baked, was the arrangement of the topping ingredients successful? Did it look appetising and suitable to be cut into single-serve portions?
5. Explain how your idea for packaging your pizza met your school's environmental policy.
6. Based on your own experience and the comments of your two taste testers, discuss any improvements you would make to either the ingredients or the production process if you were to make the pizza again.

Plain Rice

There are four simple methods for cooking plain rice, which is ideal for serving as an accompaniment. Try each method to see which you prefer.

INGREDIENTS

Rapid boil

4 cups boiling water

⅔ cup long-grain rice

Absorption

1½ cups water

1 cup long-grain rice

Rice cooker

1½ cups water

1 cup long-grain rice

SERVES TWO

METHOD

Rapid boil

1. Bring the water to the boil in a saucepan.
2. Add the rice. Gently stir with a fork to separate the grains.
3. Rapidly boil uncovered for 12–15 minutes.
4. Test to see if the rice is tender by tasting a grain.
5. Drain in a sieve and serve.

Absorption

1. Bring the water to the boil in a saucepan.
2. Add the rice. Gently stir with a fork to separate the grains.
3. Place a lid on the saucepan. Lower the heat and simmer for 12–15 minutes. Do not lift the lid during cooking.
4. Remove from the heat. Keep covered with the lid and allow to stand for 5 minutes. Toss with a fork.

Rice cooker

1. Rinse the rice and place it in the rice cooker, then add water. Cover with the lid and switch on.
2. The rice cooker will switch off when the cooking is complete. Allow the rice to rest for 10 minutes before serving. Some rice cookers have a warming element that will keep the rice warm for several hours.

Microwave

Follow the instructions on the packet of rice, as cooking times vary for different varieties. Also remember to check the instructions in your microwave manual since the power levels vary considerably between models.

WKTAN/Shutterstock.com

EVALUATION

1. Why is it necessary to stir the rice when you add it to the boiling water?
2. Explain why it is important to leave the lid on the rice when cooking by the absorption method.
3. Why is it necessary to allow the rice to stand for five minutes before serving when cooking by the absorption method?
4. Why does the rice increase in size once it has been cooked?
5. Explain why rice is a component of meals in many countries throughout the world.

9780170495714

Sushi Rolls

INGREDIENTS

4 sheets nori

Sushi Rice

2 cups Koshihikari or sushi rice

2 cups water

Sushi Su

½ cup su (rice vinegar)

2 tablespoons sugar

pinch salt

Filling (choose 1–3)

fish fingers, prepared according to packet instructions

drained canned tuna

red capsicum

shredded egg

cucumber batons

avocado

pickled radish

egg mayonnaise or Japanese-style mayonnaise

For serving

wasabi paste

pickled ginger

soy sauce

MAKES FOUR SUSHI ROLLS

METHOD

Sushi Rice

1 Place the rice in a medium saucepan. Rinse several times with cold water until the water is clear. Drain thoroughly to remove excess starch.

2 Add the water to the rice. Stand for 10 minutes.

3 Bring to the boil and boil for 1 minute. Turn the heat to low. Cover the saucepan tightly and simmer for 20 minutes.

4 Remove from the heat and allow to stand for 10 minutes. Do not uncover until ready. This will ensure the rice stays moist.

Sushi Su

1 Mix all the Sushi Su ingredients well.

2 Transfer the freshly cooked rice to a large bowl and gently toss it with the Sushi Su. The rice is now ready to be used to make the Sushi Rolls.

To assemble the dish

1 Select one, two or a maximum of three ingredients for the filling. Cut the chosen filling ingredients into thin strips.

2 Place half a sheet of nori onto a bamboo mat. Allow one edge of the nori to hang over the edge of the mat that is furthest away from you by about 1 centimetre.

3 Place a band of Sushi Rice in the centre of the nori. Spread evenly.

4 Place the filling ingredients in layers across the centre of the rice to form a long band from left to right.

5 Use the mat to roll up the sushi, rolling away from you. Press down firmly.

6 Remove the mat from the roll. Leave the roll whole or cut it into equal-sized serving portions. Serve with wasabi paste, pickled ginger and soy sauce.

Artur Kozlov/Moment/Getty Images

EVALUATION

1 Why is a short-grain rice used to prepare the Sushi Rolls?

2 Explain why you need to rinse the rice several times in cold water before using it in this recipe.

3 Why is it important to make sure the lid is not removed from the rice while it stands for 10 minutes after cooking?

4 What ingredients could be used to fill the Sushi Rolls other than those suggested in the recipe?

5 Discuss why Sushi Rolls are considered to be a healthy snack or lunch food.

Polenta Pancakes

INGREDIENTS

25 grams unsalted butter
110 millilitres water
¼ teaspoon salt
2 tablespoons quick-cook polenta
½ cup plain flour
¼ teaspoon baking powder
¼ teaspoon bicarbonate of soda
1 egg
½ cup Greek yoghurt
2 tablespoons milk
1 tablespoon honey
2 spring onions, finely sliced
30 grams additional butter for frying the pancakes

To serve

Corn Salsa (page 239)
sour cream or Greek yoghurt

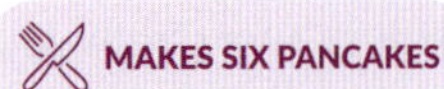
MAKES SIX PANCAKES

METHOD

1. In a small saucepan, add the butter, water and salt. Over medium heat, bring the liquid to a simmer, then pour in the polenta in a fine stream. Stir continuously over a moderate heat for 1–2 minutes, until the mixture thickens.
2. Remove from the heat and allow to cool for 15 minutes. Stir occasionally to prevent a skin from forming on the surface.
3. Sift together the flour, baking powder and bicarbonate of soda.
4. In a medium bowl, use a whisk to combine the egg and yoghurt until it is lump-free. Add the milk and honey and whisk to combine all four ingredients.
5. Pour a third of the liquid mixture into the saucepan with the cooled polenta. Whisk to loosen, then gradually add in the rest of the liquid mixture, whisking until it is fully combined with the polenta.
6. Fold in the sifted dry ingredients, then the spring onions. Cover and refrigerate while cleaning up the work area before cooking the pancakes.
7. Heat a frying pan over medium heat. And add 1 teaspoon of butter and swirl to coat the pan.
8. Pour a quarter cup of batter into the pan. It will spread.
9. Cook for 2 minutes, then use a lifter to flip the pancake and cook for 2 more minutes. Adjust the heat if necessary.
10. Transfer to a plate or oven tray and keep warm until the remainder of the pancakes are cooked. In between batches, wipe the frying pan with paper towel and add an additional half teaspoon of butter.

To serve

Create two stacks of three Polenta Pancakes each. Top with Corn Salsa and add a side serve of 1–2 tablespoons of sour cream or Greek yoghurt.

EVALUATION

1. Describe the sensory properties – appearance, aroma, flavour, texture and sound – of your Polenta Pancakes served with Corn Salsa.
2. Identify the reaction that occurs to the polenta in step 1. Explain why the polenta is the ingredient responsible for thickening the mixture.
3. Identify the main ingredients in the pancake batter that are responsible for aerating the mixture. Explain how they contribute to the light texture of the final product.
4. Describe the difference between the physical processes of whisking and folding.
5. Reflect on the success of your Polenta Pancakes served with Corn Salsa and your management of production time. Highlight the successful aspects of production and make recommendations for improvements.

9780170495714

Corn Salsa

This salsa can be wrapped in fresh flatbread, used to add colour, texture and flavour to a salad lunch box, or served with barbecued meat or fish.

INGREDIENTS

½ teaspoon cumin seeds

¼ teaspoon coriander seeds

1 corn cob

¼ red capsicum

¼ green capsicum

2 spring onions, sliced

2 teaspoons olive oil

pinch salt and pepper

SERVES ONE

METHOD

1. In a small saucepan over medium heat, dry-fry the cumin and coriander seeds until they become fragrant (about 30–60 seconds). Stir with a wooden spoon to prevent the seeds from burning.
2. Transfer the seeds to a mortar and pestle and grind into a coarse powder. Set aside.
3. Using a serrated knife, slice the corn kernels from the cob.
4. Dice the red and green capsicum into approximately 5-millimetre cubes, about the same size as the corn kernels.
5. Return the saucepan to medium heat and add the olive oil.
6. Sauté the diced capsicum for 3 minutes, then add the spring onion and sauté for a further 2 minutes. Stir occasionally and adjust heat if necessary to ensure that the vegetables do not brown.
7. Add the corn kernels and the ground spices, then cover with a tight-fitting lid. Cook for 5 minutes, stirring occasionally.
8. Remove from the heat. Rest covered for 3 minutes.
9. Season with salt and pepper.

EVALUATION

1. Explain why corn is classified as a grain food.
2. Describe how the sensory properties of the cumin and coriander seeds change during dry-frying.
3. Explain why the diced capsicum is sautéed rather than dry-fried.
4. Why is the serrated knife the best tool to remove the corn kernels from the cob?
5. Use the *Australian Guide to Healthy Eating* food selection model to explain why the Corn Salsa would rate more highly than ham and tomato sauce as a filling in a flatbread wrap.

Mark Fergus Photography

Instant Noodles in Soupy Goodness

INGREDIENTS

2 stalks broccolini

5 snow peas

⅙ red capsicum

1 packet dried instant noodles, with flavour sachet

200 millilitres water

1 egg

⅓ cup bean sprouts

To serve

2 spring onions, finely sliced

3–4 slices red chilli

2–3 teaspoons fried shallots

1–2 teaspoons soy sauce (to taste)

1 lime cheek

SERVES ONE

METHOD

1 Rinse the vegetables. Slice the broccolini into 3-centimetre pieces, keeping the florets together. Trim the tops off the snow peas and dice the capsicum.

2 Place the contents of the flavour sachet into a small saucepan. Add the water and bring to the boil, then reduce the heat to medium.

3 Add the broccolini, snow peas and capsicum and simmer for 1 minute. Add the dried noodles and simmer for 1 more minute.

4 Use some tongs to move the noodles to one side of the saucepan, then carefully crack an egg into the space. Cook for 1 minute.

5 Remove the vegetables and noodles from the saucepan and place into a serving bowl. Continue cooking the egg and soup over medium heat for 3 minutes. Don't worry if the egg is a bit misshapen.

6 Place the bean sprouts on top of the vegetables and noodles in the serving bowl.

7 Use a metal spoon to transfer the egg to the serving bowl, then pour the soup over the vegetables, noodles and egg.

8 To serve, top with your choice of finishing ingredients. Squeeze the lime cheek over the soup and eat while it is fresh and hot.

Mark Fergus Photography

EVALUATION

1 Describe the sensory properties – appearance, aroma, flavour, texture and sound – of your Instant Noodles in Soupy Goodness.

2 Explain the difference between boiling and simmering.

3 Why are short cooking times for the ingredients important in this recipe?

4 Classify the ingredients for the Instant Noodles in Soupy Goodness on a diagram of the *Australian Guide to Healthy Eating* and discuss the value of adding extra ingredients to a popular snack food.

5 Recommend two other combinations of ingredients – one protein and three or four vegetables – that could be used to create an improved version of the recipe.

9780170495714

Multigrain Batard

INGREDIENTS

1½ cups white bread flour
½ cup multigrain flour
½ teaspoon salt
¼ teaspoon sugar
1 teaspoon dried yeast
1 cup warm water
canola oil spray for greasing tray
2 teaspoons milk for glaze
poppy or sesame seeds

MAKES ONE LOAF

METHOD

1 Preheat oven to 220°C.
2 Sift the flours, salt, sugar and yeast into a large bowl and make a well in the centre.
3 Add the warm water all at once and mix to a soft dough using a spatula.
4 Turn onto a lightly floured board and knead for 8–10 minutes, until the dough is smooth and elastic.
5 Use the palm of your hand to form the dough into a rough oval.
6 Turn the dough so that the long edge is facing you. Fold one-third of the dough towards the centre and gently press along the edge. Bring the opposite long edge of the dough to the centre so that it meets the first edge and press down gently.
7 Next, fold one half on top of the other and press the dough down gently to seal the edges.
8 Lightly flour your hands and gently pat the dough out to form a plump oval. Place the loaf on a baking tray with the seam side down.
9 Score the top once with a cut lengthwise. Lightly cover with a clean tea towel or oiled plastic wrap and allow to prove in a warm place for 20 minutes.
10 Carefully brush with the milk and sprinkle with the poppy or sesame seeds.
11 Bake for 25–30 minutes or until the loaf is golden brown and sounds hollow when tapped.

Mark Fergus Photography

EVALUATION

1 Explain why it is important to sift the dry ingredients before mixing in the liquid.
2 What would happen if cold water was used instead of warm water when making the bread dough?
3 Explain why it is important to prove and knead the bread dough.
4 Describe the tests used to decide if the loaf is cooked.
5 Explain why the Multigrain Batard would be a better choice than a loaf of white bread according to the guidelines of the *Australian Guide to Healthy Eating*.

Flatbreads

INGREDIENTS

Pita Bread

1 cup unbleached flour

1 teaspoon freeze-dried yeast

½ teaspoon gluten flour

½ teaspoon salt

¼ teaspoon sugar

½ cup warm to hot water

Naan

1 cup self-raising flour

1 tablespoon plain yoghurt (live, if possible)

½ teaspoon salt

65 millilitres lukewarm water, approximately

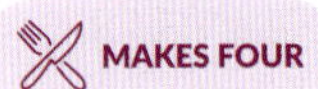
MAKES FOUR

METHOD

Pita Bread

1 Sift the dry ingredients into a bowl, add the water and mix into a soft dough.

2 Cover the surface of the dough with plastic wrap and rest it in a warm place for 10–15 minutes to prove.

3 Turn the dough onto a floured board and lightly knead. Divide into four portions.

4 Form each portion into a ball. Roll out each ball on a floured surface into a circle approximately 14 centimetres in diameter. Allow to rest on the bench for 5 minutes so that a slight skin forms.

5 Preheat griller on high.

6 Cook the Pita Breads, one at a time, under the hot griller for approximately 1–2 minutes, until each one puffs up and lightly browns. Then turn and cook the other side.

8 Allow to cool before filling with any accompaniments.

Naan

1 Put the flour, yoghurt and salt into a mixing bowl. Add the water a little at a time, working the mixture into a soft, slightly sticky dough with your fingers.

2 Cover the dough with a damp tea towel and leave it in a warm spot to ferment for 30–45 minutes.

3 Divide the dough into four portions and form each into a ball. Roll out each ball on a floured surface into an oval approximately 20–23 centimetres in length and 0.8 centimetres thick.

4 Preheat griller on high.

5 Cook the Naans, one at a time, under the hot griller until each one puffs up and is speckled with brown spots. Then turn and cook the other side. Naans cook very quickly, so keep watching them.

6 Serve warm.

9780170495714

EVALUATION

1 Draw up a table similar to the one below to record the similarities and differences in the sensory properties of the homemade flatbreads and some commercial flatbreads.

Type of flatbread	Appearance	Aroma	Flavour	Texture	Sound
Homemade pita bread					
Commercial pita bread					
Homemade naan bread					
Commercial naan bread					

2 Discuss the skills involved in the production of the flatbreads you made. What aspects of the production were difficult?

3 What are the advantages of homemade flatbreads?

4 Of all the flatbreads you tested, which product did you like best? Why?

5 Suggest a range of fillings or toppings that could be used with any of the flatbreads that you evaluated to enable the filled flatbread to successfully meet the recommendations of the *Australian Guide to Healthy Eating*.

Mark Fergus Photography

Pizza Bases

Pizza bases can be high and fluffy, like a bread crust, or they can be thin and crisp. These recipes are for the thin and crisp style. They have a higher fibre content than high and fluffy bases. Try one or the other with your favourite toppings.

INGREDIENTS

Wholemeal Pizza Base

½ cup plain flour

½ cup wholemeal flour

1 teaspoon dried yeast

¼ teaspoon salt

¼ teaspoon sugar

1 teaspoon olive oil

100–125 millilitres warm water

1 tablespoon semolina

Polenta Pizza Base

1 cup plain flour

⅓ cup fine polenta

2 teaspoons dried yeast

1 teaspoon sugar

1 teaspoon salt

⅓ cup warm milk

½ cup warm water

1 tablespoon olive oil

METHOD

Wholemeal Pizza Base

1 Preheat oven to 210°C.

2 Sift the flours, yeast, salt and sugar into a medium bowl. Add the oil to the warm water, then mix into the dry ingredients. Cover the surface of the dough with oiled plastic wrap and leave to prove for 10–15 minutes in a warm place.

3 Turn the dough onto a lightly floured board and knead lightly. Roll the dough into the shape of your pizza tray.

4 Oil the pizza tray and sprinkle it with semolina. Then oil your fingers and place the dough on the tray, spreading it to fit.

5 Add your choice of toppings and bake for 15–20 minutes.

Polenta Pizza Base

1 Preheat oven to 210°C.

2 Sift the flour, polenta, yeast, sugar and salt into a medium bowl. Combine the warm milk, water and oil, then mix into the dry ingredients. Cover the surface of the dough with oiled plastic wrap and leave to prove for 10 minutes in a warm place.

3 Turn the dough onto a lightly floured board and knead lightly. Roll the dough into the shape of your pizza tray.

4 Oil the pizza tray. Then oil your fingers and place the dough on the tray, spreading it to fit.

5 Add your choice of toppings and bake for 15–20 minutes.

EVALUATION

1 What are the three ideal environmental conditions yeast requires to grow?

2 Describe the physical changes that occur to the yeast dough during the proving process.

3 Why is it important to knead a yeast dough after the proving process?

4 Identify the ingredients in these Pizza Bases that are good sources of fibre.

5 Describe the sensory properties of a 'good' pizza base.

9780170495714

Napoli Sauce

Napoli sauce is a rich, flavoursome, concentrated tomato sauce from Italy. It is used as a base for many pizzas, and is spread over the dough base before other toppings are added. Traditionally, fresh tomatoes are used; if they are in season, they have wonderful colour and flavour. However, canned tomatoes are a suitable alternative. This recipe is suitable to spread on pizza bases. It can also be stirred through freshly cooked pasta.

INGREDIENTS

2 teaspoons olive oil

½ small onion, finely diced

1 clove garlic, crushed

1 cup canned diced tomatoes or passata

½ teaspoon dried basil

¼ teaspoon salt

½ teaspoon sugar

pepper

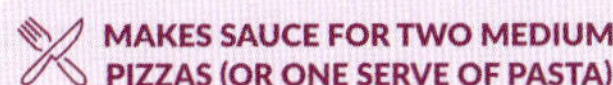

MAKES SAUCE FOR TWO MEDIUM PIZZAS (OR ONE SERVE OF PASTA)

METHOD

1. In a small saucepan, heat the oil and sauté or gently fry the onion and garlic until they are transparent.
2. Add the tomatoes or passata, basil, salt and sugar and cook uncovered until the sauce thickens.
3. Season with pepper, to taste.
4. Cool before spreading on pizza bases.

EVALUATION

1. Name the cooking process in which onion is fried until it is transparent.
2. What causes the Napoli Sauce to thicken in step 2 of the recipe?
3. Why is the Napoli Sauce cooled before it is spread on pizza bases?
4. What are some other herbs that could be used in this recipe?
5. Suggest some commercially available products that could be used on top of a pizza instead of making Napoli Sauce from scratch. Discuss the nutritional advantages of making Napoli Sauce rather than using a commercial product.

Mark Fergus Photography

9

RECIPES

Roasted Vegetable Pizza

Vegetables roasted in the oven develop a delicious brown crust, their flavour is intensified and their texture is softened, making them a tasty topping for a pizza. The Maillard reaction, dextrinisation and caramelisation occur when vegetables such as potato, sweet potato, parsnip and pumpkin are roasted. Mushrooms, eggplant, zucchini and garlic contribute a range of colour, textures and flavours to the end product.

INGREDIENTS

½ potato

150 grams sweet potato

150 grams pumpkin

¼ parsnip

¼ zucchini

¼ red capsicum

⅙ eggplant

2 mushrooms

1 clove garlic, peeled

1 tablespoon olive oil

salt and pepper

1 quantity Wholemeal or Polenta Pizza Base (page 244), prepared up to the end of step 4

½ quantity Napoli Sauce (page 245) or ⅓–¼ cup commercial equivalent

60 grams cheddar cheese, grated

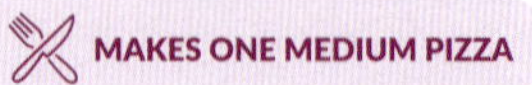

METHOD

1 Preheat oven to 210°C.

2 Wash all the vegetables except the mushrooms. Wipe the mushrooms with a clean, damp cloth.

3 Peel the potato, sweet potato, pumpkin and parsnip and cut into cubes, 1–2 centimetres in size.

4 Cut the zucchini, capsicum and eggplant into similar-sized cubes. Cut the mushrooms in half.

5 Combine all the vegetables and the clove of garlic in a bowl. Add the oil and toss together.

6 Cover a baking tray with a sheet of baking paper. Lay out the vegetables on the tray in a single layer. Season with salt and pepper.

7 Bake for 20–30 minutes or until the vegetables are tender.

8 Oil a pizza tray. If using the Wholemeal Pizza Base, sprinkle the tray with semolina. Roll out the pizza base dough thinly, transfer it to the baking tray and spread it with the Napoli Sauce.

9 Cover with the roasted vegetables, then sprinkle on the cheese.

10 Bake for 10–15 minutes or until the pizza crust is golden brown.

Mark Fergus Photography

EVALUATION

1 List the vegetables in the recipe that could be described as root vegetables.

2 What are the benefits of roasting vegetables rather than boiling them?

3 Explain why the Napoli Sauce and grated cheese are essential in making the Roasted Vegetable Pizza.

4 Why are pizzas baked in a very hot oven?

5 Discuss why the Roasted Vegetable Pizza would receive a better health rating than an Aussie pizza.

9780170495714

Calzones

A calzone, or Italian pocket pizza, is a half-moon-shaped dish with a filling folded inside a pocket of pizza crust. It is usually made as an individual serve and with a range of filling ingredients, such as meats, vegetables and cheese. The cheese is either bocconcini or mozzarella, both of which are stretchy when heated. A calzone can be baked or deep-fried. This version is baked.

INGREDIENTS

Dough

1 cup plain flour

1 teaspoon yeast

¼ teaspoon sugar

½ teaspoon salt

1 tablespoon olive oil

⅓ cup warm water

Filling

1 teaspoon olive oil

¼ medium onion, finely chopped

1 clove garlic, crushed

1 medium Roma or egg tomato, chopped, or ⅓ cup canned diced tomatoes

1 tablespoon fresh basil, chopped

2 slices mild salami, sliced

2 small bocconcini, sliced, or 2 slices mozzarella

2 teaspoons milk, for glazing

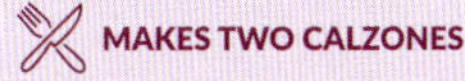

METHOD

1 Sift the flour, yeast, sugar and salt into a large bowl. Stir in the oil and warm water and mix to a dough.

2 Knead the dough on a floured surface until smooth and elastic.

3 Place the dough in an oiled bowl, cover with plastic wrap and stand in a warm place for 15–20 minutes to prove.

4 Preheat oven to 200°C.

5 Heat the oil in a small saucepan and fry the onion and garlic until soft.

6 Add the tomato or canned tomatoes and cook until the liquid has evaporated.

7 Remove from the heat and add the basil and salami. Allow the mixture to cool.

8 Turn the dough onto a floured surface, cut it in half and roll each half into the shape and size of a dinner plate.

9 Cover half of one round with half of the filling mixture. Top the mixture with half of the cheese and fold the top half over to enclose the filling. Dampen one edge with a little milk or water to help hold the dough closed.

10 Roll and twist the edges closed to seal the Calzone. Repeat to make the second Calzone.

12 Place the Calzones onto a greased tray and glaze them with milk. Cut two slits in the top of each to vent the dough. Bake for 15–20 minutes or until well browned.

Mark Fergus Photography

EVALUATION

1 Why does the calzone's design make it an ideal food to be eaten 'on the run'?

2 Explain why the filling is cooled before being wrapped in the dough.

3 What may happen if you forget to cut slits in the dough before baking?

4 Discuss any changes that you would make to the filling ingredients if you made this recipe again.

5 Classify the ingredients for the Calzones on a diagram of the *Australian Guide to Healthy Eating*. Comment on the health rating you would give this recipe.

VEGETABLES FOR GOOD HEALTH

Vegetable protein

Folate

Iron

Low in fat

Carbohydrates

Zinc

Vitamin A

High in dietary fibre

THE VALUE OF FIBRE

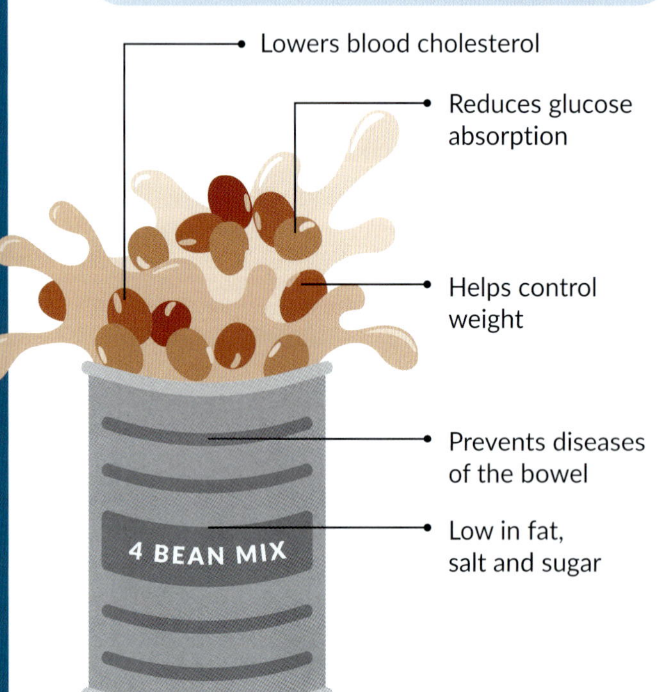

WHAT'S IN A COLOUR?

Antioxidants that help reduce the risk of cancer

Vitamin A for healthy eyes and mucous membranes

Folate and phytochemicals for protection from cancer

Antioxidants that protect body cells, reducing the risk of cardiovascular disease

Health-promoting phytochemicals

9780170495714

BENEFITS OF PROTECTED CROPPING

Higher yields; faster-growing and better-quality vegetables

Consistent supply; vegetables can be grown out of season

Reduces carbon footprint; less use of pesticides and herbicides

Protection from adverse weather, pests and diseases

Reduced water consumption

10 VEGETABLES AND LEGUMES

KEY TERMS

legumes the seeds from the *Leguminosae* family; these vegetables are eaten in their immature form as green peas and beans, and in their mature form as dried peas, beans, lentils and chickpeas

protected cropping a farming practice that involves growing food crops under, or sheltered by, an artificial structure, such as a greenhouse or polytunnel

solanine a toxin that develops when potatoes are exposed to light, resulting in a green colour on exposed surfaces

Templates
- PMI table
- Food order
- Sensory wheel
- Australian Guide to Healthy Eating

Worksheets
- Practical activity 10.1
- Practical activity 10.2
- Practical activity 10.3
- Design activity 10.1

Quizzes
- Testing knowledge 10.1
- Testing knowledge 10.2
- Testing knowledge 10.3

Nelson MindTap — To access resources above, visit **cengage.com.au/nelsonmindtap**

9780170495714

Vegetables and legumes/ beans in the *Australian Guide to Healthy Eating*

Vegetables and legumes/beans form one of the largest segments of the *Australian Guide to Healthy Eating* and should make up most of your plate. They are nutrient-dense plants that can be eaten raw or cooked, and help to make our daily diet interesting.

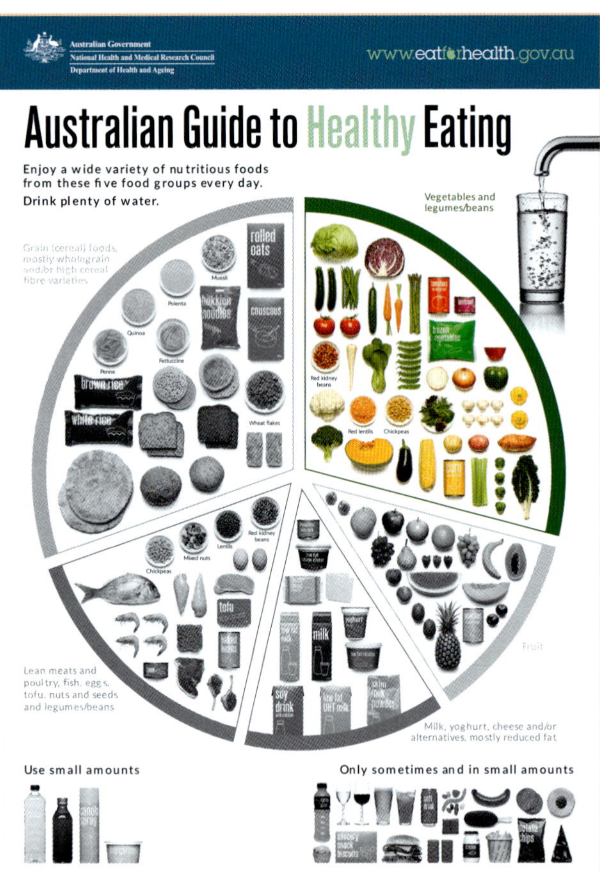

Figure 10.1 The place of vegetables and legumes/beans in the *Australian Guide to Healthy Eating*

Vegetables and legumes for good health

Vegetables and legumes are important for good health.

- Both vegetables and legumes are a valuable source of dietary fibre, which is necessary for a healthy digestive system.
- They are low in fat, and high in vitamins and minerals.
- Legumes are rich in protein, carbohydrate, iron and zinc, and therefore provide a great nutritional alternative to meat.
- Dark green leafy vegetables and orange and red vegetables supply vitamin A, which helps to repair and maintain tissue and assists with night vision.
- Dark green vegetables such as silverbeet and broccoli provide some of the iron that the body needs to help transport oxygen.
- Green vegetables and dried peas and beans are good sources of folate.
- Vegetables contain a range of other minerals, including magnesium, zinc and calcium.
- Vegetables are 70–95 per cent water, so most do not contribute many kilojoules to the daily energy intake.
- Legumes and vegetables such as potatoes and sweet potatoes are high in starch and are a good source of energy. However, when potatoes are cooked in a lot of fat, they soak it up like a sponge. Consequently, potato crisps and chips are concentrated sources of energy, and should be eaten only occasionally.
- Vegetables and legumes have a low energy density, so a diet high in a variety of these important foods will assist with weight management and reduce the risk of becoming overweight or obese.
- Eating a variety of vegetables and legumes each day may help to prevent some chronic diseases, such as heart disease and some cancers.

Figure 10.2 A range of vegetables served with hummus, which is made of chickpeas

Table 10.1 How much of the vegetables and legumes/beans group should I eat?

	Serves per day	
	12–13 years	14–18 years
Boys	5.5	5.5
Girls	5	5

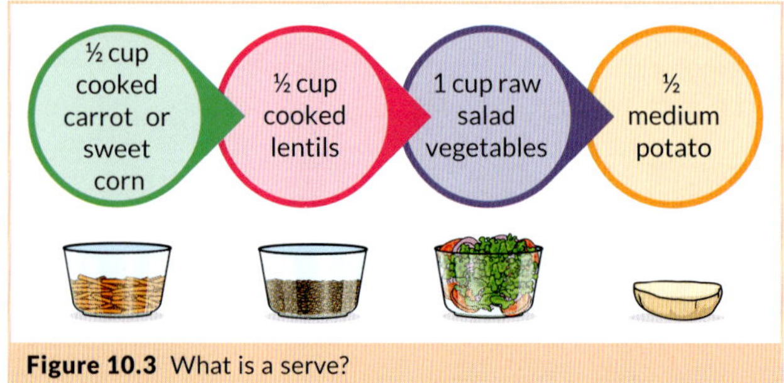

Figure 10.3 What is a serve?

9780170495714

Are we eating enough?

Data from the Australian Institute of Health and Welfare released in June 2024 shows that 96 per cent of children and adolescents aged 2–17 do not consume the recommended number of serves of vegetables each day. Similarly, 94 per cent of Australian adults aged 18 and over do not meet the requirements of the *Australian Dietary Guidelines* for vegetable intake each day. The research also showed that women are more likely to meet the recommendations for vegetable consumption than men.

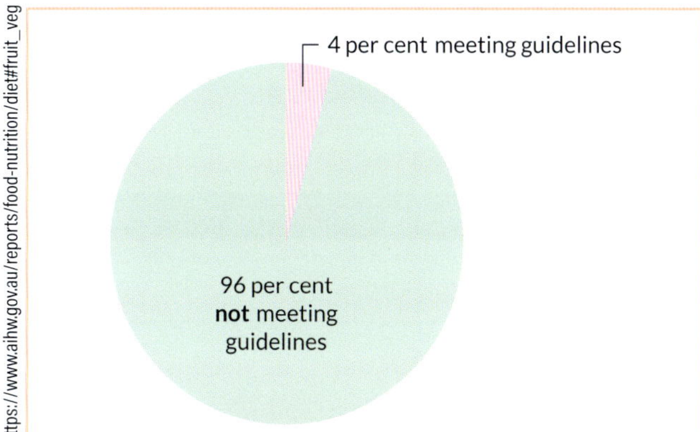

Figure 10.4 Percentage of children and adolescents aged 2–17 who meet the requirements of the *Australian Dietary Guidelines* for vegetable consumption

Why Australians do not consume the recommended serves of vegetables

Possible reasons why individuals do not consume the recommended serves of vegetables per day include the following:

- There is a lack of knowledge about the health benefits of vegetables.
- When planning meals, the protein component is often decided first and vegetables are not seen as such an important part of the meal.
- Vegetables are thought to be too expensive.
- There is minimal knowledge about the wide variety of fresh and processed vegetables available in Australia.
- Consumers are not confident in preparing and cooking vegetables.
- Some people think fresh vegetables take too long to prepare and cook.
- The sensory properties of some vegetables do not appeal to many children and so are often rejected when served as part of a meal.

Strategies to improve vegetable consumption

Several national and state government organisations have developed programs and strategies to encourage Australians to eat more vegetables. These include the Victorian Government's Better Health Channel and Nutrition Australia's Healthy Lunchbox Week.

Better Health Channel

Weblinks
Better Health Channel
Sanitarium

A vast range of information is available on the Victorian Government's Better Health Channel to help individuals learn about the importance of consuming more fruit and vegetables every day. Some of the key topics on the website provide information about the nutrients present in fruit and vegetables, with suggestions on how to cook and serve these foods to maximise their nutrient content. Parents and carers can also explore a range of fact sheets; for example, about why it is important to eat different types and colours of fruit and vegetables.

Having access to detailed information about fruit and vegetables will increase community members' knowledge about the health benefits of including these important foods in their own diet and in the diet of their family members.

Figure 10.5 A vegetable stir-fry makes a delicious meal.

National Lunchbox Week

An initiative by Nutrition Australia is National Lunchbox Week, which is held in February each year, just as students are returning to school after the summer holidays. The campaign offers ideas that aim to inspire parents and carers to create enjoyable and nourishing lunch boxes. This is valuable advice, as it is estimated that young people consume one-third of their daily food intake at school, so the foods in their lunch boxes need to contribute positively to their daily nutritional requirements.

Figure 10.6 National Lunchbox Week suggests ideas for preparing a healthy lunch box.

10

Activity 10.1

National Lunchbox Week

Visit the National Lunchbox Week website.

1 Explore the website and identify the goals and aspirations of National Lunchbox Week.

2 Explore the lunch box recipe ideas on the website. Which one would you enjoy the most in your lunch box? Explain why.

3 Read the fact sheets and guides, and select one that appeals to you. After reading the tips on your chosen fact sheet or guide, complete a PMI (Plus, Minus, Interesting) table about the information it contains.

4 Select one of the videos and watch it. Which suggestion would you be most likely to try? Justify your answer.

5 Work with a partner to design a lunch box for a child in Grade 3 or 4. In your menu planning, keep in mind the National Lunchbox Week initiative and the promotion of eating vegetables. Explain why your lunch box menu is suitable for a child in Grade 3 or 4.

Template
PMI table

Weblink
Lunchbox Week

Plus	Minus	Interesting

CASE STUDY 10.1

Effects of lowering vegetable consumption

Read the article below, and then answer the questions that follow.

Drop in vegetable consumption driving farmers out of business, harming health

Jane McNaughton, *ABC Rural*, 9 October 2024

Are you eating enough vegetables?

You are in the minority if you are, according to the Australian Bureau of Statistics.

The bureau's National Health Survey says only 6.5 per cent of Australians are eating their daily recommended dose of five serves per day, with most eating fewer than two serves and ultra-processed foods taking up more of our diets.

Dietitian and nutritionist Jemma O'Hanlon is worried that our busy lifestyles are contributing to preventable diseases, mostly due to what we are eating.

'Many of us are choosing too many unhealthy foods in our daily diets and it's really not the individual's fault,' she said.

'We're exposed to these unhealthy foods wherever we go, at the supermarket, the petrol station, the convenience store, even school canteens have a lot of unhealthy foods in there.

'Overweight and obesity, heart disease, type two diabetes and many cancers are also linked with poor diets, so this is a really serious problem.'

Bad food habits affecting farmers

The downward trend in vegetable consumption also affects the people who grow it, with many already doing it tough due to economic conditions and other industry pressures.

A recent study by peak horticulture industry body AusVeg found that more than a third of growers were considering leaving the industry in the next 12 months, citing cost increases for fuel, electricity, labour costs and fertiliser, poor retail pricing, and regulatory burdens.

...

Staying healthy more affordable

But it's not just farmers' bottom lines being affected by Australians eating less vegetables – Ms O 'Hanlon said the economy was also suffering from the side effects.

'The total healthcare spend on risk factors that are potentially avoidable is actually $24 billion,' she said.

9780170495714

'This sort of money could actually be invested [by governments] in putting towards programs and campaigns that promote our health and wellbeing.'

So why aren't we eating enough vegetables?

Ms O'Hanlon said in addition to the busy lives, there was a perception that eating healthy was too expensive.

'The good news is that veggies are not expensive. They are, in fact, very affordable and I think this a bit of a myth and misconception out there,' she said.

'In fact, fresh veggies on average are about 65 cents a serve.'

Multiple studies have found that on average, a supermarket shop of foods that meet Australia 's dietary guidelines was 7 per cent cheaper than what's in the average person's basket.

Healthy diet, healthy economy

Ms O'Hanlon said one of the most effective ways to help the waistline and the wallet, was to embrace whole foods and eat more vegetables.

'People are looking for cheaper options to make sure they can live and cover all of their expenses,' she said.

'The healthiest and the most affordable diet is the Mediterranean diet – it beats the Western diet that a lot of us are eating at the moment.

'Vegetables are rich in fibre, which nourishes our gut – in essential vitamins, minerals and antioxidants, which protect us and boost our health and wellbeing.'

The Mediterranean diet includes eating a variety of fruits, vegetables, wholegrains, legumes, fish and seafood, and minimal processed foods and red meat. …

Source: 'Drop in vegetable consumption driving farmers out of business, harming health' by Jane McNaughton on ABC Rural, 9 October 2024 https://www.abc.net.au/news/rural/2024-10-09/dropping-vegetable-consumption-hurting-farmers-and-health/104429338

iStock.com/Wavebreakmedia

Figure 10.7 Selecting food that meets the *Australian Dietary Guidelines* is a cheap and healthy option for shoppers.

Analysis

1 What percentage of Australians are not eating the recommended five serves of vegetables per day?
2 Explain why most Australians are consuming more ultra-processed foods and fewer vegetables each day, and what impact this has on their health and wellbeing.
3 Create a mind map that summarises the impact of the lower vegetable consumption on the farming community and on the national economy.
4 Explain why, from a health perspective, consuming the recommended number of serves of vegetables each day will benefit individuals.
5 Compare your usual food consumption with the makeup of the Mediterranean diet, and comment on the similarities and differences.

The importance of dietary fibre

Dietary fibre is a type of carbohydrate that is essential for good health. Foods that are high in dietary fibre have been shown to help lower blood cholesterol, reduce glucose absorption, and prevent diseases of the bowel such as constipation and diverticulitis. High-fibre foods can also help control weight because they provide greater satiety; that is, they make you feel full for longer.

Health professionals recommend that adults should consume approximately 25–30 grams of dietary fibre each day. Interestingly, most Australians do not include sufficient dietary fibre in their diets, consuming only 18–25 grams of fibre daily.

Dietary fibre is found in the cell walls of all plant foods, such as fruits and vegetables, peas, beans and cereals. Foods that have a high level of dietary fibre are usually low in fat, salt and sugar.

There are two main types of dietary fibre: insoluble dietary fibre and soluble dietary fibre.

The body is unable to digest or absorb insoluble dietary fibre, but it is nevertheless crucial to include it in daily meals because it adds bulk to the diet and helps to eliminate the waste material from the body.

By helping food pass through the digestive system more quickly, insoluble dietary fibre helps to maintain our bowel health and prevents us from becoming constipated. A diet high in insoluble dietary fibre is also known to increase the number of good bacteria, or microbiota, that live in our gut. Gut microbiota help to prevent the risk of serious diseases of the bowel, such as bowel cancer.

Insoluble dietary fibre is found in a wide range of vegetables. The table below lists the amount of insoluble dietary fibre found in some popular vegetables.

iStock.com/mustipan

Figure 10.8 Dietary fibre is found in the cell walls of all plant foods.

Table 10.2 Insoluble dietary fibre in vegetables

Vegetable	Insoluble fibre (raw) per 100 grams (g)
Beans	2.7
Broccoli	4.1
Capsicum	0.9
Carrot	3.3
Cauliflower	1.8
Celery	1.8
Cucumber	1.1
Eggplant	2.3
Lettuce	1.7
Mushroom	2.5
Onion	1.5
Parsnip	2.5
Peas	2.3
Potato	1.7
Pumpkin	1.5
Spinach	2.7
Sweet corn	4.5
Sweet potato	2.0
Tomato	1.2
Zucchini	1.6

Soluble dietary fibre is found in oatmeal, legumes and the bran from rice and barley. One of the most important functions of soluble dietary fibre is to help to reduce blood cholesterol levels. It has also been found to be helpful in managing diabetes by stabilising blood glucose levels.

Classification of vegetables

Vegetables can be classified according to the way they grow, their structure, or the part of the plant that is eaten, as shown in Figure 10.10 on the following page.

Eating the rainbow

Vegetables can also be categorised according to colour. If you choose vegetables in all the colours of the rainbow, you will consume a wide range of nutrients that are essential for good health.

iStock.com/AnaBGD

Figure 10.9 An array of colourful vegetables

What's in a colour?

Guideline 2 of the *Australian Dietary Guidelines* encourages us to enjoy 'plenty of vegetables, including different types and colours and legumes/beans'. To achieve this, over one week we should try to eat a range of vegetables.

Vegetables make meals more interesting because they contribute a variety of colours, flavours and textures. The varying colours in vegetables are due to different pigments:

- **Red:** Tomatoes, capsicum and radishes contain lycopene, which produces their red colour. Lycopene is an antioxidant that may help reduce the risk of developing some cancers.
- **Orange/yellow:** Carrots, sweet potato, sweet corn and pumpkin contain carotenoids – including betacarotene, which is converted to vitamin A. This helps to maintain healthy mucous membranes and healthy eyes.

9780170495714

- **Green:** The green colour of spinach, asparagus, broccoli, peas, green beans, avocados and green capsicum comes from chlorophyll. These vegetables also contain phytochemicals, which help fight against cancer. Green leafy vegetables are a good source of iron and folate.
- **Purple/blue:** The main pigment found in purple/blue vegetables such as beetroot, red cabbage and eggplant is anthocyanin. This is an antioxidant that helps to protect our body cells, and reduces the risk of cardiovascular disease and some cancers.
- **Brown/white:** The white flesh of cauliflower, garlic, onions, potatoes, parsnips and turnips contain health-promoting phytochemicals. Allicin, which is found in garlic, has antiviral and antibacterial properties.

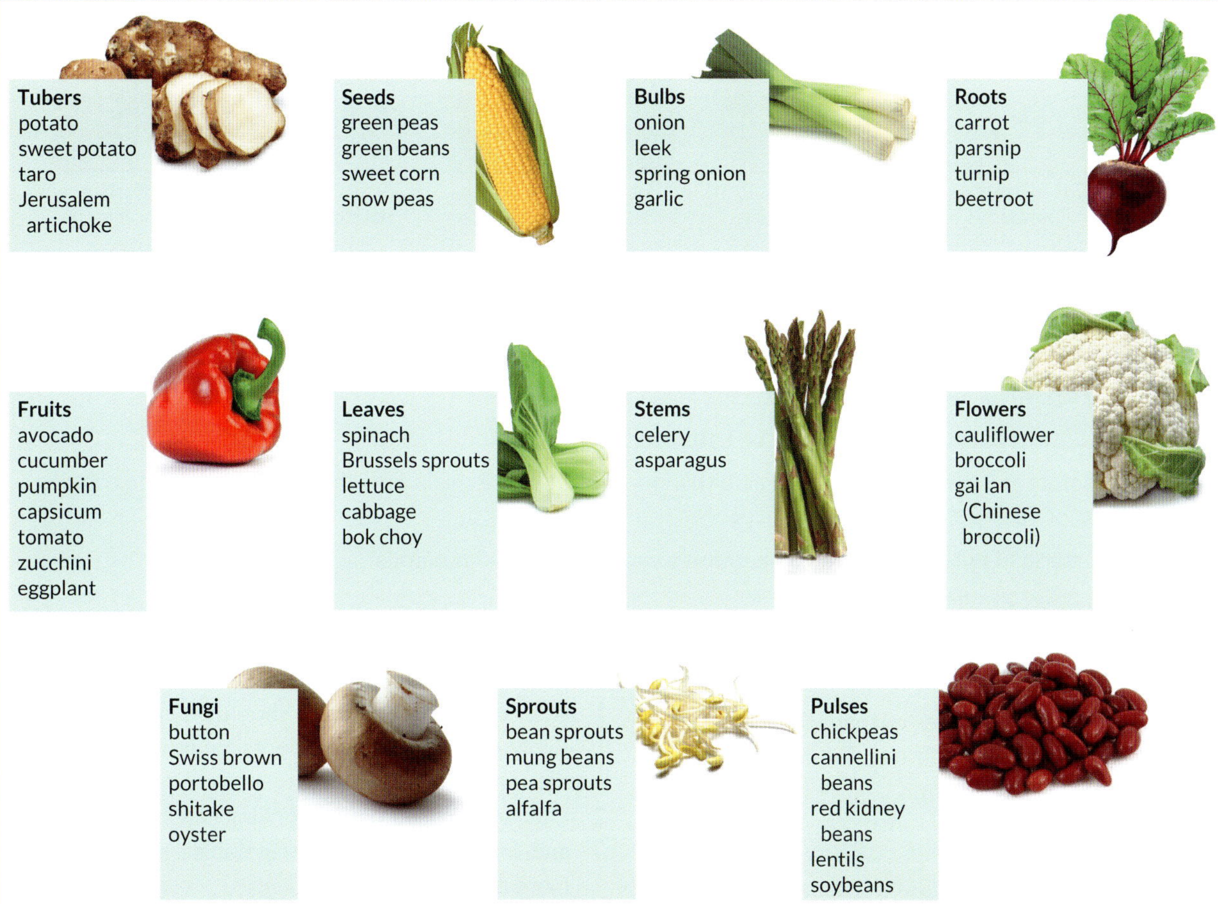
Tubers
potato
sweet potato
taro
Jerusalem artichoke
Seeds
green peas
green beans
sweet corn
snow peas
Bulbs
onion
leek
spring onion
garlic
Roots
carrot
parsnip
turnip
beetroot
Fruits
avocado
cucumber
pumpkin
capsicum
tomato
zucchini
eggplant
Leaves
spinach
Brussels sprouts
lettuce
cabbage
bok choy
Stems
celery
asparagus
Flowers
cauliflower
broccoli
gai lan (Chinese broccoli)
Fungi
button
Swiss brown
portobello
shitake
oyster
Sprouts
bean sprouts
mung beans
pea sprouts
alfalfa
Pulses
chickpeas
cannellini beans
red kidney beans
lentils
soybeans

Figure 10.10 The classification of vegetables

Top left to right: iStock.com/kaanates, iStock.com/rimglow, iStock.com/tpzijl, iStock.com/xamtiw; Middle left to right: iStock.com/Josef Mohyla, iStock.com/PotaeRin, iStock.com/stuartbur, iStock.com/Eivaisla; Bottom left to right: iStock.com/Floortje, iStock.com/4kodiak, iStockphoto.com/lleerogers

Activity 10.2

Thinking about vegetables

1 Sketch and colour in a food rainbow. Annotate the rainbow with as many vegetables as you can think of for each colour.

 a Share your food rainbow with other class members, and add to your rainbow after discussion.

 b Which colour has the greatest variety of vegetables?

 c Identify the two vegetables you like most and the two you like least in your rainbow. Compare your choices with those of other class members, and then explain why there are similarities and differences in your choices.

 d Which vegetables can be eaten raw and which are more enjoyable cooked?

 e Which colour vegetables would be a good source of iron in the diet?

 f Suggest some vegetables you could eat to boost vitamin A in your diet.

2 Look at the graph below, and answer the questions that follow.

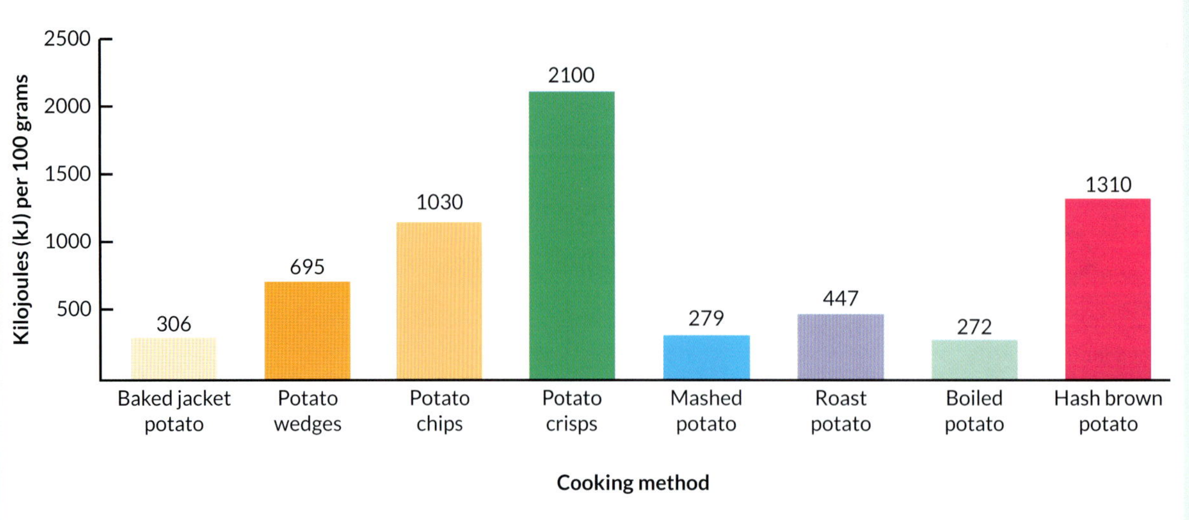

Figure 10.11 Energy value for 100-gram portions of potatoes

a Why do you think potato crisps have such a high energy value?

b Explain why hash browns should only be eaten occasionally.

c Baked potatoes have a lower energy value than roast potatoes. Explain how the differences in the cooking methods of baking and roasting may account for this.

d Why do mashed potatoes have a higher energy value than boiled potatoes?

Cooking vegetables

Preparing and cooking vegetables changes their appearance, aroma, flavour, texture and sound, and sometimes their nutrient content.

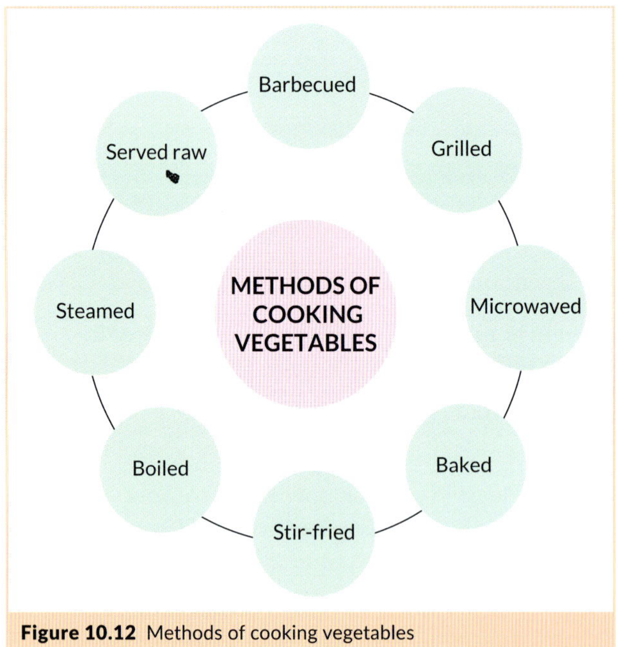

Figure 10.12 Methods of cooking vegetables

Top tips for cooking vegetables

1 After washing, vegetables should be peeled thinly, because many nutrients are found just under the skin.

2 The heat applied during cooking can intensify flavours and soften the texture of vegetables, so that they are more enjoyable to eat.

3 The processes of baking and frying cause some vegetables, such as onions and potatoes, to caramelise, and converts the reducing sugars to dextrin, making them more appealing to many people.

4 The processes of boiling and steaming should be carried out until the vegetables are just cooked to avoid the loss of water-soluble nutrients, such as vitamin C, and to prevent vegetables – particularly greens – from losing their colour.

5 Microwaving is a very quick cooking process that minimises nutrient loss and allows vegetables to maintain their colour.

9780170495714

Testing knowledge 10.1

Quiz Testing knowledge 10.1

1 Create a mind map to highlight the nutrients in different types of vegetables.
2 Explain why including vegetables in the diet is an effective strategy in maintaining a healthy weight.
3 How many serves of vegetables is it recommended that Australians should eat each day?
4 Consider the data in Figure 10.4 on page 251, regarding vegetable consumption for children and adolescents. From your experience, suggest reasons why so few children and adolescents meet the *Australian Dietary Guidelines* for vegetable consumption.
5 Outline the type of information available on the Better Health Channel that will help individuals learn about the importance of increasing their consumption of vegetables and fruit.
6 Discuss why dietary fibre is important for good health.
7 Explain the difference between soluble and insoluble fibre, and how each contributes to good health.
8 What creates different colours in vegetables? List two examples.
9 How does cooking vegetables make them more appetising to eat?
10 Why is microwaving a good method of cooking vegetables?

Fresh beans

The most common fresh bean varieties in supermarkets and markets are green beans. Other types include fresh butter beans, which are the same shape and size as green beans, and snake beans, which are used in Asian-style cooking and are very long with dark brown tips.

Some of the green bean harvest is processed so that consumers can purchase them frozen, canned or dried.

Fresh green beans should have a bright green, slightly glossy appearance, and snap cleanly when broken in half. Beans that bend, are wrinkled or have some discoloured spots have poor flavour and texture.

Fresh green beans can be stored in a plastic bag in the crisper of the refrigerator for up to four days. Fresh green beans should be cooked quickly to retain their vitamins A and C.

Figure 10.13 Green beans

iStock.com/loops7

PRACTICAL ACTIVITY 10.1

Worksheet Practical activity 10.1

Cooking green beans

Aim

To compare the flavour, texture and colour of fresh green beans cooked by different methods

Ingredients

- 30 fresh green beans (6 beans per test)
- 1 teaspoon of oil

Equipment

Study the results table on the following page and determine the equipment list required for the test.

Method

1 Wash and cut the top and tails off all the beans.
2 Use a vegetable knife to slice each bean on a 45° angle.
3 Prepare the cooking equipment for each test.

Results

	Flavour	Texture	Colour	Sound
Boil the beans in 2 centimetres of water with the lid on. Cook for 3 minutes.				
Boil the beans in 2 centimetres of water with the lid off. Cook for 3 minutes.				
Steam the beans. Cook for 3 minutes.				
Microwave the beans with 1 tablespoon of water in a covered container. Cook for 1 minute 30 seconds.				
Stir-fry the beans in 1 teaspoon of oil. Cook for 2 minutes.				

Analysis

1 Which cooking method produced beans with the best flavour?
2 Which cooking method produced beans with the best texture?
3 Which cooking method produced beans with the best colour?
4 Does the sound of biting into a fresh bean alter once it is cooked? Explain your answer.
5 Decide which cooking method would retain the most nutrients in fresh beans. Explain why.

Conclusion

What was the best cooking method to achieve good flavour, texture, colour and sound in fresh green beans?

Potatoes

Potatoes are a nutritious food, as they are a good source of carbohydrates, vitamin C, potassium and dietary fibre. They also provide some magnesium, niacin and thiamine. Potatoes are fat-free, but the methods used to cook them – such as deep-frying – can increase their energy value.

The potato is a tuber that grows underground on the roots of the plant. The leaves of the potato plant are poisonous. In Australia, potatoes come in many different skin colours, but the flesh inside the potato is usually white or cream in colour.

All potatoes should be firm, well-shaped for their variety, and blemish-free. Avoid potatoes that are wrinkled and beginning to shoot. Greening of potatoes is caused by the development of a toxin called **solanine**, which develops when potatoes are exposed to light – either natural or fluorescent. Light causes chemical changes in the pigment chlorophyll, resulting in a green colour on all exposed surfaces. Potatoes that have a green tinge should be thrown away, because the toxin solanine can cause illness.

Potatoes stored in a cool, dark, well-ventilated cupboard will keep for two to three weeks. Keeping them away from light will prevent them from greening, and storing them separately from onions will maximise their storage life.

Potato varieties are often differentiated by reference to whether they are waxy or starchy.

iStock.com/jenifoto

Figure 10.14 Potatoes

9780170495714

Table 10.3 Characteristics and uses of waxy and starchy potato varieties

Waxy potatoes		
Waxy potatoes have a high moisture content, but are low in starch. They hold their shape and remain firm when boiled, and consequently are not ideal for mashing.		
Kipfler	• Small, long and oval – sometimes called the 'finger potato' • Yellow skin and yellow waxy flesh • Best steamed	Picture Partners/Alamy Stock Photo
Desiree	• Long and oval • Pink skin and creamy yellow waxy flesh • Suited to baking, boiling and mashing, but not deep-frying	Martin Lee/ Alamy Stock Photo
Pontiac	• Round • Red/pink skin and white waxy flesh • Denser than other varieties, so requires longer cooking • An all-purpose potato – good for baking, boiling and grating	Enlightened Media/ Shutterstock.com
Starchy potatoes		
Starchy potatoes have a low moisture and sugar content, but are high in starch. They mash and bake well, but tend to collapse when boiled because of their low sugar content. They produce golden chips when fried.		
Coliban	• Round • White potato with smooth skin and floury flesh • Suited to baking and mashing, but sometimes disintegrates with boiling	Flower Studio/ Shutterstock.com
Sebago	• Round/oval • White, dry flesh • Considered an 'all-rounder', and ideal for baking, roasting, boiling, frying and mashing	iStock.com/domnicky

PRACTICAL ACTIVITY 10.2

Worksheet Practical activity 10.2

Template Sensory wheel

Scientific experiment: Investigating the best cooking method for varieties of potatoes

Aim

To determine the cooking method that produces the best eating properties in waxy and starchy potatoes

Ingredients

4 potatoes, each of a different variety: 2 waxy and 2 starchy

Equipment

- 1 steamer
- 1 saucepan with lid
- oven tray covered with baking paper
- 1 container suitable for use in microwave
- chopping board
- vegetable knife

Method

1. Select 4 potatoes, each of a different variety – two with waxy flesh and two with starchy flesh. Refer to Table 10.3 on page 259 to help you in your selection.
2. Preheat oven to 220°C and set up a steamer. Fill a saucepan with water and set it to boil.
3. Peel and cut each of the varieties of potatoes into 1-centimetre cubes.
4. Divide the potato cubes of each variety into four groups, one for each method of cooking.
5. Brush one-quarter of the potato cubes of each variety with oil. Place on a lined baking tray, and bake until tender and golden brown.
6. Steam, boil and microwave the remaining potato cubes.
7. When cooked, present samples on plates for evaluation.
8. Record your results on the table.

Results

In a table like the one below, record your observations of the sensory properties – appearance, aroma, flavour, texture and sound – for each variety of potato.

Potato variety	Baked	Steamed	Boiled	Microwaved

Analysis

1. Which cooking method allowed the most varieties of potato to retain their shape?
2. Identify the varieties of potato that performed best for:
   - baking
   - steaming
   - boiling
   - microwaving.
3. Consider the information about the varieties of potatoes in Table 10.3 on page 259. Compare this information with your results, and identify any that were significantly different. If there were differences in your results, suggest reasons for this.
4. How does each cooking method affect the nutrient content of potatoes?
5. Suggest other foods that would complement each method of cooking potatoes to make up a main meal.
6. List a safety consideration for each method of cooking potatoes.

Conclusion

1. Decide which cooking method produces the best eating properties for each variety of potato.
2. 'The sensory properties of cooked potatoes will vary depending on whether the flesh is waxy or starchy, and on the method used to cook them.' Do your results support this statement? Explain your answer.
3. Make recommendations about the selection of varieties of potatoes for future recipes you might prepare.

9780170495714

CASE STUDY 10.2

A family potato-growing business

Read the article below, and then answer the questions that follow.

K.P. Maher and Sons: Growing potatoes in Springbank, Victoria

Background

Potato growing in Victoria is historically connected to many of the Irish potato farmers who left Ireland during the potato famine. Between 1845 and 1852, the potato crop that was staple food of the Irish people was destroyed by the potato blight fungus. In the early 1850s, facing starvation and an uncertain future, the founder of K.P. Maher and Sons, John Maher, along with many other potato farmers, left Ireland for Australia. John Maher settled in Springbank, a district outside of Ballarat, which had rich free-draining soils similar to Ireland. The weather was also suitable for growing potatoes as it is usually frost-free between November and April. Today, the Maher family continue to grow potatoes in Springbank as they have for the past seven generations.

Process of growing potatoes

Planting

In Victoria, the paddocks are ploughed in September and then harrowed to smooth the surface of the soil and conserve the moisture. Farmers start planting potato crops in late October when the frosts have finished, and the soil is beginning to warm. The farmers use certified seed potatoes that are grown by specialist growers and have been tested to ensure they are disease free. On the farm, the seed potatoes are cut by machine into uniform pieces ensuring they have at least two 'eyes' or shoots. The seed potatoes are laid out to dry and then transferred to a potato planter which plants the potato pieces in rows in the prepared paddocks. A range of fertilisers such as nitrogen, phosphorous, potash, sulphur and calcium are added to the soil when the potatoes are planted.

Photos box/Shutterstock.com

Figure 10.15 Certified seed potatoes being sown

Growing

The potatoes are irrigated during their growing season, using lateral irrigators so that water is evenly applied across the crop. This method of applying water is not affected by wind because the sprinkler heads move with the wind, minimising the impact of wind drift on the way water is distributed across the crop. This is therefore a more sustainable agricultural practice as it reduces the amount of water used. Water used for irrigation on the Springbank farm is sourced from underground bores and stored in dams on the property.

iStock.com/georgeclerk

Figure 10.16 A lateral irrigator being used to water potato crops

Harvesting

The potato harvest begins in mid-March when the tops of the potato plants die down and the skins on the underground tubers set. Potatoes that are to be stored are left in the ground until the temperature of the soil cools to 16°C as this prevents the tubers from rotting. Potatoes that are processed into French fries are harvested earlier in the year.

Processing potatoes

The harvested potatoes are graded and trucked to storage facilities for sale to the fresh food market or transported to food processing plants.

K.P. Maher and Sons grow a specific variety of potatoes on their property called the 'Innovator' for McDonald's restaurants. This variety of potatoes is the ideal size and length to make into the French fries that McDonald's is renowned for – super-long chips that are crisp on the outside and fluffy on the inside. As well as supplying McDonald's with potatoes for their French fries, some of their potato crop is exported to be made into potato crisps.

Sustainable practices

To maximise the fertility of the soil, K.P. Maher and Sons use a four-year crop rotation cycle on their farm. In between their crops of potatoes, they grow canola and wheat, and also grass as feed for cattle and sheep.

Figure 10.17 Grading potatoes after harvest

iStock.com/georgeclerk

Analysis

1. Explain why many Irish potato growers emigrated to the Australia during the 1800s.
2. Outline the reasons why the Ballarat district was one of the areas in Victoria selected by Irish settlers to establish their new potato farms.
3. Explain why it is important to use certified seed potatoes when planting a new crop.
4. Discuss the benefits of using a lateral irrigator when watering a potato crop.
5. Outline the environmental factors that will determine when farmers harvest the potatoes that are to be stored before being sold into the fresh food market.
6. List the main products that potatoes grown by K.P. Maher and Sons are processed into.
7. Suggest reasons why a four-year planting rotation is used for farmland growing potatoes.

Growing vegetables sustainably

Climate change has had a detrimental impact on food production, and, as a result, many horticulturalists are now using a system of **protected cropping** to grow a reliable source of vegetables for Australian consumers. The development of protected cropping is expected to increase dramatically in Australia in the future, in response to the effects of extreme temperatures, humidity, flooding and plagues of pests such as mice and grasshoppers.

Protected cropping involves growing food crops under, or sheltered by, an artificial structure, such as a greenhouse or polytunnel. This enclosed structure protects vegetables such as tomatoes, cucumbers, green leafy vegetables, capsicums and eggplants from adverse weather, pests and diseases. It also enables farmers to create optimal growing conditions, as they can control the temperature, light and moisture inside the structures. This system is different from the traditional method of growing vegetable crops outdoors in paddocks, where they are exposed to variations in weather, and are at risk of being attacked by pests and disease.

Growing vegetables in a protected cropping environment is growing rapidly, and approximately 20 per cent of the total value of Australia's vegetable production now uses this system.

9780170495714

PATTAMA BOONSIRI/Shutterstock.com

Figure 10.18 Protected cropping is used to grow many of Australia's green leafy vegetables.

Sergey Bezverkhy/Shutterstock.com

Figure 10.19 Growing tomatoes in a protected environment

Features of a protected cropping system

- Crops are often grown in a soil-like material made up of sand, compost, sawdust and gravel, rather than directly in the soil. This allows good moisture retention.
- Hydroponic watering systems provide the plants with water and nutrients, and are automated and computer controlled. Excess water and nutrients are recycled and used several times.
- Fans and heaters are used to control the humidity. This is important because it prevents excessive humidity within the growing structure that can cause disease and increase the need for pesticides.

Protected cropping and sustainability

Protected cropping helps to safeguard against climate change and to ensure sustainability by reducing the amount of water, herbicides and pesticides required. However, operating the heating and cooling systems for optimal growing is very energy-intensive. Some farmers who use protected cropping systems produce their own solar power in order to improve their environmental sustainability and to reduce production costs.

Table 10.4 Benefits of protected cropping

Vegetable production	Environmental impact
Higher yields	Reduced carbon footprint
Faster growing times	Less use of pesticides and herbicides
Improved consistency of quality	Targeted nutrients, so less fertiliser is used
Vegetables can be grown out of season	Reduced water consumption
Consistent supply, because growing time is known	Minimal pests and diseases
	No weeds competing for nutrients
	Vegetables can be grown closer to urban settlements, so food miles are reduced

Pre-prepared salad mixes and vegetables

In the past few years, there has been a change in the way many consumers purchase their salads and vegetables at the supermarket. Until recently, if a consumer wanted to make a salad at home, they would purchase a variety of individual salad vegetables, wash and dry them at home, and then tear or slice them into a suitable serving size before tossing them in a salad bowl with a homemade dressing. Similarly, consumers would purchase a piece of pumpkin, several carrots, potatoes, red onions and zucchini to prepare at home as an accompaniment to the family roast. Often, there were leftover vegetables that had to be stored or thrown out.

Today purchasing a pre-prepared, pre-packed bag of salad ingredients from the supermarket is a very convenient and time-saving option for shoppers. The bag includes a mix of leafy, washed salad greens sufficient for a family. All the cook needs to do is simply tip the contents into a salad bowl and add a splash of salad dressing!

Figure 10.20 Salad mixes are a convenient option for families.

Consumers can also purchase a wide range of pre-prepared vegetables such as ready-to-cook roast vegetables. They can buy sweet potato or zucchini pre-cut into 'noodles', or cauliflower or broccoli processed into 'rice' – a few minutes tossed in a wok with a splash of oil, and they are ready to eat as a quick and healthy vegetable for dinner.

Figure 10.21 Pre-prepared mixed vegetables, ready to roast

For those families wanting to make healthy lunch boxes that are quick and easy to prepare, they can now purchase a variety of pre-prepared vegetables and fruit, including carrot sticks and celery sticks.

Figure 10.22 Pre-prepared fresh carrot sticks are a healthy snack.

Australian shoppers can now select from a wide variety of pre-prepared salad mixes, green leafy vegetables and other vegetables. The range of these products has increased significantly in recent years as a result of consumer demand for products that meet their specific needs.

The advantages of using pre-prepared salad mixes and vegetables

Convenience

- Many people find they have little time to prepare a meal from scratch, given their work, study and/or family commitments. Purchasing a product such as a pre-prepared salad mix to serve with a pan-fried chicken fillet or barbecued lamb chop means they can cut down the preparation time required to serve a nutritious meal for their family. Similarly, purchasing the vegetables for roasting that have already been peeled and cut into serving sizes makes meal preparation much quicker and less onerous. When making a stir-fry, the cook can simply purchase an Asian stir-fry kit that includes a sachet of sauce. There is no need to shop for different components to make the stir-fry, slice the vegetables or prepare the sauce – they can simply heat the wok, and dinner is almost ready to serve!
- Buying a packet of pre-cut carrot sticks or baby cucumbers for snacks is a time-saver when preparing school lunches.
- Pre-prepared vegetables are of benefit to people who have limited cooking skills or find peeling and chopping vegetables a tedious task. Simply purchasing vegetables that have been pre-cut or pre-prepared may encourage more people to cook at home and may lead to healthier eating habits.

Increasing the number of serves of vegetables individuals eat

- Having easy access to pre-prepared salad mixes and vegetables may encourage more people to increase their vegetable consumption, and help them meet the *Australian Dietary Guidelines* for the number of serves of vegetables they should eat each day.

Minimising food waste

It is important to minimise wasted vegetables, as the breakdown of vegetable matter gives off methane, a greenhouse gas that contributes to global warming.

- Using pre-prepared salad mixes and vegetables may help lessen food waste, particularly for people who live alone or in small households. If, for example, they want to have coleslaw with their dinner, they can buy a packet of coleslaw mix – rather than purchasing a whole cabbage and a bag of carrots to make their own coleslaw, and then not being able to use it all and throwing out the leftovers.
- People who live in apartments may not have access to composting facilities, so if they prepare their own vegetables from scratch, the peelings and offcuts may have to be put in the rubbish bin and end up being sent to landfill.

The disadvantages of using pre-prepared salad mixes and vegetables

The environmental impact of packaging materials

- Pre-prepared products, including salad mixes and vegetables, are typically packed in polypropylene

9780170495714

plastic bags or placed on polystyrene trays, covered in plastic wrap. Most of these materials are described as 'single use' plastics, and may not be recyclable or biodegradable. As it degrades, plastic packaging gives off the greenhouse gas methane, polluting the environment.

Loss of nutrient content

- Nutrients may be lost through oxidation when vegetables are peeled, cut and then stored under plastic wrap for an extended time. This may lead to a shortened shelf life that, in turn, leads to food waste.

Increased cost of pre-prepared salad mixes and vegetables

- The cost of purchasing pre-prepared vegetables and salad mixes has been shown to be considerably more than the cost of preparing them ourselves. This will impact on the family budget, especially when there are many people in the family.

  An article published in *Choice* magazine in November 2023 observed:

  > *Being time-poor can make those pre-cut trays and bags of veggies look very attractive, but our research shows they can cost you a lot. When we looked at things like carrot sticks and broccoli florets, we found you could pay around nine times as much for the convenience. Pre-cut produce can be a game changer for the elderly or people with disability, but sadly, you'll pay a premium for it.*

Source: *Choice* magazine article, '10 tips to help you save on your fresh fruit and veg' by Margaret Rafferty – 7 November 2023 https://www.choice.com.au/food-and-drink/groceries/fruit-and-vegetables/articles/tips-to-help-you-save-on-fruit-and-veg

Testing knowledge 10.2

Quiz
Testing knowledge 10.2

11 Describe the features you would look for when selecting fresh green beans to buy and the best method of storing them.

12 Explain why it is important to include potatoes in a well-balanced diet.

13 Why should potatoes that have green on their skin not be consumed?

14 Which varieties of potato would be suitable for mashing if you were making gnocchi?

15 What is 'protected cropping', and what types of vegetables are grown using this system?

16 List the features of a protected cropping system.

17 Explain the benefits to the environment of growing vegetables using protected cropping.

18 Identify two reasons why consumers find pre-prepared and pre-packaged vegetables convenient.

19 Discuss how pre-prepared and pre-packaged vegetables assist in minimising food waste.

20 Create a mind map to highlight the disadvantages of purchasing pre-prepared and pre-packaged vegetables.

Figure 10.23 What varieties of potato are suitable for making gnocchi?

PRACTICAL ACTIVITY 10.3

Comparative food testing: Comparing pre-prepared coleslaw products

Worksheet
Practical activity 10.3

Template
Sensory wheel

Aim

To compare two different types of pre-prepared coleslaw products

Method

1. List the key ingredients you would expect to find in a coleslaw.
2. Describe the sensory properties of a good coleslaw.
3. Work with a partner to develop a vocabulary list of words you could use to describe the sensory properties of coleslaw.

4 Your teacher will provide you with samples of two different types of pre-prepared coleslaw products:
   - a coleslaw mix pre-dressed with mayonnaise from the chill cabinet of the supermarket
   - a coleslaw salad kit that includes a salad dressing from the fresh vegetable section of the supermarket.
5 Prepare the coleslaw salad kit by tossing it in a bowl with the dressing.
6 Undertake a taste test of both salads. Record your results, using descriptive language, in a table like the one below. Refer to the sensory wheel on page 4 to assist you.

Results

	Appearance	Aroma	Flavour	Texture	Sound
Coleslaw mix pre-dressed with mayonnaise					
Coleslaw salad kit					

7 Refer to the labelling information on each product, and record the nutrition information for 100 grams of each type of salad in a table like the one below.

Nutrition information	Coleslaw mix pre-dressed with mayonnaise	Coleslaw salad kit
Energy (per 100 grams)		
Protein (per 100 grams)		
Total fat (per 100 grams)		
Carbohydrate (per 100 grams)		
Sodium (per 100 grams)		

Analysis

etorres/Shutterstock.com

1 Which salad had the most appealing sensory properties? Justify your decision.
2 Which coleslaw provided the lowest amount of energy, total fat and carbohydrate per 100 grams?
3 Explain why it is important to minimise the amount of energy and fat we include in our diet.
4 Which coleslaw contained the highest level of sodium? How does the amount of sodium per 100 grams of this product compare with the amount of sodium a 15-year-old should consume on a daily basis?
5 What type of packaging material is used to contain each product? What impact will the use of these packaging products have on the environment?

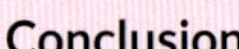

Conclusion

If you were going to buy pre-prepared coleslaw for a meal, which would you purchase? Justify your decision based on:
- the sensory properties of the two products you taste tested
- the nutritional properties of the two products
- the type of packaging used for each product
- the extent to which each product met your expectations for a pre-prepared salad.

Reducing food waste going to landfill

It is estimated that one-third of all food produced in Australia goes to waste. According to Steven Lapidge, Chief Executive of the federally funded Fight Food Waste Cooperative Research Centre, 'Australia's food waste would fill the MCG 10 times each year or would fill semi-trailers stretching from Perth to Sydney'. Wasted food is not only a financial issue, but also a danger to our environment.

If vegetable waste goes to landfill, it decomposes anaerobically (without oxygen) and releases

9780170495714

greenhouse gases, primarily methane. It is estimated that at least 3 per cent of Australia's greenhouse gas emissions come from organic matter decomposing in landfills. This is of concern, because methane affects the quality of the air we breathe and contributes to global warming and climate change. When the decomposing organic matter from food waste mixes with other metals and chemicals that are disposed of in landfill, a toxic sludge is created. The resulting toxins can leech into water supplies, impacting the water that humans and animals drink.

Planning meals

- Decide what meals you are going to cook
- Make a list of ingredients required for these meals
- Check the pantry and refrigerator to determine what food you already have and what you need to purchase
- Make a final shopping list before leaving home

Purchasing food

- Stick to your shopping list
- Buy loose vegetables rather than packaged ones, so you only purchase the amount you need

Cooking food

- Calculate the amount of food you need to cook carefully, so as to avoid leftovers
- If there are leftovers, store them correctly to avoid them going 'off'

Disposing of food waste

- If you're able to do so, set up a home compost system or worm farm
- Dispose of food scraps in green bins for regular collection by local councils
- Give food scraps to community gardens for their composting systems, if they are happy to receive them

Figure 10.24 Strategies to reduce food waste at home

iStock.com/svetikd

Figure 10.25 A home compost system

OzHarvest

Weblink
OzHarvest

OzHarvest is a national charity that fights food waste, and rescues surplus food to provide meals and food packs for people who are at risk of food insecurity. To help individuals to minimise food waste, OzHarvest has developed strategies such as Use It Up tape. People can use this tape to mark shelves of their refrigerator or pantry to show food items that need to be used up first. It can also be used on individual items. A recent impact study conducted by OzHarvest has shown that this tape is a practical and effective way to avoid food waste, and that households that use it have reduced their food waste by up to 40 per cent.

Good & Fugly

Good & Fugly aims to reduce food waste created on farms. It was launched in Sydney in 2020 and now also operates in Melbourne. The company makes up weekly boxes of fruit and vegetables that have been rejected by supermarkets, and delivers them from the farmers directly to people who subscribe to this service. Recipe cards are also included, to help people produce delicious meals.

Figure 10.26 OzHarvest's Use It Up tape shows which food needs to be used up first.

Figure 10.27 Good & Fugly's small fruit and veg box

The fruit and vegetables may have been rejected by supermarkets because they do not meet their specifications, as they may be too small, not perfectly and uniformly formed or shaped, or have minor scratches or blemishes on the skin.

It is estimated by Good & Fugly that 25 per cent of all produce never leaves the farm because it is 'ugly'. Farmers usually dispose of this fresh produce as there is no market for it, but Good & Fugly has come to the rescue of both the farming community and the environment. Excess produce is sent to OzHarvest, where it is used to provide meals that give food security for people in need.

Good & Fugly states that 'every purchase with Good & Fugly delivers a triple promise: it's good for you, the planet and our hard-working farmers'.

Weblink Good & Fugly

Figure 10.28 'Ugly' carrots rejected by supermarkets are still suitable to eat.

Legumes

Legumes are classified in the *Leguminosae* family, and include the leaves, stems and pods of the plant. Plants in this family produce seeds within a pod; for example, fresh green peas, snow peas and fresh green beans. Legumes are a very nutritious food as they are high in protein and fibre. Peanuts and soybeans are also classified in the *Leguminosae* family as they produce seeds within a pod.

When the seedpods of these plants mature on the plant, they become dry and are then known as 'pulses'. Dried peas, beans, lentils and chickpeas are normally referred to as pulses as they have been allowed to mature on the vine and are harvested when they are dry.

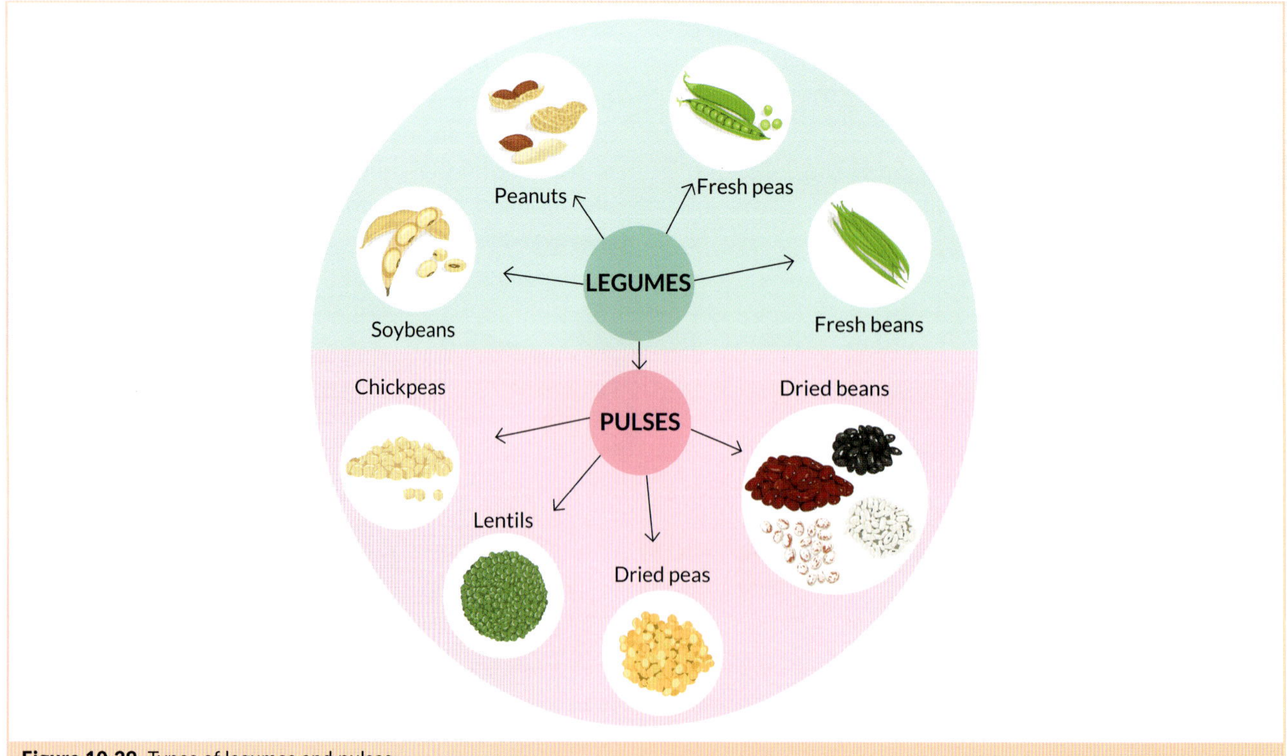

Figure 10.29 Types of legumes and pulses

9780170495714

Figure 10.30 The nutritional properties of legumes

Nutrition

Legumes are a good source of plant protein, dietary fibre, carbohydrates, B group vitamins, zinc, iron, potassium and folate.

As they are rich in soluble fibre, they form a thick gel when eaten, which slows down the digestive process, reducing the absorption of 'bad cholesterol'. This process assists heart health and lowers blood sugar levels by slowing the absorption of sugar. Insoluble fibre passes through the digestive tract adding bulk, which keeps the colon healthy and reduces the risk of colon cancer.

Because of their nutritional properties, particularly their protein content, legumes are found in two segments of the *Australian Guide to Healthy Eating*:

- vegetables and legumes/beans
- lean meats and poultry, fish, eggs, tofu, nuts and seeds, and legumes/beans.

Growing legumes sustainably

Legumes can play a valuable role in sustainable agriculture as they help farmers reduce the use of chemical fertilisers. Legume crops have small nodules on their root systems that contain nitrogen-fixing bacteria. These bacteria use the nitrogen from the air to 'fix' the nitrogen in the nodules the plants use while growing. When the plant dies, the fixed nitrogen is released into the soil, making it available to other plants and improving the health of the soil, reducing the need for the addition of chemical fertilisers. When farmers rotate their cereal crops such as wheat with leguminous crops such as chickpeas or lentils, the soil is 'topped up' with nitrogen.

Overuse of chemical fertilisers pollutes the air and drinking water and can affect soil health by depleting other nutrients in the soil.

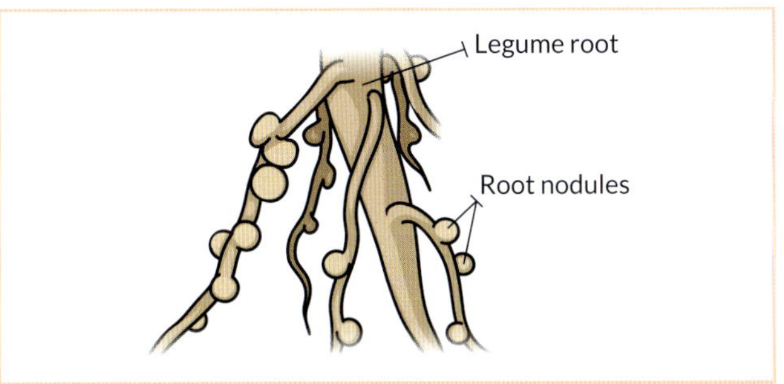

Figure 10.31 Nitrogen-fixing bacteria on the roots of legume plants 'fix' nitrogen in the nodules.

Recent research has shown that legume crops are not as heavily reliant on water as other crops, so these crops are more sustainable in low rainfall areas.

iStock.com/Grigorenko

Figure 10.32 Growing chickpeas

Testing knowledge 10.3

Quiz
Testing knowledge 10.3

21 Approximately what proportion of all food produced in Australia goes to waste?
22 Explain how food decomposes in landfill.
23 Why does food sent to landfill cause an environmental problem, and how does it impact on the health of Australians?
24 Describe the problem that occurs when organic waste mixes with metals and chemical waste in landfill.
25 Identify four ways kitchen food waste can be reduced.
26 What is one environmental advantage of the Good & Fugly food strategy?
27 Explain the difference between legumes and pulses.
28 Outline why legumes are a valuable food to include regularly in the diet.
29 Why do many farmers use a legume crop in their crop rotation system when growing other crops?
30 What impact can the overuse of chemical fertilisers have on the environment?

Critical and creative thinking 10.1

Compare the nutrient content of fruit and vegetables

1. Select four vegetables to compare. Each of the vegetables must be from a different classification.
2. Identify the key vitamins and minerals each of the vegetables contains.
3. Identify the amount of fibre present in each vegetable.
4. Explain how the vegetables are similar and different with respect to the nutrients they contain.
5. Make recommendations about how frequently people should eat the vegetables you have compared, and explain why they are important to good health.

DESIGN ACTIVITY 10.1

Worksheet Design activity 10.1

Templates Food order

Sensory wheel

Australian Guide to Healthy Eating

All Things Potatoes café

Ballarat is one of Victoria's major potato-growing areas. It has productive soil and access to reliable irrigation water – two important resources for a bountiful potato crop.

Design brief

The Spud family have decided to expand their potato-growing business in Ballarat and open a café in the centre of the city to showcase the wonderful dishes that can be produced from the humble potato. It will be called the All Things Potatoes café. Ballarat has a diverse population, and tourists come from around the world, so the Spud family want the lunch menu for the café to reflect potato dishes from many cultures, and celebrate how different cultures incorporate potatoes into their everyday food. While potato chips will not feature, the lunch menu will highlight at least 'five ways with potatoes' to demonstrate the versatility of potatoes as a main or side dish. There will only be one chef and an assistant working in the kitchen of the café, so each dish should be able to be prepared and cooked in under one hour.

	Idea 1	Idea 2	Idea 3	Idea 4
Recipe and source				
Cultural origin				
Served as main/side dish				
Variety of potato used				
Other main ingredients				
Cooking methods used				
Preparation and cooking time				
Thinking notes on why the idea was selected				

9780170495714

Investigating and defining

1 Based on the solution requirements and constraints in the design brief, develop five design criteria to evaluate the success of a lunch dish showcasing potatoes.

2 Use the internet and recipes from home cooks to research the types of recipes from different cultures that include potatoes and can be served as a main or side dish for lunch. The preparation and cooking time should be approximately one hour. You could begin your research by using the recipe index of this book to explore a range of potato recipe ideas you might not have tried before.

3 Record four recipes that appeal to you and meet the needs of the brief in a table like the one on the previous page. Remember to include the sources of the recipes.

Generating and designing

1 Create two designed solution options for a lunch item to showcase potatoes at the All Things Potatoes café.

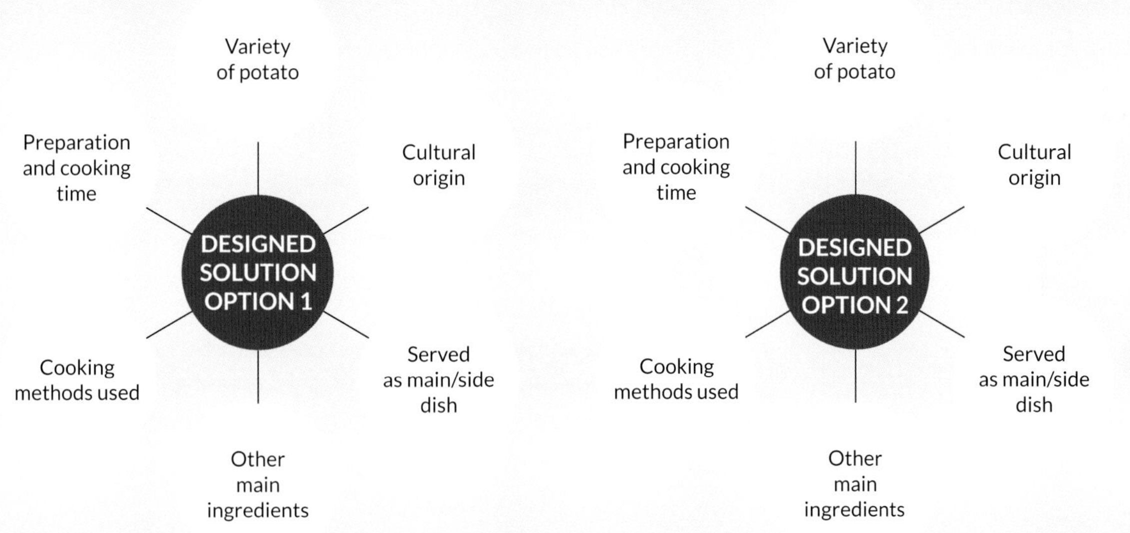

2 Select the option that you prefer and justify your choice. Explain why you did not select the other option.

3 Write up a new recipe for the potato dish ready for production.

Planning and managing

1 Prepare a food order.

2 Annotate a copy of your recipe to indicate the tools and equipment needed in production and to highlight key safe work practices.

Producing and implementing

1 Prepare the product.

2 Take a photo after plating your potato dish and before taste testing. Include it with your evaluation.

Evaluating

1 Respond to the design criteria you developed.

2 Describe the sensory properties – appearance, aroma, flavour, texture and sound – of your potato dish.

3 Discuss the effectiveness of your time management for this production.

4 What recommendations would you make to improve the quality of the recipe, if you were to make it again?

5 Classify the ingredients of your potato dish on a diagram of the *Australian Guide to Healthy Eating*. Comment on how well the dish meets the recommendations of this food selection model. Suggest other food items or recipes that could accompany your potato recipe and improve the nutritional value of the meal.

Broccoli, Chickpea and Tomato Salad

INGREDIENTS

10 almonds

½ cup canned chickpeas, drained

olive oil spray

200 grams broccoli, cut into small florets

1 tablespoon tahini

½ lemon, juiced

1 teaspoon olive oil

1 tablespoon coriander, finely chopped

1 tablespoon parsley, finely chopped

½ cup cherry tomatoes, halved

SERVES TWO

METHOD

1. Preheat oven to 180°C. Line a baking tray with baking paper.
2. Place the almonds on the lined baking tray and bake for 6 minutes. Remove the almonds from the oven tray and place on a chopping board. Allow to cool. Chop roughly.
3. Rinse the chickpeas. Pat dry thoroughly with paper towel.
4. Place the chickpeas on the lined baking tray. Spray lightly with the olive oil spray.
5. Roast for 10 minutes or until crisp.
6. Bring a medium saucepan of water to the boil. Add the broccoli florets and cook for 2 minutes. Drain and refresh in cold water. Drain again.
7. Place the tahini, lemon juice, olive oil, coriander and parsley in a small food processor and blend until combined. Add 1 tablespoon of warm water and process again until the mixture has a smooth and a creamy consistency.
8. Place the blanched broccoli, chickpeas and cherry tomatoes in a serving bowl and spoon over the tahini dressing. Season with salt and pepper and scatter with the chopped almonds.

Mark Fergus Photography

EVALUATION

1. Describe the sensory properties – appearance, aroma, flavour, texture and sound – of your Broccoli, Chickpea and Tomato Salad.
2. Compare the sensory properties of the unroasted and roasted chickpeas.
3. Discuss why it is important to refresh the broccoli florets in cold water.
4. Explain how including this salad as part of their evening meal would assist a young person aged 12–13 years to meet the recommended number of serves of vegetables for their age.
5. Classify the ingredients for the Broccoli, Chickpea and Tomato Salad on a diagram of the *Australian Guide to Healthy Eating*. Explain how well this dish meets the guidelines of this food selection model. Suggest a protein-based recipe you could serve to accompany the salad to make a healthy lunch or evening meal.

9780170495714

Potato and Thyme Baked Tortilla

INGREDIENTS

200 grams kipfler or red or white chat potatoes

1 small red onion

1 tablespoon extra virgin olive oil

1 clove garlic, finely chopped

¼ teaspoon saffron threads

1 teaspoon thyme leaves, finely chopped

⅛ teaspoon salt

pinch pepper

125 millilitres water

4 eggs

To serve

1 tablespoon chopped chives or parsley

1 tablespoon sour cream or crème fraiche

2 cherry tomatoes, halved

METHOD

1 Wash the kipfler or chat potatoes well and pat dry. Thinly slice into rounds.

2 Peel the onion, cut in half lengthwise, then thinly slice into half-moons.

3 Preheat oven to 180°C. Line an 18-centimetre loaf tin or foil container with baking paper.

4 Heat the oil in a frying pan over low heat and add the potato, onion, garlic, saffron and thyme. Stir occasionally until the potato is starting to soften – approximately 8–10 minutes.

5 Add the salt, pepper and water to the frying pan and increase the heat to a simmer. Cook until the water has nearly evaporated and the potatoes are just tender – approximately 10 minutes.

6 Transfer the potato mixture to a large bowl to cool.

7 In a medium bowl, lightly beat the eggs. Do not aerate them.

8 Add the eggs to the potatoes and toss gently, taking care not to break up the vegetables.

9 Transfer the mixture to the lined loaf tin or foil container and bake until just set – approximately 15 minutes.

10 Cool to room temperature.

To serve

1 Use scissors to finely cut the chives.

2 Remove the tortilla from the baking tin and slice into serving portions.

3 Top each portion with a teaspoon of sour cream or crème fraiche and sprinkle with some chives or parsley. Serve with the cherry tomatoes on the side.

4 Serve with a salad and fresh bread for a light meal.

iStockphoto.com/esseffe

EVALUATION

1 Describe the sensory properties – appearance, aroma, flavour, texture and sound – of your Potato and Thyme Baked Tortilla.

2 Describe how frying and simmering change the sensory properties of the potatoes.

3 Identify the functional role of the eggs in the tortilla and describe how their properties change during baking.

4 Recommend two salads that would complement the Potato and Thyme Baked Tortilla and justify your selection.

5 Classify the ingredients of the Potato and Thyme Baked Tortilla, including those of one of your salad ideas and some fresh bread, on the *Australian Guide to Healthy Eating* and discuss how well this meal meets the recommendations of the food model.

Yakitori Chicken Skewers

INGREDIENTS

2 serves cooked rice

Yakitori Sauce

¼ cup cooking sake

¼ cup tamari or light soy sauce

3 tablespoons mirin

2 teaspoons brown sugar

1 clove garlic, crushed

1 teaspoon (15 grams) fresh ginger, grated

Chicken Skewers

4 chicken thigh fillets

4 spring onions

olive oil spray

To serve

Green Vegetable Stir-fry (see recipe on the following page)

SERVES TWO

METHOD

1 Soak 4 bamboo skewers in water for 10 minutes.

2 Prepare the rice (see page 236) and keep warm.

Yakitori Sauce

1 In a small saucepan, combine the cooking sake, tamari or light soy sauce, mirin, brown sugar, garlic and ginger.

2 Sir well over a moderate heat to dissolve the sugar. Reduce the heat to low and simmer uncovered for approximately 10 minutes until the sauce has reduced by a third. Turn off the heat and allow to cool to room temperature.

Chicken Skewers

1 Cut the chicken into 2-centimetre pieces. Cut the spring onion into batons the same length as the chicken pieces.

2 Preheat griller to high.

3 Thread the chicken and spring onion pieces alternately onto the bamboo skewers. Spray with olive oil spray.

4 Grill the first side of the chicken skewers for approximately 4–5 minutes.

5 Brush the skewers with a little of the yakitori sauce, then turn them over and brush a little more yakitori sauce on the second side.

6 Grill the second side for 3–4 minutes.

To serve

Serve the Yakitori Chicken Skewers on a bed of rice accompanied by the Green Vegetable Stir-fry and the remaining yakitori sauce.

Mark Fergus Photography

9780170495714

Green Vegetable Stir-fry

INGREDIENTS

4 pieces broccolini

½ zucchini

1 bok choy

8 snow peas

2 tablespoons vegetable oil

1 centimetre fresh ginger, finely diced

2 tablespoons vegetable stock

pinch sugar

salt to taste

METHOD

1 Cut the broccolini into 3-centimetre pieces. Keep the stems and florets separate.
2 Cut the zucchini into bite-size pieces.
3 Trim the base of the bok choy, then separate the leaves and wash them. Cut the white stems from the green leaves and cut the stems in half. Coarsely shred the green leaves. Keep the stems and leaves separate.
4 Trim the snow peas and cut in half.
5 Prepare a bowl of iced water and place it in the sink.
6 Half fill a medium saucepan with water and bring to the boil. Blanch the broccolini stems for 2 minutes, then add the broccolini florets and boil for 2 minutes. Drain and immediately plunge the stems and florets into the bowl of iced water to stop the cooking process. Drain well.
7 Heat a wok over high heat and add the vegetable oil. Add the most dense vegetables to the wok first – start with the broccolini, followed by the snow peas, zucchini and bok choy stems.
8 Add the ginger and stir-fry for 30 seconds, then pour over the stock and bring to the boil.
9 Add the shredded bok choy leaves, sugar and salt and toss the vegetables until the leaves begin to wilt.
10 Serve immediately.

EVALUATION

1 Why does the recipe suggest soaking the bamboo skewers in water before using them?
2 Explain one important food safety rule to observe when preparing the chicken.
3 Identify and describe the changes that occur to the chemical and physical properties of the chicken when it was grilled.
4 Explain why, when stir-frying the green vegetables in step 7, it is recommended that you cook the most dense vegetables first.
5 Classify the ingredients for the Yakitori Chicken Skewers served with Green Vegetable Stir-fry on a bed of rice on a diagram of the *Australian Guide to Healthy Eating*. Write a paragraph to explain how well this meal meets the recommendations of this food selection model.

Orange Vegetable Dhal

INGREDIENTS

150 grams butternut pumpkin
150 grams sweet potato
1 medium carrot
25 grams butter
2 teaspoons olive oil
½ brown onion, finely diced
1 garlic clove, crushed
2 centimetres fresh ginger, peeled and grated
2 teaspoons ground coriander
2 teaspoons ground turmeric
2 teaspoons garam masala
2 teaspoons curry powder
½ stick cinnamon
10 curry leaves, fresh or dry
1 bay leaf
1 cup coconut milk
½ cup (100 grams) split red lentils
1 cup vegetable stock
½ cup water
200 grams canned diced tomatoes
2 tablespoons palm sugar
1 tablespoon lemon juice
salt and pepper

To serve

½ cup natural yoghurt
2 tablespoons picked coriander leaves
½ long red chilli, finally sliced
fresh flatbread (page 242) or rice (page 236)

SERVES TWO

METHOD

1. Peel the butternut pumpkin, sweet potato and carrot and cut into 1-centimetre pieces.
2. Place the butter and oil in a medium-to-large saucepan and heat to medium-low. Add the onion and sauté with the lid on for 6-8 minutes. Stir occasionally. When the onion is translucent, add the garlic and ginger and stir for 30 seconds.
3. Add the diced orange vegetables, spices, curry leaves, cinnamon stick and bay leaf and stir over the heat for 2 minutes until they become fragrant.
4. Deglaze the saucepan by adding the coconut milk and stirring the bottom of the saucepan to ensure that all the spices are completely incorporated into the liquid.
5. Rinse the lentils in cold water.
6. Add the rinsed lentils, vegetable stock, water, tomatoes and palm sugar to the saucepan. Stir well.
7. Bring to the boil, then reduce the heat and simmer for approximately 20 minutes. Stir occasionally. When the vegetables are tender, and the lentils have become soft and creamy, remove from heat.
8. Discard the cinnamon stick and bay leaf. Add the lemon juice and season to taste with salt and pepper.
9. Transfer to a serving bowl. Top with yoghurt, coriander leaves and chilli slices.
10. Serve as a meal with flatbread or rice or as part of a shared table.

EVALUATION

1. Describe the sensory properties – appearance, aroma, flavour, texture and sound – of your Orange Vegetable Dhal.
2. Explain how deglazing the saucepan in step 4 contributes to the flavour of the dish.
3. Outline the nutritional value of including orange vegetables in the diet.
4. Explain why including lentils in a recipe is a strategy of some people who follow a vegetarian diet.
5. Classify the ingredients for the Orange Vegetable Dhal on a diagram of the *Australian Guide to Healthy Eating*. Discuss how well this dish meets the recommendations of this food selection model.

9780170495714

Mark Fergus Photography

Shepherd's Pies

Mashed potato makes a delicious topping on these pies, and is much lower in fat than traditional pastry.

INGREDIENTS

Filling

2 teaspoons oil

½ onion, finely diced

125 grams minced steak

⅓ carrot, grated

1 teaspoon parsley, finely chopped

1 tablespoon tomato paste

1 tablespoon tomato sauce

¼ cup beef stock

½ teaspoon Worcestershire sauce

¼ teaspoon mixed herbs

pepper

Topping

2 large potatoes

2 tablespoons milk

2 teaspoons butter

pinch salt

40 grams cheese, grated

MAKES TWO PIES

METHOD

1. Peel the potatoes and cut into even-sized pieces. Place in a saucepan with sufficient water to cover. Cook with the lid on for approximately 15 minutes or until tender.
2. Heat the oil in a medium saucepan. Add the onions and fry lightly.
3. Add the minced steak and cook until brown. Stir constantly to prevent the meat clumping together.
4. Add the carrot, parsley, tomato paste, tomato sauce, beef stock, Worcestershire sauce, herbs and pepper. Simmer for approximately 5 minutes or until the mixture thickens slightly.
5. Preheat oven to 200°C.
6. Drain the potatoes. Mash and add milk, butter and a pinch of salt. Mix well.
7. Divide the meat mixture between two foil containers and spread the mashed potato on top. Sprinkle with grated cheese.
8. Place on an oven tray and bake for 10 minutes or until golden brown.

EVALUATION

1. Describe the sensory properties – appearance, aroma, flavour, texture and sound – of your Shepherd's Pies.
2. Why is it important to cut the potato into even-sized pieces when preparing it to make mashed potato?
3. Why are milk and butter added to mashed potato?
4. Identify two safe work practices you followed during production.
5. Classify the ingredients for the Shepherd's Pies on a diagram of the *Australian Guide to Healthy Eating*. Comment on the health rating you would give this recipe. What are the health benefits of using potato rather than pastry in this pie?

Mark Fergus Photography

9780170495714

Spicy Potatoes with Chickpeas

This recipe, served with rice, is high in complete protein and makes a delicious meal for vegetarians.

INGREDIENTS

1 medium waxy potato (150 grams)

150 grams sweet potato

1 tablespoon oil

½ onion, finely diced

1 clove garlic, crushed

1 teaspoon ground cumin

½ teaspoon garam masala

½ teaspoon ground coriander

½ teaspoon ground fennel

¼ teaspoon ground turmeric

pinch of cayenne pepper

220 grams canned diced tomatoes

½ cup vegetable stock

⅓ cup frozen peas

½ cup (100 grams) chickpeas, rinsed and drained

To serve

Plain Rice (page 236)

Tzatziki Dip (page 22)

pappadams

SERVES TWO

METHOD

1. Peel the potato and sweet potato and cut into 2-centimetre cubes.
2. Heat the oil in a medium saucepan and sauté the onion and garlic until soft but not brown.
3. Add the spices and cook for 30 seconds.
4. Add the diced potato and sweet potato and lightly toss to coat in the spices.
5. Stir in the undrained tomatoes and vegetable stock.
6. Bring to the boil. Reduce the heat and simmer for approximately 30 minutes or until the potato is tender.
7. Stir in the peas and chickpeas and heat through.
8. Serve with Plain Rice accompanied by Tzatziki Dip and pappadams.

EVALUATION

1. What is the purpose of frying the spices for 30 seconds in step 3?
2. Identify the ingredients in the Spicy Potatoes with Chickpeas and the rice that are good sources of protein for vegetarians.
3. Why are the frozen peas and canned chickpeas only heated through in step 7 and not added with the potatoes in step 4?
4. Which part of the production was the most successful and which part did you find the most challenging? Why?
5. Classify the ingredients for the Spicy Potatoes with Chickpeas served with Plain Rice, Tzatziki Dip and pappadams on a diagram of the *Australian Guide to Healthy Eating*. Comment on the nutritional value of this meal.

Mark Fergus Photography

Corn Fritters with Tomato and Avocado Salsa

INGREDIENTS

Cumin Salt

½ teaspoon cumin seeds

1 teaspoon salt flakes

Tomato and Avocado Salsa

½ ripe avocado

6 cherry tomatoes or ½ tomato, chopped

⅛ red onion, finely diced

1 tablespoon coriander, chopped

1 tablespoon lime juice

¼ green chilli, seeds removed, finely chopped

⅛ teaspoon Cumin Salt (see above)

Corn Fritters

2 cobs fresh corn

¼ red capsicum, finely diced

2 spring onions, finely sliced

2/3 cup (100 grams) flour

½ teaspoon baking powder

1/8 teaspoon Cumin Salt (see above)

1 tablespoon parmesan cheese, finely grated

50 grams tasty cheese, coarsely grated

2 tablespoon flat leaf parsley, chopped

1 egg

70 millilitres milk

fresh pepper

2 tablespoons olive oil

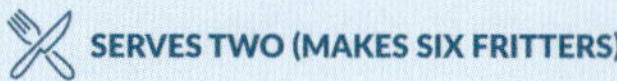

METHOD

1. To make the Cumin Salt, place the cumin seeds and salt flakes in a dry frying pan and warm over medium heat until fragrant. Remove from heat, tip into a mortar and grind into a coarse powder.
2. To make the Tomato and Avocado Salsa, peel the avocado, dice into 1 centimetre cubes and combine with the tomato, red onion, coriander, lime juice, green chilli and Cumin Salt. Gently combine the ingredients and set aside.
3. Boil the corn cobs for 6 minutes. Drain. Slice the kernels from the cobs and break them up into small pieces.
4. Sift the flour, baking powder and Cumin Salt into a large bowl. Stir through the corn kernels, cheeses and parsley.
5. In a small bowl, whisk the egg and milk together. Add the fresh pepper. Stir into the flour mixture to form a thick batter.
6. Heat 1 tablespoon oil in a frying pan over medium heat. When hot, add 2 tablespoons of batter to form one fritter. Make another two fritters in the same pan and cook for approximately 3–4 minutes on each side or until golden. Set aside and keep warm. Add the remainder of the oil and cook another three fritters.
7. Arrange the Corn Fritters on two plates and top with the Tomato and Avocado Salsa. Serve with extra Cumin Salt on the side.

Additional serving suggestion

Top with poached or fried eggs and rashers of bacon.

EVALUATION

1. Describe the sensory properties – appearance, aroma, flavour, texture and sound – of your Corn Fritters with Tomato and Avocado Salsa.
2. Explain the safety procedures you followed when cutting the ingredients in this recipe and cleaning the knives that you used.
3. Explain why frying food is a cooking method that should only be used occasionally.
4. List the ingredients in the recipe that contain starch and those that contain protein. Identify the processes that cause the changes to the structure of starch and protein when the fritters are cooked.
5. Classify the ingredients for the Corn Fritters with Tomato and Avocado Salsa on a diagram of the *Australian Guide to Healthy Eating* and explain how well it meets Guideline 2.

9780170495714

Mark Fergus Photography

Spicy Baked Beans and Eggs

INGREDIENTS

2 tablespoons olive oil

1 leek, white part only, thinly sliced

2 cloves garlic, crushed

½ green chilli, seeds removed, thinly sliced

1 teaspoon coriander seeds

400 grams canned cannellini beans, rinsed and drained

1½ cups canned diced tomatoes

1 tablespoon tomato paste

½ teaspoon ground allspice

½ teaspoon sugar

¼ teaspoon salt

pinch pepper

200 millilitres water

3 teaspoons lemon juice

2 eggs

1 tablespoon flat leaf parsley, chopped

4 slices sourdough bread, toasted, to serve

SERVES TWO

METHOD

1 Preheat oven to 200°C. Grease two ovenproof ramekins (about 1½ cup capacity each).

2 Add the oil to a medium saucepan. Sauté the leek, garlic and chilli over a medium heat for about 5 minutes or until softened.

3 Roughly crush the coriander seeds in a mortar.

4 Smash half of the cannellini beans with the back of a wooden spoon.

5 Add the tomatoes, tomato paste, crushed coriander seeds, allspice, crushed beans, sugar, salt, pepper and water to the leek mixture. Bring to the boil, then reduce the heat and simmer for 10–12 minutes or until the mixture is thick. Stir occasionally. Remove from the heat and stir in the remainder of the beans.

6 Stir through the lemon juice. Divide the bean mixture between the two ramekins.

7 Make a shallow hole in the centre of the bean mixture in each ramekin. Crack an egg into each hole. Bake for 10–15 minutes or until the eggs are cooked to your liking.

8 Remove from the oven and top with the chopped parsley.

9 Serve with the toasted sourdough bread.

Mark Fergus Photography

EVALUATION

1 Describe the sensory properties of your Spicy Baked Beans and Eggs. Consider each element of the recipe in your discussion of appearance, aroma, flavour, texture and sound.

2 Describe how the cooking method of sautéing changes the characteristics of the leek, garlic and chilli in step 2.

3 In step 5, the vegetables are boiled and then simmered. Explain how these cooking methods influence the final sensory properties of the dish.

4 Identify the process that occurs when the eggs are baked.

5 Classify the ingredients for the Spicy Baked Beans and Eggs on a diagram of the *Australian Guide to Healthy Eating* and discuss the extent to which this dish meets Guideline 2.

9780170495714

Carrot Cake

INGREDIENTS

1 large carrot

½ cup walnuts or sultanas

½ cup oil

¾ cup brown sugar

2 eggs

1 cup self-raising flour

½ teaspoon cinnamon or mixed spice

icing sugar for dusting

METHOD

1 Grease and line a 20-centimetre ring tin or small loaf tin. Preheat oven to 180°C.
2 Grate the carrot and roughly chop the walnut pieces, if using. Place the carrot and walnut pieces or sultanas in a large mixing bowl.
3 Add the oil, sugar and eggs to the mixing bowl and stir.
4 Sift in the self-raising flour and spice and mix well with the other ingredients.
5 Spoon the mixture evenly around the ring tin or into the loaf tin.
6 Place in the oven and bake for 20–25 minutes. The cake is ready when it has turned a pale golden colour, it has begun to shrink away from the sides of the tin and a skewer comes out clean and dry.
7 Turn the cake onto a cooling rack.
8 Dust the cake with icing sugar.

EVALUATION

1 Describe the sensory properties – appearance, aroma, flavour, texture and sound – of your Carrot Cake.
2 Why is it important to sift the flour and spice?
3 Outline three other tests you could use to check if the cake was done other than testing it with a skewer.
4 Explain why the oil that is used in this recipe is considered to be a 'healthier' fat than butter.
5 Classify the ingredients of the Carrot Cake on a diagram of the *Australian Guide to Healthy Eating*. Explain why, according to the Eat for Health program, carrot cake is not considered a healthy snack even though it contains some vegetables.

Mark Fergus Photography

MEAT FOR GOOD HEALTH

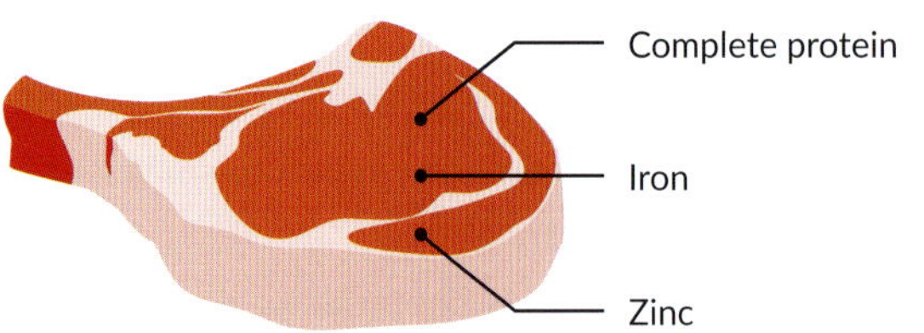

POULTRY FOR GOOD HEALTH

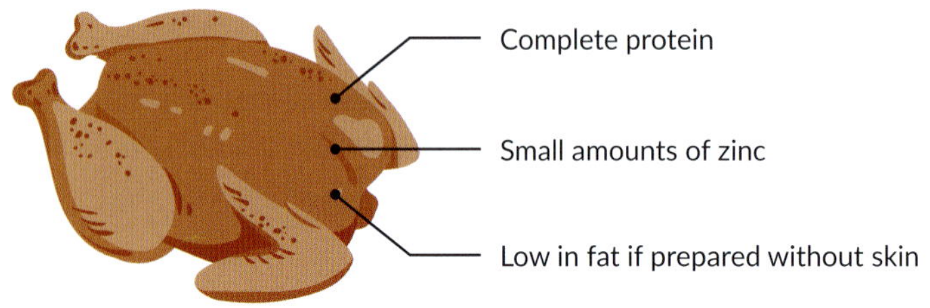

FISH FOR GOOD HEALTH

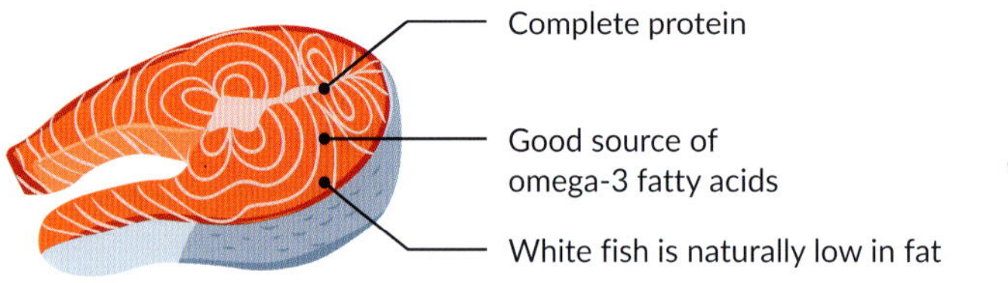

EGGS FOR GOOD HEALTH

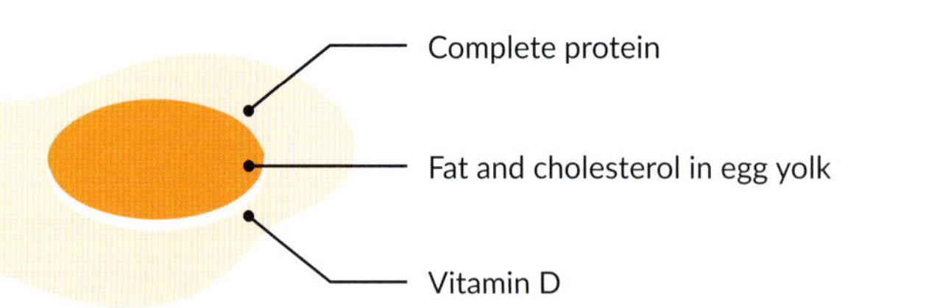

NUTS FOR GOOD HEALTH

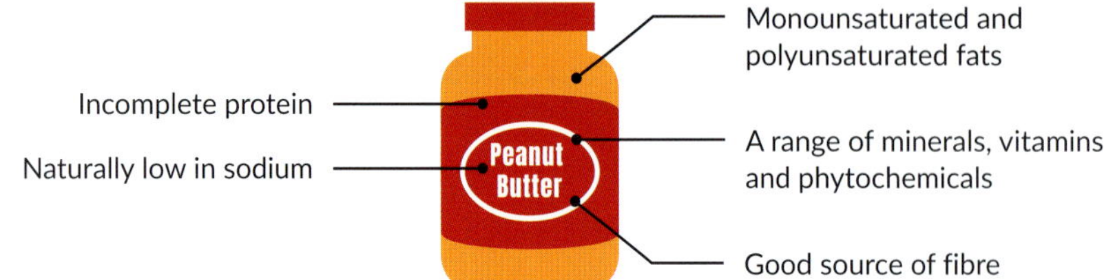

9780170495714

MEAT, POULTRY, FISH AND EGGS

SUSTAINABLE AND ETHICAL ISSUES

Cattle burp large amounts of methane, a greenhouse gas; feeding cows seaweed helps cut methane emissions

Hens kept in cages have little space to move or stretch their wings

Free-range hens are free to forage for insects, to perch and to bathe in dust

Aquaculture was developed to prevent over-fishing of some endangered fish species, but significant environmental impacts can occur; these include pollution of surrounding waterways by fish waste

Bee populations have suffered high mortality rates due to exposure to pesticides on almond orchards

KEY TERMS

aquaculture the breeding, rearing and harvesting of fish and shellfish in coastal marine waters, open oceans and fresh-water systems

connective tissue the tissue in meat that links and holds together the muscles

feedlot production where cattle are kept in pens for 180–360 days and fed a grain-based, high-energy diet of wheat, barley, sorghum and canola seed

marbling the even distribution of deposits of fat cells in red muscle tissue

modified atmosphere packaging (MAP) a method of packaging that causes change in the levels of gases inside a package in order to extend the shelf life of a product

muscle fibres cells that are bound into thin sheets of connective tissue; these bundles then form groups to create muscles

pasture-fed cattle beef cattle that spend their lives grazing on pasture

sustainable fishing the practice of leaving enough fish in the ocean so that the fish population can remain productive and healthy

Templates
- Food order
- Decision table
- Sensory wheel
- Australian Guide to Healthy Eating

Worksheets
- Practical activity 11.1
- Practical activity 11.2
- Design activity 11.1
- Design activity 11.2

Quizzes
- Testing knowledge 11.1
- Testing knowledge 11.2
- Testing knowledge 11.3

To access resources above, visit **cengage.com.au/nelsonmindtap**

9780170495714

Lean meats, poultry, fish and eggs and their alternatives in the *Australian Guide to Healthy Eating*

This segment of the *Australian Guide to Healthy Eating* encompasses a wide variety of foods, including all types of lean meats, poultry, fish, eggs, tofu, nuts, seeds and legumes/beans.

These foods are a valuable source of many nutrients and are important for good health.

Figure 11.1 The place of lean meats, poultry, fish and eggs and their alternatives in the *Australian Guide to Healthy Eating*

The lean meats group of foods for good health

The foods in this group are protein-rich and are a good source of B group vitamins, and minerals such as iron and zinc.

The *Australian Guide to Healthy Eating* recommends that one to three serves of meat or meat alternatives should be eaten each day.

Fish and seafood – especially oily fish such as salmon, trout, sardines and mackerel – provide a valuable source of long-chain omega-3 polyunsaturated fatty acids.

Legumes contain many of the nutrients found in animal sources and therefore have been placed in this group as well as in the vegetables group. For those people who do not wish to consume animal foods, including legumes, tofu, nuts and seeds in their diet is a cost-efficient means of obtaining an adequate intake of protein and other nutrients.

Nuts and seeds provide protein, essential fatty acids and a range of minerals and vitamins, especially vitamin E. However, smaller serving sizes are recommended due to their more concentrated kilojoule content.

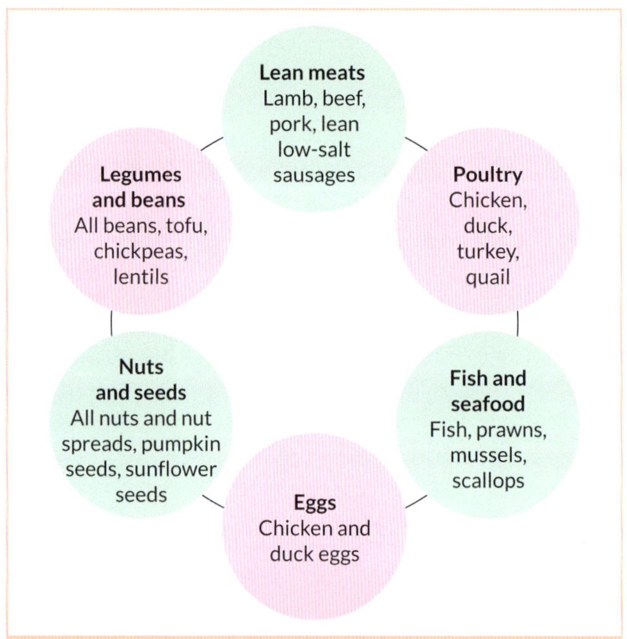

Figure 11.2 Foods in the lean meats segment of the *Australian Guide to Healthy Eating*

Table 11.1 How much of the lean meats group should I eat?

	Serves per day	
	12–13 years	14–18 years
Boys	2.5	2.5
Girls	2.5	2.5

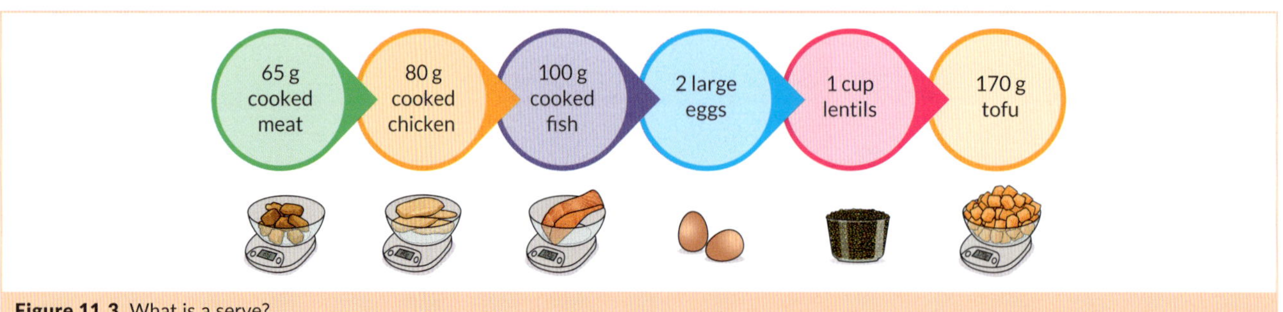

Figure 11.3 What is a serve?

9780170495714

Meat

The term 'meat' refers to the body tissues of animals, eaten as food. Meat is an important part of the diet of many Australians. The meat we eat generally comes from cattle (beef and veal), sheep (lamb), pigs (pork) and poultry (chicken, turkey and duck). Beef, lamb and pork are often described as 'red meat' and poultry as 'white meat'. In this section, we focus on red meat. Poultry is discussed in the next section (see page 292).

Meat for good health

Lean meat is an important food as it contains a high proportion of protein, which is essential for growth, and for repairing and maintaining body tissue in active people. A 100-gram serving of lean meat – which is about the size of the palm of the hand – gives our body over half the protein it needs each day. The protein found in all meat, including poultry, is 'complete'; that is, it has all the essential amino acids needed by the body in the correct proportions required for growth and tissue repair. Amino acids are the molecules or 'building blocks' that combine in different combinations to form different proteins.

Red meat, especially lean cuts, contain readily absorbed forms of iron – specifically, haem iron – and zinc. Iron helps oxygen to move around the body, and is important in producing energy. Zinc is vital in helping our body to utilise carbohydrates and protein, and to heal wounds. Meat is also a good source of niacin, riboflavin and some thiamine. These B group vitamins are essential for our body to utilise nutrients in food.

The Heart Foundation recommends limiting the amount of unprocessed beef, lamb, pork and veal we eat to less than 350 grams per week; that is, one to three lean red meat meals per week. This advice is based on research that demonstrates a link between consuming red meat and the development of cardiovascular disease.

Diets rich in lean red meat can still be low in saturated fat and not adversely affect plasma cholesterol levels. Try to choose lean cuts of meat that have little or no visible fat. Highly processed meats such as bacon, sausages and salami are high in fat and salt. These types of meat are in the 'only sometimes and in small amounts' section of the *Australian Guide to Healthy Eating*.

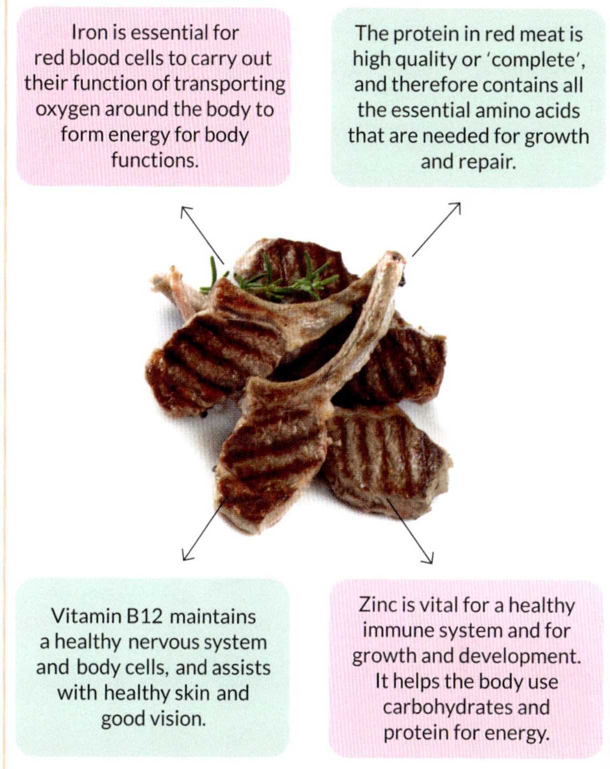

Figure 11.4 Nutrients in red meat

iStock.com/robynmac

The composition of meat

Meat is made up of 22 per cent protein, 3 per cent fat and 75 per cent water. The **muscle fibres** that form meat are cells bound by thin sheets of connective tissue. These bundles of fibres are organised in groups to form individual muscles, which are anchored to the bone by connective tissue. The longitudinal structure of the muscles forms the grain of meat. Muscle fibres are small when the animal is young and the muscles are not yet well developed. As the animal grows and exercises, the muscles enlarge, particularly those that are most frequently used, such as in the neck and legs.

Connective tissue links the muscles and holds them together. It is found between muscle fibres and between whole muscles. The more connective

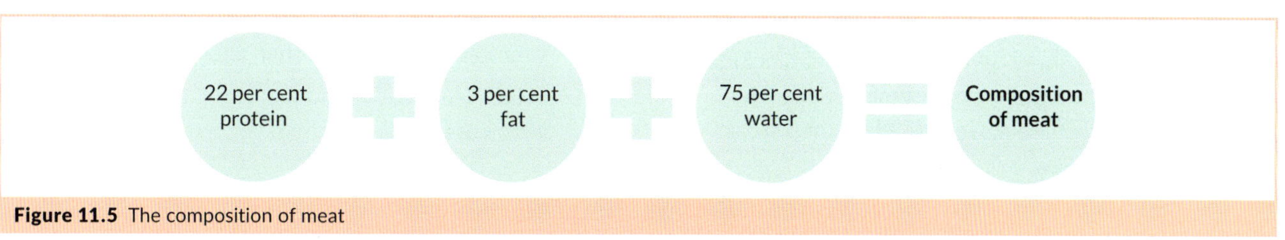

Figure 11.5 The composition of meat

tissue the meat cut contains, the tougher the meat will be. Tougher cuts of meat usually come from the leg, shoulder and forequarter of the animal, because these are the parts of the animal that receive the most exercise. When connective tissue is heated in a liquid, the insoluble collagen becomes gelatin, and the tissue becomes tender to eat.

Cuts of meat with a great deal of connective tissue are best if they are cooked slowly in a moist environment, because this softens the meat. Wet methods of cooking – such as stewing, braising and casseroling – tenderise tougher cuts of meat such as blade steak, chuck steak, topside steak, and beef and lamb shanks.

Fat tissue surrounds muscle tissue and is also incorporated in it. **Marbling** is the term used to describe the distribution of fat cells in red muscle tissue. Fat has an important role in the sensory properties of flavour and texture when meat is cooked.

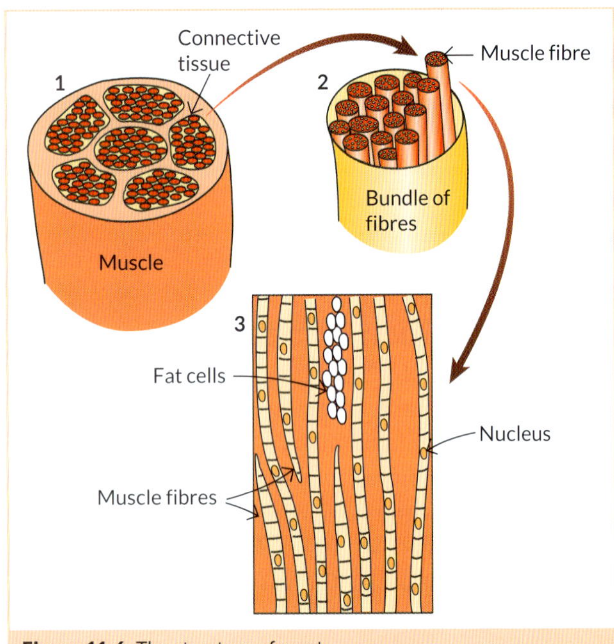

Figure 11.6 The structure of meat

More tender cuts of meat are found around the ribs and back of the animal. These cuts have little connective tissue, and can therefore be cooked by dry radiant heat methods, such as grilling, frying, roasting and barbecuing. Tender cuts of meat are easiest to use for making quick and easy everyday meals. The meat is most tender if it is carved across the grain, so that the muscle bundles are short and therefore easier to chew. Tender cuts of meat include middle loin lamb and pork chops, lamb cutlets, lamb and pork fillets, fillet steak, Scotch fillet, rump steak and porterhouse steak.

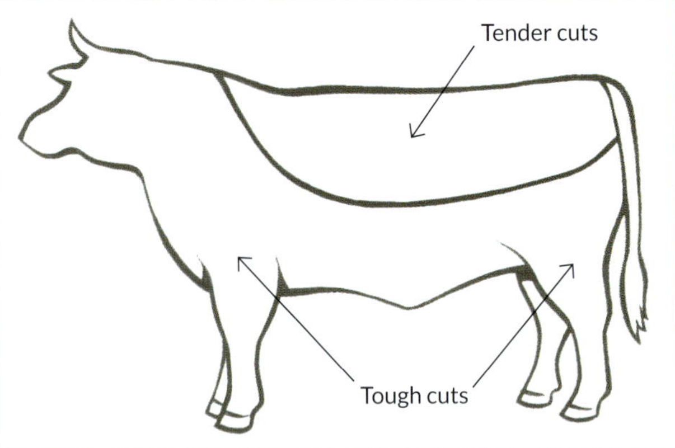

Figure 11.7 Tender and tough cuts of meat

Activity 11.1

Cuts of beef and lamb

Visit the websites of Australian Beef and Australian Lamb. Access the charts of cuts of meat on each website, and then answer the following questions.

1. Identify three possible cuts of beef that are suitable to grill on a barbecue.
2. Explain why meat from the shin is unsuitable for grilling.
3. Name a cut of boneless beef suitable to roast for a family of four.
4. Identify the cut of beef you would purchase to prepare a stir-fried beef and vegetable dish.
5. You have purchased a jar of curry simmer sauce to make a beef curry. Which beef cut would you purchase to prepare this dish? Explain why.
6. Discuss the advantages of roasting and serving a butterflied leg of lamb compared with a traditional leg of lamb containing the bone.
7. Why is the boned shoulder of lamb a popular cut in some households?
8. Identify a suitable cooking method for cooking lamb shanks and explain why the method would make this cut of meat enjoyable to eat.
9. Outline the nutritional advantages of serving a lamb tenderloin rather than a loin chop.
10. Which lamb cut would you choose to make a lamb casserole? Why?

9780170495714

Packaging meat for sale

Meat that is purchased from the supermarket is usually packaged by being either overwrapped or vacuum packaged.

When meat is overwrapped, it is placed on a tray and covered with plastic wrap to prevent any dust entering the package or any physical contamination of the meat. This method of packaging is called **modified atmosphere packaging (MAP)**, as it modifies the levels of oxygen, nitrogen, carbon dioxide (CO_2) and water vapour inside the package in order to extend the food's shelf life. Meats packed this way look more acceptable in the chiller displays of supermarkets, because they retain their bright red colour.

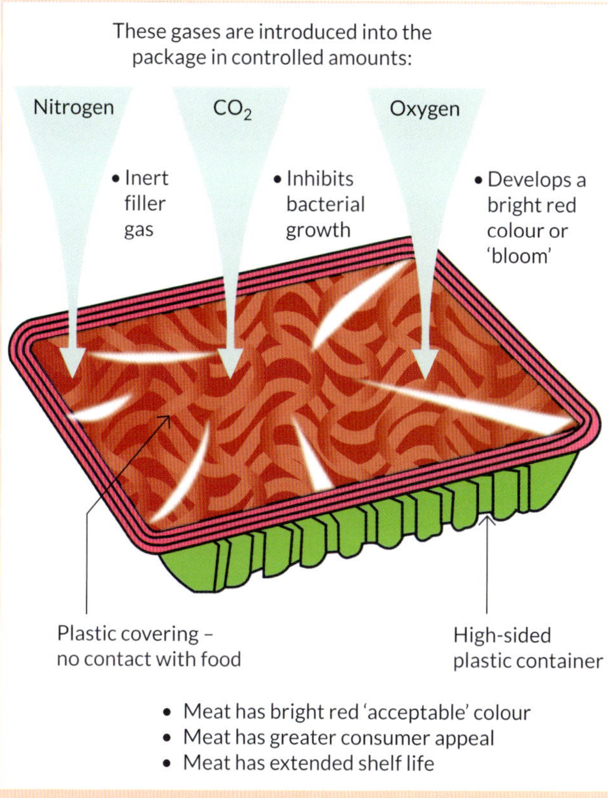

Figure 11.8 Modified atmosphere packaged minced beef

Vacuum packaging excludes oxygen entirely. The air in the package is mechanically sucked out to create a vacuum, so that the plastic package fits tightly around the meat before sealing. However, meat that is vacuum packaged may look darker in colour and less appealing than fresh meat. To return the meat to its natural red meat colour, it should be removed from the packaging for about 30 minutes before cooking so that it can 'breathe'.

Beef: From paddock to plate

Beef cattle that spend their lives grazing on pasture in Australia are called **pasture-fed cattle**. In northern Queensland, cattle graze on large cattle stations, on native pasture. In the southern states, such as in Victoria, cattle graze on smaller farm holdings that are often sown with introduced pasture and fodder crops. Whether the cattle graze on large cattle stations or a smaller farm holding, the supply of pasture depends on either an adequate supply of rainfall or on irrigation. However, during the colder winter months, the growth of pasture slows, therefore affecting the supply of meat. Another problem associated with producing pasture-fed beef is that the farmer has little control over how much grass the cattle eat. The weight and growth of the animals depend on the season and the amount of feed available. Meat from pasture-fed cattle is considered to have a complex, robust flavour, and has a yellowish fat colour.

Another method of producing beef for Australian consumers is **feedlot production**. In this system, the young cattle initially graze on pasture before being sent to a feedlot for 180–360 days. In the feedlot, the cattle are kept in pens, where they are fed a grain-based, high-energy diet that consists of wheat, barley, sorghum and canola seed. The controlled diet means that the farmer can control the growth and weight of the cattle. The meat that is produced is marbled; that is, the fat is evenly distributed throughout the muscle fibres. Feedlot production ensures that a supply of high-quality beef is available all year, particularly during winter, both for domestic markets and for export, particularly to Japan.

Bloomberg/Getty Images

Figure 11.9 Cows eat from a trough at a cattle feedlot.

Activity 11.2

Beef from paddock to plate

Create a table similar to the one below to summarise the key features of each production system in dot points. Use colour to highlight the similarities between each system.

Pasture-fed cattle	Feedlot cattle

Sustainable and ethical farming of cattle, sheep and pigs

As caretakers of the land, most Australian cattle, pig and sheep farmers are committed to producing beef, pork and lamb sustainably and ethically. They work hard to leave the land, waterways, vegetation and soils in better condition for future generations, and to keep their animals safe and well.

Environmental issues

One of the key environmental issues facing livestock producers is reducing the amount of methane produced by ruminant animals such as cattle and sheep. Research has shown that cattle and sheep produce over 10 per cent of Australia's total greenhouse gas emissions. One strategy being developed to reduce the amount of methane that cattle and sheep emit into the atmosphere is to add a small amount of asparagopsis, a red-coloured seaweed, into the feed given to animals. The asparagopsis prevents a specific hormone in the gut of animals from forming during digestion, leading to a dramatic reduction in the amount of methane the animals produce.

Damsea/Shutterstock.com

Figure 11.10 Added to the feed given to cattle and sheep, asparagopsis has been proven to help reduce the amount of methane these animals emit into the atmosphere.

The density or number of livestock that are grazed on pasture is another environmental issue that must be managed by farmers. Over-grazing can be caused by having too many animals on the property and/or leaving them in the same paddock for too long. Over-grazing can affect soil structure, as it can lead to the soil becoming compacted. This can have a major impact on the ability of the soil to store carbon, and may prevent the plants that make up the pasture from regenerating.

Ethical issues

Ensuring the welfare of their animals is a key concern for livestock producers. To raise healthy animals, farmers must ensure that the animals have adequate space, have access to plenty of pasture, and have their healthcare needs met promptly. It is also important for livestock producers to minimise the amount of antibiotics they use to ensure that antibiotic-resistant strains of disease among the animals do not occur.

The method of intensive indoor farming used to produce most of the pork for the Australian market is an ethical issue that concerns many consumers. Approximately 90 per cent of pigs used to produce pork meat for Australian consumers are housed in indoor production systems throughout their lives. Australian pork producers argue that this system allows them to protect the animals from extreme weather conditions, reduces the risk of disease, and enables them to manage the health and wellbeing of the animals.

However, others argue that this type of pork production is not ethical, as it does not take animal welfare into consideration or allow the animals to express their natural behaviours, such as wallowing in mud baths or foraging for food.

The methods used to house breeding sows is of greatest concern. The sow is kept in a farrowing crate while she is giving birth and until the piglets are weaned at about four weeks of age. This crate protects the piglets from being crushed by their mother. The crate is very confined – it only allows enough space for the sow to stand up and lie down, but not enough space for her to turn around. The piglets are separated from the sow by an open metal barrier, which allows them to suckle and feed. A key concern is that confining sows in farrowing crates causes them to become stressed and does not allow them to interact naturally with their piglets.

A more ethical practice is free-range pig farming, where pigs are allowed to roam in a paddock with access to a shelter. Only 5 per cent of Australian pork production is currently free-range.

Skyrta Olena/Shutterstock.com

Figure 11.11 A farrowing crate

9780170495714

Activity 11.3

Ethical issues in the use of farrowing crates

Visit the website of the RSPCA and explore the information in the section 'What are the animal welfare issues with farrowing crates for sows?' Then answer the following questions.

1 Write a definition of a farrowing crate.
2 Draw up a knowledge or mind map to identify four key animal welfare concerns with the use of farrowing crates. Provide an example of each of the welfare concerns you have identified.

Activity 11.4

Red meat, green facts

Explore the Australian Good Meat website, and then answer the following questions.

1 List three facts presented on the Australian Good Meat website about how the Australian red meat and livestock industry is becoming more environmentally sustainable by reducing greenhouse gas emissions.
2 Explain how the red meat and livestock industry plans to become carbon neutral by 2030.
3 Outline three ways in which livestock producers are working towards improving biodiversity on their properties.
4 List two ways in which livestock producers are reducing the amount of water used in raising beef cattle.
5 Describe one strategy that Australian livestock farmers use to ensure that their animals are treated humanely.

Testing knowledge 11.1

1 Create a mind map to highlight the nutrients in the lean meats group that are considered important for good health.
2 Outline the recommended amount of meat or meat alternatives that should be consumed each week according to the *Australian Guide to Healthy Eating*.
3 Explain why the protein found in meat is described as 'complete', and why this type of protein is essential for our health and wellbeing.
4 Identify the key nutrients, other than protein, found in red meat and explain their functions in the body.
5 Explain why the Heart Foundation recommends limiting the amount of unprocessed red meat we consume each week.
6 Discuss the differences between muscle fibres, connective tissue and fat tissue in meat.

7 What is modified atmosphere packaging (MAP)? Explain why this packaging system is used to package fresh meat.
8 Outline one strategy livestock producers are developing to reduce the amount of methane produced by ruminant animals such as cattle and sheep.
9 Explain how farmers can ensure the sustainability of their land when grazing cattle and sheep.
10 Why is raising pigs in indoor intensive systems considered to be unethical by some consumers?

Poultry

'Poultry' is the term used to describe any domesticated birds that are used as food. Chicken, turkey, duck, quail and pheasant are all part of the poultry group, of which chicken is the most popular to eat.

Today's chickens are descendants of wild fowl that roamed the dense jungles of primeval Asia. It was only after the Second World War that chicken became reasonably priced – before this, it was expensive and was served only as a roast meal on special occasions. Modern production methods have reduced the cost of this versatile food, and today it is readily available in food stores.

Poultry consumption in Australia has risen considerably over the past few decades due to its affordability, availability and convenience. It is a versatile protein food as it can be enhanced with the flavours and cooking methods of different world cuisines. There has also been a growing awareness of its nutrient value, further increasing its popularity.

Poultry for good health

Like other meats, chicken is a very good source of complete protein, and therefore helps to build and repair body tissue. However, it contains marginally less iron than red meats such as beef and lamb. Chicken also contains the B group vitamins thiamin and riboflavin, as well as some zinc. It has the same amount of protein as red and other white meat, but less saturated fat and more polyunsaturated fat. Chicken is often thought to be a good substitute for red meats because if it is prepared without the skin, it is very low in fat.

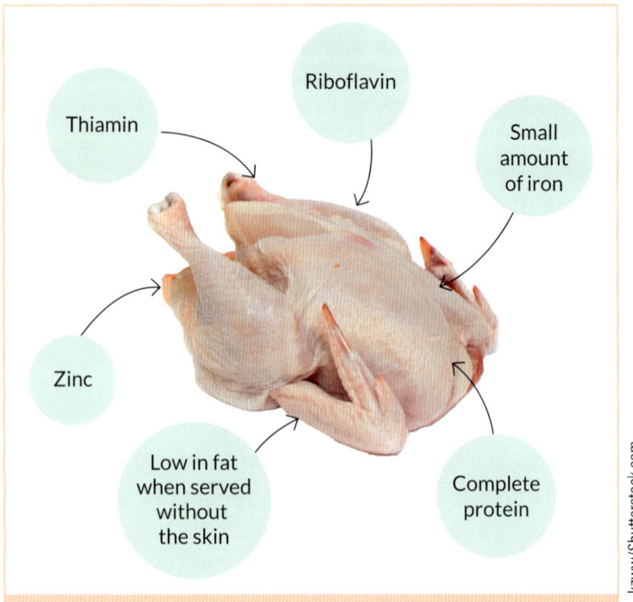

kzww/Shutterstock.com

Figure 11.12 Nutrients in poultry

Cuts of chicken

Chicken can be purchased whole for roasting or in cuts such as breast fillets, thighs, drumsticks, drumettes, cutlets, wings and winglets. Minced chicken is another option for preparing quick, simple everyday meals.

Today, there is a great variety of convenient fresh, chilled or frozen chicken products available for sale, such as chicken stroganoff and green chicken curry. Jars of sauces and packets of flavourings that can be added to chicken to make quick, easy meals also line the supermarket shelves.

iStock.com/fcafotodigital

Drumsticks

- Cuts: drumsticks, mince
- Cooking methods: oven roast, casserole

iStock.com/chengyuzheng

Legs

- Cuts: Maryland, thighs, drumsticks, drumettes
- Cooking methods: bake, casserole, barbecue

iStock.com/Timmary

Wings

- Cuts: wings, winglets, spare ribs
- Cooking methods: pan-fry, oven roast, barbecue

JohnGollop/E+/Getty Images

Whole chicken

iStock.com/Martina_L

Breast

- Cuts: fillets, tenderloins, cutlets
- Cooking methods: stir-fry, grill, pan-fry, oven roast, barbecue

iStock.com/Andrii Zastrozhnov

Thighs

- Cuts: fillets, Maryland, cutlets, mince
- Cooking methods: bake, stir-fry, casserole, barbecue

Figure 11.13 Cuts of chicken

9780170495714

Activity 11.5

Using chicken in everyday meals

Undertake an online search using the key words 'how to joint a chicken'. (The Taste website has a useful video.) Review the information you find, and then answer the following questions.

Weblink
How to joint a chicken

1 Select two cuts of chicken. Describe their physical properties and recommended cooking methods.
2 Search for recipes that use the chicken cuts you have selected.
3 Draw a table like the one below and complete it using information from the website/s.

	Chicken cut 1	Chicken cut 2
Description of the chicken cut		
Physical properties of the chicken cut		
Name of a recipe that uses this chicken cut		
Main cooking method used in the recipe		
Four main pieces of equipment required to complete the recipe		
One health and safety issue to consider in the preparation or production of the recipe		

Sustainable and ethical farming of chicken meat

Chicken meat has become a very popular food item for many Australian families. The chickens that are used as a source of meat are kept in large barns where they are free to roam throughout the day. The barns protect the poultry from predators, and are well ventilated to ensure that the temperature and air quality are kept at optimum levels.

Figure 11.14 Chickens bred for meat are kept in barns.

iStock.com/sansubba

Environmental issues

Information presented by Our World in Data demonstrates that poultry meat produces less greenhouse gas emissions per kilogram of food product than other livestock used as a source of meat for human consumption.

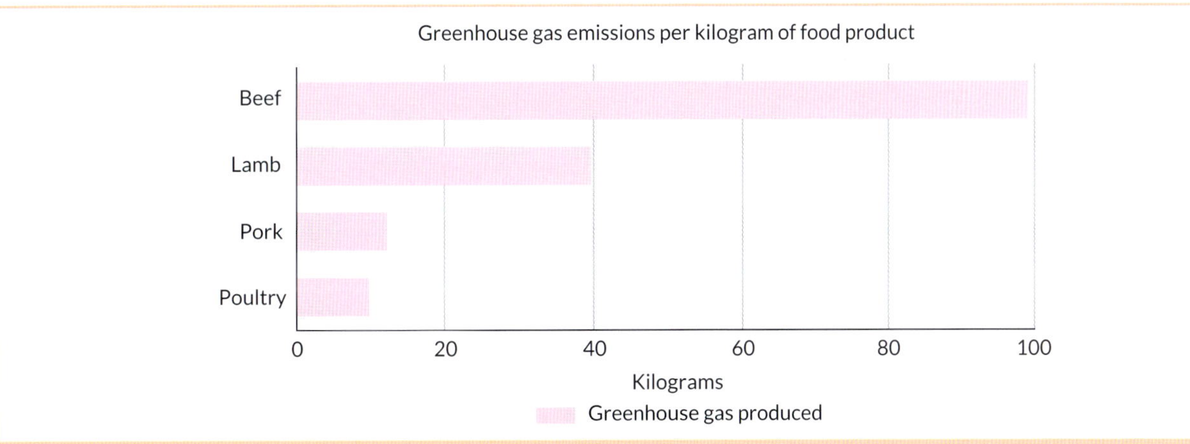

Figure 11.15 Poultry meat produces less greenhouse gas emissions per kilogram than other livestock.

Ethical issues

A number of ethical issues are associated with the production of chicken meat for human consumption. Chickens that are bred to produce meat are genetically selected to have a larger body mass and to grow more quickly than chickens used to produce eggs. These chickens reach their ideal size for slaughter by four to six weeks of age. However, this rapid growth can lead the chickens to suffer from leg deformities, as their skeleton cannot support their larger body weight, and can result in higher levels of mortality.

Another ethical concern is that some poultry meat producers keep large numbers of birds in the barns. These high stocking densities can restrict the chickens' natural behaviours, such as perching, foraging and dust bathing. Chickens can also become stressed and develop serious health issues if they do not have adequate lighting, good ventilation and clean litter covering the floor.

The methods used to stun and slaughter chickens also give rise to ethical concerns. All birds are stunned before they are slaughtered. Ideally, birds are stunned using controlled-atmosphere or gas stunning. However, some facilities use an electrical stunning system, which is far less humane.

Fish

Fish is a great ingredient for everyday meals because it contains a small amount of connective tissue with short muscle fibres, so it cooks very quickly and is usually tender. Because fish has such delicate flesh, it is best to cook it just before you want to serve it.

Fish can be purchased in a variety of ways, such as whole, as fillets, as cutlets or canned. Fish is quick and easy to cook, and can be baked, grilled, pan-fried, deep-fried, poached, steamed, stewed, stir-fried, microwaved, barbecued or cooked in foil. Be mindful when cooking fish that too many strong flavouring ingredients can overpower the delicate flavour.

Fish for good health

Health professionals encourage us to include up to two fish meals in our diets each week. Fish is a very healthy choice for meals because it is an excellent source of complete protein, and most varieties are naturally low in fat. Fish also provides important omega-3 fatty acids, which are essential for brain and eye development. In addition, eating fish on a regular basis is thought to reduce the risk of childhood asthma and to help halve the likelihood of heart attack. However, fish does not contain any calcium, unless you eat the bones in canned fish such as salmon or sardines.

It is now recognised that, due to the pollution present in many of the world's rivers and oceans, larger fish may be contaminated with heavy metals such as mercury, so it is wise to minimise the consumption of fish such as shark (flake), swordfish and barramundi.

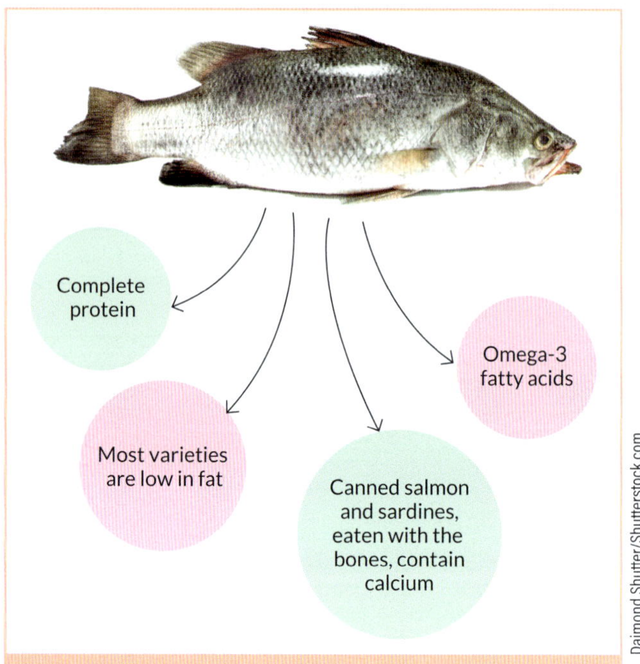

Figure 11.16 The nutrients in fish

Sustainable and ethical farming of fish

Australia has established a reputation as a supplier of safe, high-quality seafood. **Sustainable fishing** means leaving enough fish in the ocean so that the fish population can remain productive and healthy. It also means respecting aquatic habitats so that other species of marine animals can remain alive and healthy, and ensuring that people who depend on fishing can maintain their livelihoods.

Seafood is a nutritionally rich food, and, as the world's population grows, traditional methods of catching wild fish cannot meet the increase in global demand and natural stocks of seafood decline. To overcome this problem, many countries around the world have developed facilities to farm fish. **Aquaculture** involves breeding, rearing and harvesting fish and shellfish in coastal marine waters, open oceans and fresh-water systems.

Two main types of aquaculture are carried out in Australia – marine and fresh-water. The marine fish industry is carried out mainly in South Australia and Tasmania in in-shore and off-shore sea cages or sea pontoons. The main species farmed are blue fin tuna, Atlantic salmon, yellowtail kingfish and barramundi. The fresh-water fishing industry has many small farms throughout Australia, and uses intensive tank rearing systems and pond and dam systems. The main species farmed are Murray cod, silver perch and eels.

9780170495714

Environmental issues

While aquaculture practices are seen to be the solution to over-fishing of some endangered species, they have created significant environmental impacts. In Tasmania's Macquarie Harbour, where large numbers of salmon are raised in enclosed sea cages, high levels of water pollution have been found in surrounding waters. This pollution is caused by a build-up of faecal matter or fish waste, unused fish food that can lead to an overload of nutrients in the water, and antibiotic contamination. The sea water around these sites has also been depleted of oxygen, and 'dead' zones on the sea floor – where no marine life can live – have developed.

Ethical issues

One of the key concerns linked to aquaculture facilities is that the low oxygen levels in the water, combined with higher water temperatures, have led to a large number of salmon dying from asphyxiation as a result of a lack of oxygen. Another ethical issue is that seals are drawn to the aquaculture pontoons to feed on the salmon, as this is one of their natural food sources. However, the fish farming companies see the seals as predators, and use underwater explosives and projectiles to scare them away. The Australian Marine Conservation Society has also reported that in Tasmania's Macquarie Harbour, 'over 2000 fur seals were sedated and transported over land to the other side of Tasmania. This attempt to remove seals from around salmon cages, resulted in their release on the opposite corner of the State only to have many swim back.'

iStock.com/LKR Photography

Figure 11.17 Aquaculture sea cages

Choosing sustainable seafood

The Australian Marine Conservation Society has produced an online resource and app for consumers wishing to purchase sustainable seafood. *GoodFish: Australia's Sustainable Seafood Guide* considers stock status, bycatch and discards, habitat and ecosystem, and how the seafood is caught, giving each seafood one of three 'traffic light' ratings: 'Better choice', 'Eat less' and 'Say no'.

- **Better choice** indicates that the species is not over-fished and has been caught using techniques that have a low environmental impact. Species in this category include sand whiting, wild-caught coral trout and farmed barramundi.
- **Eat less** indicates wild-caught species that have been caught using methods that damage the habitat and produce bycatch. Wild-caught snapper and blue grenadier are in this group.
- **Say no** indicates wild-caught species that may be over-fished, or whose capture involves significant bycatch of threatened or protected species, or whose production causes environmental pollution. Wild-caught barramundi and prawns, and farmed Atlantic salmon are in this category.

Activity 11.6

Using the GoodFish app

Weblink GoodFish

Aim

To use the GoodFish app to make a responsible choice of fish for a family meal

Method

Salmon and barramundi are two types of fish that are popular to use when preparing a family meal.

1 Download the GoodFish app onto your device or visit the 'Sustainable Seafood Guide' page on the GoodFish website.

2 Use the information on GoodFish to complete the table below. For each species of fish, list two facts from GoodFish to explain why the fish is listed in this category.

Results

	Type of salmon	Type of barramundi
Better choice		
Eat less		
Say no		

GoodFish, https://goodfish.org.au/

Figure 11.18 The GoodFish app icon

11

Analysis

1 Were any of the species over-fished? If so which ones?

2 Did catching any of the fish prove harmful to other marine species? Explain your results.

3 Were any of the fish species farmed? If so, what impact has this had on the environment and on other marine species?

4 Explain why some species of salmon and barramundi are rated as a 'Better choice' rather than being rated as 'Eat less' or 'Say no'.

Conclusion

Identify which species of fish you would select for a family meal based on its sustainability, and explain why this would be a good choice.

Cooking meat, poultry and fish

Meat, poultry and fish are cooked in order to make them safe to eat and easier to chew and digest, and to destroy any harmful microorganisms. Methods of cooking can be divided into two groups – dry methods and moist methods.

The dry methods of cooking are fast, and best suited to tender cuts of meat, poultry or fish. These cuts have enough water in their tissue to enable the conversion of the protein collagen to gelatin. Therefore, these cooking methods use little-to-no liquid, and the time taken to cook the meat, poultry or fish depends on the size and thickness of the cut. Examples of these methods are pan-frying, stir-frying, grilling, barbecuing and roasting.

The moist methods of cooking are slower, and best suited to less tender cuts of meat and poultry. These cuts do not have enough water in their tissues to convert collagen to gelatin. Therefore, the long, slow, moist cooking softens the connective tissue that makes meat tough. Examples of these methods are casseroling, pot-roasting, braising and stewing.

Stepanek Photography/Shutterstock.com

Figure 11.19 Tougher cuts of meat are suitable to use in a casserole.

Changes that occur during cooking

During the cooking process, a number of changes occur to the physical and sensory properties of meat, poultry and fish.

Texture

- When heat is applied, the protein in the meat, poultry or fish is denatured by coagulation. The protein food shrinks as some of the water is expelled, and the texture becomes firm.
- Cooking improves the palatability of meat, poultry and fish by making it easier to chew.
- Connective tissue is tenderised and the collagen is converted to gelatin, if moisture is present. This method breaks down the protein in the muscle and connective tissue.
- Fat melts, giving meat, poultry and fish a crisp, brown surface when exposed to dry heat.

Colour

- When dry heat is applied, such as in grilling, frying or roasting, the surface of the meat will become golden brown as a result of the Maillard reaction.
- The internal colour of meat changes from red to brown.
- The flesh of poultry changes from pale pink to white.
- The inner flesh of white fish becomes white and opaque.

Flavour and aroma

- Flavour or extractives are squeezed out of meat, poultry and fish and onto the surface when exposed to dry heat, providing flavour and aroma.
- Fat also adds flavour when meat, poultry and oily fish are cooked.

Nutrient value

When meat, poultry and fish are cooked, a number of changes occur to their nutritive value:

- if overcooked, the protein becomes tough and indigestible
- fat-soluble vitamins A, D, E and K are not affected by cooking
- water-soluble vitamins (B group) are lost through cooking.

How to tell if it is cooked

Meat

The meat should be browned on the outside. Pierce the meat with a skewer – the juices should be pinkish to clear.

Poultry

Pierce the poultry with a skewer in the thickest part of the flesh – the juices should run clear. The chicken meat should no longer be pink. The skin should be golden brown and slightly crispy.

9780170495714

Fish

The flesh of white fish should turn white and flake apart easily with a fork.

Top tips for grilling

Grilling is a quick and healthy method of cooking, and is ideal for preparing family meals in a hurry.

1 For a griller separate to the oven, leave the griller door open during cooking to prevent heat from building up inside the appliance. For a grill inside the oven, grill with the door closed or follow the manufacturer's instructions.
2 Trim visible fat from meat or poultry.
3 Use tongs for turning food during grilling.
4 If meat starts to curl while it is cooking, cut the curled-up edges with the point of a sharp knife.
5 Test to see if meat or poultry is cooked by gently pressing with blunt tongs. Do not cut the meat or poultry, as this allows the juices to escape. Fish will flake apart with a fork.
6 Soak bamboo skewers for kebabs in water before threading to prevent them from burning during grilling.
7 If meat, poultry or fish is marinated with honey or sauces containing sugar, heat the grill to medium rather than high to prevent the marinade from burning.

Top tips for stir-frying

Stir-frying is a very quick method of cooking in a wok over intense heat. Tender cuts of meat, chicken and/or vegetables should be cut into equal bite-sized pieces prior to the cooking process. Meat and chicken can be marinated before stir-frying, but the marinating liquid should be drained off before cooking to prevent stewing.

1 Prepare meat strips by trimming off any fat and slicing across the grain. Strips should be around 5–8 centimetres long. Cut chicken pieces in equal sizes so they cook evenly.
2 Add a small quantity of oil to the wok and swirl to coat its sides and base. Heat oil until smoke point is almost reached.
3 Drain off excess marinade, add the meat or chicken pieces and stir-fry for 2–3 minutes. The meat or chicken pieces should sizzle when added to the hot oil. Sear in small batches to prevent the meat or chicken from shedding its juices and stewing, and consequently toughening. Allow the wok to heat up again between batches.
4 Remove the meat or chicken from the wok and stir-fry the vegetables separately – firm vegetables first, then softer or leafy varieties. Return everything to the wok to warm before serving.

Figure 11.20 Stir-fry cooking

Top tips for frying

1. Always have the oil hot enough so that when the food is added, it sizzles. Test with a cube of bread – it should turn golden in 30 seconds. Food soaks up cold oil like a sponge.
2. Make sure all portions of food are dry before placing them in the hot oil. This prevents spitting.
3. Fry only small portions of food – they will cook quickly and more evenly.
4. Drain fried food well before serving.
5. Take care when frying because hot fat and oil reach very high temperatures. Use oven mitts to protect your hands.

Cavan Images/Alamy Stock Photo

Figure 11.21 Chicken thigh is a tender cut of poultry that is suitable for frying.

Activity 11.7

Cooking safely with meat, poultry and fish

Work in small teams to prepare a digital notice board or fact sheet to inform consumers how to cook safely with protein foods, such as meat, poultry and fish. The digital notice board or fact sheet will be available in supermarkets and most butchers. Your digital notice board or fact sheet should include:

- appropriate headings
- information about how to select good-quality meat, poultry or fish
- information about safely transporting meat, poultry or fish home (see Chapter 2)
- information about strategies for storing meat, poultry or fish safely (see Chapter 2)
- instructions for preparing and cooking meat, poultry or fish
- diagrams or photos, to help make the message clear.

Testing knowledge 11.2

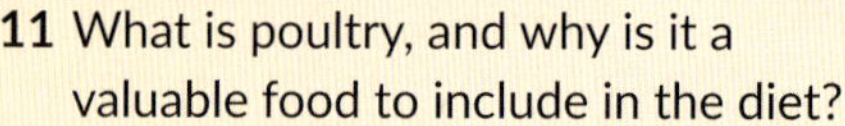

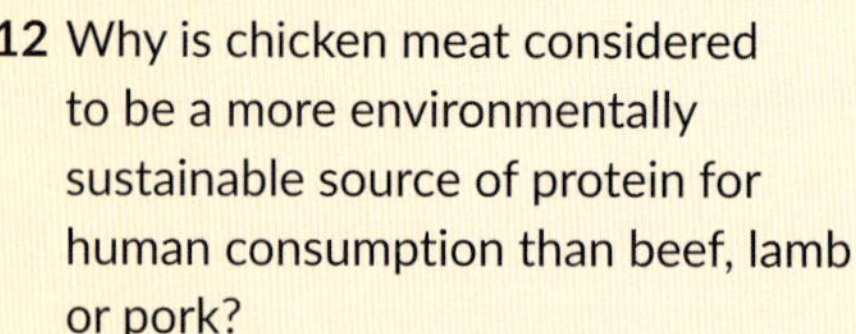

Quiz
Testing knowledge 11.2

11. What is poultry, and why is it a valuable food to include in the diet?
12. Why is chicken meat considered to be a more environmentally sustainable source of protein for human consumption than beef, lamb or pork?
13. Outline two ethical issues associated with the production of chicken meat.
14. Explain why fish can be cooked very quickly.
15. Create a mind map to highlight why fish is a very healthy choice for meals.
16. What is meant by the term 'sustainable fishing'? Provide one example of how this can be achieved.
17. Outline one environmental and one ethical issue associated with the use of aquaculture in Tasmanian waters.
18. Explain how dry methods of cooking differ from moist methods of cooking.
19. Describe the impact of cooking on the nutrient value of meat, poultry and fish.
20. Create a diagram to demonstrate the changes that occur to meat and poultry when it is cooked.

Eggs

An egg is the reproductive cell of a female bird that contains nutrients to support the development of the new embryo.

Eggs can form the basis of a cooked breakfast or a quick and easy family meal. They are also an important ingredient in many recipes, assisting with multiple processes used in food preparation.

Eggs for good health

Eggs are a popular source of animal protein as they are low in cost and therefore more affordable than meat products, and can be easily prepared in a variety of ways. Like other sources of animal protein, eggs are rich in vitamins and minerals – most notably vitamin D. One egg fulfills 40 per cent of our daily vitamin D requirement, which is essential for calcium absorption. However, eggs do not contain any vitamin C, and the amount of carbohydrate they contain is also very minimal.

The fat and cholesterol that are present in eggs are only found in the egg yolk. Because eggs contain cholesterol, they were once considered to be bad for

9780170495714

our heart health, but recent research indicates that eggs do not have a negative impact on our heart. However, health professionals recommend that people with type 2 diabetes or those who need to lower their LDL (low-density lipoprotein) cholesterol (or 'bad' cholesterol) levels should eat no more than seven eggs per week.

Table 11.2 Approximate nutrient content of a 55-gram egg

Energy	297 kilojoules
Water	37.7 grams
Protein	6.35 grams
Fat	5.05 grams
Carbohydrate	0.15 grams
Cholesterol	187.5 milligrams
Sodium	66.5 milligrams
Potassium	57 milligrams
Calcium	19.5 milligrams
Iron	0.8 milligrams
Zinc	0.45 milligrams

Components and structure of an egg

The structure of an egg is complex. It has an external shell and an inner membrane to protect the developing embryo from changes in the external environment. The yolk containing the egg cell is surrounded by both thin and thick albumen, or egg white. The chalaza or cord anchors the yolk to the shell and keeps it centred. The air sac is important, as it will provide the hatching chicken with its first gulp of air.

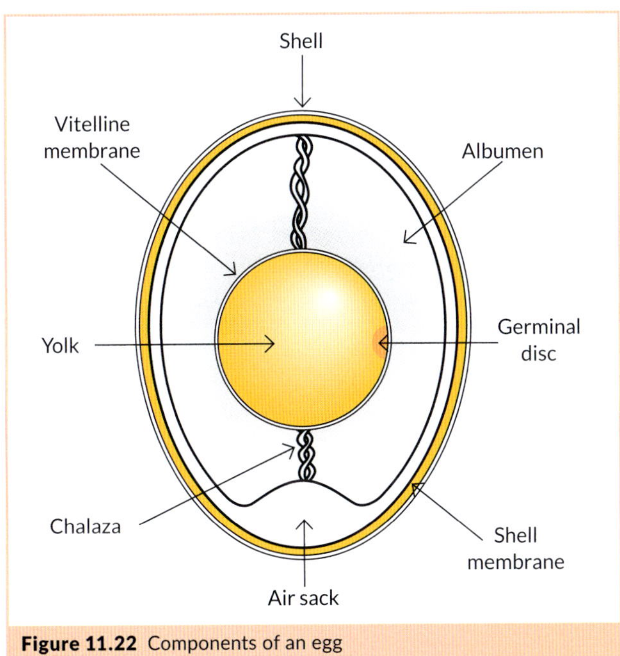

Figure 11.22 Components of an egg

PRACTICAL ACTIVITY 11.1

Scientific experiment: Investigating the structure of eggs

Worksheet
Practical activity 11.1

1. Collect an egg, a small plate and a skewer.
2. Break the egg onto the plate, making sure not to break the yolk.
3. Carefully examine both the shell and the egg.
4. See if you can locate all of the components of the egg shown in Figure 11.22. Place a tick next to each component as you identify it.
5. What do you think is the purpose of the shell membrane?
6. Carefully pierce the egg yolk with your skewer. What happens to the yolk?
7. Explain how the yolk keeps its round shape.
8. What do you think is the role of the chalaza in an egg?
9. What do you think is the purpose of the germinal disc in an egg?

Testing the freshness of eggs

Eggs can be tested for freshness by filling a bowl with cold water and gently lowering the egg into the water.

- The egg is fresh if it stays on the bottom of the bowl.
- The egg is stale if it floats to the surface of the bowl.

CORDELIA MOLLOY/Science Photo Library

Figure 11.23 Testing the freshness of eggs – fresh eggs sink and stale eggs float.

Purchasing eggs

When purchasing eggs, always check the best-before date to ensure that they are fresh. Eggs are usually purchased in a carton, where they sit in individual spaces, point-down so that the yolk remains in the centre of the white. Remember that eggs are a perishable food, so they must be refrigerated to ensure that they are safe to eat.

Figure 11.24 Always check the best-before date on an egg carton to ensure that you buy eggs that are fresh.

Methods of egg production

Egg farming is the process of raising hens to produce eggs for human consumption. Australians eat almost 19 million eggs every day.

In Australia, there are three egg farming systems.

Free-range eggs

The hens are free to roam and forage for food outside for at least eight hours each day. The hens' food, water and nests are housed in sheds, where they are protected from predators. There has been a significant increase in the number of consumers who purchase free-range eggs, and they now make up approximately 56.5 per cent of all eggs sold in supermarkets.

William Edge/Shutterstock.com

Figure 11.25 Free-range hens can roam outside during the day.

Organic eggs

Organic eggs are also produced in a free-range environment where the hens are able to roam in paddocks during the day. However, unlike the feed fed to hens in a traditional free-range system, their feed is free from pesticides, herbicides and synthetic fertilisers. They do not receive any antibiotics. Organic eggs only make up about 2 per cent of all eggs sold in supermarkets.

Figure 11.26 Organic eggs do not contain any pesticides or herbicides.

Barn-laid or cage-free eggs

The hens are free to roam, perch and socialise in a large, climate-controlled sheds, and are not confined to cages. They are able to spread their wings and stretch out, but they do not have access to the outdoors. The hens have nest boxes to lay their eggs in. The sheds are highly automated and the eggs roll gently from the nesting box along a conveyor belt to the packing sheds. Approximately 10 per cent of all eggs sold in supermarkets are barn-laid or cage-free eggs.

iStockphoto.com/Galdric

Figure 11.27 Hens confined to a barn do not have access to the outdoors.

Cage eggs

The hens are confined to small, barren wire cages in large climate-controlled sheds. Each hen only has a space approximately the size of an A4 sheet of paper. The cages – which are known as 'battery cages' – are stacked in multiple tiers, with conveyor belts between each tier to collect the eggs and remove the bird manure automatically. In 2023, Australia's state and territory agriculture ministers endorsed new Australian Animal Welfare Standards and Guidelines for Poultry, which include phasing out barren battery cages by 2036.

9780170495714

Mai.Chayakorn/Shutterstock.com

Figure 11.28 These hens are confined to small wire cages.

Sustainable and ethical issues in egg production

The demand for eggs has increased in recent years, as they are widely available and are a budget-friendly form of protein. However, the production of all forms of animal protein, including eggs, brings sustainability and ethical challenges, such as those relating to environmental pollution, a reduction in natural resources and animal welfare.

Sustainable issues in egg production

Compared with other protein foods such as beef, lamb and pork, eggs have a relatively low environmental footprint. However, many egg producers are constantly looking for ways to further improve their sustainability by improving feed production, minimising their energy use, and reducing the impact of poultry manure produced on farm.

One strategy being implemented on many poultry farms to address these issues is the development of fully automated 'smart sheds'. These sheds allow farmers to control environmental conditions inside the shed – such as temperature, air flow and light – to ensure the optimum conditions for their hens. They can also monitor and control feed availability and water consumption based on the health and productivity of the hens. Solar panels fitted on the roof of the sheds are a source of renewable energy. Poultry producers are also developing strategies to recycle and repurpose poultry manure, both on and off farm.

Ethical issues in egg production

Sound animal welfare standards are critical to ensure the sustainability of the egg industry. Hens are clever, inquisitive and social animals, and need to be free to roam and spread their wings. Ideally, they should have access to the outdoors to socialise, scratch and dust bathe, and have a perch on which to roost at night. Hens kept in barren, wire battery cages cannot express their natural behaviour, as they have little-to-no space to move or stretch their wings. Because of their lack of space, they also cannot escape the aggression of other hens.

In recent years, consumers have become increasingly concerned about animal welfare issues including cage egg production systems. As a result, many consumers are purchasing free-range and barn-laid or cage-free eggs with lower hen-density production ratios, rather than eggs produced using intensive production systems. Egg producers have responded to this demand by moving to barn-laid or cage-free systems and by providing hens with furnished cages, including a perch, a nest box and increased space. Supermarkets have also responded to consumer demand, so many more free-range and cage-free eggs are now available on the supermarket shelves.

For many years, the RSPCA has been advocating for an end to battery cage egg production. Its current campaign, 'Bye Bye Battery Cages', aims to encourage state and territory governments to implement the new Australian Animal Welfare Standards and Guidelines for Poultry before the required date of 2036. The RSPCA argues that hens will continue to suffer unnecessarily in battery cages if changes are not urgently made.

Weblink
RSPCA's 'Bye, bye battery cages' campaign

Nuts and seeds

Nuts and seeds are both nutritious foods that should be included as a part of a well-balanced diet. Nuts are the edible kernel of dried fruits from some trees and plants. They are really the seeds of a plant enclosed in a hard, brittle or woody tough shell. Similarly, many vegetables and plants produce edible seeds, such as pumpkin seeds, chia seeds, sesame seeds and sunflower seeds.

Christopher Boswell/Shutterstock.com

Figure 11.29 Growing almonds uses large volumes of water, which has a negative impact on the environment.

Both nuts and seeds can be eaten as a snack food, or used as a garnish or flavouring ingredient. Nuts can be tossed through a salad or stir-fry, or used ground as a substitute for flour in a cake; while seeds can be used as a nutritious addition to muesli or fruit smoothies, or as a topping on muffins and bread rolls.

Nuts and seeds can add variety and a concentrated source of nutrients to everyday meals.

Classification of nuts

Nuts are grouped or classified according to how they grow: on a tree, under the ground or as the seed of a fruit.

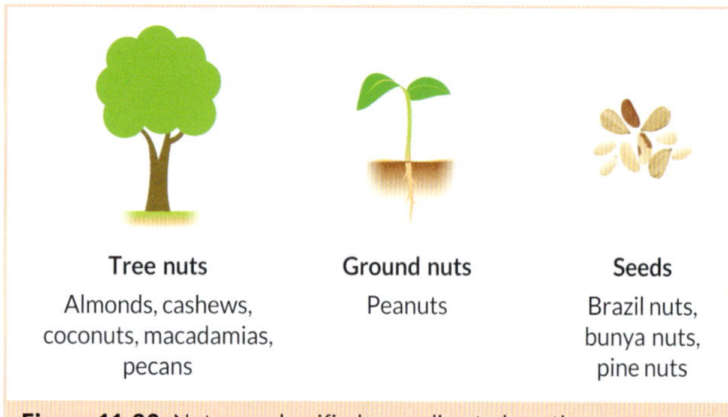

Figure 11.30 Nuts are classified according to how they grow.

Nuts and seeds for good health

Nuts are a healthy plant food, as they provide protein, healthy monounsaturated and polyunsaturated fats, and a range of minerals, vitamins and phytochemicals. They are also a good source of fibre and are naturally low in sodium. Because of their nutritional properties, nuts are often eaten by people who follow a vegetarian or vegan diet. It is recommended that we include a handful – about 30 grams – of nuts in our diet every day. Nuts are best enjoyed raw or toasted and unsalted, as too much salt can increase blood pressure.

A small proportion of people are allergic to nuts and, should they ingest them, the consequences can be life threatening. They may be allergic to one or more tree nuts, peanuts only, or both tree nuts and peanuts. Further information on nut allergies can be found in Chapter 8.

iStock.com/WRS Photos

Figure 11.31 A range of nuts and seeds

Like nuts, seeds provide a concentrated source of important nutrients. Sesame seeds are a good source of protein, omega-6, vitamins and fibre, while chia seeds are renowned for their high concentration of omega-3 fatty acids, dietary fibre, protein and calcium. Recent research suggests that including chia seeds in the diet has many health benefits, especially in helping to reduce cholesterol levels and the risk of cardiovascular disease.

Environmental sustainability of nut production

Almond production has increased significantly in Australia in recent years, and almond orchards are now one of the largest horticultural crops grown in South Australia. However, almond orchards use significantly more irrigated water than other crops. Almond trees require approximately 12.5 million litres of water per hectare which is the equivalent of approximately 7000 litres of water per kilogram of shelled almonds. This high level of water use has a major impact on the environment, as most of these orchards grow along the Murray River and draw water from this river system. Evidence shows that the Murray–Darling Basin system is already under stress, and drawing irrigated water from this system – especially in drought years – will be detrimental to its environmental sustainability.

Another environmental concern is the impact that almond production has on bee populations. Beekeepers have recorded high mortality rates among bee populations that are used to pollinate almond orchards, most likely as a result of exposure to pesticides.

Nut spreads and milks

Nuts are commonly used to make spreads such as peanut butter, cashew butter and almond butter. They are also widely used as an alternative to milk made from dairy cows. Plant-based milks – such as almond milk, coconut milk,

Figure 11.32 Plant-based milks are growing in popularity.

9780170495714

macadamia milk and cashew milk – are growing in popularity, and supermarkets and health food stores now stock a wide range of these milks for sale, both fresh and in long-life varieties.

Plant-based milks have become more popular as they are dairy-free and suitable for people who are allergic to cow's milk or who are lactose intolerant, or those following a vegan diet.

When choosing a plant-based milk, check the label to ensure that it has similar nutritional properties to dairy milk. This will mean that it has been fortified with calcium and has adequate protein for vegans. Also be aware that many of these milks contain added sugar or sugar syrup; aim to purchase unsweetened or organic varieties.

CASE STUDY 11.1

Ranking different 'milks'

Read the article below, and then answer the questions that follow.

Oat, soy, almond, coconut, dairy: A dietitian rates eight 'milks' from healthiest to hell no

Susie Burrell, *The Age*, 3 July 2023

Long gone are the days when the biggest milk decision you needed to make was full-cream or light. Now, we have a range of plant-based 'milks' with sales forecast to grow by as much as 50 per cent in the next four years. While they aren't technically milk, they're taking the place of cow milk in many Australians' diets.

iStock.com/photoguns

Figure 11.33 Alternative milks can be made from coconut, oats, soy beans, almonds and more.

Full-cream dairy milk

Full-cream milk is a whole, natural food that contains 8 g of protein and close to 300 mg of calcium per 250 ml glass, as well as 10 g of total fat, 6 g of which is saturated. With the rise in heart disease rates in Australia over the past 30 to 40 years, it has been recommended that adults help to reduce their overall intake of saturated fat, and as such, swap to a reduced or low-fat milk, which contains far less fat with all the same nutritional benefits associated with consuming an animal-based milk.

- **Pros:** a whole, natural food, rich in key nutrients including protein and calcium; good option for families
- **Cons:** relatively high in saturated fat
- **Rating:** 6/10

Low-fat dairy milk

Contrary to popular belief, low-fat milk does not contain more lactose – a naturally occurring sugar – than full-cream milk. It is simply lower in fat and calories per serve, containing between 0–4 g of fat per 250 ml serve.

- **Pros:** a lower fat, lower calorie dairy milk; good for those with high cholesterol
- **Cons:** lacks the rich mouthfeel of full-cream milk
- **Rating:** 10/10

Soy milk

Soy milk, the plant-milk alternative that is closest nutritionally to dairy milk, contains 8–10 g of protein per 250 ml serve. Nutritionally, it is lower in saturated fat than dairy milk, with most of its fat coming from polyunsaturated fat, and there are regular, reduced-fat and low-fat soy milks available to suit your preference. The majority of readily available soy milks are fortified with good amounts of the essential nutrients typically found in dairy milk, including calcium. Be aware though: soy milks can also have sugars and/or vegetable oils added.

- **Pros:** low in saturated fat, high in protein; the best option nutritionally for vegetarians and vegans
- **Cons:** has a distinct flavour
- **Rating:** 8/10

Oat milk

Of all plant-based milks, it is oat milk that has attracted plenty of attention over the past couple of years, likely thanks to its creamy taste and the fact that it sounds healthy. Made from a mix of oats, water and oat flour, oat milk is much lower in protein than dairy milk, with just 4 g on average per serve, and little to no natural calcium. While some varieties may add calcium, not all do, and the formulated milk product still lacks the range of nutrients, such as phosphorus and magnesium working synergistically, the way they do in cow's milk.

Oat milk contains significantly more carbohydrate than all other milks, including nut and soy.

- **Pros:** rich, creamy taste, low in calories. Best option for families looking for a plant-based milk alternative.
- **Cons:** lacks natural calcium and protein, can contain added sugars
- **Rating:** 5/10

Almond milk

Popular on paleo and other low-carb regimens, unsweetened almond milk contains few calories per serve and no sugars. It's also a good source of the key nutrient vitamin E, which helps cell regeneration in the body.

The biggest issue with almond milk is that it is naturally low in protein and calcium, so always choose an almond milk with added calcium, to reap the bone health benefits, and be careful of products that contain added sugars, as this bumps up sugar and calorie content significantly.

- **Pros:** low in calories and carbs, good option for low-carb and keto diets
- **Cons:** offers few nutrients naturally, especially if unfortified with calcium and protein
- **Rating:** 3/10

Rice milk

Popular with allergy sufferers, rice milk can be a good option for individuals with a nut allergy, who prefer a vegan or vegetarian alternative to animal milk. Generally made from brown rice and water, rice milk, like almond milk, can be exceptionally low in protein and calcium, so look for fortified varieties. Rice milk can contain significantly more sugars and calories than almond milk.

- **Pros:** low in calories and carbs; may be the best option for those with food allergies
- **Cons:** offers few nutrients naturally, especially if unfortified with calcium and protein
- **Rating:** 2/10

Coconut milk

Made from a mix of water and coconut flesh, coconut milk as a dairy replacement is still an emerging product, with only a couple of supermarket options currently available. Relatively high in fat, with 10 g per serve, as well as up to 10 g of sugars per cup, the current range is less likely to contain added calcium – and with exceptionally low levels of protein, nutritionally there are much better milks and alternatives you can choose instead.

- **Pros:** can be enjoyed on a keto regimen
- **Cons:** no positive nutritional attributes, extremely high in saturated fat
- **Rating:** 0/10

Susie Burrell is an accredited practising dietitian and nutritionist.

Analysis

1. List the dairy, plant and nut milks in order, according to Susie Burrell's rating.
2. Justify why low-fat dairy milk is a better choice to pour over breakfast cereal than full-cream dairy milk or oat milk.
3. Explain why soy milk is the best option nutritionally for vegetarians and vegans.
4. Outline the reasons why Susie Burrell gives coconut milk such a low rating.
5. Develop a logical argument that could be used in a promotional campaign to encourage people to swap to a plant- or nut-based 'milk'.

iStock.com/Tatiana Atamaniuk

Figure 11.34 Soy milk

9780170495714

PRACTICAL ACTIVITY 11.2

Worksheet Practical activity 11.2

Template Sensory wheel

Sensory analysis: Taste testing plant-based milk

1 Taste test a range of different types of plant-based milks. Select three or four different types, such as:

- soy milk
- oat milk
- coconut milk
- almond milk
- rice milk.

Focus on the sensory properties – appearance, aroma, flavour and texture – of each type of plant-based milk. Record your results in a table like the one below.

Plant-based milk type		Appearance	Aroma	Flavour	Texture
1					
2					
3					
4					

2 Rate the plant-based milks from the one you liked the most to the one you liked the least. Justify your ratings, using your analysis of the sensory properties.

3 Examine the label of each plant-based milk and record the kilojoule, fat, sugar and calcium content per 100 grams in a table like the one below.

Plant-based milk type		Content per 100 grams			
		Energy (kJ)	Fat	Sugar	Calcium
1					
2					
3					
4					

4 Draw conclusions about the plant-based milk with the highest energy content and the plant-based milk with the lowest energy content, and the amount of fat and sugar present in each plant-based milk.

5 Compare the calcium content of each type of plant-based milk. Develop a logical argument to explain the variation in the calcium content you have observed in each type.

6 With a partner, brainstorm and record why there is such a large variety of plant-based milks available today.

7 If you were responsible for buying plant-based milk for your family, which type would you choose? When writing your response, consider the sensory properties, nutritional value and the way that plant-based milk is used in your home.

Legumes and beans

Legumes have many of the same nutritional properties as meat, poultry, fish, eggs, nuts and seeds, and so have been placed in this section of the *Australian Guide to Healthy Eating*, as well as in the vegetables group. Legumes such as chickpeas and lentils are often made into burgers and patties, and used as a substitute for similar meat products. More detailed information on legumes and beans can be found in Chapter 10.

iStock.com/a-lesa

Figure 11.35 Chickpea and lentil patties

11

Testing knowledge 11.3

21 Why have eggs become a popular source of animal protein for many families?

22 Create a mind map to highlight why eggs should be included as a part of a well-balanced diet.

23 Outline a simple method you could use to test if an egg is fresh.

24 Compare the production of cage eggs and free-range eggs.

25 Explain how 'smart sheds' can improve the environmental sustainability of egg production.

26 Outline one ethical issue associated with the production of eggs and one strategy some egg producers are using to address this issue.

27 What does the RSPCA hope to achieve through its 'Bye Bye Battery Cages' campaign?

28 Explain why people are encouraged to include a handful of nuts in their diet every day.

29 Outline the reasons that we should try including sesame seeds and chia seeds in our meals.

30 Why have plant-based milks become popular with some consumers?

Quiz
Testing knowledge 11.3

Critical and creative thinking 11.1

Comparison of the environmental impact of cow's milk and almond milk

1 Research the environmental impact of the production of cow's milk and almond milk. Focus your research on:
- the amount of land required to produce a litre of each type of milk
- the impact of production on soil quality
- the amount of water use in production
- the amount of greenhouse gas emissions
- the disposal of waste.

2 Write a summary paragraph to compare the impact of the production of each type of milk on the environment.

DESIGN ACTIVITY 11.1

Worksheet
Design activity 11.1

Templates
Food order

Sensory wheel

Australian Guide to Healthy Eating

A gourmet burger

Burgers are a popular food to purchase from food trucks at farmers' markets.

Design brief

Design a gourmet burger that incorporates a protein ingredient and a range of vegetables and flavourings that would be suitable to sell from a food truck at a farmer's market to customers who are health conscious. The farmers' market promotes environmental responsibility among its store holders, so the burger must be packaged in a material that is sustainable.

Investigating and defining

1 Develop four design criteria that cover the solution requirements and constraints outlined in the design brief to evaluate your gourmet burger.

2 Investigate the types of meat-based and plant-based protein ingredients that would be suitable to include in a burger.

3 Undertake an online search of other ingredients that could be included in burger recipes. Focus your research on vegetables, flavouring ingredients and bread products.

4 Complete a table like the one on the following page by identifying four ingredients suitable to include in each category.

5 Research two ideas for sustainable packaging that could be used to serve the burgers from the food truck.

9780170495714

Protein ingredients	Vegetables	Flavourings	Breads

Generating and designing

1 Using the information you have gained from your investigation into burger ingredients, develop two options for burgers that could be sold from a food truck at the farmers' market.

2 Annotate each burger option, highlighting the advantages and disadvantages of each. Based on your evaluation of each option, select your designed solution.

3 Using the basic recipe for Zucchini Burgers (page 309) as a guide to the proportion of ingredients required, write out the recipe for your new burger.

Planning and managing

1 Prepare a food order.

2 Describe two health and safety practices you will need to follow when preparing your burger. One safety issue must focus on the management of raw ingredients during preparation and the other must focus on cooking the burger.

Producing and implementing

Produce the burger you have designed, following the health and safety practices you have identified.

Evaluating

1 Respond to your four design criteria in detail to determine the success of your burger.

2 Describe the sensory properties – appearance, aroma, flavour, texture and sound – of your burger.

3 Evaluate the flavour of your burger. Did the flavours complement each other?

4 What changes or modifications would you make to your recipe if you were to make the burger again?

5 Classify the ingredients for your burger on a diagram of the *Australian Guide to Healthy Eating*. Discuss whether your burger meets the guidelines of this food selection model and is a healthy option to sell from a food truck at the farmers' market.

DESIGN ACTIVITY 11.2

A meal using minced meat or a plant-based mince substitute

Beef, lamb, pork, chicken and plant-based ingredients can all be minced to create a versatile product that can be used in sauces; as a filling in pies, pastries, crepes and pasta; and in balls, patties and loaves. Mince can be fried, baked, grilled, steamed or stewed to make a variety of recipes suitable for serving as a meal.

Worksheet
Design activity 11.2

Templates
Decision table

Sensory wheel

Australian Guide to Healthy Eating

Design brief

1 Write your own design brief based on the '5 Ws'. The theme is a main meal in which minced meat or a plant-based mince substitute is the hero.

- Who – who will be eating the meal
- What – the occasion at which the meal will be served, and the time available to prepare and serve it
- When – the time of day or season when the meal will be eaten
- Why – why the meal will be eaten
- Where – where the meal will be served

2 Formulate the sentences or statements you developed in step 1 into a paragraph that will become your design brief.

Investigating and defining

1. Use the solution requirements and constraints in your design brief to develop five design criteria to evaluate your meal.
2. Research and select two recipes that use minced meat or a plant-based mince substitute, that are suitable for a main meal and can be prepared in the time identified in your brief.

Generating and designing

1. Develop a concept sketch of the presentation of each design option. Annotate each option to include:
   - recipe title
   - list of main ingredients
   - major cooking methods
   - major processes
   - references.
2. Use a decision table similar to the one below to select the designed solution. Justify your choice.

<table>
<tr><th colspan="2">Decision table</th></tr>
<tr><td colspan="2">Decision to be made:</td></tr>
<tr><td>Option 1
Advantages:
Disadvantages:</td><td>Option 2
Advantages:
Disadvantages:</td></tr>
<tr><td colspan="2">Designed solution:</td></tr>
<tr><td colspan="2">Justify your decision based on the solution requirements and constraints in your design brief:</td></tr>
</table>

Planning and managing

1. Calculate the ingredients required to prepare the recipe for the number of people you are serving.
2. Write up your new recipe.
3. Identify three relevant health and safety issues you will need to consider when preparing and cooking your main meal.

Producing and implementing

1. Prepare the product using safe work practices.
2. Note any modifications or changes you made during the production of the recipe.

Evaluating

1. Respond to your design criteria in detail to determine the success of your meal.
2. Describe the sensory properties – appearance, aroma, flavour, texture and sound – of your meal.
3. What was the most challenging part of the production? Outline how you managed this challenge.
4. If you were to produce this meal again, what changes would you make to improve it?
5. Classify the ingredients of your meal on a diagram of the *Australian Guide to Healthy Eating*. Comment on the nutritional value of this meal.

9780170495714

Zucchini Burgers

While burgers are a favourite fast food purchased by many people, they are very simple to make at home. One of the benefits of making your own burgers is that you can vary the type of protein ingredient used in the patty. You can use traditional minced beef, minced chicken or even a whole chicken breast, or a combination of equal proportions of beef and pork mince for a very moist and tasty burger. Or you can substitute a plant-based mince if you are looking for a vegetarian option. Adding a vegetable such as grated zucchini or even grated carrot improves the nutritional properties of the burgers and helps you to meet the five serves of vegetables we should aim to eat each day. Homemade burgers are also lower in saturated fat, salt and sugar than their fast-food cousins and are therefore more nutritionally balanced.

INGREDIENTS

¼ medium zucchini

100 grams lean minced steak or plant-based mince

¼ onion, grated

½ small clove garlic, crushed

½ teaspoon soy sauce

2 teaspoons beaten egg

2 tablespoons fresh breadcrumbs

1 tablespoon flour

2 tablespoons oil, for frying

2 hamburger buns

2 lettuce leaves

½ tomato, sliced

METHOD

1. Grate the zucchini and allow to stand for 5–10 minutes. Squeeze the grated zucchini to remove as much moisture as possible.
2. Combine the grated zucchini, minced steak or plant-based mince, onion, garlic, soy sauce, egg and breadcrumbs. If the mixture is too wet, add extra breadcrumbs.
3. Shape into two round patty shapes and dust with flour.
4. Barbecue on an oiled plate or fry in a frying pan until the meat juices run clear. Cook plant-based burgers for approximately 4 minutes on each side.
5. Assemble on buns with lettuce and tomato slices.

iStockphoto.com/-lvinst-

EVALUATION

1. Describe the sensory properties – appearance, aroma, flavour, texture and sound – of your Zucchini Burgers.
2. Why is it important to squeeze the moisture from the grated zucchini before adding it to the hamburger mixture?
3. Why is the onion grated and not diced in this recipe?
4. Explain the function of the egg in the hamburger mixture.
5. Explain why, according to the *Australian Guide to Healthy Eating*, we should eat burgers – especially those purchased from a fast-food outlet – 'only sometimes and in small amounts'.

Baked Meatballs and Spaghetti

Everyone loves a bowl of pasta, especially when topped with a simple tomato sauce and sprinkled with parmesan cheese. The addition of another favourite – meatballs – to the pasta creates a delicious, satisfying meal perfect for dinner or a family get-together. The meatballs can be prepared with any type of minced meat, or you could use a plant-based mince instead. Using a wholemeal pasta instead of the traditional spaghetti and serving the dish with a green leafy salad would improve its nutritional profile.

INGREDIENTS

140 grams spaghetti

Meatballs

1 slice stale bread

2 tablespoons milk

200 grams minced beef, minced chicken or plant-based mince

¼ onion, grated

½ egg, lightly beaten

1 tablespoon grated parmesan cheese

pinch cayenne pepper

pinch salt and pepper

Tomato Sauce

1 cup tomato passata or sugo sauce

1 cup canned diced tomatoes

¼ cup water

2 teaspoons sugar

2 teaspoons basil, finely chopped

2 teaspoons parsley, finely chopped

To complete the dish

2 tablespoons grated parmesan cheese

SERVES TWO

Mark Fergus Photography

METHOD

1 Fill a large saucepan two-thirds full with water and bring to the boil. Add the spaghetti and cook until al dente, stirring once or twice. Drain. While the spaghetti is cooking, prepare the meatballs.

Meatballs

1 Preheat oven to 200°C. Lightly oil a baking tray.

2 Trim the crusts from the bread and tear into small pieces. Place in a small bowl, add the milk and allow the bread to soak for 5 minutes. Squeeze out excess milk.

3 Combine the minced meat or plant-based mince, onion, egg, parmesan cheese, cayenne pepper, salt and pepper and the soaked bread. Mix well.

4 Using about 2 teaspoons of the mince mixture per meatball, shape into small balls and place on baking tray. Bake for approximately 15 minutes or until firm and lightly browned.

Tomato Sauce

Combine the passata or sugo sauce, canned tomatoes, water, sugar, basil and parsley in a medium saucepan and simmer for 10 minutes, so the sauce can thicken a little.

Completing the dish

1 Place the cooked spaghetti, tomato sauce and meatballs in a small casserole dish. Gently stir to mix the sauce through the dish.

2 Sprinkle with extra parmesan cheese and bake for 10–15 minutes or until a light golden brown.

EVALUATION

1 Describe the sensory properties – appearance, aroma, flavour, texture and sound – of your Baked Meatballs and Spaghetti.

2 What is the role of the soaked bread in the meatballs?

3 Discuss the preparation and nutritional benefits of baking the meatballs instead of frying them.

4 Describe a test you could use to determine whether the meatballs were cooked.

5 Write a critique of the Baked Meatballs and Spaghetti recipe to justify its inclusion on the National Healthy School Canteens list of recommended menu items.

9780170495714

Beef and Noodle Stir-fry

INGREDIENTS

200 grams rump steak, thinly sliced

Marinade

2 teaspoons oil

1 tablespoon oyster sauce

1 tablespoon soy sauce

1 tablespoon dry sherry or rice wine vinegar

Noodle Stir-fry

200 grams fresh hokkien noodles

2 tablespoons oil, for frying

½ onion, cut into wedges

3 centimetres fresh ginger, finely sliced

1 clove garlic, finely sliced

⅓ carrot, julienned

1 piece broccoli, cut into florets

¼ red capsicum, julienned

½ zucchini, cut into batons

4 snow peas, tops removed

¼ cup water or stock

1 tablespoon bamboo shoots, sliced

1 tablespoon water chestnuts, sliced

1–2 tablespoons satay sauce or sweet chilli sauce

SERVES TWO

METHOD

1 Combine the steak, oil, oyster sauce, soy sauce and dry sherry or rice wine vinegar and marinate in the refrigerator for 20–30 minutes.

2 Place the noodles in a large bowl, cover with boiling water and leave to stand for about 5 minutes.

4 Drain the marinade from the steak and discard the marinade. Heat 1 tablespoon of the oil in a wok on high.

5 Fry the onion for 30 seconds, stirring, then add half of the steak. Cook for 1 minute, stirring, then add the remainder of the steak. Stir-fry until almost cooked. Remove from the heat and keep warm.

6 Wipe out the wok and heat the remaining oil on high.

7 Add the ginger and garlic to the wok, then the carrot, broccoli, capsicum, zucchini and snow peas, allowing 30 seconds between the addition of each vegetable.

8 Add the water or stock, bamboo shoots and water chestnuts. Cook until the vegetables are just tender.

9 Drain the noodles.

10 Return the steak and onion to the wok, then stir through the noodles and satay or sweet chilli sauce.

EVALUATION

1 Describe the sensory properties – appearance, aroma, flavour, texture and sound – of your Beef and Noodle Stir-fry.

2 Describe the best way to cut meat to ensure that it is tender and easy to chew after being stir-fried.

3 Explain why the beef slices were marinated, and describe the functional role of each ingredient in the marinade.

4 Identify two safety factors you took into consideration when preparing the Beef and Noodle Stir-fry.

5 Classify the ingredients for the Beef and Noodle Stir-fry on a diagram of the *Australian Guide to Healthy Eating* and explain how well this recipe meets the guidelines of this food selection model.

Mark Fergus Photography

Spicy Stir-fried Pork and Green Beans

This dish is based on a traditional recipe from the province of Sichuan in the south-west of China. The Shaoxing wine, chilli bean sauce and rice vinegar give the sauce a spicy, piquant flavour. Traditionally, the green beans are charred over a high heat until blistered, but we have steamed them here for a healthier option. If fresh green beans are unavailable, you could swap them for another crunchy green vegetable, such as broccoli florets or broccolini. For a vegetarian or vegan option, you can substitute plant-based mince for the minced pork.

INGREDIENTS

- ⅔ cup jasmine rice, to serve
- 2 tablespoons vegetable oil
- 4 spring onions, thinly sliced on an angle
- 2 tablespoons ginger, finely grated
- 2 cloves garlic, crushed
- 250 grams minced pork or plant-based mince
- 3 tablespoons soy sauce
- 3 tablespoons Shaoxing wine
- 2 teaspoons chilli bean sauce
- 2 teaspoons caster sugar
- 250 grams green beans, topped and tailed
- 90 millilitres chicken or vegetable stock
- 3 teaspoons rice vinegar
- 3 teaspoons sesame oil

METHOD

1. Cook the jasmine rice by the absorption method according to the instructions on page 236. Keep covered with the saucepan lid to keep warm.
2. Half fill the base of a steamer with water and place on low heat.
3. Heat the oil in a wok or large frying pan.
4. When the oil has reached medium-to-high heat, add the spring onions (retaining a small portion for serving), ginger and garlic and stir-fry for approximately 30 seconds.
5. Add the minced pork or plant-based mince and fry, breaking up any clumps with a wooden spoon.
6. Continue to cook the minced pork or plant-based mince for 2–3 minutes.
7. Combine the soy sauce, Shaoxing wine, chilli bean sauce and sugar in a small bowl and stir to remove any lumps. Add to the minced pork or plant-based mince mixture and stir through.
8. Bring the water in the steamer to the boil, add the green beans and steam for 4 minutes or until just tender.
9. Add the stock to the minced pork or plant-based mince mixture and simmer for 3–4 minutes or until the liquid has reduced by half.
10. Add the steamed beans, rice vinegar and sesame oil to the minced pork or plant-based mince mixture and toss quickly to combine.
11. To serve, place the cooked jasmine rice in the base of a bowl and top with the Spicy Stir-fried Pork and Green Beans. Sprinkle with the reserved spring onions.

Mark Fergus Photography

EVALUATION

1. Describe the sensory properties – appearance, aroma, flavour, texture and sound – of your Spicy Stir-fried Pork and Green Beans.
2. What other green vegetables could replace the green beans in this recipe?
3. Outline two safety rules to follow when cooking on the cooktop.
4. Suggest other cereal-based foods other than rice that could be served with the Spicy Stir-fried Pork and Green Beans.
5. Classify the ingredients for the Spicy Stir-fried Pork and Green Beans served with jasmine rice on a diagram of the *Australian Guide to Healthy Eating*. Explain how well this meal meets the recommendations of this food selection model.

9780170495714

Thai Fishcakes

Fishcakes are a popular street food found throughout Thailand. This recipe uses canned tuna, but fresh, firm-fleshed white fish could be substituted. These fishcakes are quick and easy to make and, when served in a lettuce cup, make a delicious, spicy, healthy snack.

INGREDIENTS

185 grams canned tuna (in brine)

1 teaspoon Thai green curry paste

¼ cup coconut cream

2 tablespoons cornflour

1 egg

2 tablespoons coriander, chopped

1 cup fresh breadcrumbs

2 tablespoons peanut oil

lettuce cups and sweet chilli dipping sauce, to serve

SERVES TWO

METHOD

1. Drain the tuna well.
2. Place the tuna in a food processor with the Thai green curry paste, coconut cream, cornflour, egg and coriander.
3. Process until just combined – do not over-process or allow the mixture to become paste-like.
4. Add the breadcrumbs and combine in the food processor; if the mixture is too wet, add more breadcrumbs.
5. Shape a tablespoon of the mixture into a small patty. Repeat until you have used all the mixture.
6. Rest the fishcakes in the refrigerator for 10 minutes to firm up.
7. Heat the oil in a frying pan and cook the fishcakes for 2–3 minutes or until lightly browned all over. Turn once only.
8. Drain on absorbent paper before serving.
9. Serve in lettuce cups with sweet chilli dipping sauce.

EVALUATION

1. Describe the sensory properties – appearance, aroma, flavour, texture and sound – of the Thai Fishcakes served in lettuce cups with sweet chilli dipping sauce.
2. What is the purpose of adding fresh breadcrumbs to the mixture?
3. List the important safety issues to consider when frying the fishcakes.
4. Explain why it is important to drain the fishcakes on absorbent paper before serving.
5. Identify the benefits of including fish in a well-balanced diet.

Mark Fergus Photography

Spicy 'Shake and Bake' Chicken with Potato Latkes

Tossing the chicken drumettes – the portion of the wing that attaches to the body of the chicken – in a spicy coating creates a delicious crunchy texture when they are baked in the oven. Potato latkes are a type of potato pancake that are traditionally associated with Jewish cuisine. Other vegetables such as grated carrot, zucchini or onion can be added to the mixture to increase the vegetable content. Serving the chicken and latkes with steamed broccoli or green beans makes a very nutritious family meal.

INGREDIENTS

Spicy 'Shake and Bake' Chicken

3 tablespoons rice flour
2 teaspoons Cajun spice
2 teaspoons olive oil
2 teaspoons soy sauce
8 chicken drumettes
olive oil spray

Potato Latkes

2 large potatoes
¼ onion
2 tablespoons plain flour
1 egg
salt and pepper
15 grams butter
2 tablespoons oil

To serve

steamed broccoli or green beans

METHOD

Spicy 'Shake and Bake' Chicken

1. Preheat oven to 180°C. Line a baking tray with baking paper.
2. Place the rice flour and Cajun spice in a plastic bag.
3. Mix together the olive oil and soy sauce and lightly brush over the chicken drumettes.
4. Place the drumettes in the plastic bag with the rice flour and Cajun spice. Shake well to coat.
5. Place the coated chicken drumettes onto the baking tray. Lightly spray with the olive oil spray.
6. Bake in the preheated oven for 20–25 minutes or until tender and juice runs clear when a skewer is placed in the thickest part of a chicken drumette.

Potato Latkes

1. Peel and grate the potatoes. Firmly squeeze the grated potato to remove as much water as possible. Place in a medium bowl.
2. Peel and grate the onion and mix with the grated potato.
3. Add the flour and egg, then season with salt and pepper. Mix until smooth.
4. Heat the butter and oil in a frying pan over medium heat. Test the heat by dropping a small cube of bread into the pan – if the butter and oil mix is hot, the bread will sizzle and brown immediately.
5. Drop ¼ cup of the potato mixture into the hot frying pan and pat down with a fork to form a small latke. Repeat with the remaining potato mixture – it should make four latkes.
6. Lightly fry for approximately 8 minutes until the bottom of the latkes are brown and crisp. Gently turn over and cook the other side for a further 8 minutes until brown and crisp and the potato is cooked through.
7. Remove from the frying pan and drain on absorbent paper.
8. Serve with the Spicy 'Shake and Bake' Chicken and steamed broccoli or green beans.

9780170495714

EVALUATION

1 Describe the sensory properties – appearance, aroma, flavour, texture and sound – of your Spicy 'Shake and Bake' Chicken with Potato Latkes.
2 Why are olive oil and soy sauce brushed over the chicken drumettes before they are coated in the rice flour and Cajun spice?
3 List one important safety rule to observe when frying the latkes.
4 Which part of the production was the most successful and which part did you find the most challenging? Why?
5 Classify the ingredients for the Spicy 'Shake and Bake' Chicken with Potato Latkes served with steamed broccoli on a diagram of the *Australian Guide to Healthy Eating*. Discuss how well this meal meets the recommendations of this food selection model.

Mark Fergus Photography

11 RECIPES

Oyakodon

Oyakodon is a Japanese chicken and egg rice bowl. It is classic homestyle cooking – a comfort food served in one bowl.

Note:
To cook two serves together, use a medium frying pan – approximately 25 centimetres.
To cook one serve, halve the ingredients and use a small frying pan – approximately 20 centimetres.

INGREDIENTS

- 2 small skinless chicken thighs (approximately 300 grams in total)
- 1 tablespoon cooking sake
- ½ brown onion
- ½ cup dashi (Japanese soup stock) or chicken stock
- 2 tablespoons mirin
- 2 tablespoons soy sauce
- 2 teaspoons caster sugar
- 4 eggs, at room temperature

To serve

- 2 portions of cooked rice
- 2 spring onions, sliced on the diagonal
- 60 grams baby spinach leaves
- shichimi togarashi (Japanese spice) (optional)

METHOD

1. Prepare the rice using the absorption method (see page 236) and keep the lid on the pan so that it stays warm and is ready for serving.
2. Trim any fat from the chicken thighs, then cut along the grain into 2-centimetre strips. Next, with the knife nearly parallel to the chopping board, slice the chicken strips across the grain into slivers. This technique makes the chicken cook faster and more evenly.
3. Place the chicken pieces in a bowl, add the sake and toss. Rest for 5 minutes.
4. Peel the onion and slice lengthwise into strips approximately 5 millimetres wide. Slice the spring onions diagonally and set aside for serving.
5. Place a medium frying pan on the cooktop but do not turn on the heat yet.
6. Add the dashi, mirin, soy sauce and caster sugar to the frying pan and stir to dissolve the sugar. Place the sliced onions in a single layer, ensuring that they are evenly distributed. Then add the chicken pieces in an even layer over the onions.
7. Turn the heat to medium and bring the liquid to a simmer, then turn the heat to low. Cook uncovered for 5 minutes until the chicken pieces are no longer pink. Turn the chicken pieces over halfway through cooking. The sauce will reduce and the flavour will intensify.
8. Crack the eggs into a medium bowl and use chopsticks to cut the egg yolks and whites so that they are marbled and not fully mixed.
9. Increase the heat to medium, then drizzle the eggs in a circular pattern over the onions and chicken pieces. Avoid pouring to the edge of the pan, as this can overcook the eggs.
10. Cook until the eggs are just set.

To serve

Spoon the steamed rice into 2 bowls. Place the spinach leaves on top of the rice, then use a large spoon to place a portion of the chicken, onions and egg on top. Take care not to break up the omelette too much. Spoon a little of the sauce into each bowl to soak into the rice. Sprinkle over the spring onions and some shichimi togarashi (if desired). Eat while hot.

9780170495714

EVALUATION

1. Describe the sensory properties – appearance, aroma, flavour, texture and sound – of your Oyakodon.
2. Explain why detailed attention is given to slicing the chicken thighs.
3. Identify the function of sake in this recipe and explain how it contributes to the sensory qualities of the final dish.
4. What component of the eggs is responsible for the change in the physical properties when the eggs are cooked? Describe the changes in colour and structure of the eggs.
5. Classify the ingredients for the Oyakodon on a diagram of the *Australian Guide to Healthy Eating* and discuss how well this meal meets the recommendations of this food selection model.

Mark Fergus Photography

11 RECIPES

Fried Egg

INGREDIENTS

1 tablespoon vegetable oil
cube of bread (to test the oil)
1 egg

METHOD

1 Add the oil to a small frying pan and place over medium heat for approximately 30 seconds.
2 Add the cube of bread to the oil. Once it becomes golden brown, the oil is hot enough to cook the egg. Remove the cube of bread from the frying pan.
3 Carefully crack the egg into the hot oil.
4 Cook for 2–3 minutes or until the white is set and the yolk is slightly set.
5 Use an egg slice to remove the egg from the frying pan. Drain it on paper towel before serving.
6 Allow the frying pan to cool before washing.

EVALUATION

1 Why is it important to preheat the oil in the frying pan before adding the egg?
2 Why is it desirable to drain the egg on paper towel before serving?
3 Identify and explain the main safety rules that should be followed when frying eggs.
4 Discuss why nutritionists recommend boiling and poaching as the preferred methods of cooking eggs.
5 Suggest some foods or recipes to serve with the fried egg to make a delicious meal that includes at least four of the recommended *Australian Guide to Healthy* Eating food groups.

iStock.com/milanfoto

9780170495714

Scrambled Eggs

INGREDIENTS

2 eggs
4 tablespoons milk
pinch salt
pinch pepper
1 teaspoon butter
warm toast, to serve
1 teaspoon parsley, chopped, to serve

SERVES ONE

METHOD

1. Whisk the eggs, milk, salt and pepper in a small bowl until well combined.
2. Melt the butter in a small frying pan.
3. Pour in the egg mixture and gently stir over a low heat until the mixture thickens but is still soft. Do not overcook the eggs.
4. Serve with the warm toast and sprinkle with the chopped parsley.

EVALUATION

1. Describe the sensory properties – appearance, aroma, flavour, texture and sound – of your Scrambled Eggs.
2. Why is it important to stir the eggs gently while they are cooking?
3. Identify the process that is responsible for eggs thickening and explain how this occurs when cooking the Scrambled Eggs.
4. What effect would overcooking have on the texture of the eggs?
5. Consider the recommendations of the *Australian Guide to Healthy Eating* food selection model and suggest some other foods that could be served with the Scrambled Eggs to create a healthy breakfast.

Mark Fergus Photography

Korean-style Beef Bowl

The preparation of this meal requires careful planning and management. There are quite a few steps to produce this recipe successfully, so it is important to be well organised. Make sure that you and your partner read through the recipe carefully and have all of the ingredients prepared – including cooking the plain rice to serve as a base for the Korean minced beef – before you start.

INGREDIENTS

Toppings

1 tablespoon sesame seeds

1 medium carrot

1 teaspoon vegetable oil

1 spring onion

Korean minced beef

1 tablespoon soy sauce

1 tablespoon mirin

1 tablespoon gochujang paste

1 teaspoon brown sugar

1 teaspoon sesame oil

3 teaspoons vegetable oil

½ onion, finely diced

2 cloves garlic, crushed

1 teaspoon ginger, peeled and grated

300 grams minced beef

½ long red chilli, seeds removed and thinly sliced

2 spring onions, thinly sliced

½ teaspoon salt

black pepper

To serve

1 quantity cooked plain rice (page 236)

METHOD

Toppings

1. Toast the sesame seeds in a small saucepan over a low heat for approximately 5 minutes, or until lightly toasted. Remove from the saucepan and set aside to cool.
2. Using a vegetable peeler, slice the carrot lengthwise into fine ribbons.
3. Heat the vegetable oil in a frying pan over a medium heat. Add the carrot ribbons and stir-fry for 1–2 minutes until just wilted. Remove from the frying pan and set aside.
4. Slice the spring onion diagonally.

Korean minced beef

1. Prepare the flavouring sauce for the minced beef. Combine the soy sauce, mirin, gochujang paste, brown sugar and sesame oil in a small bowl and stir well.
2. Add the vegetable oil to the frying pan and sauté the onion for 2–3 minutes or until soft and just starting to brown. Add the garlic and ginger and stir-fry for a further 1 minute or until aromatic.
3. Add the minced beef to the pan, increase the heat to medium-high and stir until it breaks up and changes colour and is no longer pink.
4. Stir in the flavouring sauce and mix thoroughly. Add the sliced chilli and spring onions and season with salt and pepper. Simmer for a further 1–2 minutes to develop the flavour. Remove from the heat.

To serve

Place the cooked plain rice in two serving bowls and pile the Korean-style beef mixture on top. Top with the stir-fried carrot ribbons, sliced spring onion and toasted sesame seeds.

9780170495714

EVALUATION

1 Describe the sensory properties – appearance, aroma, flavour, texture and sound – of your Korean-style Beef Bowl.
2 List two rules for cooking safely on a cooktop.
3 Identify and describe the changes that occurred to the physical properties of the minced beef when it was cooked.
4 Outline the stages in the Australian food system that vegetables such as carrots and onions would go through to reach a supermarket in a fresh condition.
5 Classify the ingredients for the Korean-style Beef Bowl on a diagram of the *Australian Guide to Healthy Eating*. Write a paragraph to explain how well this recipe meets the recommendations of this food selection model.

Mark Fergus Photography

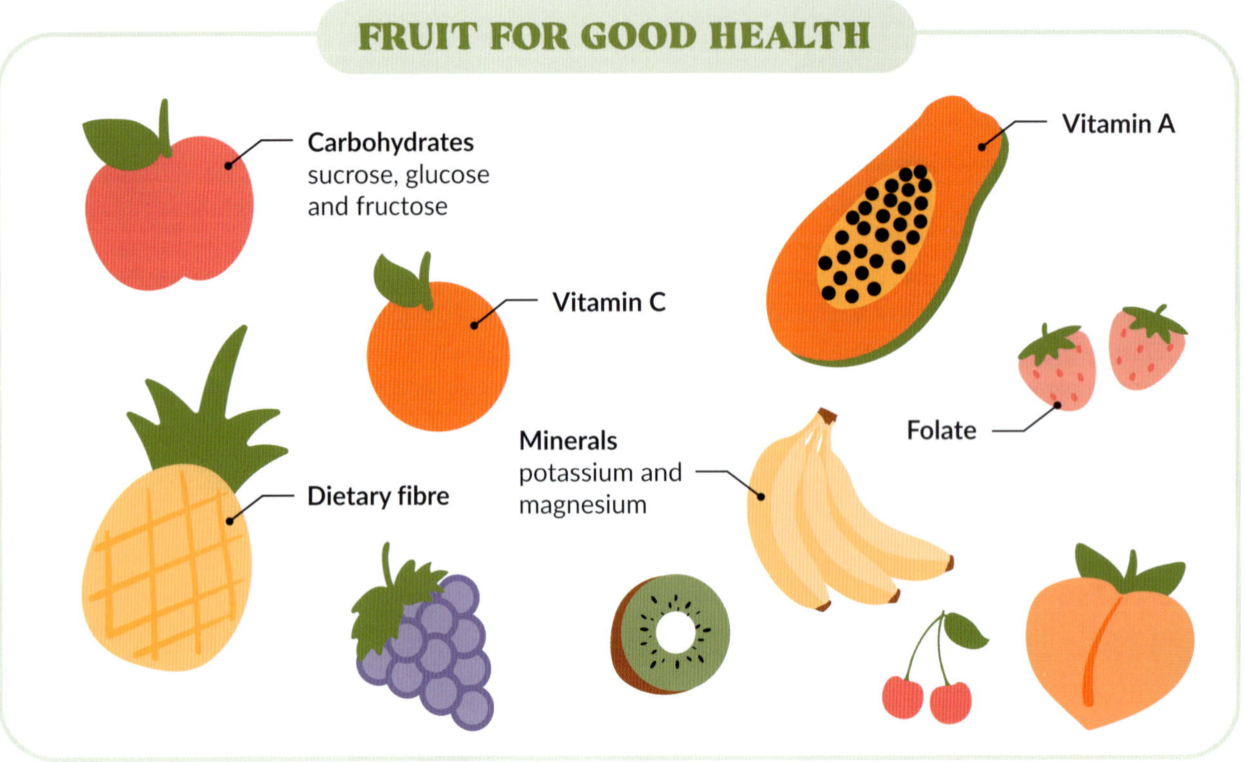

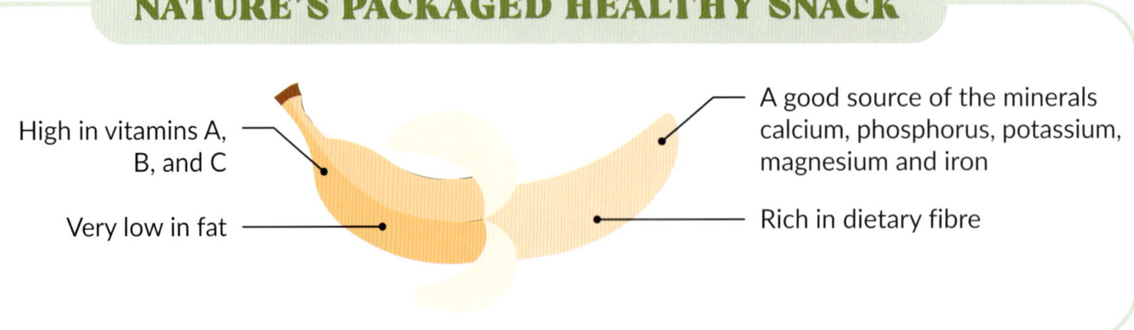

9780170495714

GROWING, PACKAGING & PROCESSING FRUIT SUSTAINABLY

Minimise the use of fertilisers and pesticides

Use efficient drip systems to water crops and prevent run-off

Minimise water use by recycling water during fruit processing

Reduce landfill by repurposing fruit waste for animal food

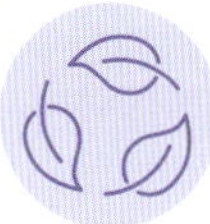
Promote new plastic-free packaging for fruit

12 FRUIT

KEY TERMS

citrus fruit a fruit family that includes oranges, mandarins and lemons; their peel has a distinctive aroma and their flesh is divided into segments; their flavour ranges from acidic to sweet

enzymatic browning a process that occurs when the enzymes in cut or peeled fruit are exposed to oxygen in the air and cause browning

in season the time of year when a fruit or vegetable has its best sensory properties

pome fruit fruit that have crisp, juicy flesh surrounding a core that contains seeds; for example, apples and pears

sustainable farming farming practices that maintain the land's productivity so that it will be available for future generations

Templates
- Decision table
- Food order
- PMI table
- Sensory wheel
- Australian Guide to Healthy Eating

Worksheets
- Practical activity 12.1
- Design activity 12.1

Quizzes
- Testing knowledge 12.1
- Testing knowledge 12.2
- Testing knowledge 12.3

To access resources above, visit **cengage.com.au/nelsonmindtap**

9780170495714

Fruit in the *Australian Guide to Healthy Eating*

Fruit is often described as a perfect snack food. It comes in its own biodegradable package, it is convenient to eat, there are many varieties, and it is a good source of a wide range of nutrients. For this latter reason, it represents a significant proportion of the *Australian Guide to Healthy Eating*.

Figure 12.1 The place of fruit in the *Australian Guide to Healthy Eating*

Fruit for good health

Fruit is a very good source of a wide range of nutrients:

- **Carbohydrates:** Fruits contain disaccharides, sucrose and monosaccharides – such as glucose and fructose – that give fruit its sweet taste. Unripe fruits contain starches and, as the fruit ripens, the starch is converted to sugars. Ripe fruits have a higher concentration of sugars. Dried fruit is more energy-dense than fresh fruit.
- **Vitamins:** Fresh fruits – particularly citrus fruit, berries and some tropical fruits – are rich in vitamin C and folate. Vitamin C is valuable in normalising the function of the immune system. Orange/yellow fruits also provide carotene (vitamin A), which is thought to enhance immune function.
- **Minerals:** Potassium and magnesium are found in fruit, and are linked to lower blood pressure. Magnesium is also necessary for normal muscle function.
- **Dietary fibre:** Fruit provides a valuable source of dietary fibre, both in the flesh and particularly in the edible skins. The fibre in fruit helps food move through the digestive tract, and reduces constipation and the risk of some cancers, particularly colorectal cancer.

Figure 12.2 Fruit provides a wide range of nutrients.

Fruit is a valuable food to be eaten each day, as it may help to reduce the risk of chronic diseases, such as heart disease, stroke and some cancers. As noted above, there is also evidence that minerals found in fruit, such as potassium and magnesium, are linked to lower blood pressure.

Overall, fruit is low in energy (kilojoules), and people who consume fruit will feel fuller for longer because fruit has a high fibre and water content. The inclusion of fruit in our daily diet reduces the risk of overeating, which can cause weight gain.

Are we eating enough fruit?

Data released by the Australian Institute of Health and Welfare (AIHW) in June 2024 reveals that only 64 per cent of children and adolescents aged 2–17 consume the recommended number of serves of fruit each day. Concerningly, as we grow older, the amount of fruit we consume decreases even further. The data from the AIHW also indicates that fewer than 44 per cent of adults aged 18 and over consume the recommended two serves of fruit each day.

Table 12.1 How much fruit should I eat?

	Serves per day	
	12–13 years	14–18 years
Boys	2	2
Girls	2	2

9780170495714

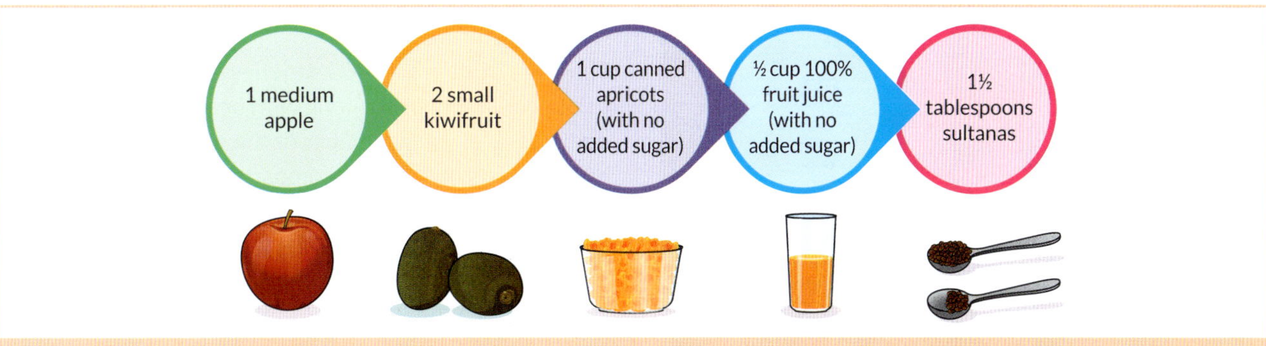

Figure 12.3 What is a serve?

Adding more fruit to our diet

Fruit is best and most nutritious when eaten fresh and raw. It can be added to cereal for breakfast, it can form part of a salad for lunch, or it can be eaten at the end of a meal. Fruit can be used as a base for desserts such as baked apples or fruit crumbles, and can be stewed or poached. Fruit such as strawberries, blueberries and raspberries can be mixed into or used to decorate pancakes or drop scones. Grated apple or chopped banana can be folded into muffins.

Frozen or canned fruit can be used to provide greater convenience and variety, especially when the fruit is not in season; however, if choosing processed fruit, choose fruit that is canned in fruit juice without any added sugar. Dried fruits may be eaten, but these are higher in kilojoules. They may also stick to the teeth and increase the risk of tooth decay.

Classification of fruit

Most fruit originates from flowers, which contain the seeds of the plant.

Buying fruit in season

Fruit matures with the seasons; as the weather changes – for example, from summer to autumn – some fruits reach their peak. When they are **in season**, they have their best flavour, texture and aroma, and, because they are plentiful, they are available at the best price.

Figure 12.4 The classification of fruit

Top left to right: iStock.com/Roman Samokhin, iStock.com/Everyday better to do everything you love, iStock.com/jerryhat; Middle left to right: iStock.com/ansonsaw, iStock.com/xxmmxx; Center: iStock.com/ MarcoFood

Jopiejusuf/Shutterstock.com

Figure 12.5 A wide variety of fruit is available at fresh food markets and farmers' markets.

Testing knowledge 12.1

Quiz Testing knowledge 12.1

1 Why is a portion of the *Australian Guide to Healthy Eating* allocated to fruit?

2 Explain why ripe fruit has a sweet flavour.

3 Why should we only eat small amounts of dried fruit?

4 Copy and complete the following table to identify the function of each of the following nutrients, which are present in fruit, in the body.

Nutrient	Function in the body
Vitamin C	
Vitamin A	
Potassium	
Magnesium	

5 Explain why it is important to eat both the flesh and skin of fresh fruit, such as apples.

6 Create a mind map of the benefits of including fruit in the diet each day.

7 Explain why, from a health perspective, health professionals would be concerned that our consumption of fruit decreases as we age.

8 Suggest a strategy that a secondary school student could follow to consume the number of serves of fruit recommended each day in the *Australian Dietary Guidelines*.

9 Create a diagram to demonstrate how an individual could include more fruit in their diet.

10 Discuss the benefits of buying fruit when it is in season.

Apples

There are more than 7000 varieties of apple grown throughout the world. Along with pears, nashi and quince, apples are a member of the **pome fruit** family, which means they have a compartmentalised core that contains seeds, and their flesh is crisp and juicy.

Figure 12.6 Parts of an apple

Nutrition

Apples are one of nature's best snacks. The firm skin keeps the crisp, juicy flesh in good condition for days, and ensures that the apple is easily transported in a lunch box or bag without refrigeration.

The saying 'an apple a day keeps the doctor away' reflects the fact that apples are a very healthy food, as their flesh is high in dietary fibre in the form of a soluble dietary fibre called pectin. Apples also contain some dietary fibre in their skin. Dietary fibre is important for good health because it helps to keep the digestive system working. In addition, apples contain a small number of vitamins and minerals. A medium-sized eating apple contains about 270 kilojoules and is a filling snack.

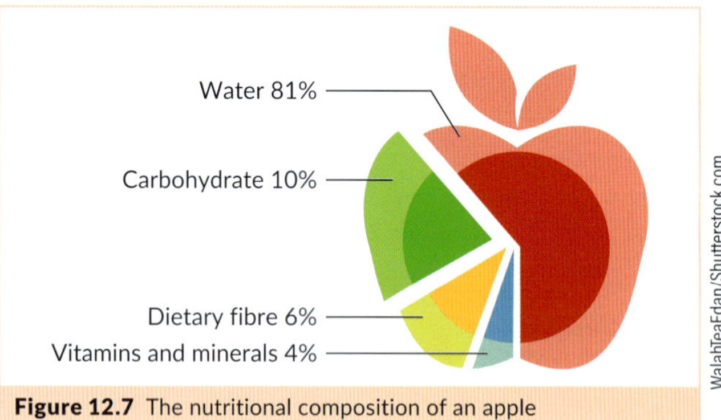

WalahTeaEdan/Shutterstock.com

Figure 12.7 The nutritional composition of an apple

Growing apples in Australia

Apples are grown in most Australian states; however, the majority are grown in Victoria and Tasmania. Approximately 43 per cent of Australia's apple crop is grown in the Goulburn Valley, the Yarra Valley, Harcourt, Gippsland and the Mornington Peninsula in Victoria.

9780170495714

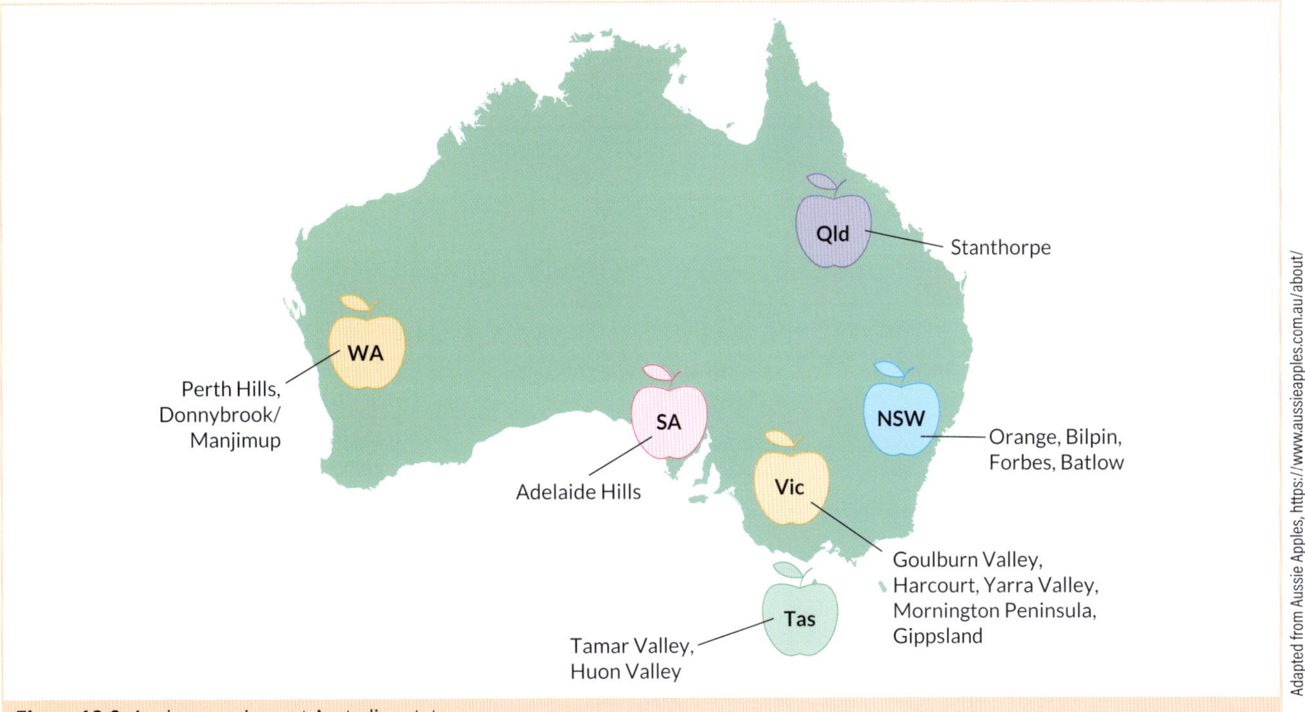

Figure 12.8 Apples grow in most Australian states.

Apples are in season during autumn and winter, when their flavour, texture and colour is at their peak and there are a wider range of apple varieties available for consumers to buy.

Australia is one of the world's healthiest fruit-growing environments, because its isolation and strict biosecurity laws ensure that it is free from many of the world's pests and diseases.

Apple varieties

Australia grows a wide variety of apples. Granny Smith and Red Delicious are two of the most popular varieties, and these make up a large proportion of the market.

Research scientists have been developing new apple varieties that are beginning to appear in our supermarkets. They are responding to consumers' desire for apples with enhanced sensory properties – such as a sweeter flavour or a crisper texture – or apples that are resistant to enzymatic browning, such as Cosmic Crisp®. Other new varieties, such as Kissabel®, have been bred to have a red-coloured flesh similar to a crab apple. Newer varieties of apple – which include Pink Lady, Royal Gala, Fuji, Jazz, Bravo and Kanzi – are trademarked and grown under licence by individual orchardists.

Australia has a history of cultivating new apple varieties, including the Granny Smith apple, which was first grown in New South Wales by Maria Ann Smith in 1860 and is now popular all over the world. It has a green 'greasy' skin and a tart flavour, and is in season from mid-April to May. It is popular for cooking and for processing into apple products. The Pink Lady, which was bred in Western Australia by John Cripps in 1973, was originally called the 'Cripps Pink'. Its flesh is crisp and firm, and has a flavour similar to that of Golden Delicious. Pink Ladies are available from May onwards.

The Australian Fuji was bred in Japan and is a late-season apple, available from April onwards. It has a dense flesh and a sweet, distinctive flavour, and its skin is a blushed pink-and-red colour. The Royal Gala originated in New Zealand and is an early-season apple, usually available from February onwards. It has a dense, crisp texture and is sweeter than Red Delicious.

Figure 12.9 Cosmic Crisp® is one of the new varieties of apples available to Australian consumers.

Table 12.2 Popular apple varieties grown in Australia

	Variety		
	Royal Gala	Golden Delicious	Jazz®
Image	iStock.com/MahirAtes	iStock.com/bergamont	iStock.com/karimitsu
Flesh	Firm, crisp, sweet, juicy	Crisp, creamy, flavoursome, sweet, juicy	Firm, dense, crunchy with a tangy, sweet flavour, similar to peaches and melon
Colour	Red blush to solid red over a golden background	Greenish to golden yellow, occasionally with a slight pink blush	Pink/red flush over a light green background
Size	Medium	Medium	Small to medium
Availability	March–September	Late March–July	April–October
Comments	A very attractive apple with excellent eating quality	An internationally popular apple, always in demand	A distinct new variety – a cross between Royal Gala and Braeburn
	Red Delicious	Granny Smith	Fuji
Image	iStock.com/Dimitris66	iStock.com/LauriPatterson	iStock.com/Everyday better to do everything you love
Flesh	Medium sweetness, crisp, juicy, white	Hard, crisp, tart flavour, white	Very sweet, firm, crisp, juicy
Colour	Solid bright red with slight stripe	Green to greenish yellow, occasionally with a slight pink blush	Predominantly blushed dull pink/red, with some russet yellow stripes
Size	Medium to large	Medium	Medium, but can be small
Availability	March–December	Harvest March–April, available all year	April–October
Comments	An internationally popular apple, always in demand	Australia's own, world-renowned green apple	Very sweet 'honey core', slightly sweeter than other varieties
	Pink Lady™	Ambrosia®	Rockit™
Image	iStock.com/chengyuzheng	iStock.com/Suzifoo	HSTUDIO99/Shutterstock.com
Flesh	Firm, sweet, crisp, juicy, creamy white flesh	Light, crisp, juicy texture, creamy white flesh	Sweet, juicy, crisp texture
Colour	Pink or light red blush over yellow-green background	Orange to red blush with some yellow	Deep red-pink over a yellow background
Size	Large	Medium	Miniature/small
Availability	Late April–February	March–October	April–late August
Comments	A unique and very popular apple that is delicious to eat	Keeps well in the refrigerator and makes a great snack	Thin skin, small core, excellent for children's lunch boxes

9780170495714

PRACTICAL ACTIVITY 12.1

Worksheet
Practical activity 12.1

Sensory analysis: Varieties of apples

Aim

To compare the sensory properties of different varieties of apples

Method

1. Select four different varieties of apples. Wash and dry them before the comparison.
2. Draw up a table like the one below, and record the appearance of each variety of apple before slicing it for taste testing.

Results

Apple variety	Appearance		Aroma (e.g. sweet, tart, fruity, tangy, fresh)	Flavour (e.g. sweet, sour, sharp, tart, bland)	Texture (e.g. crisp, crunchy, soft, floury, firm, juicy, dry)	Sound (e.g. crunchy, crisp, dull, juicy, brittle)	Rating 5 = like a lot 4 = like 3 = OK 2 = dislike 1 = dislike a lot
	Colour (e.g. red, yellow-green, pink, mottled, striped, speckled)	Shape (e.g. round, oval, flat, big, small, uneven)					

Analysis

1. Why is it always advisable to wash fruit before eating?
2. Identify the variety of apple that you liked best. Explain why.
3. Which variety of apple did you like least? Explain why.
4. Which sensory property is the most important to you when selecting an apple to eat raw?
5. Outline the qualities that would be the most important if you wanted to cook with an apple.
6. Discuss why today's apple growers supply the market with a wide variety of apples.
7. What is meant by 'in season' when discussing fruit?
8. List the months when the apple varieties you compared are in season.

Conclusion

Which variety of apple do you prefer to eat raw? Justify your answer – consider all of the sensory properties of your preferred apple.

Photoongraphy/Shutterstock.com

Figure 12.10 Which variety of apple do you prefer to eat raw?

12

Commercial storage

Australians like to eat apples all year round, so distributors use cold storage to allow supply to continue between harvests. Cold storage is a form of refrigeration that maintains the fruit between 0 and 1°C, with humidity around 85 per cent. This system prevents over-ripening of the fruit, and maintains the crispness and freshness of the apples.

Activity 12.1

Research task: Australian apples

Aussie Apples is a website funded through the Hort Innovation Apple and Pear Fund. It is a grower-owned, not-for-profit research and development corporation for Australian horticulture. It provides consumers with a wide range of information about Australian apples and explains why they are such a great snack food.

Visit the Aussie Apples website to broaden your knowledge about the apple industry in Australia, and then complete the research task below.

1 Go to the 'About' page and research answers to the following questions:
   - a When is the picking/harvesting season for apples in Australia?
   - b What are the features of a good-quality apple?
   - c Explain why the preferred method for storing apples is cold storage.
   - d Outline how wax on the skin influences the eating quality of apples.

2 Go to the 'Health' page and research answers to the following questions:
   - a Read the nutrition information about apples. Taking that information into account, explain why there is some truth in the statement 'an apple a day keeps the doctor away'.
   - b Read the 'Top 5 crunch facts' about apples by Dr Joanna McMillan.
     - i Create a diagram to summarise the 'Top 5 crunch facts'.
     - ii Explain how you know the information on this webpage is from a reliable source.

Weblink
Aussie Apples

3 Go to the 'Recipes' page and select one recipe from each group that you would like to try. Read the recipe. Then, using your knowledge of preparing and cooking food, describe how you imagine the final sensory properties of each recipe would be. Record your responses in a table similar to the one below.

	Recipe title	Reason for selection	Final sensory properties
Breakfast			
Savoury and salads			
Snacks and hacks			
Sweets and treats			

CASE STUDY 12.1

Growing apples in Australia

Read the case study below, and then answer the questions that follow.

Montague Farms: Australian apple growers

About Montague Farms

Montague Farms was established in 1948 when it was founded by William Montague. Since that time three generations of the Montague family have operated the business growing a wide variety of apples, stone fruit, pears, citrus and grapes.

Montague Farms are a leading producer of Australian apples. They continue to supply Australians with their favourite varieties including Granny Smith, Pink Lady, Royal Gala and Red Delicious. However, in recent years, several new types of apples have been developed though

9780170495714

crossbreeding traditional varieties. Some of the new apples that have become available in our fresh food markets and supermarkets include Jazz, Ambrosia and Envy. These new varieties of apples are grown by apple orchardists under licence across six Australian states. All the apples grown by Montague Farms are hand-picked to minimise bruising or damage to the fruit.

In Victoria, Montague Farms have orchards in the Goulburn Valley, Yarra Valley and Harcourt. Not only do these areas have the ideal growing conditions, but they are close to their consumer markets helping to minimise their environmental footprint. As well as selling fruit throughout Australia, Montague farms export to Asia, Europe and North America.

The environmental sustainability strategy of Montague Farms

In 2021, Montague Farms opened a new fruit-processing facility in Narre Warren North, Victoria. The new washing, grading, packing and distribution facility processes over 277 million apples annually. The new facility has been designed to be environmentally sustainable and to reduce the amount of carbon dioxide (CO_2) they emit into the atmosphere.

Montague Farms has developed a wide range of environmental sustainability strategies including:

- installing banks of solar panels to reduce their reliance on the energy grid
- new energy-efficient design features in the new fruit-processing building
- minimising water use by collecting, treating and recycling water used in fruit processing on site
- repurposing any fruit waste as feed for pigs and worm farms
- minimising waste that is sent to landfill with the aim for there to be no waste in the future
- repurposing waste from fruit production into fruit juice, fruit pastes and apple cider
- developing new plastic-free packaging for all their fruit products that is biodegradable, recyclable or made from compostable materials
- minimising the use of synthetic fertilisers in their fruit orchards
- increasing the types of biodiversity on the property by setting aside land for conservation

Growing Australian apples

WINTER

- Apple trees are dormant; they conserve energy for the growing season
- Trees are pruned to encourage new growth and allow maximum sunshine to reach the fruit
- New trees are planted

SPRING

- Budburst occurs and leaf buds begin to shoot
- Apple trees produce blossom that will develop into fruit buds
- Bees pollinate the flowers, and small apples start to form; beehives may be placed in apple orchards to encourage pollination
- Fruit trees are fertilised to maximise the nutrients available for growth

SUMMER

- Apples rapidly grow to full size and develop their colour
- Reflective sheets are placed under the trees to reflect the sun, so the fruit colours evenly
- Some of the apple crop is thinned to ensure best-quality fruit is produced
- Trees are netted to protect from hail, birds and wind damage

AUTUMN

- Apples are ready to pick; different varieties ripen in different months
- The fruit is harvested by hand to prevent damage or bruising
- Apples are washed, graded, stored in controlled conditions, and packed for distribution and sale

Figure 12.11 The growth of apples, by season

Analysis

1. List the popular varieties of apples that Montague Farms grows and supplies to Australian consumers.
2. Explain how new varieties of apples are developed, and identify three new varieties of apples that have recently come onto the market.
3. Describe how the apples grown by Montague Farms are harvested and why this process is used.
4. How does pruning the apple trees assist apple production?
5. Explain the role of bees in producing an apple crop and why this is a crucial stage in the growth cycle of an apple.
6. What is the purpose of placing reflective sheets under the apple trees during summer?
7. How do orchardists protect their crop as the fruit develops and why is this important?
8. Create a mind map of the strategies Montague Farms is using to ensure their environmental sustainability.
9. Access the Montague Farms website.
   a. Click on the 'Fruit' tab, then select 'Apples'. Choose one variety of apple and describe its sensory properties.
   b. Watch the video about the grower of the Smitten, Jazz, Ambrosio or Eve variety, and complete a PMI (Plus, Minus, Interesting) table about the grower and the apples they produce.

Template PMI table

Weblink Montague Farms

Cooking with apples

Apples are a very versatile fruit that, while delicious eaten raw, can be used in a wide variety of recipes, including cakes, muffins and desserts. They can also be added to salads or savoury dishes.

One concern with apples when they are peeled and prepared for cooking is that they may sometimes turn brown. This is called **enzymatic browning**. Enzymatic browning is a process that occurs when the enzymes in cut or peeled fruits are exposed to oxygen in the air. It can be prevented by covering the cut fruit with water or sprinkling it with lemon juice.

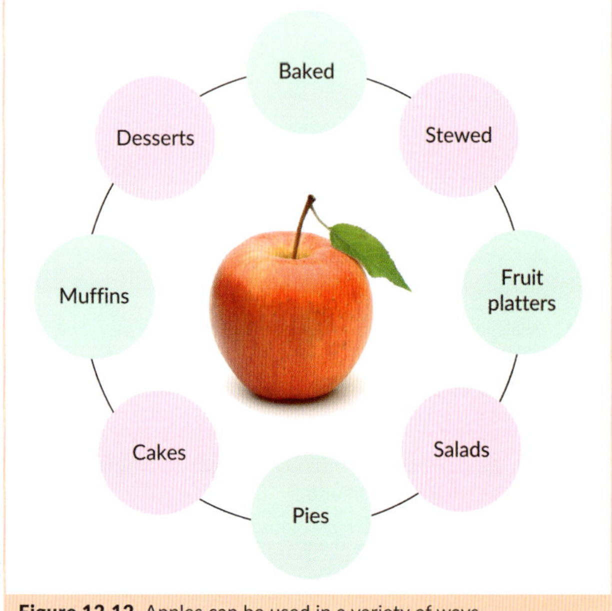

Figure 12.12 Apples can be used in a variety of ways.

Figure 12.13 Apple tart is a popular dessert using apples.

Oranges

Oranges – along with lemons, mandarins, grapefruit, limes, cumquats and pomelos – are part of the **citrus fruit** family. They have a distinctive aroma and their flesh is divided into segments. The flavour of citrus fruit ranges from acidic to sweet.

Citrus fruit trees are evergreen, and grow in subtropical climates. To grow, they require rich, well-drained soil with a good water supply. The most suitable citrus-growing areas in Australia are in the irrigation regions of the Murray and Murrumbidgee Rivers, on the plains in New South Wales, in Central Queensland, and on the outskirts of Perth in Western Australia.

9780170495714

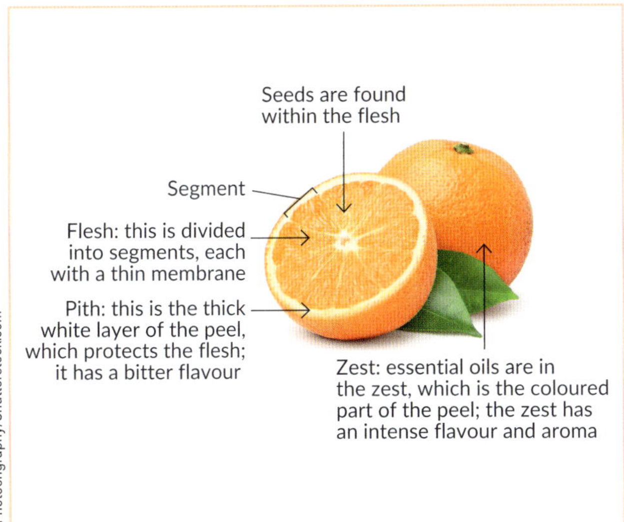

Figure 12.14 Parts of an orange

Orange varieties

The two most commonly available varieties of orange in Australia are the Valencia and the navel. Valencia oranges are considered best for juicing, and are available throughout summer and autumn. They have a smooth skin, which often has a green tinge. The navel orange is the most popular orange to eat fresh because it is almost seedless, and is easy to peel and break into segments. It is available from May to December.

Nutrition

Oranges are naturally high in vitamin C as well as being a good source of folic acid, potassium and dietary fibre. Vitamin C helps in the healing process if we have been ill or injured. It helps to strengthen body tissues and bones, and assists in the absorption of iron.

Activity 12.2

Field to Feast: Oranges

Go to SBS OnDemand and access the video 'Field to Feast: Oranges', featuring Thanh Truong – the 'Fruit Nerd'.

The video focuses on Australian orange growers. It examines the properties of navel and Valencia oranges, and the variety that is best for juicing. It also explains how the fruit is distributed. After watching the video, answer the following questions.

1 Explain how to select a good-quality orange when shopping at the supermarket or a farmers' market.

2 What is the best way to store oranges at home to maintain their freshness? Identify the chemical property of oranges that assists in lengthening their shelf life.

3 After listening to the presenters discuss the sensory properties of navel oranges, compare the sensory properties of two varieties of navel oranges: Washington, which is the most common, and Cara Cara. Record your descriptions in a table similar to the one below.

Sensory property	Washington navel	Cara Cara navel
Appearance • skin • flesh		
Aroma		
Flavour • sugar level • acid level		
Texture		
Sound		

4 Oranges are often picked with a green skin tinge. Identify the process they undergo during packing to change the skin to an even, bright orange and make them more acceptable to consumers.

5 Explain why the Mildura region is ideal for growing oranges.

6 Complete a flow chart that identifies the key stages in moving oranges through the logistics hubs that supply fresh fruit to the supermarket distribution centres (DCs) and other markets.
7 What is the timeline for moving oranges from the farmer to the distribution centres around Australia?
8 Describe the difference in appearance of the oranges that are exported, those sold in the domestic market and those made into juice.
9 Describe the unique growing features of Valencia orange trees.
10 Explain why Valencia oranges are best suited to the production of fruit juice, compared to navel oranges.
11 Describe the process of producing fruit juice from concentrated orange juice.
12 What is pasteurisation and why does orange juice undergo this process?
13 Outline the difference in the sensory properties of fresh orange juice, short-life juice and long-life juice. What accounts for the differences?
14 When oranges are processed and the juice extracted, the peel of the oranges becomes a waste product. Explain how this waste product is used to prevent it from going to landfill.

Testing knowledge 12.2

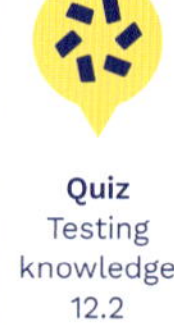

Quiz Testing knowledge 12.2

11 Explain how fruit may be classified into specific groups.
12 How does the old saying 'an apple a day keeps the doctor away' still relate to good health today?
13 Outline when apples are in season in Australia and why they are at their best at this time of the year.
14 Explain why new varieties of apples, such as Jazz or Cosmic Crisp, are available on supermarket shelves.
15 Select one apple variety from Table 12.2 on page 328 and explain why it appeals to you.
16 Many varieties of apples are available to consumers all year round. Outline the process that apple producers use to enable many varieties of apples to be available when they are not in season.
17 What is enzymatic browning and how can it be prevented?
18 Describe the main differences between a Valencia orange and a navel orange.
19 Identify the section of an orange that contains the zest.
20 Explain how the zest improves the sensory properties of an orange cake.

Fruit juices and drinks

Many people find that fruit juice is a tasty and refreshing drink; however, there are a number of nutritional concerns linked to consuming fruit juice on a regular basis. When fruit is juiced, the cell structure of the fruit is broken down and most of the dietary fibre is removed. Therefore, a valuable component of the fruit is lost when it is juiced rather than eaten whole. Another problem with consuming fruit juice on a regular basis is that it is very acidic and can lead to the breakdown of tooth enamel, which results in tooth decay.

Figure 12.15 Orange juice

iStock.com/BoardingINow

9780170495714

While many beverages are sold as fruit drinks, a large number of these products only contain a small percentage of real fruit juice. Many also have artificial colouring and flavourings added to them, along with added sugar to make them resemble real fruit juice. It is important to read the label carefully when choosing fruit drinks.

Comparing whole fruit and fruit juice

When we eat an orange, we eat the highly fibrous inner membrane and pulp. These sections are the primary source of the orange's flavonoids. These flavonoids work together with various nutrients, such as vitamin C. When an orange is juiced, these sections – and consequently the flavonoids – are lost.

Biting into a crisp apple involves us eating the skin of the apple as well as the flesh. The skin of the apple is exposed to the sun and, because of this, various coloured pigments such as carotenoids, anthocyanins and flavonoids are formed. As well as giving an apple its characteristic colours, these pigments have valuable nutritional properties.

In addition to the skin, the apple pulp is a valuable source of fibre and contains phytochemicals. When an apple is juiced the fibre is removed. Three apples are needed to produce one glass of apple juice. The resulting juice has the fibre removed, and a clear, slightly coloured juice remains. This juice is a concentrated form of energy and it is easy to consume too much.

One apple weighs approximately 200 grams, and contains 3 grams of fibre and approximately 200 kilojoules. It takes about 10 minutes to eat a whole apple.

One 250-millilitre glass of apple juice contains 0 grams of fibre and approximately 520 kilojoules. It is equivalent in energy value to three apples, but only takes about two minutes to drink, as it has no skin and no flesh to chew.

Bananas

Bananas are another of nature's cleverly packaged, healthy snacks. In fact, they are sometimes described as 'brain food' because they are high in potassium. This is a major nutrient required in the process of carrying the trillions of messages that nerves move around the human body. Foods that are good sources of potassium help concentration at school and home, as well as assisting in our performance in physical activities.

Nutrition

Bananas are low in fat and are a good source of kilojoules, making them a great high-energy food. As bananas ripen, their starches are converted to sugar, which the body can turn into energy quickly. Bananas are an excellent source of the minerals potassium, calcium, magnesium, phosphorus and iron. They are also high in vitamins A, B and C. Bananas are rich in dietary fibre, which means we feel full after eating them and so don't feel like eating other snack foods that are high in fat, salt or sugar. The other benefit of dietary fibre is that it helps to absorb water in the intestines, so the digestive process is more efficient.

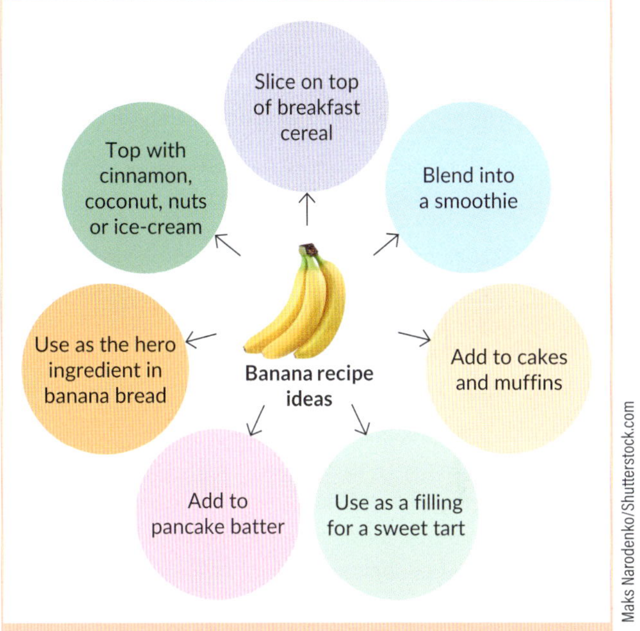

Figure 12.16 Bananas can be used in a variety of ways.

Maks Narodenko/Shutterstock.com

Storage

Allow bananas to ripen at room temperature, out of direct sunlight. Once ripe, bananas can be placed in a brown paper bag and stored in the crisper of a refrigerator for two to three days. Even though this turns the skin black, the flesh inside will remain delicious to eat.

Growing bananas using sustainable practices

Bananas grow in tropical regions on large, palm-like plants. The banana palm is not a tree, because it does not have a woody trunk. The banana palm develops a flower stem containing female flowers, from which the bananas grow. A single banana is called a 'finger'; several bananas – about 12 – make a 'hand'; and several hands form a 'bunch'. The most popular variety of banana grown in Australia is the Cavendish, which is long and thin. Lady Finger bananas are short and plump, and have a very distinctive flavour and texture.

Figure 12.17 Bananas growing on a banana palm

Some farmers are now producing bananas using **sustainable farming** practices. These practices maintain the land's productivity so that it will be available for future generations. When growing bananas in a sustainable manner, farmers use reduced amounts of fertilisers and pesticides on their crops, which has considerable benefits for the environment and reduces production costs. Instead of using chemicals to increase production, sustainable methods follow the natural growing cycle of bananas, which in turn improves the quality and fertility of the soil. Banana crops are also watered by efficient drip systems that prevent run-off and contamination of nearby waterways.

While these sustainable farming practices slow down the growing process, producers believe that bananas grown in this way have a firmer texture, a sweeter flavour and an extended shelf life. Before the bananas are sent to market, farmers dip their ends in red wax to indicate that they have been grown using sustainable farming practices.

Figure 12.18 Red wax-tipped bananas

CASE STUDY 12.2

Bananas for energy

Read the article below, and then answer the questions that follow.

How many bananas does it take to fuel the Australian Open?

Billie Eder, *The Age* – 12 January 2024

The most important thing a professional tennis player has in their bag isn't a racquet or a change of outfit.

It's a banana.

The 120-gram superfood is the snack of choice for most tennis players, and across the 15 days of the 2024 Australian Open, about 7800 bananas will be eaten in the gym, during practice and on court at Melbourne Park.

Take out the ball kids, officials and the athlete's entourage, and about 5000 will be consumed by the athletes.

If you lay 7800 standard size bananas (15–20 cm) end-to-end, and took out the bend, it would total between 1.17 km and 1.56 km, the equivalent of three-and-a-bit laps around the MCG.

According to Jo Shinewell, the national performance dietitian at Tennis Australia, there's no better snack an athlete can have.

'Number one, they are the most carbohydrate rich in comparison to all the other fruits, but they're also portable and easy. They can put them in their tennis bag and take them on court rather than having to cut it up before. So there's the portability of it as well,' Shinewell says.

'When you are [playing] for longer than 60 to 90 minutes at what we would call a moderate to high intensity, at that 90 minute mark, essentially

9780170495714

based on research, your fuel levels are getting close to depletion. So, if you're going to be going longer than that 90 minutes, you need to top yourself up with carbohydrate while you're active.'

The biggest indicator your fuel levels need a top-up is fatigue. When fatigue kicks in, moving around the court becomes harder, Shinewell says, and other areas of performance are negatively impacted, such as spatial awareness and decision-making.

The average length of a men's grand slam match last year was two hours and 54 minutes, according to *The Athletic*, but at the extreme end of the spectrum, matches can stretch to more than six hours.

...

'The longer you go, the more you're dependent on the external fuel that you're putting in,' Shinewell says.

'We generally say, from that 90 minutes where they need fuel, depending on the size of your body, you would be looking at having somewhere between 30 and 60 grams of carbohydrate per hour ... so if you had a banana and a gel or some sports drink or something else, you would be hitting somewhere in that mark.'

Bananas are good because they have a low glycaemic index (GI), which means they hit the bloodstream gradually, helping to sustain energy levels, while sport gels have a high GI, which means they kick in faster.

The other benefit of bananas is potassium.

'In terms of our sweat, we need what we call electrolytes, particularly sodium and potassium, to maintain a fluid balance within the body, and that's why sports drinks have got both sodium and potassium,' Shinewell says.

Stock Connection Blue/Alamy Stock Photo

Figure 12.19 Almost 8000 bananas are consumed by tennis players over the 15 days of the Australian Open.

'Having the potassium in the banana is going to be an added benefit to that fluid balance. Plus, potassium helps with ... blood vessel contractions, so it helps relax the blood vessels a bit as well which is a bonus.'

Of course, every athlete is different, and while bananas have mass appeal, some athletes like alternate snacks.

2023 US Open champion Coco Gauff is known for taking a container of fruit salad on court. Watermelon, rockmelon, pineapple and grapes are the usual favourites, which parents Candi and Corey pack for her.

'It's just fruit. My mum prepares it before the matches, and sometimes my dad does too,' Gauff told the US Today Show. 'I don't know if it's like a stress thing, and also they say to always eat to give you energy.' ...

Analysis

1 Annotate a diagram of a banana to highlight the reasons why Jo Shinewell says 'there's no better snack an athlete can have'.

2 Why do athletes such as professional tennis players need to top up their carbohydrate intake when playing for longer than 60–90 minutes?

3 Explain why having a low glycaemic index (GI) makes bananas a great snack food for tennis players.

4 Bananas are an excellent source of potassium. Outline the importance of potassium in the diet of an athlete.

5 Justify why bananas are a better snack food than a sandwich made with white bread and jam.

12

Activity 12.3

Fruity snacks

1 Develop a list of foods that students in your class eat as snacks.

2 Classify each item on a diagram of the *Australian Guide to Healthy Eating*. How many items are in the fruit section of this food selection guide?

3 Explain why fruit is an excellent snack food in terms of nutrition and convenience.

4 List four examples of processed foods and drinks that contain fruit and could be consumed as snacks.

5 Are any of the items you selected in question 4 high in fat or sugar? How do you know this? Where would you place them on a diagram of the *Australian Guide to Healthy Eating*?

6 Crunching and chewing on an apple helps to reduce dental plaque. What other health benefits are there in eating an apple?

Template
Australian Guide to Healthy Eating

Testing knowledge 12.3

21 Outline two health concerns linked to consuming fruit juice.

22 How do fruit drinks differ from fruit juice? Explain why it is important to understand the difference between them.

23 Justify why it is more beneficial to eat a whole apple than drink a glass of apple juice.

24 Bananas are a good source of potassium. Explain why it is important to include foods high in potassium in the diet.

25 Create a diagram to promote the nutritional properties of bananas.

26 Explain whether or not adding banana to a cake mixture would change the classification of the cake in the *Australian Guide to Healthy Eating*.

27 Why are bananas classified as a tropical fruit?

28 What are the benefits to the environment of using sustainable farming methods when growing bananas?

29 Discuss the benefits to a farmer of using sustainable farming practices in their business.

30 How does growing bananas sustainably change the properties of the fruit?

Quiz
Testing knowledge 12.3

Critical and creative thinking 12.1

Comparing fruit

Select two fruit types to compare from two different fruit classifications.

1 List the characteristics of each type of fruit, including:
- physical structure
- sensory properties
- nutrient value
- storage requirements
- seasonality.

2 Discuss the similarities and differences between the two fruits you have compared, and create a diagram to record your comparison.

9780170495714

DESIGN ACTIVITY 12.1

Promoting fresh Australian-grown apples and oranges

Worksheet
Design activity 12.1

Templates
Decision table

Food order

Sensory wheel

Australian Guide to Healthy Eating

Weblink
A better choice

Background information: 'A Better Choice'.

'A Better Choice' is a national initiative designed by Fresh Markets Australia and the Central Markets Association of Australia 'to educate consumers about the quality, service and freshness of produce at their local fruit and vegetable shop'. Shopping locally means fresher produce and greater variety, and consumers have easy access to the business owners, who are passionate about their industry and understand it – from grower, to market, to plate. The program supports more than 800 independent fruit and vegetable retailers across Australia, who together supply 15 per cent of the fresh produce sold each year.

'A Better Choice' is updating the recipe section of its website and is looking for new ideas to promote fresh Australian-grown apples and oranges.

Design brief

Design a recipe that promotes fresh Australian-grown apples or oranges. The recipe can have a sweet or savoury focus, and should be suitable to serve for two people as part of a shared table at a family gathering. A photograph of the finished dish and a description of its nutritional properties, based on the recommendations of the *Australian Guide to Healthy Eating* will be included with the recipe on the 'A Better Choice' website.

Investigating and defining

1. Explore the 'A Better Choice' website to gain a deeper understanding of aims of the program and its target market.
2. Select the type of fruit – apples or oranges – that will be the focus of your designed solution.
3. Based on the solution requirements and constraints in the design brief, develop three design criteria to evaluate the success of your designed solution.
4. Research varieties of Australian-grown apples or oranges that will be in season at the time of production. Select two varieties of the fruit selected.
5. Research recipes that will include the selected fruit and can be served as part of a shared table at a family gathering. Remember to record the source of the recipe. These recipes will become your designed solution options.
6. Investigate the nutritional properties of the selected fruit.

Generating and designing

Complete a decision table like the one below.

Selected fruit	Designed solution option 1	Designed solution option 2
Recipe title		
Source of recipe		
Savoury or sweet		
Suitability for a family gathering		
Nutritional properties of the recipe compared with the recommendations of the *Australian Guide to Healthy Eating*		
Advantages of the recipe as a designed solution option		
Disadvantages of the recipe as a designed solution option		
Selected designed solution, with justification		

Planning and managing

1 Prepare a food order.
2 Annotate a copy of your designed solution recipe to indicate the tools and equipment needed in production and to indicate key safe work practices.

Producing and implementing

1 Produce your designed solution fruit recipe for part of a shared table at a family gathering.
2 Photograph the product before taste testing.

Evaluating

1 Respond to the design criteria you developed.
2 Describe the sensory properties – appearance, aroma, flavour, texture and sound – of your designed solution fruit recipe for part of a shared table at a family gathering.
3 Discuss the effectiveness of your time management for the production.
4 Classify the ingredients of your fruit recipe on a diagram of the *Australian Guide to Healthy Eating*. Discuss how well the dish meets the guidelines of this food selection model.
5 What recommendations would you suggest to improve the quality and nutritional properties of the recipe if you were to make it again?

9780170495714

Pear and Parmesan Salad with Lemon Chicken Bites

INGREDIENTS

Lemon Chicken Bites

2 slices white bread, crusts removed

⅓ cup milk

200 grams minced chicken

2 tablespoons parmesan cheese, grated

1 clove garlic, crushed

salt and pepper

1 tablespoon parsley, chopped

zest of 1 small lemon

1 tablespoon oil for frying

Pear and Parmesan Salad

2 cups of mixed salad leaves, such as rocket leaves and other lettuce varieties

¼ cup pecan nuts, chopped in half lengthwise

2 tablespoons spring onions, chopped

1 firm pear

Dressing

2 tablespoons olive oil

1 tablespoon cider vinegar

salt and ground black pepper, to taste

1 tablespoon parmesan cheese, shaved, to garnish

Mark Fergus Photography

METHOD

Lemon Chicken Bites

1. Soak the slices of bread in milk for a few minutes, then squeeze to remove excess moisture.
2. Combine all of the ingredients except the oil in a medium bowl.
3. Work the mixture together by hand, until well combined.
4. Wet your hands and roll the mixture into walnut-sized balls. Refrigerate while preparing the salad and the dressing.
5. Heat the oil in a frying pan, add the balls of chicken mixture and fry, turning gently until cooked through.
6. Drain the Lemon Chicken Bites on absorbent paper and allow to cool.

Pear and Parmesan Salad

1. Place the salad leaves, pecans and spring onions in a large bowl.
2. Just before serving, peel the pear, cut it into quarters and then into slices. Add the slices to the salad.
3. To prepare the dressing, place the olive oil and vinegar in a lidded jar, add the seasonings, then put on the lid and shake to combine.
4. Add the dressing to the salad and toss to combine.
5. Divide the Pear and Parmesan Salad into two portions and place on serving plates. Place half of the Lemon Chicken Bites on top of each. Garnish with the shaved parmesan cheese.

EVALUATION

1. Describe the sensory properties – appearance, aroma, flavour, texture and sound – of your Pear and Parmesan Salad with Lemon Chicken Bites.
2. Identify and describe the process that occurs to the protein in the chicken mince when the meatballs are fried.
3. What sensory properties would be altered if the pear was cut early in the preparation of the dish, rather than just before serving?
4. How could you treat the pear if you wished to prepare it in advance?
5. Evaluate the nutritional properties of your Pear and Parmesan Salad with Lemon Chicken Bites by classifying the ingredients on a diagram of the *Australian Guide to Healthy Eating*. Comment on whether this recipe is suitable to serve as a healthy meal for children and families.

12 RECIPES

Watermelon and Cherry Tomato Salad

INGREDIENTS

Dressing

1 tablespoon extra virgin olive oil

1 teaspoon balsamic vinegar

2 teaspoons lemon juice

1 teaspoon honey

½ clove garlic, crushed

Salad

500 grams seedless watermelon (skin on)

100 grams cherry tomatoes

½ small Lebanese cucumber

½ small red onion, finely sliced

⅓ cup pitted kalamata olives, halved lengthwise

½ cup mint, coarsely chopped

½ cup flat leaf parsley, coarsely chopped

To garnish

50 grams Australian feta cheese

SERVES TWO

METHOD

1. Prepare the dressing by placing all of the ingredients in a small screw-top jar and shaking well to combine.
2. Remove the skin from the piece of watermelon, then cut the flesh into 2.5-centimetre pieces. Don't worry if the pieces are different shapes.
3. Cut the cherry tomatoes into halves.
4. Cut the Lebanese cucumber in half lengthwise and use a teaspoon to scoop out the seeds.
5. Cut thin slices across the cucumber to create half-moon shapes.
6. Combine the watermelon, cherry tomatoes, Lebanese cucumber, red onion, olives, mint and parsley in a large bowl.
7. Shake the dressing again, then drizzle over the fruit and vegetables and toss gently to combine.
8. Garnish the salad with crumbled feta cheese.

EVALUATION

1. Describe the sensory properties – appearance, aroma, flavour, texture and sound – of your Watermelon and Cherry Tomato Salad.
2. Explain why the dressing used in this recipe is a temporary emulsion.
3. List the different types of knives you used to cut the ingredients for this recipe. Explain why each one was the best tool for the task.
4. Suggest another vegetable that could be added to the salad to increase the number of serves of vegetables that it would provide. Justify your choice.
5. With reference to the recommendations of the *Australian Guide to Healthy Eating*, justify whether you think the Watermelon and Cherry Tomato Salad served with Pork Schnitzels (see page 343) would be a healthy main meal for a family.

Mark Fergus Photography

9780170495714

Pork Schnitzels

Schnitzel is a favourite meal for many people. This is a very versatile recipe, as the technique of coating food in egg and breadcrumbs can be used with a variety of ingredients, such as chicken breast fillets, thin slices of beef or veal, or thick slices of eggplant.

INGREDIENTS

2 pork loin medallions (approximately 100 grams each)

2 tablespoons plain flour

salt and pepper

1 egg

1 tablespoon milk

½ cup dry breadcrumbs

2 tablespoons grated parmesan cheese

2 tablespoons oil

SERVES TWO

METHOD

1 Carefully remove any fat or silver membrane from around the edges of the pork loin medallions.
2 Cover the medallions with a sheet of baking paper or plastic wrap and use a rolling pin to press them to an even thickness of about 1 centimetre.
3 Place the flour on a sheet of paper towel and season with salt and pepper.
4 Beat the egg and milk together in a small bowl. Place on a flat plate.
5 Combine the breadcrumbs and parmesan cheese in a small bowl. Place on a sheet of paper towel.
6 Coat each medallion in the seasoned flour, then brush with the egg mixture and coat in the breadcrumb mixture. Press the crumbs on firmly.
7 Coat each medallion in a second layer of egg and breadcrumbs. Press the crumbs on firmly.
8 Rest the crumbed schnitzels in the refrigerator for at least 15 minutes. This allows the moisture to be evenly distributed in the crumbing mixture and helps to prevent it from falling off the food during cooking.
9 Heat the oil in a frying pan over a low heat. Place the crumbed schnitzels in the hot oil and cook for approximately 3–4 minutes or until brown on the bottom.
10 Carefully turn the schnitzels over and cook on the second side for a further 3–4 minutes or until brown on the bottom.
11 Drain on paper towel.

EVALUATION

1 Describe the sensory properties – appearance, aroma, flavour, texture and sound – of your Pork Schnitzels.
2 Explain why it is important to ensure that the meat is an even thickness before coating it.
3 Why are the schnitzels coated in flour before being brushed with the egg mixture?
4 Describe two safety rules to follow when frying food.
5 Explain why, according to the recommendations of the *Australian Guide to Healthy Eating*, fried food such as Pork Schnitzels should be eaten 'only sometimes and in small amounts'.

Vegetable Lasagna

INGREDIENTS

Vegetable Sauce

2 teaspoons olive oil

½ onion, finely diced

1 clove garlic, crushed

100 grams mushrooms, sliced

½ zucchini, grated

200 grams canned diced tomatoes

1 tablespoon tomato paste

¼ carrot, grated

¼ cup celery, finely diced

¼ red capsicum, finely diced

½ Granny Smith apple, peeled and grated

¼ teaspoon dried oregano

¼ teaspoon dried basil

¼ teaspoon dried rosemary

White Sauce

1½ cups milk

⅛ onion, finely diced

1 clove

pinch of nutmeg

30 grams butter

2½ tablespoons flour

salt and pepper

To assemble the dish

3 fresh lasagna pasta sheets

2 slices mozzarella cheese

1 tablespoon parmesan cheese, grated

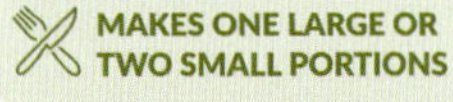
MAKES ONE LARGE OR TWO SMALL PORTIONS

METHOD

Vegetable Sauce

1 Heat the oil over a medium heat in a large saucepan and sauté the onion and garlic. Cook for 1–2 minutes but do not brown. Add the mushrooms and cook until they are soft.

2 Add the remaining ingredients to the saucepan and cook for 3 further minutes, stirring. Remove from the heat. The sauce will be sloppy.

White Sauce

1 Place the milk, onion, clove and nutmeg in a small saucepan and heat until simmering. Remove from the heat and allow to cool.

2 Strain the cooled, flavoured milk into a measuring jug.

3 Melt the butter in a small saucepan and blend in the flour. Cook for 30 seconds, taking care not to brown. Remove from the heat.

4 Gradually stir in the flavoured milk. When blended and lump-free, return to the heat, stirring constantly. Cook until boiling, then remove from heat.

5 Season with salt and pepper.

To assemble the dish

1 Preheat oven to 180°C.

2 Line a baking dish with one pasta sheet, then alternately layer the vegetable sauce, white sauce, mozzarella and more pasta. Finish with the white sauce and sprinkle with the parmesan cheese.

3 Bake for 20 minutes or until golden brown.

EVALUATION

1 Describe the sensory properties – appearance, aroma, flavour, texture and sound – of your Vegetable Lasagna.

2 How does the grated apple add to the flavour of the vegetable sauce?

3 Identify the steps in the recipe that help to ensure that the white sauce is lump-free, and explain how this sauce thickened.

4 List the ingredients that contributed to the browning of the Vegetable Lasagna during baking. What process occurs during baking?

5 Write a paragraph that outlines why the Vegetable Lasagna would be considered a healthy meal.

Mark Fergus Photography

9780170495714

Stewed Apples

INGREDIENTS

2 Granny Smith apples

¼–⅓ cup water

1 tablespoon sugar

1 clove

SERVES TWO

METHOD

1. Wash and peel the apples and cut into quarters.
2. Remove the cores and slice the apples thinly.
3. Place all the ingredients in a small saucepan. Stir to coat the apple with the other ingredients.
4. Bring to the boil, place a lid on the saucepan and reduce the heat to a simmer. Stir occasionally.
5. Cook until the apple is tender.

EVALUATION

1. Explain the meaning of the term 'stewing'.
2. Which knife would be most suitable for cutting and slicing the apples? Describe the safest way to slice apples.
3. Why is it important to keep the lid on when stewing apples?
4. List some other fruits that could be stewed successfully.
5. Explain why the saying 'an apple a day keeps the doctor away' is appropriate nutritional advice.

udra11/Shutterstock.com

Fruit Crumble

INGREDIENTS

1 cup fruit – stewed, canned or freshly sliced

2 tablespoons caster sugar or brown sugar

2 tablespoons coconut or almond slivers

2 tablespoons rolled oats or fresh breadcrumbs

pinch nutmeg

pinch cinnamon

2 tablespoons self-raising flour (white or wholemeal)

20 grams butter

SERVES TWO

METHOD

1 Preheat oven to 180°C. Lightly butter two ovenproof ramekins.

2 Divide the fruit evenly between the ramekins.

3 Mix all the dry ingredients in a bowl.

4 *Either* melt the butter in a microwave and stir into the dry ingredients *or*, using the fingertips, rub the cold butter into the dry ingredients until the mix resembles fresh breadcrumbs.

5 Sprinkle the fruit with the crumble topping. The surface of the crumble should be rough, not smooth.

6 Bake for 15–20 minutes or until golden brown.

EVALUATION

1 Describe the sensory properties – appearance, aroma, flavour, texture and sound – of your Fruit Crumble.

2 List four examples of fresh fruits that would be suitable to use in a crumble recipe.

3 Identify two safe work practices you followed when using the oven.

4 Classify the ingredients for the Fruit Crumble on a diagram of the *Australian Guide to Healthy Eating*. Rate the Fruit Crumble as 'healthy', 'reasonably healthy' or 'not very healthy'. Justify the rating you have given this dish.

Mark Fergus Photography

9780170495714

Free-form Fruit Tart

Light, buttery pastry filled with a luscious fresh fruit makes a beautiful dessert.

INGREDIENTS

Pastry

½ cup self-raising flour

2 tablespoons plain flour

30 grams butter, chilled and cubed

2 teaspoons caster sugar

¼ teaspoon cinnamon

1–2 tablespoons iced water

extra sugar for sprinkling

Filling

½ egg white, lightly beaten

2 teaspoons fine polenta or semolina

200 grams apricots, nectarines or peaches, peeled pitted and sliced

or

1 Granny Smith apple, peeled, cored and thinly sliced

2 teaspoons sugar

To serve

cream or ice-cream

METHOD

Pastry

1 Process the flours, butter, sugar and cinnamon in a food processor until the mixture resembles fine breadcrumbs.

2 Add the iced water and process until the mixture starts to come together.

3 Turn the dough onto a lightly floured bench and bring together until smooth. Do not over-knead.

4 Flatten the dough into a disc, cover with plastic wrap and rest in the refrigerator for 20–30 minutes.

5 Roll the disc into a 3-millimetre-thick round, approximately 20 centimetres in diameter. Turn the pastry as you roll so that it does not stick. Transfer to a lightly oiled oven tray.

To assemble and bake the tart

1 Preheat oven to 200°C.

2 Brush the pastry circle with egg white and sprinkle with polenta.

3 In a small bowl, toss the fruit in the sugar.

5 Place the sugared fruit in the centre of the pastry, leaving a 4–5-centimetre border.

6 Fold the border of the pastry over the fruit, pleating to fit.

7 Brush the pastry with more of the egg white and sprinkle with the extra sugar.

8 Bake for 15–20 minutes.

9 Serve warm with cream or ice-cream.

Mark Fergus Photography

EVALUATION

1 Describe the sensory properties – appearance, aroma, flavour, texture and sound – of your Free-form Fruit Tart.

2 What are the advantages and disadvantages of making a free-form tart compared with a tart baked in a traditional flan or pie tin?

3 What are the advantages and disadvantages of using a food processor to make the pastry?

4 Why is the pastry rested in the refrigerator?

5 Classify the ingredients for the Free-form Fruit Tart on a diagram of the *Australian Guide to Healthy Eating*. Explain how well this dish meets the recommendations of this food selection model.

Fruit Galettes with Custard Powder Sauce

INGREDIENTS

Fruit Galettes

1 sheet butter puff pastry (approximately 22 × 22 centimetres)

40 grams butter, at room temperature

½ cup ground almonds

4 tablespoons demerara or raw sugar

½ teaspoon vanilla extract

1 fresh yellow peach, 1 Pink Lady apple or 8 canned apricot halves

1 tablespoon milk

Custard Powder Sauce

1 tablespoon custard powder

1–2 teaspoons white sugar

200 millilitres milk

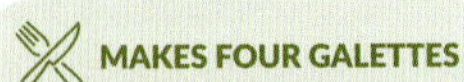

METHOD

Fruit Galettes

1. Separate a sheet of pastry from the frozen packet and defrost in the refrigerator until ready to use.
2. Preheat oven to 200°C. Line an oven tray with baking paper.
3. In a medium bowl, use a wooden spoon to soften the butter, then add the ground almonds, 1 tablespoon demerara or raw sugar and vanilla. Mix well and set aside.
4. Prepare the fruit:

   Yellow peach – cut the peach in half, remove the stone and slice thinly. Try to retain the natural curve of the fruit.

   Pink Lady apple – cut the apple into quarters and slice thinly. Try to retain the natural curve of the fruit.

   Canned apricot halves – drain the apricot halves and pat dry with paper towel.
5. Remove the pastry sheet from the refrigerator and use a sharp knife to cut it into four squares, approximately 11 × 11 centimetres each. Lay the pastry squares on the lined oven tray.
6. Divide the butter mixture into four portions and place on the centre of each pastry square. Spread it out a little, but leave a 2-centimetre band of pastry on all sides.
7. Arrange the fruit over the top of the butter mixture, brush the borders with milk and sprinkle the remaining demerara or raw sugar equally over the four galettes.
8. Bake for 12–15 minutes or until the pastry is puffed and golden.
9. Serve with Custard Powder Sauce. Alternatively, you could serve the galettes with ice-cream or cream.

Custard Powder Sauce – on the cooktop

1. In a small bowl, blend the custard powder and sugar with 2 tablespoons of the milk.
2. In a small saucepan, heat the remaining milk until simmering. Remove from heat.
3. Pour the hot milk onto the blended custard powder and stir until combined.
4. Transfer the custard mixture to the saucepan, then return to the cooktop. Bring to the boil over a low-medium heat, stirring constantly.
5. Cook for 1 minute, then remove from heat.

9780170495714

METHOD

Custard Powder Sauce – in the microwave

1 In a small microwave-safe bowl, blend the custard powder and sugar with 2 tablespoons of the milk.
2 Gradually add the remaining milk. Whisk or stir well.
3 Microwave on high for 2 minutes. Remove and whisk to remove any lumps.
4 Microwave on high for a further 1–2 minutes or until the mixture thickens.
5 Remove from the microwave and whisk again.

EVALUATION

1 Describe the sensory properties – appearance, aroma, flavour, texture and sound – of your Fruit Galettes with Custard Powder Sauce.
2 Identify the two processes that cause the pastry to brown when the galette is baked.
3 Identify the process that occurs to the Custard Powder Sauce during cooking.
4 Predict the outcome to the sensory properties of the Custard Powder Sauce if it is not whisked or stirred.
5 Classify the ingredients of the Fruit Galettes with Custard Powder Sauce on a diagram of the *Australian Guide to Healthy Eating*. After considering the distribution of the ingredients on this food selection model, recommend how often this dessert should be eaten.

Mark Fergus Photography

MILK PRODUCTS

NUTRIENTS IN MILK PRODUCTS

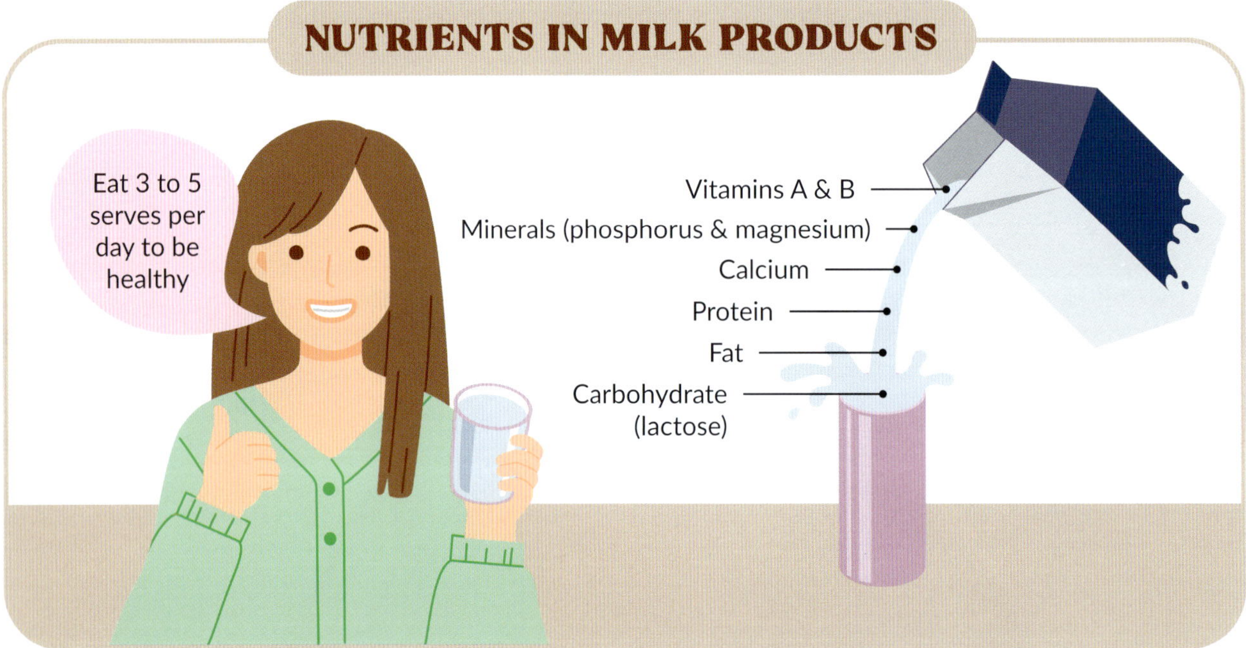

SUSTAINABILITY OF MILK PRODUCTION IN AUSTRALIA

CHALLENGES FOR DAIRY FARMERS

- Drought
- High water costs
- High fodder costs
- Low milk farmgate prices

PRODUCTION OF GREENHOUSE GASES

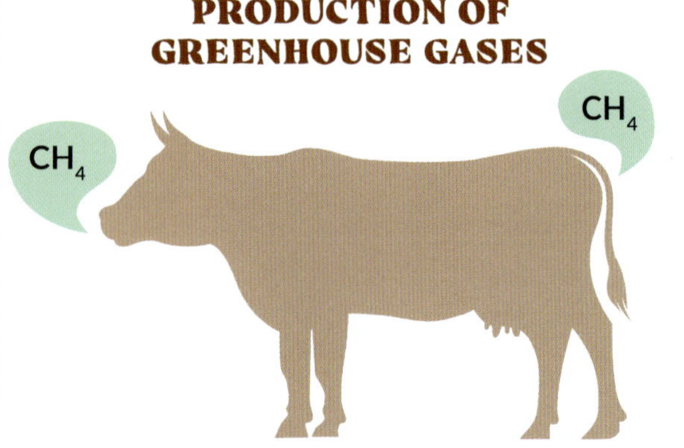

9780170495714

PROCESSING OF MILK

PASTEURISATION

Makes milk safe and extends shelf life

HOMOGENISATION

Evenly distributes fat through milk

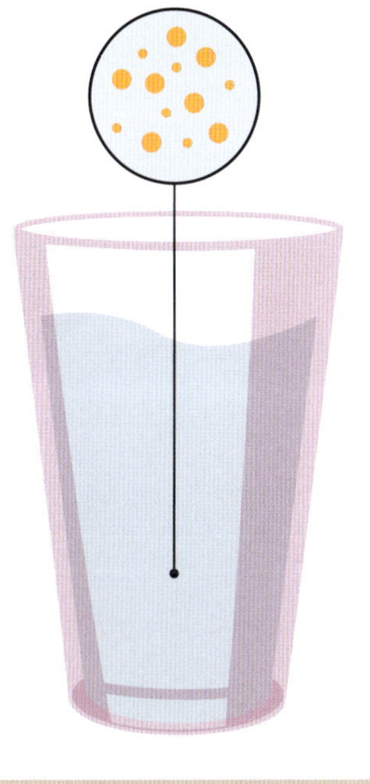

13 MILK, YOGHURT AND CHEESE

KEY TERMS

casein the protein present in the curds of milk

cheese a product made from casein, the protein present in the curds of milk that are separated from the water and lactose, or whey

homogenisation a process that breaks the globules of fat within milk into minute particles so that the cream does not rise to the surface

lactose the sugar (carbohydrate) present in milk

lactose intolerance the reduced ability to digest milk sugars, due to insufficient amounts of the enzyme lactase

osteoporosis a medical condition that occurs when calcium is lost from the bones, making them very fragile and easily broken

pasteurisation a process that destroys pathogenic, or disease-causing, bacteria and also extends the shelf life of milk

UHT milk a type of milk that has undergone ultra-high temperature processing of heating the milk to approximately 135°C for 2–3 seconds

Templates
- SWOT analysis table
- Food order
- Sensory wheel
- Australian Guide to Healthy Eating

Worksheets
- Practical activity 13.1
- Practical activity 13.2
- Practical activity 13.3
- Design activity 13.1

Quizzes
- Testing knowledge 13.1
- Testing knowledge 13.2
- Testing knowledge 13.3

Nelson MindTap

To access resources above, visit **cengage.com.au/nelsonmindtap**

9780170495714

Milk, yoghurt and cheese in the *Australian Guide to Healthy Eating*

Health professionals recommend that we include a variety of dairy products such as milk, yoghurt and cheese or their alternatives – such as nut-based milk – in our diet each day, as they are essential for good health. Based on this advice, these food products make up a significant proportion of the *Australian Guide to Healthy Eating*.

Australian Government
National Health and Medical Research Council
Department of Health and Ageing

www.eatforhealth.gov.au

Australian Guide to Healthy Eating

Enjoy a wide variety of nutritious foods from these five food groups every day.
Drink plenty of water.

Vegetables and legumes/beans

Fruit

Milk, yoghurt, cheese and/or alternatives, mostly reduced fat

Lean meats and poultry, fish, eggs, tofu, nuts and seeds and legumes/beans

Use small amounts

Only sometimes and in small amounts

Figure 13.1 The place of milk, yoghurt and cheese in the *Australian Guide to Healthy Eating*

Milk, yoghurt and cheese for good health

It is important to include milk, yoghurt and cheese or their alternatives in our diet each day as they are an excellent source of a wide range of nutrients, including protein, fat and carbohydrates (lactose). Milk is high in the mineral calcium, which is essential for bone development, and is a much higher source of this nutrient than most other foods. Milk is also a source of other vitamins and minerals, especially vitamins A, B1 (thiamine), B2 (riboflavin), and B12 (cobalamin). In addition, milk provides the minerals phosphorus and magnesium, which help calcium to be absorbed into the body. Milk is composed of 87 per cent water.

Including a variety of milk, yoghurt and cheese products in the diet each day can help to reduce the risk of developing health conditions, including high blood pressure, heart disease, stroke, type 2 diabetes and some types of cancer.

Figure 13.2 Milk, yoghurt and cheese are essential parts of a healthy and well-balanced diet.

iStock.com/Anna Puzatykh

Dietary advice released by the Heart Foundation in January 2025 states that healthy Australians – those people who do not have any risk factors for heart disease – can choose to eat unflavoured full-fat milk, yoghurt and cheese rather than reduced-fat options if they prefer. The body of research that underpins this advice indicates that milk, yoghurt and cheese have a 'neutral' effect on heart health. However, it is important that people who have high cholesterol or have been diagnosed with heart disease select unflavoured reduced-fat milk, yoghurt and cheese products to reduce the amount of saturated fat in their diet.

The *Australian Dietary Guidelines* also recommends that, when selecting foods for a healthy diet, we should all minimise the amount of butter, cream and ice-cream we eat, as these products are 'eat sometimes' foods and are more likely to increase the level of LDL (low-density lipoprotein) or 'bad' cholesterol in our blood. The dietary advice from the Heart Foundation supports this recommendation.

Figure 13.3 Ice-cream is a delicious dairy food but, like butter and cream, it should only be eaten sometimes.

iStock.com/kirin_photo

9780170495714

Table 13.1 How much of the milk, yoghurt and cheese group should I eat?

	Serves per day	
	12–13 years	14–18 years
Boys	3.5	3.5
Girls	3.5	3.5

Figure 13.4 What is a serve?

Lactose intolerance

Approximately 4 per cent of the Australian population suffer from **lactose intolerance**. Far more women than men have an intolerance to lactose. People who have a lactose intolerance lack the enzyme lactase, or have it in insufficient quantities in their system, and are therefore unable to digest **lactose**, which is the sugar in milk. The result is that the bacteria that live in the colon cause the lactose to ferment and produce carbon dioxide (CO_2). This causes the bowel to retain water and leads to symptoms that include bloating and diarrhoea.

Milk

Cow's, goat's, sheep's and camel's milk are the types of milk most commonly consumed by human beings. Cow's milk is the most popular source of milk in Australia. The Australian Bureau of Statistics reports that we consume approximately 267 grams of milk products per day.

Milk processing

After milk has been taken from the animal and before it can be sold on the commercial market, it undergoes pasteurisation to make it safe. It may also be homogenised.

Pasteurisation

Milk is a protein food, and is the perfect medium for the growth of microorganisms. It is therefore easily contaminated. The most common method of treating milk to make it safe to consume is to heat it to 72°C for 15 seconds, and then to cool it rapidly to 4°C. This process of **pasteurisation** is referred to as 'high-temperature, short-time pasteurisation' and was developed in the 1860s by Louis Pasteur, a French chemist and microbiologist. It destroys all disease-causing bacteria and extends the shelf life of milk, while having a limited effect on the flavour.

Figure 13.5 The nutrient composition of different animal milks

Homogenisation

The **homogenisation** of milk involves breaking the globules of fat within milk into tiny particles so that the cream does not rise to the surface. The process of homogenisation involves forcing the milk through a very fine nozzle at a high pressure. Homogenisation breaks the fat molecules into uniform particles that are about one-quarter of their original size. The texture and mouth feel of the milk is the same for the whole contents of the container.

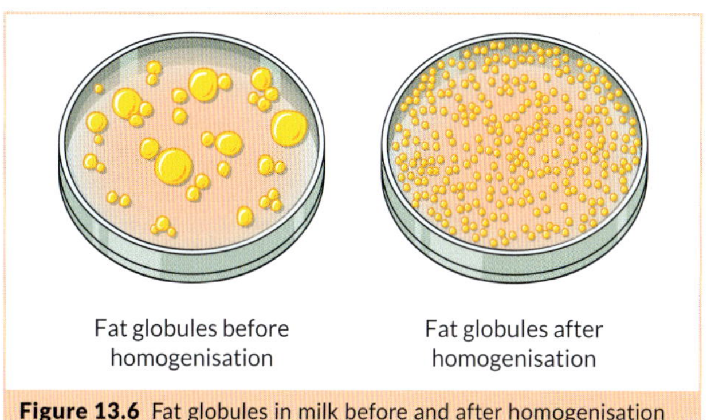

Figure 13.6 Fat globules in milk before and after homogenisation

Ultra-high temperature (UHT) milk

UHT milk has undergone ultra-high temperature processing, which involves heating the milk to approximately 135°C for 2–3 seconds. It is then packaged using aseptic packaging. Aseptic packaging is a process whereby the food product and the package are sterilised separately and then brought together in a sterile environment.

UHT milk has similar sensory properties to regular milk. It is a long-life milk that can be stored for three to six months without refrigeration until it is opened.

Types of milk

A wide variety of milk is available to consumers.

- Full-cream milk contains approximately 4 per cent fat.
- Reduced-fat milk contains about half the fat (less than 2 per cent) of full-cream milk. It has the fat-soluble vitamins A and D added to it to replace the nutrients that are lost when the fat is removed during processing.
- Low-fat milk contains less than 1.5 per cent fat and is fortified with vitamins A and D.
- Skim milk has the lowest proportion of fat (no more than 0.15 per cent). Skim milk, too, has the fat-soluble vitamins A and D added to it to replace the nutrients that are lost when the fat is removed during processing.

Mark Fergus Photography

Figure 13.7 A variety of dairy milks is available to consumers.

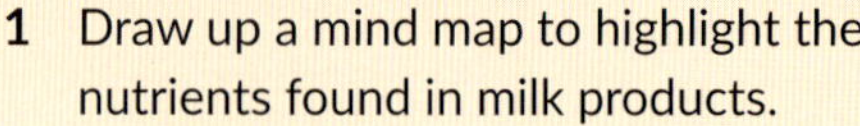

Testing knowledge 13.1

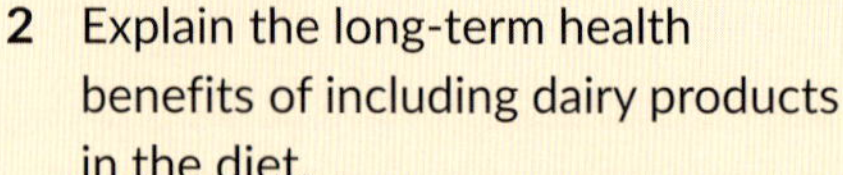

Quiz
Testing knowledge 13.1

1. Draw up a mind map to highlight the nutrients found in milk products.
2. Explain the long-term health benefits of including dairy products in the diet.
3. Summarise the latest advice from the Heart Foundation about including full-fat dairy products in the diet.
4. Explain the meaning of the term 'lactose intolerance' and the impact of this condition on people's health.
5. Design a breakfast menu for a 12–13-year-old boy or girl to demonstrate how they could consume at least two serves of dairy foods.
6. Describe the main differences in the nutritional composition of milk produced from cows and camels.
7. Outline the process used to pasteurise milk, and explain why pasteurisation is an important process in milk production.
8. Describe how milk is homogenised, and explain why milk undergoes homogenisation before being sold to consumers.
9. Identify and describe the process that milk undergoes to extend its shelf life to six months.
10. Discuss how skim milk differs from full-cream milk.

Challenges facing Australian dairy farmers

At the end of the 2024 financial year, the Australian Department of Agriculture, Fisheries and Forestry estimated that 5 per cent of Australian farm businesses were classified as dairy farms (4502 farms).

Most dairy farms are concentrated in the high-rainfall coastal areas and irrigated inland areas of Victoria, New South Wales, Tasmania and south-eastern Queensland. Some dairying also takes place in South Australia and Western Australia. In recent decades, the viability of many of these dairy farms is becoming more and more challenging, given ongoing drought conditions, increasing production costs and milk-processing costs, and the fact that the prices that are paid to dairy farmers by supermarkets are well below the cost of production.

9780170495714

iStock.com/Chris Gordon

Figure 13.8 Australian dairy cows

DROUGHT
Significant drought conditions have affected Queensland, New South Wales, north-west Victoria and eastern South Australia.

HIGH WATER COSTS
Water prices have increased significantly in recent years. Some farmers faced a 420 per cent increase in the price of water over the previous year.

HIGH FODDER COSTS
Drought conditions have led to a shortage of grain and hay, forcing dairy farmers to pay excessively high prices for fodder to feed their cattle.

LOW MILK PRICES
In September 2024, Coles and Woolworths reduced the price of fresh milk by 5 cents, to $1.55 per litre – the first time since 2011 that the cost of milk in major Australian supermarkets has decreased.

Figure 13.9 The challenges facing Australian dairy farmers

CASE STUDY 13.1

The effects of the drop in milk prices

Read the article below, and then answer the questions that follow.

Supermarkets drop milk price for the first time since 2011 as farmers fear return of 'milk war'

Fiona Broom and Warwick Long, *ABC Rural*, 8 October 2024

In short:

The cost of supermarket brand milk has gone backwards for the first time since 2011.

The move comes as Australia's big two supermarkets face scrutiny over discount pricing practices.

What's next?

Dairy farmers say the price drop could signal a new milk price war between the retailers, and threaten producers' livelihoods.

The cost of milk in major Australian supermarkets has gone backwards for the first time since 2011 amid growing scrutiny over discount pricing.

Coles and Woolworths reduced the price of its generic fresh milk last month by 5 cents, to $1.55 per litre.

Discount supermarket chain Aldi held firm at $1.60, before matching the price in October.

It's the first time, supermarket milk prices have gone backwards since January 2011, when Australia's big three supermarkets slashed their prices to $1 a litre.

'We are very concerned where this is heading,' said Dairy Farmers Victoria president, Mark Billing.

Mr Billing runs a 400-head dairy farmer in Larpent, in south-west Victoria.

He said the supermarkets' approach devalues the product and could push farmers to the financial brink.

Coles said it had not changed the price it paid dairy farmers, while Woolworths said it is passing on savings from milk processors to customers.

Milk wars 2.0

Australian Dairy Farmers, the peak body for dairy farmers, told the ABC that retailers view milk as a 'discount leader' – a cheap item that draws in customers.

While 1-litre milk dropped by 5 cents per litre, two-litre bottles have gone from $3.10 to $3.00, and 3-litre bottles from $4.50 to $4.35.

Mr Billing said the discount was significant.

'It's extremely frustrating, particularly in the environment we're in . . . margins are really, really tight, and it's putting a lot of pressure on farm management,' he said.

The price reduction comes as Coles and Woolworths face legal action from the consumer watchdog over pricing practices.

The Australian Competition and Consumer Commission (ACCC) has alleged the companies misled consumers with discount pricing claims on hundreds of products, and has launched legal action against both companies.

Its latest report revealed customers don't trust 'sale price' claims at the grocers.

Mr Billing has questioned the timing of the price drop.

'We just need to understand what the strategy is,' he said.

'I'd love to be able to hear what their thinking is behind these reductions.'

Lower prices 'passed on to customers'

Retailers are able to drop their prices as farmers are being paid less by the milk processors, according to the supermarkets.

'Changes in the farmgate milk price flow directly into the price we pay processors for their finished product – if the milk price goes up we pay more, and when it goes down we pay less,' a Woolworths spokesperson said.

'We've reduced the price of Woolworths branded milk to pass on the savings we're receiving from our processors.'

The current reduction is not an all-time low for the product, however.

In 2022, Woolworths sold its supermarket brand milk for $1.35 for one litre, before it increased the price to $1.60.

Coles said it bought direct from farmers for its own brand milk, and that it had not changed the price it was paying producers.

'We introduced a direct sourcing model for our own brand milk in 2019 to ensure we could provide fair, competitive and guaranteed farmgate prices to dairy farmers directly,' a Coles spokesperson said.

An Aldi spokesperson said the supermarket remained committed to fair pricing for suppliers.

'We remain focused on reviewing pricing for the whole dairy category on an ongoing basis,' the spokesperson said.

Long, hard road

The price drop comes as milk production in Australia is at its lowest in 30 years as dairy imports are increasing.

Pricing uncertainty since the outbreak of the $1-litre 'milk wars' has left farmers exhausted, Mr Billing said.

'Farmers are a pretty resilient bunch, but a piece of wire is resilient too – you can only bend it so many times before it breaks,' he said.

'As it gets devalued, then it can be a little bit soul destroying when you've got a milk price drop.'

Around the turn of the century, there were 7409 Victorian dairy businesses.

There are now 2796.

Last year in Victoria, Australia's biggest milk producing state, 8 per cent of its dairy farmers quit the business.

Analysis

1. Evaluate the validity of this article. Consider the reliability of the sources of information used in the article, the context and the presentation of evidence.
2. Draw conclusions about why the authors published this article.
3. Hypothesise the impact of the information presented in this article on the future of the Australian dairy industry.
4. Formulate two recommendations you could present to the Federal Minster for Agriculture or Dairy Australia to ensure the sustainability of the Australian dairy industry.

The environmental impact of milk production

Dairy farming is one of Australia's main rural industries and, like other forms of primary food production, it has a major impact on the environment.

Greenhouse gas emissions

One of the most significant ways in which the dairy industry impacts on the environment is through the production of greenhouse gases that contribute directly to climate change. Dairy cows add substantial amounts of greenhouse gases to the atmosphere because when

9780170495714

cows digest their food, they produce methane (CH_4). Of all the emissions from dairy farms, 60–70 per cent are methane. The dung and urine that dairy cows excrete contain high levels of nitrous oxide (N_2O), a greenhouse gas that is known to be far more harmful to the environment than CO_2.

Water use

Dairy farms use large amounts of water – approximately 12 per cent of our national water consumption – to produce the milk we all enjoy. Dairy farmers use water to irrigate the pasture and grow the grain that they need to feed their dairy cows, as well as to provide their animals with drinking water. Water is also essential to ensure that the milking sheds are kept clean and hygienic. According to Sustainable Table, an environmental not-for-profit organisation, it takes approximately 800 litres of irrigated water to produce 1 litre of milk.

Soil health

While organic and biodynamic dairy farmers use natural products such as compost and manure to improve the health of the soil and to grow the pasture they need to feed their animals, most dairy farmers rely on synthetic fertilisers and herbicides. However, the use of such fertilisers and herbicides can impact on soil health by destroying many of the microbes that live in the soil, reducing its ability to trap carbon and to grow healthy pasture and crops.

Another concern is that unless grazing is well managed, the soil can be compacted by over-grazing.

Figure 13.10 It takes approximately 800 litres of irrigated water to produce 1 litre of milk.

This reduces the soil's capacity to grow healthy crops and to prevent erosion if pasture cover is reduced.

Dairy farmers must also be aware of the importance of carefully managing the waste stream – especially the dung and urine produced by their dairy herds – as the synthetic fertilisers it contains can wash off or leach into rivers and streams, polluting the waterways.

Sustainable milk production

Dairy farmers and dairy food processors have been aware for years of the necessity to develop more sustainable industry practices in the face of a changing climate. A 2016 report titled *Dairy's (Climate) Changing Future*, prepared by The University of Melbourne, highlights the

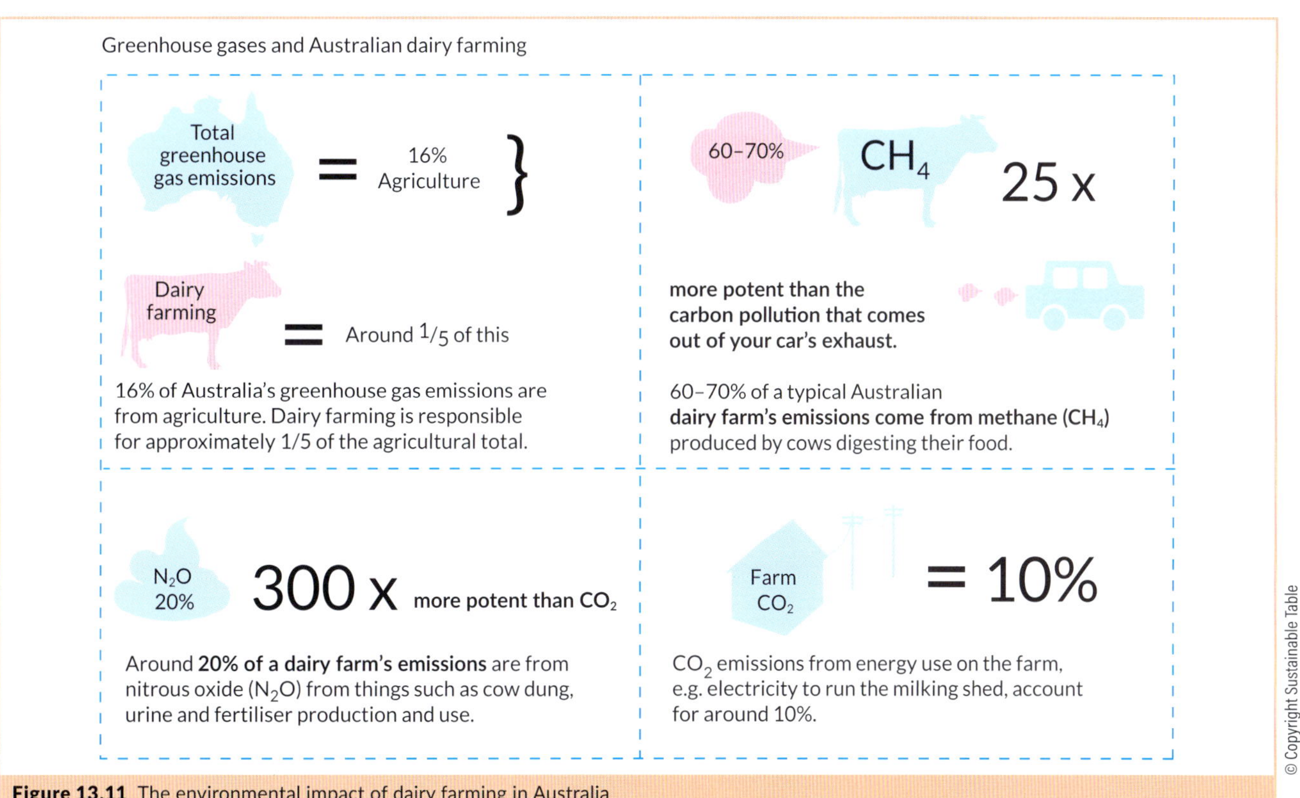

Figure 13.11 The environmental impact of dairy farming in Australia

way Australia's changing climate will impact on the dairy industry. According to this report:

> *By 2040, farmers will have to deal with warmer temperatures and more extreme weather events, while more variable rainfall will see seasons shift and feeding strategies altered. Summers will extend well beyond the usual summer period and dry spells will last longer.*

This will affect animal welfare and pasture growth and as a result, milk production.

In an effort to address the issue of sustainability, the Australian dairy industry has developed the Australian Dairy Industry Sustainability Framework. One of the key features of this framework is to implement strategies to reduce the impact of the dairy industry on the environment.

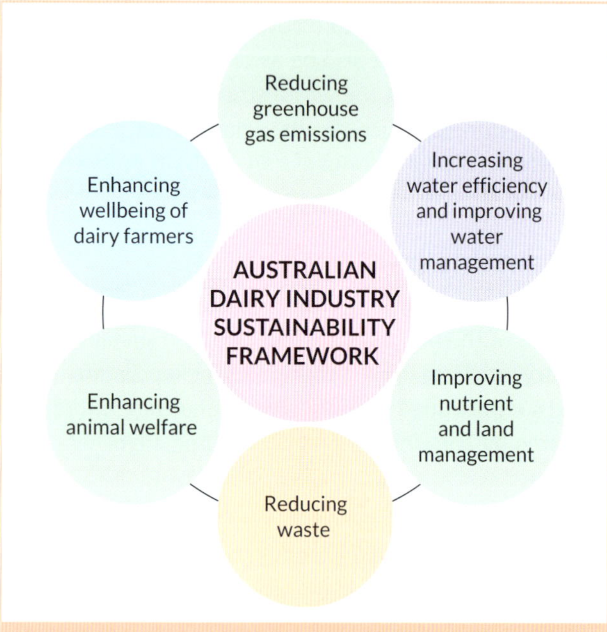

Figure 13.12 Key features of the Australian Dairy Industry Sustainability Framework

Strategies to reduce greenhouse gas emissions

- Aim to reduce greenhouse gas emissions across the industry by 30 per cent.
- Breed dairy cows that produce lower levels of methane in their gut.
- Improve the variety of grains and pastures used to feed animals to reduce the amount of methane and nitrous oxide the animals excrete.
- Provide tools such as the 'Dairy Climate Toolkit' for dairy farmers so that they become more informed and can measure and reduce the emissions on their farms.
- Use environmentally friendly sources of electricity such as wind power, where possible.
- Convert organic waste from dairy processing into power and treated water.

Strategies to increase water efficiency and improve water management

- Reduce the amount of water consumed by dairy companies by 30 per cent.
- Recycle water from the dairy sheds.
- Have all farmers monitor water consumption.
- Have all farmers establish and implement a water security management plan.

Strategies to improve nutrient and land management

- Exclude all stock from waterways.
- Manage and maintain the vegetation along riverbanks.
- All dairy farmers to complete and implement a soil and nutrient management plan.
- Ensure no further net deforestation occurs.
- Implement and document a biodiversity action plan.

Strategies to reduce waste

- Aim for dairy processors to divert all waste sent to landfill by 2030.
- Recycle all polyethylene material used to wrap the silage or fodder used to feed dairy cows.
- Ensure that packaging for all dairy products is compostable, reuseable or recyclable; for example, the polyethylene plastic lining in milk cartons has been replaced with cartons made from renewable resources, including wood fibres and sugar cane.

The Australian dairy industry continues to promote and support research and initiatives to advance the sustainability of the industry and combat the effects of climate change. To learn more, visit the 'Climate and environment' page on the Dairy Australia website.

Weblink Dairy Australia

Figure 13.13 A recyclable milk carton

9780170495714

Milk alternatives

For people who have an allergy or intolerance to dairy products, or for those who wish to avoid consuming animal products for ethical or environmental reasons, there are now a wide range of plant-based products available on the supermarket shelves. These milk alternatives include 'milk' made from soy, almond, coconut, rice and oats.

- **Soy milk** has a high protein content and is quite high in fat. Many soy milks are fortified with calcium.
- **Almond milk** contains monounsaturated fats rather than saturated fat. It is low in kilojoules and contains a moderate amount of protein. Unlike cow's milk, almond milk does not contain any calcium, but it is often fortified with this nutrient.
- **Coconut milk** is very high in saturated fat but does not contain any protein or calcium.
- **Rice milk** is suitable for most people with allergies. It contains very little fat and no saturated fat. It is low in protein, but may be fortified with calcium. Rice milk has a naturally high sugar content.
- **Oat milk** contains only small amounts of fat and saturated fat, and a medium level of protein. Not all oat milk products are fortified with calcium. People who are gluten-intolerant may be allergic to oat milk.

Figure 13.14 Plant-based 'milk' products

Testing knowledge 13.2

11 Using the Y diagram below, discuss the relationship between ongoing drought conditions and the cost of water and fodder for dairy farmers.

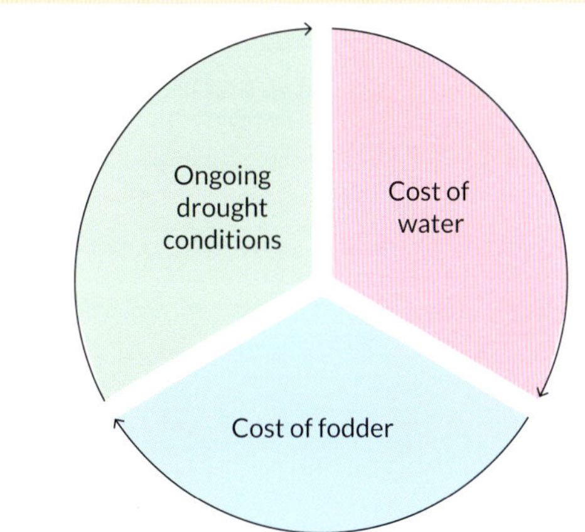

12 Explain how low supermarket milk prices impact on the sustainability of dairy farms.

13 Describe how dairy cows can contribute to greenhouse gas production.

14 Copy the diagram below and use it to complete a summary of the uses of water on a dairy farm.

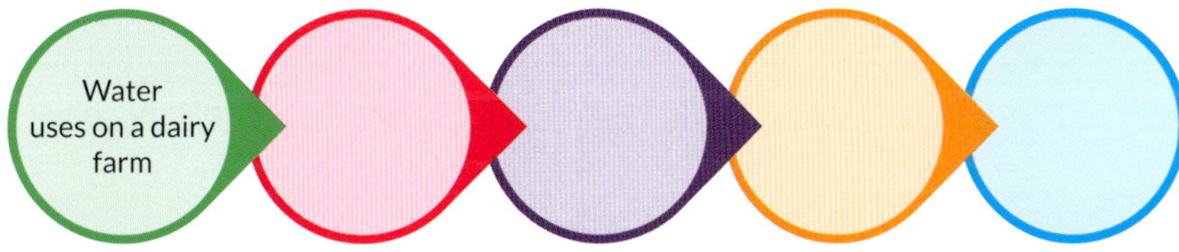

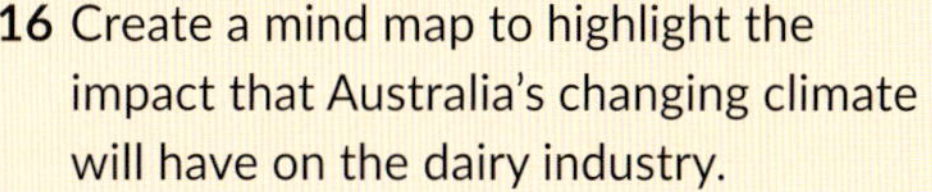

15 Outline two strategies dairy farmers can use to improve the soil health on their farms.

16 Create a mind map to highlight the impact that Australia's changing climate will have on the dairy industry.

17 Explain how dairy farmers could increase the sustainability of their industry by introducing strategies to:

- reduce their greenhouse gas emissions
- reduce waste.

18 Outline two strategies dairy farmers could use to improve nutrient and land management.

19 Select two different types of milk alternatives and explain why some consumers would prefer to consume these rather than cow's milk.

20 Explain why coconut milk may have a detrimental impact on the health of some consumers.

9780170495714

Yoghurt

What is yoghurt?

The word 'yoghurt' is thought to have originated from the Turkish word *yoğurmak*, which means 'to thicken'. To make yoghurt, two cultures are added to whole or skim milk to begin the fermentation process: *Lactobacillus bulgaricus* and *Streptococcus thermophilus*. These cultures convert the lactose (the sugar in milk) to lactic acid, which coagulates the protein and creates the thick consistency typical of yoghurt.

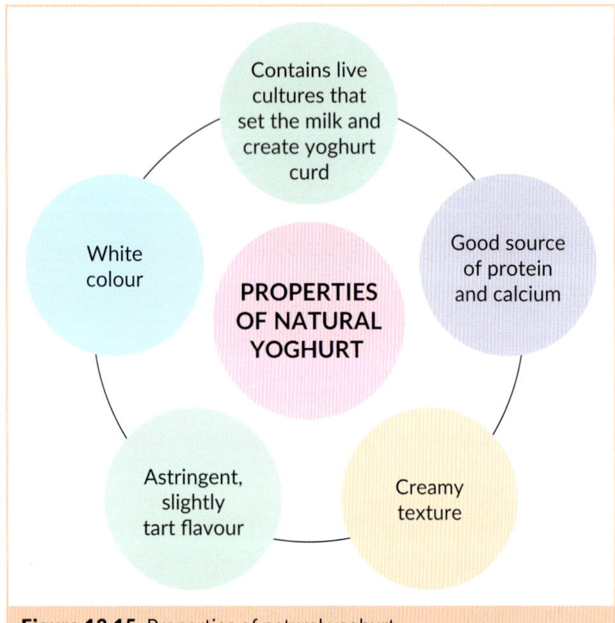

Figure 13.15 Properties of natural yoghurt

In most countries, yoghurt is made from cow's milk, but it can also be made from the milk of other animals, especially goats and sheep. Today, dairy-free yoghurt-style products made from soy and coconut milk are also available.

Nutritional properties of yoghurt

Yoghurt is considered to be a nutrient-rich food, as it is an excellent source of protein and calcium. While most dairy products are not considered suitable for people who are lactose-intolerant, the bacteria that is used to set the milk – or create the yoghurt 'curd' – converts the sugar in milk (lactose) to lactic acid, making it suitable for lactose-intolerant individuals. Yoghurt that contains the probiotic cultures *Lactobacillus acidophilus*, *Bifidus* and *Lactobacillus casei* promote health by increasing the level of intestinal flora – or 'good' bacteria – that live in the digestive system.

Figure 13.16 Yoghurt is an excellent source of protein and calcium.

PRACTICAL ACTIVITY 13.1

Worksheet Practical activity 13.1

Template Sensory wheel

Taste testing yoghurt

1 Taste test a range of different types of vanilla yoghurt. Select four different types of yoghurt, such as:

- full-fat
- reduced-fat
- soy milk
- coconut milk
- goat's milk
- sheep's milk
- almond milk
- yoghurt in a pouch.

Focus on the sensory properties – appearance, aroma, flavour, texture and sound – of each type of yoghurt. Record your results in a table like the one below.

Yoghurt type		Appearance	Aroma	Flavour	Texture	Sound
1						
2						
3						
4						

9780170495714

2 Rate the yoghurts from the one you liked the most to the one you liked the least. Justify your ratings, using your analysis of the sensory properties.

3 Examine the label of each yoghurt, and record the kilojoule, fat, sugar and calcium content per 100 grams.

Yoghurt type		Nutritional content per 100 grams			
		Energy (kJ)	Fat	Sugar	Calcium
1					
2					
3					
4					

4 Draw conclusions about the yoghurt with the highest energy content per 100 grams and the yoghurt with the lowest energy content in relation to the amount of fat and sugar present in each yoghurt.

5 Compare the calcium content of each type of yoghurt. Develop a logical argument to explain the variation in the calcium content you have observed between each type of yoghurt.

6 With a partner, brainstorm and make a list of reasons why there is such a large variety of yoghurts available today.

7 If you were responsible for buying yoghurt for your family, which type would you choose and explain why? When writing your response, consider the sensory properties, nutritional value and the way that yoghurt is used in your home.

Labneh

Labneh is a delicious soft cheese that is made by straining yoghurt through fine muslin. It originated in the Middle East. Labneh is low in kilojoules, spreads easily and is a great substitute for cream cheese. While it is widely available in markets and supermarkets, labneh is easy to make at home.

iStock.com/Mizina

Figure 13.17 Labneh can be used to make a dip.

PRACTICAL ACTIVITY 13.2

Making your own labneh

Worksheet Practical activity 13.2

Ingredients

1 kilogram Greek-style natural yoghurt

300 millilitres olive oil

3 garlic cloves, bruised

3 × strips lemon rind, 2-centimetres thick

2–3 small sprigs fresh rosemary

Method

1 Spoon the yoghurt into a fine sieve lined with muslin and place over a bowl. Make sure the sieve does not touch the bottom of the bowl.

2 Cover the bowl with plastic wrap and place in the fridge overnight to drain. If you prefer a thicker labneh, allow the yoghurt to drain for a longer period. The longer you let it drain, the thicker the labneh will be.

3 Roll the labneh into walnut-sized balls.

4 Place the balls of labneh in a jar. Add the garlic, lemon rind and rosemary, and cover with the olive oil.

Cheese

Cheese is made from **casein**, the protein present in the curds of milk that are separated from the water and lactose, or whey. It is a concentrated form of milk – it takes 1 litre of milk to make 100 grams of cheese. Like milk, cheese is considered to be an important food to include in the diet as it contains a wide variety of essential nutrients, particularly calcium, protein, iodine, vitamin A, vitamin D, riboflavin, vitamin B12 and zinc.

The most commonly used milk for cheese making in Australia is from cows and goats. In parts of the world that have a harsher climate and terrain, the milk from animals that are better suited to these conditions – such as reindeer, yak, horses and water buffalo – is frequently used.

Australians consume approximately 13 kilograms of cheese per person every year, the most popular of which is cheddar-style cheese.

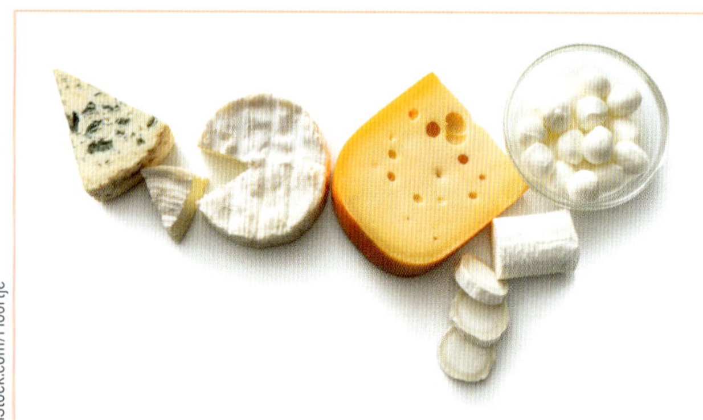

iStock.com/Floortje

Figure 13.18 Cheese contains a wide variety of essential nutrients.

Classification of cheese

Like most other food types, cheese can be classified in a variety of ways:

- the type of milk used to make the cheese; for example, cow's, goat's or sheep's milk
- the way in which the cheese has been ripened; that is, whether it has been surface-ripened, such as Gruyère cheese, or interior-ripened, such as Swiss, cheddar and blue cheeses
- the fat content; that is, full-cream or low-fat
- the firmness or texture of the cheese; that is, hard, semi-soft, soft or fresh. This is the most common method used to classify cheese. The firmness or texture of cheese is determined by the water content, with soft and fresh cheeses having a moisture content of up to 80 per cent and hard cheeses as little as 30 per cent.

Table 13.2 Examples of different types of cheese

Types of cheeses	Examples
Fresh cheeses	Cream cheese, cottage, ricotta, mozzarella, bocconcini, mascarpone
Soft white cheeses	Brie, Camembert
Semi-soft cheeses	Edam, Havati, Colby
Hard cheeses	Cheddar, Jarlsberg, Gouda, Gruyère, Emmenthal, Parmesan, Romano
Processed cheeses	Cheese slices, cheese sticks, flavoured cheeses
Goat's milk cheese	Soft, surface-ripened, unripened

PRACTICAL ACTIVITY 13.3

Worksheet Practical activity 13.3

Comparing cheeses

Cheese is a great source of calcium. For example, as shown in Table 13.3 on page 364, 30 grams of cheddar or tasty cheese contains approximately 240 milligrams of calcium.

Your teacher will arrange to have a variety of different types and brands of tasty cheese products for your class to taste test and compare; for example, a block of tasty cheese, tasty cheese slices, cheese sticks or cheese wedges, and a tasty cheese dip with crackers.

Figure 13.19 A range of different tasty cheese products

9780170495714

Aim

To compare the sensory properties and calcium and sodium content of a variety of tasty cheese products

Method

1 Place labelled samples of each cheese product on a small plate. Create a copy of the table below.
2 Taste test each cheese and fill in the first five rows of the table.
3 Read the labels on the packets to find out how much calcium and sodium is contained in each type of cheese.
4 Complete the table.

Results

	Sample 1	Sample 2	Sample 3	Sample 4
Type of tasty cheese product				
Colour (e.g. creamy white, pale yellow, dull yellow, intense yellow)				
Flavour (e.g. milky, salty, strong, mild)				
Mouth feel or texture (e.g. smooth, creamy, granular, dry)				
Comment (e.g. what you thought of the flavour)				
Milligrams of calcium per 100 grams				
Milligrams of sodium per 100 grams				
Your rating out of 10				

Analysis

1 Which cheese had the best flavour? Explain your choice.
2 Which cheese had the best mouth feel? Explain your choice.
3 Which cheese contained the highest amount of calcium?
4 Identify the cheese with the lowest sodium content and the cheese with the highest sodium content. How did the sodium content impact on the sensory properties of the cheese?
5 Compare the packaging used for each type of cheese product. Which products used the most packaging? Discuss the impact of this packaging on the waste stream and the environment.

Conclusion

1 Which cheese did you prefer overall? Explain your choice.
2 After considering all the cheese samples, which one would you recommend to your friends to eat as a morning or afternoon snack? Justify your decision. Remember to take into account the calcium and sodium content of the products, as well as the packaging used.
3 Recommend two other high-calcium foods that people who do not like cheese could eat for a snack.

Osteoporosis

Osteoporosis is a major health concern for the Australian population. Like many other health issues, it is a condition that only becomes evident in the later stages of life. Osteoporosis affects over 1 million Australians and is more common in both women and men over the age of 50.

Osteoporosis occurs when calcium is lost from the bones, making them very fragile and easily broken. A normal bone has a strong outer shell, but osteoporosis causes the outer shell of the bone to become thin. The internal structure of the bone is also affected, and, instead of having a strong, mesh-like structure, the bone develops large holes, making it thin and weak. The most common osteoporosis-related fractures occur in the hips, spine, pelvis and wrists. People who have osteoporosis often suffer from severe pain, especially in their backs. Height loss and a stooped appearance are other side effects of osteoporosis.

Healthy bones

While osteoporosis does not usually become evident until late adulthood, it is during adolescence and early adulthood that we can take steps to avoid developing this debilitating condition. To minimise our risk of developing osteoporosis, it is essential to develop peak bone mass during our teenage years. Peak bone mass can be achieved by combining a diet high in calcium with significant weight-bearing exercise during adolescence, when we are rapidly growing. Peak bone mass is usually achieved by the age of 18, but we can continue to gain bone mass until the age of around 30.

Protein, calcium, phosphorus and vitamin D are the nutrients that are essential for healthy bones. During adolescence, we need larger amounts of calcium to provide for the considerable increase in the bone structure of our body. Boys aged 8–11 require approximately 800 milligrams of calcium daily, but between the ages of 12 and 15, their need for calcium increases by approximately 50 per cent – to 1200 milligrams daily. This is greater than the amount of calcium needed by girls, mainly because boys will generally grow to be taller than girls. Girls aged 8–11 require slightly more calcium than boys of the same age (900 milligrams daily) because they start their growth spurt earlier; however, they need only 1000 milligrams daily between the ages of 12 and 15.

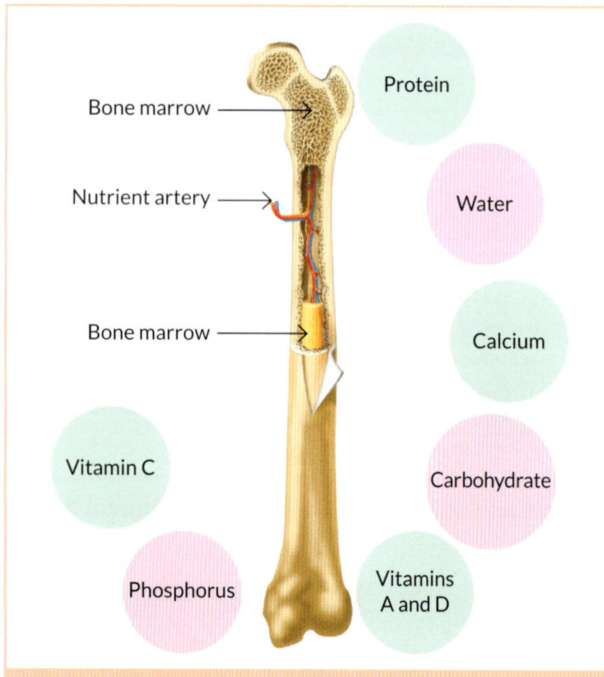

Figure 13.20 Nutrients needed for bone growth

Selecting foods high in calcium

Many adolescents and older Australians do not consume enough calcium to meet their daily needs. Calcium is found in a wide variety of foods, but some calcium sources are better than others. Remember that some foods and additives work against the absorption of calcium, preventing it from passing from the food to the body during digestion. The foods most likely to work against calcium absorption are fats, the fibre in cereals, and some green leafy vegetables. Salt, caffeine (in cola, tea and coffee) and phosphates, which are added to processed foods and drinks, also increase the amount of calcium that is lost through urine.

Table 13.3 The best food sources of calcium

Food source	Amount	Milligrams of calcium
Whole milk	1 cup (250 mL)	310
Fortified milk	1 cup (250 mL)	438
Evaporated milk	1 cup (250 mL)	658
Skim milk	1 cup (250 mL)	310
Chocolate-flavoured milk	1 small carton (300 mL)	348
Fortified soy drink	1 cup (250 mL)	295
Yoghurt	1 small carton (200 g)	255
Cheddar or tasty cheese	1 slice (30 g)	240
Edam cheese	1 slice (30 g)	288
Swiss cheese	1 slice (30 g)	320
Parmesan cheese	40 grams	460
Milk chocolate	6 squares (30 g)	73
Canned salmon (eaten with the bones)	½ cup (125 mL)	325
Sardines	5 small	285
Dried figs	5	150
Baked beans	½ cup (125 mL)	47
Carrot	½ cup (125 mL)	23
Spinach or silverbeet	½ cup (125 mL)	40
Parsley	1 tablespoon (15 mL)	20
Honeydew melon	1 cup (250 mL)	64
Almonds	¼ cup (60 mL)	95

Figure 13.21 Foods high in calcium

Carey Jaman/Shutterstock.com

9780170495714

Testing knowledge 13.3

Quiz
Testing knowledge 13.3

21 Outline the key steps used in the production of yoghurt.

22 Why is yoghurt considered to be a food suitable for people who are lactose-intolerant?

23 What is labneh and why is it considered to be a good substitute for cream cheese?

24 Write a definition for cheese and explain why cheese is considered to be a valuable food to include in our diet.

25 Explain the key differences between fresh cheeses and hard cheeses. Give two examples of each type of cheese.

26 Describe the effect of osteoporosis on the bones.

27 What is 'peak bone mass'? How is it achieved?

28 Why do boys who are 12–15 years old need more calcium than girls of the same age?

29 Make a list of the factors that prevent the body from absorbing calcium properly.

30 Identify three foods that are excellent sources of calcium.

Critical and creative thinking 13.1

Issues in milk consumption

Template
SWOT analysis table

1 Develop a logical argument for an increase in the price of milk compared to the price of bottled water.

2 Prepare a SWOT analysis (Strengths, Weaknesses, Opportunities, Threats) of almond milk. Consider the environmental, health and economic impacts of the production and use of almond milk.

Figure 13.22 Dairy foods deliver nutrients that are essential for healthy bones, especially during adolescence.

triloks/E+/Getty Images

DESIGN ACTIVITY 13.1

Worksheet
Design activity 13.1

Templates
Food order

Sensory wheel

Australian Guide to Healthy Eating

Showcasing Australian dairy ingredients

International Dairy Week is an expo of events held each January at Tatura Park, which is in the heart of dairy country in the Goulburn Valley region of Victoria.

The daily cooking demonstrations that are held in the exhibition hall are always booked out, as visitors are keen to learn about the latest products and food trends that celebrate Australian milk, cheese and yoghurt.

Design brief

One of the daily demonstrations will focus on quick and healthy workday dinners that include milk, cheese and/or yoghurt. The recipes must reflect the principles of the *Australian Guide to Healthy Eating*, so a variety of different-coloured vegetables must be part of the dinner menu, either in the main meal or as a side dish.

The organisers of International Dairy Week have invited members of the community to submit recipes, accompanied by a photograph of their meal, via social media. The creators of the selected recipes will be acknowledged each day and will receive a dairy hamper.

Investigating and designing

1 Based on the solution requirements and constraints in the design brief, develop five design criteria to evaluate the success of your quick and healthy workday dinner that includes milk, cheese and/or yoghurt.

2 Research recipes that meet the requirements of the design brief. The dinner could be a single recipe or a main dish with side dishes. The style of the dinner could be:
- a soup or salad
- a pasta or rice dish
- a grill, pan-fry or stir-fry
- a tray bake.

3 List four recipe ideas that appeal to you in a table like the one below. Remember to add the source of the recipe and the reason why you decided to include each one.

Recipe name and source, and reason for inclusion		Milk, cheese and/ or yoghurt	Vegetables (types, colours, sensory properties)	Style of dinner
1				
2				
3				
4				

Generating and designing

1 Create two quick and healthy workday dinner designed solution options.

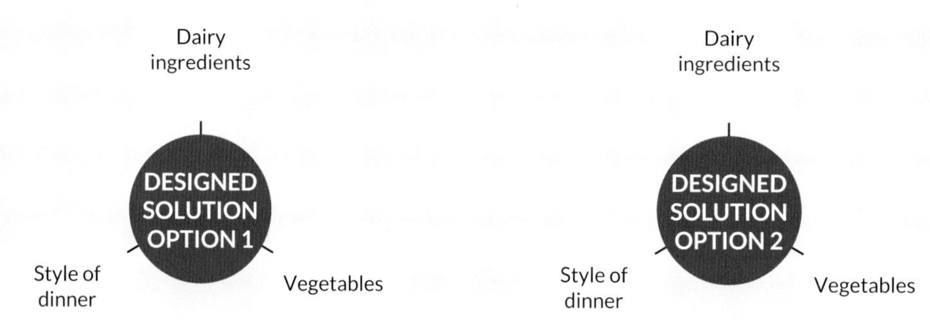

2 Select the designed solution option you prefer and justify your choice. Explain why you did not select the other option as your designed solution. Remember to refer back to the requirements of the design brief when discussing your decision-making.

3 Prepare a copy of your recipe ready for production.

Planning and managing

1 Prepare a food order.

2 Annotate a copy of your recipe to indicate the tools and equipment needed in the production and key safe work practices.

Producing and implementing

1 Produce your quick and healthy workday dinner that includes milk, cheese and/or yoghurt.

2 Photograph the meal before taste testing.

Evaluating

1 Respond to the design criteria that you developed.

2 Describe the sensory properties – appearance, aroma, flavour, texture and sound – of your quick and healthy workday dinner that includes milk, cheese and/or yoghurt.

3 Discuss the effectiveness of your time management for the production.

4 What recommendations would you make to improve the quality of your workday dinner if you were to make it again?

5 Classify the ingredients of your recipe on a diagram of the *Australian Guide to Healthy Eating*. Discuss how well the meal meets the recommendations of this food selection model.

9780170495714

Cheese and Broccoli Nuggets

INGREDIENTS

1 head broccoli (approximately 350–400 grams)

2 spring onions, finely sliced

90 grams cheddar cheese, grated

½ cup dried breadcrumbs

2 tablespoon parsley, chopped

⅛ teaspoon salt and pepper

1 egg, lightly beaten

spray olive oil

To serve

½ cup Greek-style yoghurt

1 tablespoon kasundi (optional)

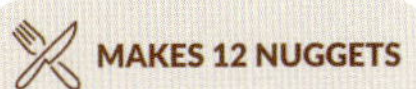
MAKES 12 NUGGETS

METHOD

1. Preheat oven to 200°C. Line a baking tray with baking paper.
2. Cut the head of broccoli into florets. Trim the stem, then cut across the trunk into slices.
3. Half fill a large saucepan with water and bring to the boil. While the water is heating, fill a large bowl with cold water and add some ice cubes so that it is ready to refresh the broccoli.
4. Blanch the broccoli pieces in the boiling water for 1 minute. Do not cover the saucepan, so that the broccoli will retain its bright colour.
5. Drain and refresh the broccoli in the iced water. When cool, drain well.
6. Set up a food processor. Add the drained broccoli, spring onions, cheese, dried breadcrumbs, parsley, and salt and pepper. Pulse the ingredients until they are finely chopped. Add the beaten egg and pulse until it is just incorporated into the mixture.
7. Turn the mixture into a medium bowl.
8. Using 1 heaped dessert-spoonful of mixture (about 40–45 grams) per nugget, shape into 12 nuggets.
9. Place the nuggets on the lined baking tray and spay with olive oil.
10. Bake for 20 minutes or until golden brown.
11. Serve with a side of Greek-style yoghurt and some kasundi for a touch of spice, if desired.

Mark Fergus Photography

EVALUATION

1. Describe the sensory properties – appearance, aroma, flavour, texture and sound – of your Cheese and Broccoli Nuggets.
2. Why is the broccoli only blanched for 1 minute?
3. List the ingredients that contribute to the brown colour of the nuggets after baking. Identify the process that is responsible for the browning.
4. Classify the ingredients for the Cheese and Broccoli Nuggets on a diagram of the *Australian Guide to Healthy Eating* and discuss how well this dish meets the recommendations of this food selection model.
5. Do you think the Cheese and Broccoli Nuggets would appeal to primary school children as a snack food? Justify your answer.

Summer Pasta Salad

Ricotta is a fresh, unripened cheese that has a mild flavour and does not melt when heated. It simply holds its shape when added as part of the topping on this Summer Pasta Salad. Ricotta has no rind and has a soft, smooth and moist texture. Being a fresh cheese, it has a short shelf life. The delicate milky flavour of ricotta complements the subtle flavours of the vegetables and herbs in the salad.

INGREDIENTS

½ teaspoon salt

180 grams dried penne or similar pasta

1 green zucchini

1 clove garlic

125 grams ripe cherry tomatoes

2 tablespoons extra virgin olive oil

½ cup baby spinach leaves

To serve

100 grams ricotta, crumbled

½ cup basil leaves, torn

8 black kalamata olives, halved (optional)

METHOD

1. Fill a large saucepan with water, add the salt and bring to the boil.
2. Add the pasta to the boiling water. Stir gently to ensure that the pasta does not stick to the bottom of the saucepan. Two minutes before the recommended cooking time on the packet is reached, test the pasta. It should be al dente.
3. Retain ¼ cup of the pasta cooking water, drain the pasta and set aside.
4. Cut the zucchini in half lengthwise, then slice into half-moons about 1-centimetre thick.
5. Peel and thinly slice the garlic.
6. Cut the cherry tomatoes in half.
7. Add the oil to a frying pan and turn the heat to medium-high. Add the zucchini and fry for about 5 minutes until lightly browned. Stir occasionally.
8. Add the garlic and tomatoes and fry until the tomatoes have softened. Stir occasionally. Reduce the heat if the zucchini slices are starting to lose their shape.
9. Add the cooked pasta, pasta cooking water and baby spinach leaves to the frying pan and toss to combine.
10. Season with salt and pepper.
11. Transfer the warm vegetable pasta to serving bowls and top with the crumbled ricotta, torn basil leaves and halved olives (if desired).

Mark Fergus Photography

EVALUATION

1. Describe the sensory properties – appearance, aroma, flavour, texture and sound – of your Summer Pasta Salad.
2. Identify the process that takes place when pasta is cooked.
3. What process occurs when the zucchini slices begin to brown during frying?
4. Recommend another type of cheese that could be added to the recipe to intensify the cheesy flavour.
5. Suggest some protein food recipes that could be served with the Summer Pasta Salad so that the resulting main meal can meet the recommendations of the *Australian Guide to Healthy Eating* food selection model.

9780170495714

Cheese and Spinach Pastries

INGREDIENTS

60 grams frozen spinach, defrosted

2 spring onions

50 grams ricotta

50 grams feta cheese

1 egg

pinch nutmeg

4 shakes black pepper

8 sheets filo pastry

¼ cup olive oil

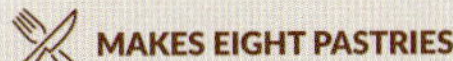
MAKES EIGHT PASTRIES

METHOD

1 Preheat oven to 200°C. Brush an oven tray with melted butter or line it with baking paper.
2 Squeeze the moisture from the defrosted spinach.
3 Finely slice the spring onions and mix with spinach, ricotta, feta cheese, egg, nutmeg and black pepper in a medium bowl.
4 Use a metal spoon to divide the mixture into eight portions within the bowl.
5 Lay one piece of filo pastry on the bench and brush lightly with oil. Fold into thirds, lengthwise.
6 Place one portion of the filling mixture on the lower edge of the pastry. Fold up the pastry to form a right-angled triangle, then continue to fold triangles the length of the pastry.
7 Place the finished pastry on the oven tray and brush with oil. Repeat the folding process to make eight pastries.
8 Bake for approximately 10 minutes or until a pale golden colour.

EVALUATION

1 Describe the sensory properties – appearance, aroma, flavour, texture and sound – of your Cheese and Spinach Pastries.
2 How were you able to prevent the filo pastry from drying out while you were preparing the Cheese and Spinach Pastries?
3 Explain the function of the egg in the filling of the Cheese and Spinach Pastries.
4 Classify the ingredients for the Cheese and Spinach Pastries on a diagram of the *Australian Guide to Healthy Eating* and discuss how well this dish meets the recommendations of this food selection model.
5 Explain why this recipe would be a better choice as a healthy snack than a sausage roll.

Mark Fergus Photography

9780170495714

Light Spinach and Ricotta Cannelloni

INGREDIENTS

Tomato Sauce

¼ onion, finely diced

2 cloves garlic, crushed

1 tablespoon olive oil

200 millilitres tomato passata, sugo or puree

¼ teaspoon sugar

pinch salt

Spinach and Ricotta Filling

50 grams frozen spinach, defrosted

120 grams ricotta

1 tablespoon grated parmesan cheese

⅛ teaspoon grated nutmeg

black pepper

White Sauce

10 grams butter

15 grams cornflour

150 millilitres low-fat milk

½ cup flat leaf parsley, finely chopped

To assemble the dish

2 fresh lasagna sheets

1 tablespoon grated parmesan cheese

Mark Fergus Photography

METHOD

Preheat oven to 180°C.

Tomato Sauce

1 In a small saucepan, sauté the onion and garlic in the olive oil.
2 Add the tomato passata, sugo or puree, sugar and salt and simmer for 3 minutes, then remove from the heat. Pour into an ovenproof dish or foil container (20 × 16 × 5 centimetres).

Spinach and Ricotta Filling

1 Squeeze the moisture from the defrosted spinach.
2 Combine the spinach, ricotta, parmesan, nutmeg and pepper in a medium bowl and mix well.

White Sauce

1 In a small saucepan, melt the butter over low heat, add the cornflour and stir for 30 seconds. Remove from the heat and gradually stir in the milk, ensuring that all the lumps have been removed.
2 Return the saucepan to the heat and bring to the boil, stirring all the time. Remove from the heat and stir through the chopped parsley. Set aside.

To assemble the dish

1 Cut the lasagna sheets in half and divide the spinach and ricotta filling between the four pieces of pasta. Spread with the filling and roll up. Place the four filled cannelloni onto the tomato sauce in the ovenproof dish or foil container.
2 Pour over the white sauce and sprinkle with the parmesan cheese.
3 Bake for 20 minutes.

EVALUATION

1 Describe the sensory properties – appearance, aroma, flavour, texture and sound – of your Light Spinach and Ricotta Cannelloni.
2 Identify the process that takes place when making the white sauce and explain how it changes the viscosity of the mixture.
3 Identify the ingredients in this recipe that are a good source of calcium.
4 What types of cheese are ricotta and parmesan? Describe the similarities and differences between the properties of these two cheeses.
5 Classify the ingredients for the Light Spinach and Ricotta Cannelloni on a diagram of the *Australian Guide to Healthy Eating*. Explain how well this dish meets the recommendations of this food selection model.

9780170495714

Cheesy Pasta and Broccolini Bake

INGREDIENTS

150 grams penne pasta

2 teaspoons olive oil

½ onion, finely diced

2 rashers bacon, finely diced

3 stems broccolini, cut into 3-centimetre lengths

250 grams ricotta

1 egg

125 millilitres full-cream milk

¼ cup parmesan cheese, grated

1 tablespoon parsley, chopped

salt and pepper

40 grams cheddar cheese, grated

SERVES TWO

METHOD

1 Preheat oven to 180°C.

2 Lightly grease two small ovenproof dishes or foil containers.

3 Bring a large saucepan of water to the boil.

4 Add the penne to the boiling water. Stir once or twice. Cook the penne for approximately 12 minutes or until al dente. To test for al dente, bite a piece of pasta. It should be firm, but not hard. You should not be able to feel or see the hard centre. Drain and keep warm.

5 Add the olive oil to a non-stick frying pan over medium heat, then fry the onion and bacon for 4–5 minutes or until soft and just beginning to brown.

6 Half fill the base of a steamer with water and bring to the boil. Place the broccolini pieces into the top of the steamer and steam for 4 minutes.

7 Mix the ricotta, egg, milk, parmesan cheese, parsley and salt and pepper in a medium bowl until well combined.

8 Place a quarter of the pasta in the base of each of the two ovenproof dishes. Spread with half the onion and bacon mixture and half the broccolini.

9 Top with a quarter of the ricotta and egg mixture. Divide the remaining penne between the two dishes. Pour the remaining ricotta and egg mixture over the top.

10 Sprinkle with the cheddar cheese.

11 Bake for 20–25 minutes or until golden on the top.

Mark Fergus Photography

EVALUATION

1 Describe the sensory properties – appearance, aroma, flavour, texture and sound – of your Cheesy Pasta and Broccolini Bake.

2 Why is it important to add the penne to boiling water? What process is taking place in the penne when it is boiled?

3 Explain why an egg is added to the sauce in this recipe.

4 Refer to Table 13.3 on page 364 that shows sources of calcium. Calculate the amount of calcium this recipe would provide for one person.

5 Classify the ingredients for the Cheesy Pasta and Broccolini Bake on a diagram of the *Australian Guide to Healthy Eating* and discuss how well this dish meets the recommendations of this food selection model.

Lamb Kofta Wraps with Herbed Yoghurt

INGREDIENTS

Lamb Koftas

2 tablespoons fine burghul

1 tablespoon olive oil

¼ red onion, finely diced

1 clove garlic, crushed

200 grams lean minced lamb

½ egg

½ teaspoon ground cumin

½ teaspoon ground coriander

½ teaspoon dried mint

¼ teaspoon chilli flakes

½ lemon, zested and juiced

⅛ teaspoon salt

small quantity of extra olive oil, for frying

Herbed Yoghurt

2 tablespoons mint leaves

2 tablespoons coriander leaves

2 tablespoons flat leaf parsley

½ clove garlic, crushed

2 teaspoons extra virgin olive oil

1 tablespoon tahini

60 grams Greek yoghurt

½ lemon, juiced

salt and pepper

To serve

2 wholegrain flatbreads

1 Lebanese cucumber, finely sliced

2 leaves iceberg lettuce, shredded

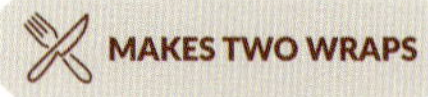
MAKES TWO WRAPS

Mark Fergus Photography

METHOD

Lamb Koftas

1. Place the burghul in a small bowl, cover with cold water and soak for 10 minutes. Drain well through a fine sieve.
2. Add the olive oil to a small frying pan and sauté the onion and garlic over a medium heat for 3–4 minutes or until softened. Do not brown. Cool.
3. Combine the minced lamb, egg, drained burghul, cooked onion and garlic, cumin, coriander, mint, chilli flakes, lemon zest and juice and salt in a medium bowl. Mix well.
4. Shape the lamb mixture into 10–12 walnut-size oval balls. Cover and refrigerate for 15–20 minutes.
5. Heat the extra oil in the frying pan over medium heat and fry the Lamb Koftas, turning regularly to brown them evenly. Drain on paper towel.

Herbed Yoghurt

1. Roughly chop the herbs, then add them to a small food processor along with the garlic and olive oil. Process to combine.
2. Add the tahini, yoghurt and lemon juice and pulse to combine.
3. Season with salt and pepper.

To serve

Serve the Lamb Koftas with the wholegrain flatbreads, Herbed Yoghurt, sliced Lebanese cucumber and shredded iceberg lettuce.

EVALUATION

1. Describe the sensory properties – appearance, aroma, flavour, texture and sound – of your Lamb Kofta Wraps with Herbed Yoghurt.
2. Identify the process that occurs during the frying of the koftas. Explain how the changes in the sensory properties occur.
3. What is the main ingredient in tahini and why is it included in the meat section of the *Australian Guide to Healthy Eating*?
4. Discuss why serving a yoghurt-based sauce with the other components of the recipe is valuable to the diets of adolescent people.
5. Classify the ingredients for the Lamb Kofta Wraps with Herbed Yoghurt on a diagram of the *Australian Guide to Healthy Eating* and explain how well this dish meets the recommendations of this food selection model.

9780170495714

Lemon Olive Oil Cake

INGREDIENTS

melted butter, for greasing cake tin

2 tablespoons self-raising flour, for dusting cake tin

½ cup (100 grams) caster sugar

1 lemon, zested

2 eggs

100 millilitre extra virgin olive oil (mild flavour)

½ cup (125 grams) full-fat Greek yoghurt

1 cup (125 grams) self-raising flour

¼ teaspoon fine salt flakes

1 tablespoon icing sugar for dusting

METHOD

1. Brush the walls and base of a 20-centimetre ring tin generously with melted butter. Line the base of the tin with baking paper, then brush the paper with more melted butter. Add the flour and move the tin around to coat it with a fine layer of flour. Tap out the excess flour. Set the tin aside.
2. Preheat oven to 180°C (fan-forced).
3. Place the sugar and lemon zest in a large bowl. Use your fingertips to rub the zest into the sugar. The mixture will become damp and pale yellow.
4. Add the eggs to the bowl and, using a hand-held beater, beat until the mixture is foaming and light in colour. It is ready when you can lift the beater and create a figure 8 in the mixture that does not immediately disappear.
5. Add the olive oil and beat until it is incorporated, then beat in the yoghurt.
6. Sift the flour and salt into the bowl and beat until they are just combined. Do not overbeat. If any flour is still visible, use a large metal spoon or spatula to fold it into the batter, rather than beating.
7. Pour the cake batter into the prepared tin and bake for approximately 20–25 minutes. Test with a skewer – it will come out clean when the cake is ready.
8. Allow the cake to cool in the tin for 5 minutes, then turn it out onto a cake rack to finish cooling.
9. Use a fine sieve to dust the cake with icing sugar before serving.

EVALUATION

1. Describe the sensory properties – appearance, aroma, flavour, texture and sound – of your Lemon Olive Oil Cake.
2. Explain how rubbing the zest into the sugar impacts on the flavour of the finished cake.
3. Create a mind map to illustrate the role of the eggs in making the Lemon Olive Oil Cake.
4. Why is it important to allow the cake to rest in the tin for 5 minutes before turning it out onto a cake rack?
5. Explain why, according to the recommendations of the *Australian Guide to Healthy Eating*, cakes and biscuits should be eaten 'only sometimes and in small amounts'.

WHY MINIMISE JUNK FOODS?

MONOUNSATURATED FATS

Monounsaturated fats are a good source of omega-3 and omega-6 fatty acids

SATURATED FATS

Saturated fats are high in cholesterol

TRANS FATS

Trans fats raise the risk of heart disease

WHY CHOOSE FAIRTRADE CHOCOLATE?

9780170495714

TIPS FOR MAKING YOUR OWN HEALTHY SNACKS

Adjust the recipe to reduce the amount of sugar the snack contains

Use a vegetable-based spread for a healthier option

Swap white flour for wholemeal flour to increase the fibre content

Add fruit or vegetables to improve the nutritional properties

Use herbs and spices for a more interesting flavour

14 ONLY SOMETIMES!

KEY TERMS

discretionary foods foods that are high in kilojoules and do not provide any of the essential nutrients needed for good health

Fairtrade a movement developed to assist farmers, including cocoa farmers, to produce their crops in a sustainable manner and to have equity in trading their products

free sugar any sugar that is added to ultra-processed foods such as confectionery, fruit juices, sugary drinks, cakes and pastries

monounsaturated fats fats found in olives, olive oil, avocados and nuts that have been shown to reduce blood cholesterol levels

saturated fats fats found mainly in foods of animal origin, such as meat, cheese and butter, that are linked to raised cholesterol levels; coconut oil and palm oil are also high in saturated fats

trans fats 'bad' fats that can lead to serious health concerns and should be avoided; they are found mainly in hydrogenated vegetable oil used by food manufacturers in processed and fast foods

Templates
- Decision table
- Food order
- Production plan
- Sensory wheel
- Australian Guide to Healthy Eating

Worksheets
- Practical activity 14.1
- Practical activity 14.2
- Practical activity 14.3
- Practical activity 14.4
- Design activity 14.1

Quizzes
- Testing knowledge 14.1
- Testing knowledge 14.2

To access resources above, visit **cengage.com.au/nelsonmindtap**

9780170495714

Only sometimes and in small amounts

Foods that are high in saturated fats, added sugar or salt are often described as **discretionary foods** because they are not essential to good health. As a result, they are placed at the bottom of the *Australian Guide to Healthy Eating*, underneath the five key food groups needed for good health.

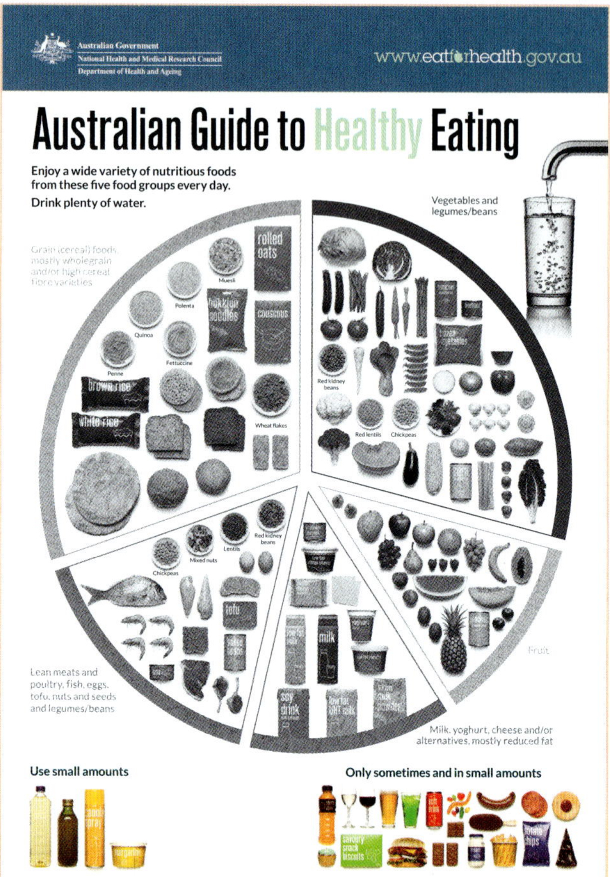

Figure 14.1 The place of discretionary foods in the *Australian Guide to Healthy Eating*

Why use small amounts or only sometimes?

Australian Dietary Guideline 3 states that we should limit foods 'containing saturated fat, added salt [and] added sugars'. However, including some unsaturated fats in our diet is important because they are a source of essential fat-soluble vitamins. The *Australian Guide to Healthy Eating* recommends that we use small amounts of polyunsaturated and monounsaturated spreads and oils in our diet as a source of these important nutrients.

According to the Eat for Health program, it is necessary to limit the amount of discretionary foods we include in our diet as they are high in kilojoules and do not provide any of the essential nutrients needed for good health. Another major concern is that discretionary foods – such as cakes, biscuits, doughnuts, soft drinks and burgers – often displace other more nutritious foods from the diet. Research shows that many children get almost 41 per cent of their energy needs from discretionary foods.

Visit the Eat for Health website and access the 'Average recommended number of serves calculator' to explore this topic more.

Weblink Eat for Health

Table 14.1 What amount of unsaturated spreads and oils can be included in a healthy diet?

Age	Serves per day
Children 3–12 years	1
Children 12–13 years	1½
Adolescents 14–18 years	2

Figure 14.2 What is a serve of unsaturated spreads and oils?

Table 14.2 How much of the discretionary foods group can be included in a healthy diet?

Age	Serves per day
Children up to 8 years	No more than ½ serve
Children and adolescents who are more active	0–2½
Older adolescents who are still growing and/or are very active	0–3

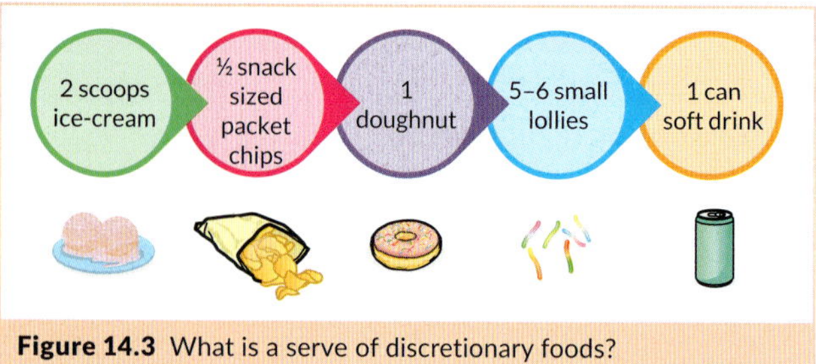

Figure 14.3 What is a serve of discretionary foods?

9780170495714

Fat

Fat is one of the main nutrients that provides people with energy. While it is true that we should limit the amount of fat we consume, not all fats are bad. We need some fat in our diet, because it provides the essential fat-soluble vitamins A, D, E and K. These vitamins are vital for many aspects of good health, including encouraging healthy skin and bone growth, and preventing blood clotting. Rather than eliminating fat completely from our diet, it is more important to select foods that contain good fats.

Monounsaturated fats

Monounsaturated fats are the 'good' fats. They can be found in olives, olive oil, avocados and nuts. Consuming foods such as oily fish – for example, tuna or salmon – is also encouraged, because they contain the essential fatty acids omega-3 and omega-6.

iStock.com/JulijaDmitrijeva

Figure 14.4 Foods containing monounsaturated fats

Saturated fats

Saturated fats are 'bad' fats. We should try to limit the amount of these fats that we consume. Saturated fats are found in animal foods such as meat and dairy products, as well as in palm oil and coconut oil.

Saturated fat can be a problem in the diet because it is high in cholesterol and can lead to heart disease. Many processed foods and fast foods – including biscuits, cakes, pastries, burgers, pizza, fried foods, potato chips and processed meats – are extremely high in saturated fat, and, if eaten in excess, can lead to overweight and obesity.

iStock.com/LauriPatterson

Figure 14.5 Hamburgers and chips contain saturated fats.

Trans fats

Like saturated fats, **trans fats** are 'bad' fats. They can lead to serious health concerns and should be avoided. Trans fats increase the level of the 'bad' low-density lipoprotein (LDL) cholesterol in our bloodstream. However, they also have the added problem of reducing the body's 'good' high-density lipoprotein (HDL) cholesterol, which helps to protect us from heart disease.

While trans fats are found naturally in small amounts in dairy products and meat, their main source in our diet is from the hydrogenation of vegetable oils. Food manufacturers use these oils in the manufacture of processed foods and fast foods, such as pastries, chicken nuggets, hamburgers and fried foods.

Phairoh chimmi/Shutterstock.com

Figure 14.6 Foods containing trans fats

Table 14.3 Characteristics, sources and health impacts of fats and oils

	Monounsaturated fats	Saturated fats	Trans fats
Characteristics	• Liquid at room temperature • Mainly vegetable sources • Fat-soluble vitamins A, D, E and K	• Solid at room temperature • Animal and vegetable sources • Concentrated energy source: 1 g = 37 kJ	• Semi-solid at room temperature • Unsaturated fats that behave like saturated fats because of their chemical structure
Sources	• Olive oil, peanut oil, canola oil	• Butter, cheese, cream, egg yolks, fat on meat • Coconut and palm oil, chocolate, hydrogenated fats	• Hydrogenated vegetable oils, processed foods
Health impacts	• Lower levels of blood cholesterol	• Increase the risk of developing cardiovascular disease • Raise levels of blood cholesterol	• Raise levels of blood cholesterol by increasing bad LDL cholesterol • Reduce good HDL cholesterol, increasing the risk of heart attack and heart disease

Sugar

Sugar, like other forms of carbohydrate, is one of the body's main sources of energy. Sugar can occur naturally in foods; for example, as fructose in fruit and as lactose in milk. Some vegetables also contain sugar in the form of glucose.

iStock.com/franny-anne

Figure 14.7 Fruit contains the natural sugar, fructose.

However, most of the sugar that Australians consume is added as 'free sugar' during food processing. **Free sugar** is any sugar that is added to ultra-processed foods such as confectionery, sugary drinks, cakes and pastries. These sugars are described as 'free' because they are not contained inside the cells of the food we eat, as they are, for example, in fruit, vegetables and milk. The sugar found in these foods is consumed along with the other nutrients the food naturally contains, such as fibre, protein or calcium. However, when fruit is processed into fruit juice and the structure of the fruit is broken down, the sugar becomes 'free'.

Food manufacturers add sugar to many food products to improve the flavour and colour, and to increase the product's bulk and viscosity. Ultra-processed foods – such as cakes, biscuits, ice-cream, pastries, flavoured yoghurt, breakfast cereals and soft drinks – all contain high levels of added sugar. Even savoury products such as tomato sauce, frozen meals, pasta sauces and canned soups contain significant amounts of added sugar.

Consumers often find it difficult to identify whether sugar has been added to a processed food, making it difficult to make healthy and informed food choices. There are over 40 types of ingredients that are a form of sugar in processed food, including agave nectar, barley malt, cane sugar, caster sugar, demerara, fruit juice concentrate, palm sugar, panela, powdered sugar and rapadura. Therefore, it is important to read the food label on a processed food carefully and check the nutritional panel with the amount of sugar in it before placing the product in your shopping trolley.

While the *Australian Dietary Guidelines* state that it is essential to limit foods containing sugar, a small amount of sugar can add interest, variety and enjoyment to our daily diet. Foods that contain natural sugars, such as fruit and milk, provide the body with additional nutrients and are called 'nutrient-dense'. In contrast, many ultra-processed snack foods that contain free sugar provide little nutritional benefit and contain only 'empty kilojoules'.

Health professionals recommend that only 10 per cent of our total energy intake should come from sugar. Excess consumption of ultra-processed foods containing free sugar can contribute to excess weight gain and lead to obesity and tooth decay.

Fat or sugar: What's the difference?

Many people are confused about whether there is a difference between fat and sugar and seem to think that they are the same thing! The truth is, they are completely

9780170495714

Figure 14.8 A range of food products containing significant amounts of sugar

iStock.com/monticello

different. Fat and sugar come from different food sources, and have different functional properties and flavours. Fat is a lipid, which is another name for fat, while sugar is a form of carbohydrate. Carbohydrates, including sugars, are more easily absorbed into the body, but any that are not used to produce energy are stored in the body as fat.

Food manufacturers often combine fat and sugar in food products to make them more appealing to consumers. If you tried to eat a tablespoon of fat or sugar on its own, you would find it very unappetising. But when fat and sugar are combined, they become very palatable – and in some cases quite blissful. Just think about eating chocolate!

Food manufacturers combine fat and sugar in many food items – such as chocolate, sweet biscuits, cakes, muffins and even hamburgers – to make them appealing to consumers. This means that the sales of these items increase and, as a result, so do the manufacturers' profits. However, the problem for the consumer is that eating too many foods that are high in fat and sugar can add a large number of unwanted kilojoules to our daily intake.

Salt

Salt is a mineral that is made up of sodium and chloride. These two electrolytes are important for good health as they help to regulate the amount of blood circulating in our body and the fluids moving in and out of our tissues. They also help our nerves to transmit messages. Children aged 9–13 years need 400–800 milligrams of sodium per day, while adolescents aged 14–18 years need 460–920 milligrams per day. However, sodium can be bad for our health if we consume more than our body needs.

The average Australian consumes approximately 10 grams – or more than two teaspoons – of sodium per day. The problem with consuming too much sodium is that it can lead to poor health outcomes, including increased blood pressure and therefore increased risk of developing a stroke. A high sodium intake can also lead to heart and kidney disease, and increase our risk of developing osteoporosis.

Most people are unaware of the amount of salt or sodium they consume as it can be 'hidden' in many of the processed foods we enjoy. Some of the processed foods that are high in added salt include bread and bread rolls, processed meat such as ham and bacon, canned soups and vegetables, frozen meals, pizza, tomato sauce, soy sauce and some breakfast cereals. The Eat for Health program recommends that we should always read the label on processed foods and select those that have no more than 120 milligrams of sodium per 100 grams, and definitely no more than 400 milligrams per 100 grams.

Eating out can also be a problem, as many of the foods we buy in fast-food restaurants – such as hamburgers, pizza, pasta, fish and chips, and some Asian foods – can be especially high in sodium.

Figure 14.9 Fish and chips are high in sodium when sprinkled with sea salt.

TheLiftCreativeServices/Shutterstock.com

Testing knowledge 14.1

1. Outline the place of discretionary foods in the *Australian Guide to Healthy Eating*.
2. Create a mind map to highlight the reasons why the amount of discretionary foods included in the diet should be limited.
3. Explain why it is important to include some fat in our diet. List four sources of monounsaturated fats.
4. Outline the health problems associated with eating discretionary foods, such as cakes, pastries, burgers, pizza, fried foods and potato chips.
5. What are trans fats and why are they considered to be bad for our health?
6. Identify the main types of sugar that can occur naturally in food.
7. Explain why food manufacturers add sugar to many food products. List six names that can be given to sugar added to processed food.
8. Explain why health professionals recommend we limit our consumption of ultra-processed snack foods.
9. Outline the main differences between fat and sugar.
10. Copy and complete the diagram below by listing the important functions of salt in the body and the problems associated with an over-consumption of salt.

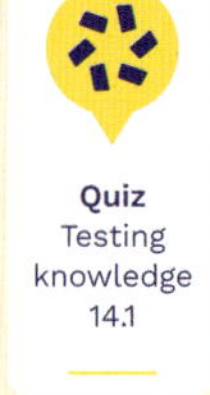

Quiz Testing knowledge 14.1

The importance of salt for good health

Problems with the over-consumption of salt

Shining a light on snack foods and sugary drinks

One of the common characteristics of being an adolescent is that you can constantly feel hungry. Adolescence is one of the periods of most rapid physical growth, during which your body demands food to satisfy its need for energy. Sometimes, having three meals a day just isn't enough!

A food habit that is common with many teenagers is regular snacking. The problem is that snack foods and sugary drinks may come to replace more nutritious foods in the diet. Snack foods and sugary drinks are generally low in important nutrients, but are usually high in fat, salt and/or sugar, and consequently provide a poor source of fuel for the body. They are also high in kilojoules and can contribute to an increase in weight. Therefore, it is important to remember that snack foods and sugary drinks should be considered as 'extra foods' that should only be eaten sometimes and in small amounts.

While it may seem that snacking can be a health hazard, it is possible to snack on foods that are lower in fat, sugar and salt, such as a piece of fruit, a handful of nuts, rice paper rolls or a salad wrap.

beats1/Shutterstock.com

Figure 14.10 Snack foods are usually high in fat, salt and/or sugar and therefore in kilojoules.

9780170495714

PRACTICAL ACTIVITY 14.1

Worksheet
Practical activity 14.1

Product analysis: What's in snack food?

Many snack foods that students eat during study breaks are high in fat, sugar and salt. These ingredients can come in many forms and are included in countless manufactured products.

The appearance of the terms in Table 14.4 on a label show that the food contains these ingredients, or their equivalents.

Table 14.4 How to read the ingredient list to choose healthier foods

Saturated and other added fats	Sugars	Salt
Note: terms such as 'oven fried' and 'baked' or 'toasted' imply that fat has been used during food preparation	Note: look for ingredients ending in '-ose' or '-tol'	
• beef fat • butter • shortening • coconut • coconut oil or palm oil • copha • cream • dripping • lard • mayonnaise • sour cream • vegetable oils and fats • hydrogenated oils • full-cream milk powder • egg (cholesterol) • mono-, di- or triglycerides	• brown sugar • corn syrup • deionised fruit juice • dextrose • disaccharides • fructose • fruit juice concentrate/fruit paste • glucose • golden syrup • honey • lactose • malt • maltose • mannitol • maple syrup • molasses • monosaccharides • raw sugar • sorbitol • sucrose • xylitol	• baking powder • booster • celery salt • garlic salt • sodium • meat or yeast extract • onion salt • monosodium glutamate (MSG) • rock salt • sea salt • seasoning • sodium bicarbonate • sodium metabisulphate • sodium nitrate/nitrite • stock cubes

1 Collect a range of snack foods, such as single serves of potato crisps, savoury shapes, muesli bars, chocolate bars, noodles and low-fat yoghurt.

2 Analyse each product's label and record the following information in a table like the one below.

Name of snack food:			
Serving size:			
	Types of fat	Types of sugar	Types of salt
Total per serve			
Total per 100 grams			

3 Identify the types of fat, sugar and salt most commonly used in the snack foods you investigated.
4 Discuss why food manufacturers use a wide range of fats, sugars and salts in the preparation of snack foods.
5 Explain why it is important that consumers look at the total amount of these ingredients, rather than rely on their identification in the ingredients list.
6 When comparing the fat, sugar and salt content of snack foods, explain why it is important to refer to the amount per 100 grams, rather than rely on the amount per serve.
7 Why is it important to look at the nutrient content of the whole food, rather than make your decision based on one nutrient alone?
8 Draw conclusions about whether including this information about the different names for fat, sugar and salt on a label is a useful tool for consumers when selecting snack foods in the future. Justify your answer.

PRACTICAL ACTIVITY 14.2

Product analysis: Comparing snack foods

Aim

To compare the nutrient content of a range of snack foods

Method

1 Identify three different snack foods that you enjoy. You can select any of the items from the diagram below, or any other foods that you like to snack on.
2 Identify a commercial brand of each of the snack foods you have selected.
3 Research the company's website for information on each of the characteristics listed in the table on the following page. Alternately, you can use the Australian Food Composition Database on the Food Standards Australia New Zealand (FSANZ) website.

Worksheet Practical activity 14.2

Weblink Australian Food Composition Database

9780170495714

4 Record your findings in a table like the one below.

Results

Characteristics	Items to be compared		
	Snack food 1	Snack food 2	Snack food 3
Fat content			
Salt content			
Sugar content			
Kilojoule content			

Analysis

1 When comparing the nutrient content of foods, is it more accurate to compare the portion size or the content per 100 grams of each food? Justify your answer.

2 Discuss the similarities and differences for each characteristic of the three snack foods you have compared.

3 Explain the implications for your long-term health of consuming each of these snack foods on a regular basis.

Conclusion

After comparing the fat, salt, sugar and kilojoule content of the three snack foods, what recommendations would you make to consumers with regard to the long-term health effects of selecting these types of snack foods?

Activity 14.1

How many snack/junk foods do you consume on a typical day?

Visit the website of the LiveLighter campaign and access the Junk Food Calculator.

1 Complete the Junk Food Calculator quiz to calculate the amount of snack/junk food you consume each week.

2 Record the amount of fat, sugar and salt you consume each week in snack/junk foods in a table like the one shown. Compare your snack/junk food consumption with the recommendations of the *Australian Dietary Guidelines*.

3 Draw conclusions about the impact on your long-term health of continuing to consume this level of snack/junk food on a weekly basis with reference to the amount of fat, salt and sugar they contain.

4 Examine the recipes for healthy snacks and drinks on the LiveLighter website. Select one recipe and explain why this would be considered a healthy snack.

Weblink
LiveLighter: Junk Food Calculator

	My weekly consumption	*Australian Dietary Guidelines* recommended consumption
Fat		
Sugar		
Salt		

CASE STUDY 14.1

Promoting junk food

Read the article below, and then answer the questions that follow.

Junk food is promoted online to appeal to kids and target young men, our study shows

Tania Northcott and Christine Parker, The Conversation, 12 July 2024

The Australian government has been investigating whether we should ban unhealthy food advertising online, and how it could work. In the United Kingdom, a ban on unhealthy food and drink advertising online will start in October 2025.

We recently used the Australian Ad Observatory to investigate targeted junk-food ads on Facebook in Australia. Our study finds that unhealthy food and drinks are promoted in ways designed to appeal to parents and carers of children, and children themselves. Additionally, young men in our study were being targeted by fast-food ads.

Kids, young people and parents should be aware of the strategies online advertisers use to normalise unhealthy eating patterns. We should all demand a more healthy digital environment.

What did we see in the ads?

The Australian Ad Observatory has created the world's largest known collection of the targeted ads people encounter on Facebook. Our 1909 volunteers have donated 328 107 unique ads from their social media feeds.

We searched the database for ads promoting the top-selling unhealthy food and drink brands. These are 'discretionary' or 'sometimes' foods that tend to be high in fats and sugars. They include fast-food meals, confectionery, sugary drinks and snacks.

We also looked at online food delivery companies because of their popularity on digital platforms. They play a likely role in promoting unhealthy foods.

…

Fast-food giants KFC and McDonald's combined accounted for roughly 25% of all unhealthy food ad observations. Snack and confectionery brands, like Cadbury, featured in a third of the ad observations. Soft drink brands such as Coca-Cola were promoted in 11% of observations.

About 9% of ads promoted online food delivery companies, and typically promoted fast-food options. Other advertisers we might not think of as junk food brands, such as Coles supermarkets and 7-Eleven convenience stores, also regularly promoted junk foods.

The power of junk food

The vulnerability of children to junk food ads is well established. Children's exposure to food marketing has been associated with what types of food they prefer and ask their parents to purchase. When they develop preferences for unhealthy foods, this contributes to unhealthy habits and related health concerns.

But it's not only children who are susceptible to unhealthy food marketing. Junk food advertising also shapes the food norms and attitudes of young people aged 18 to 24.

Our study shows junk food advertising is disproportionately served to young people, especially young men. Young men are seeing a much higher proportion of fast food ads (71%) compared to the sample overall (50%), suggesting fast food is marketed to them more aggressively. Many ads promoted special 'app-only' deals, including free delivery, especially for fast food.

The 'halo effect'

We also found examples of ads aimed at busy parents, painting fast food as something that saves parents time, quietens children and feeds families.

Even though Facebook accounts are available only to people 13 and over, junk-food ads still use child-oriented themes, such as characters and games. Many appear to be designed to appeal directly to children. This included ads promoting 'healthy' foods, such as vegetables, in kids' meals.

The most insidious marketing tactics we found connect junk foods, and the brands synonymous with junk foods, to wholesome or popular activities. This creates a 'halo effect'.

For example, many ads use 'sports-washing' to associate unhealthy foods with healthy sports activities or pleasurable spectator sports.

9780170495714

Sports in junk-food marketing can appeal to a broad audience, including young people.

Unhealthy food advertising should be banned

Last week a Parliamentary Inquiry into Diabetes in Australia repeated calls for the government to restrict the marketing and advertising of unhealthy food to children on television, radio, in gaming and online.

The federal government should soon issue its report on how best to limit unhealthy food marketing to children. Our study supports the government's proposal to ban *all* unhealthy food and drink advertising online.

We also recommend the government should include all types of promotions. This includes ads from online food delivery companies, supermarkets and sports clubs that cross-promote unhealthy foods.

Analysis

1 Evaluate the validity of the article. Consider the reliability of the sources of information used in the article, the context and the presentation of evidence.

2 Explain why the authors are urging the Australian Government to ban unhealthy food advertising online.

3 Hypothesise why young men are the target of junk food advertising.

4 Explain what is meant by the 'halo effect.'

5 After reading the article, do you agree with the authors that there should be a ban on advertising unhealthy food online? Justify your response.

Rethink sugary drinks

Just like snack foods, sugary drinks, including soft drinks, energy drinks, fruit drinks, sports drinks and cordial are described as 'empty kilojoules' as they are high in sugar but do not provide the body with any other important nutrients. Many young Australians consume a large amount of sugar-sweetened beverages, especially sugar-sweetened soft drinks. This can be problematic, as the large number of kilojoules they consume can lead to weight gain.

Figure 14.11 Many young Australians consume a large amount of sugar-sweetened beverages.

Activity 14.2

Sugary drinks

Access the Victorian Government's Better Health Channel website.

1 Go to the 'Soft drinks, juice and sweet drinks – limit intake' page (search for 'soft drink') and read the full fact sheet about sugary drinks.

a Are you surprised by some of the facts presented? Justify your answer.

b Categorise the facts presented in a PMI (Plus, Minus, Interesting) table. Compare your results with a partner and discuss the similarities and differences in your groupings.

c Explain why the consumption of sugary drinks is seen as an important issue by the Better Health Channel.

Weblink Better Health Channel

2 Examine the 'Rethink Sugary Drinks' infographics on the next page (Figure 14.12). Discuss how this information might inspire, challenge or encourage a person to reduce the sugary drinks they consume as part of their everyday food intake.

Figure 14.12 Facts about sugary drinks from the Rethink Sugary Drink website

© Rethink Sugary Drink

Preparing your own healthy snack foods

We all enjoy eating snacks, including delicious baked products: they can give us a real lift, even if only fleetingly! These foods are usually sweet in nature – although some have a savoury flavour – and are often high in fat and/or sugar, so they should only be eaten as special treats or indulgences, and for celebrations.

Making your own snacks and baked products at home means that you can adjust the recipe to:

- reduce the sugar content or change the type of shortening – for example, to a vegetable-based spread – to make a healthier option
- increase the fibre content by substituting one ingredient in the recipe for another that has a similar function; for example, by using wholemeal flour instead of white flour
- alter the texture; for example, by cooking vegetables for less time so that they will be crunchier
- alter the flavour; for example, by adding more herbs and spices for a stronger flavour.

Figure 14.13 Rice paper rolls are easy to make and are a delicious and healthy snack.

iStock.com/GMVozd

9780170495714

Pastry

Many popular snack foods – such as pies, pasties, sausage rolls and sweet or savoury tarts – are made with pastry. Pastry is a simple mixture of flour, fat and a small amount of liquid. Flexible sheets of raw pastry can be used to wrap, cover or contain single ingredients or fillings with a mixture of ingredients. The golden crust of pastry has a crisp or flaky texture, and usually provides contrast to the soft filling it protects.

WRAP A FILLING	COVER A FILLING	CONTAIN A FILLING
pasties sausage rolls spinach triangles	chicken and leek pie apple pie cherry pie	quiche Lorraine meat pie lemon tart

Figure 14.14 Functions of pastry

Pastry products are very filling and satisfy hunger because they are high in fat. Pastry contains carbohydrates, from the flour, and a high proportion of fat, which places it in the 'only sometimes and in small amounts' group of foods in the *Australian Guide to Healthy Eating*.

Functional ingredients in pastry

As noted above, pastry is a mixture of flour, fat and liquid.

- Plain flour with a low gluten content is generally used for making pastry. Minimising the amount of gluten in the dough helps to prevent the pastry from becoming tough.
- Butter is considered to be the best form of fat to use in pastry (rather than margarine, for example) because it has good flavour and keeping qualities. It is usually rubbed into the flour so that it coats the granules of starch, resulting in a crisp, textured pastry.

iStock.com/Imagehit

Figure 14.15 Savoury tarts make a delicious lunch.

- A small amount of water is used in most pastry types; this should always be added cold to avoid causing the fat to melt. An egg yolk or a squeeze of lemon juice is added to some pastries to make them more tender.

Savoury pastry products are glazed with egg and milk; sweet pastries are glazed with sugar syrup to provide a shiny, brown finish when cooked.

Types of pastry

The ratio (or amount) of butter or fat to flour varies depending on the type of pastry being made. The method used to combine the ingredients in pastry also influences the sensory properties of the final product.

Table 14.5 Characteristics of different types of pastry

Shortcrust pastry	Puff pastry	Filo pastry	Choux pastry

• Contains flour and shortening, usually butter • Butter is added by rubbing it into the flour • Has a short, crumbly texture • Used for making pasties, meat pies and fruit pies	• Contains flour and shortening, usually butter • Butter is added by rubbing in and layering • Has a light, flaky texture and rich, buttery flavour • Used for sausage rolls, pies and vanilla slices	• Contains very thin layers of flour-and-water dough • Each layer is brushed with melted butter or oil • Has a dry, flaky texture • Used for spinach triangles and apple strudel	• Contains flour, water, butter and eggs • Dough is cooked, and then the eggs are beaten in • Has a crisp, light texture • Puffs up and increases in volume during baking • Used for eclairs, profiteroles and cream puffs

Right to left: iStock.com/ZAKmac, Robert Downer/Alamy Stock Photo, juefraphoto/Shutterstock.com, Elena Zajchikova/Shutterstock.com

Top tips for making pastry

1 Prepare pastry in a cool environment to prevent the butter from melting.
2 Knead the pastry dough only lightly and as little as possible to minimise the development of gluten.
3 Add liquid only gradually, because the amount required will vary depending on the flour's ability to absorb moisture.
4 Rest the pastry in the refrigerator to allow it to become more evenly hydrated and for the gluten to relax. This process makes the pastry easier to roll out and prevents it from shrinking during baking.
5 If the pastry becomes too soft to work with, simply return it to the refrigerator until it becomes firm again.
6 Pastry is cooked at a high temperature (200°C) to ensure that the fat is quickly absorbed into the flour.

PRACTICAL ACTIVITY 14.3

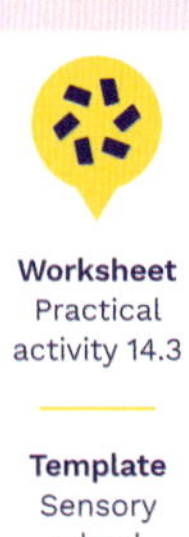

Worksheet Practical activity 14.3

Template Sensory wheel

Comparative food testing: Comparing a commercially processed fast food and an equivalent homemade product

Aim

To compare the physical and sensory characteristics and the nutritional properties of a commercially processed fast food and an equivalent homemade product

Equipment

- 1 quantity of homemade sausage rolls
- commercial sausage rolls
- 2 oven trays

Method

1 Prepare the Sausage Rolls from the recipe on page 398. Note: The homemade sausage rolls should be approximately the same size as the commercial sausage rolls.
2 Place the commercial sausage rolls on an oven tray and heat according to the manufacturer's instructions.
3 Complete a comparison of the energy, fat, sugar and sodium content and the physical and sensory properties of both types of sausage rolls. Use a table similar to the one below. Refer to the sensory wheel on page 4 for additional words to assist you.

Results

		Homemade sausage rolls	Commercial sausage rolls
Quantitative measures	Weight (in grams) of one sausage roll		
	Height and length (in centimetres) of one sausage roll		
	Colour of cooked product		
	Preparation time		
Nutritional properties per 100 grams	Energy kJ		
	Total fat, grams		
	Sugar, grams		
	Sodium, milligrams		
Qualitative measures	Appearance		
	Aroma		
	Flavour		
	Texture or mouth feel		
	Sound		
	Overall appeal (5 = like a lot, 1 = dislike a lot)		

9780170495714

Analysis

1 Describe the similarities and differences in the weight, height/length and colour of the homemade and commercial sausage rolls.
2 Did the time involved in the preparation of the homemade sausage rolls detract from their overall appeal?
3 Identify the sausage roll that had the most appealing appearance and explain why.
4 Which sausage roll had the most appealing aroma?
5 Discuss the flavour profile of each sausage roll and explain which was the most appealing.
6 Which sausage roll had the most appealing texture and sound?
7 Estimate which product had the highest salt content. Explain your decision.
8 Which product is likely to be higher in dietary fibre? Why?
9 Use your knowledge of the energy value of ingredients and the details on the product label of the commercial sausage rolls to estimate which sausage roll had the highest energy value.
10 Which product would be preferable to include in a healthy diet? Why?

Mark Fergus Photography

Figure 14.16 Homemade sausage rolls

Conclusion

After analysing your results, which product you would prefer to eat again? Justify your decision with reference to the physical and sensory characteristics and the nutritional properties of each product.

Chocolate

The word 'chocolate' comes from the Aztec language and means 'bitter water'. Chocolate is produced from cocoa beans, which are picked and then fermented, dried, roasted and cracked to remove the cocoa butter from the shell. This is then refined into chocolate liquor. Other ingredients such as sugar and milk powder are added to the liquor to make the product we know as chocolate.

Nutritional properties of chocolate

Whether it is a chocolate frog, a Tim Tam, a Mars bar or a Snickers bar, a chocolate treat is something that most people enjoy. We associate the delicious smooth, velvety, melt-in-the-mouth taste of chocolate with a sense of comfort, reward and celebration. However, chocolate is included in the 'only sometimes and in small amounts' section of the *Australian Guide to Healthy Eating*, as it is high in sugar and saturated fat and therefore in kilojoules. Consequently, an over-indulgence in this delicious treat can lead to weight gain if we are not careful.

Some academic studies have suggested that eating a small amount of dark chocolate each day is good for heart health; however, other nutrition experts dispute this claim.

There are many grades of chocolate and, usually, the higher the percentage of cocoa butter, the better the texture, aroma and flavour. In compound chocolate, vegetable oils are substituted for cocoa butter. This product has different sensory properties from chocolate that contains cocoa butter, but is sometimes easier to use when preparing decorations for a cake or dessert.

Table 14.6 Characteristics of cooking chocolate

Chocolate type	Characteristics and uses
Chocolate melts	Can be melted and set hard at room temperature
	Used for making decorations, dipping strawberries, coating biscuits and moulding
Chocolate bits	Hold their shape when baked
	Used for products that are baked in the oven, such as choc chip biscuits, chocolate cakes and muffins
	Not suitable to melt on the cooktop or in the microwave
Block cooking chocolate (e.g. Plaistowe)	Melts to a smooth, silky texture
	Used for making cakes, desserts and biscuits, and for coating and decorating

iStock.com/bhofack2

Figure 14.17 Brownies are a popular chocolate treat.

Fairtrade chocolate

Cocoa beans are widely grown throughout the world, especially in tropical climates in West Africa, South America and Asia. Most (73 per cent) of the world's cocoa crop is grown in African countries.

Cocoa beans are also grown in Fiji, Samoa, the Solomon Islands and Vanuatu. More recently, Timor-Leste has begun to develop cocoa bean plantations. However, while cocoa production is an important source of income in these Pacific Island countries, they are only able to produce small volumes in comparison with the African market. This makes it almost impossible for small farmers to compete with large multinational producers, and they often struggle to sell sufficient cocoa beans to make a sustainable living.

The **Fairtrade** movement was developed to assist farmers, including cocoa farmers, to produce their crops in a sustainable manner and to have equity in trading their products. The Fairtrade movement assists those small farmers who face a wide range of problems – none greater than that the low prices they receive for their crops, which may not even cover their costs of production. Farmers need to purchase tools, fertilisers and pesticides, as well as provide food and clothing for their families. Many farmers lack the education or financial stability to be able to work in other, more lucrative occupations. An increase in child and slave labour in cocoa-producing West African countries is another major problem facing many communities.

The introduction of the Fairtrade movement has meant that many small farmers can make a real living and plan for their future. There are now over 140 000 cocoa farmers around the world involved in this movement. It enables farmers and workers in the developing world to achieve better prices and decent working conditions, and to receive fair terms of trade. Fairtrade farmers, including those in Timor-Leste, also ensure the sustainability of their environment by using organic methods of production. In addition, buying fair trade products assists many families by giving them access to education, a safer environment, and improved healthcare and nutrition.

You can easily identify a product that uses Fairtrade ingredients by looking for the Fairtrade logo.

Mark Fergus Photography

Figure 14.18 Fairtrade chocolate

PjrStudio/Alamy Stock Photo

Figure 14.19 The Fairtrade logo

9780170495714

PRACTICAL ACTIVITY 14.4

Product analysis and sensory analysis: Sensory and chemical comparison of commercial chocolate

Worksheet
Practical activity 14.4

Aim

To analyse the sensory and chemical properties of various types of dark chocolate to determine the best-quality product

Ingredients

- dark compound chocolate
- dark couverture or cooking chocolate
- dark eating chocolate

Method

Draw a table like the one below.

1 Record the ingredients listed on the label of each product.
2 Record a description of the gloss for each sample.
3 Break a sample of each chocolate and record the sound of the snap. Was the sound sharp or dull?
4 Taste a sample of each chocolate. Let the sample melt on your tongue rather than chew it to experience all of its sensory properties.
5 Give each chocolate an overall rating.

	Dark compound chocolate	Dark couverture or cooking chocolate	Dark eating chocolate
Ingredients			
Gloss			
Snap			
Mouth feel			
Flavour			
Overall rating (5 = excellent, 3 = OK, 1 = unsatisfactory)			

Analysis

1 Discuss the similarities and differences between the ingredients in each type of chocolate.
2 What information does the order in which the ingredients appear on the food label provide?
3 Identify the ingredients that appear in some products but not in others. What reasons can you suggest for this?
4 How much fat is in each type of chocolate? What type of fat is it?
5 Which chocolate scored highest in terms of its sensory properties? Why?
6 Compare the sensory properties of a chocolate high in cocoa butter and a compound (vegetable fat) variety.

Conclusion

With other members of the class, develop a list of characteristics that denote good-quality chocolate and explain which properties influenced your decision.

Top tips for cooking with chocolate

ffolas/Shutterstock.com

Figure 14.20 Chocolate can be melted in a bowl over simmering water.

1 When melting chocolate, grate or chop the chocolate into small pieces first.
2 Make sure there is no water in the bowl before melting chocolate; water or steam will affect the chocolate's consistency and may cause it to 'seize'.
3 Chocolate can be melted in a bowl over simmering water, or in a microwave according to the manufacturer's instructions. Keep the heat as low as possible when melting chocolate to prevent it from becoming granular in texture. If using a microwave, use a low wattage for short bursts of time, and stir the chocolate frequently.
4 Do not overheat or heat the chocolate for longer than necessary, as it may burn.
5 Melt only small quantities of chocolate at a time.
6 Equal quantities of chocolate and cream can be used to make chocolate ganache as a filling or topping for cakes.
7 Set and store chocolate at room temperature. Do not refrigerate; this will cause the chocolate to sweat and lose its gloss.
8 Store chocolate away from direct heat; this causes blooming – a white discolouration on its surface.

iStock.com/stphillips

Figure 14.21 Chocolate ganache makes a delicious decoration on a cake.

Testing knowledge 14.2

Quiz
Testing knowledge 14.2

11 Justify why it is important for teenagers to limit the amount of snack foods they eat on a regular basis.
12 Outline two reasons why we should limit the amount of sugary drinks we consume.
13 List three strategies you could use to increase the health profile of baked products.
14 Create a diagram to highlight the benefits of using pastry to make snack foods.
15 Explain why pastry products are included in the 'only sometimes' section of the *Australian Guide to Healthy Eating*.
16 Justify why it is better to use butter when making pastry rather than margarine.
17 Explain why it is important to:
- rest pastry in the refrigerator before rolling out and/or baking.
- bake pastry at a high temperature.

18 Outline the nutritional reasons it is important to limit the amount of chocolate we consume.
19 Identify the ingredient in chocolate that is a key indicator of quality.
20 Create a mind map to demonstrate the benefits to small farmers of belonging to the Fairtrade movement.

Critical and creative thinking 14.1

Design an infographic to demonstrate the importance of Australian Dietary Guideline 3: 'Limit intake of foods containing saturated fat, added salt [and] added sugars'.

9780170495714

DESIGN ACTIVITY 14.1

Savoury snacks

A coding club, #appsandgames, is starting up an after-school program at your school. The club is aimed at Year 7–9 students and will run for two hours one afternoon a week during Terms 2 and 3.

Worksheet
Design activity 14.1

Templates
Decision table
Food order
Production plan
Sensory wheel
Australian Guide to Healthy Eating

Design brief

The organisers of the coding club know that the students will be hungry at the end of the day, and have decided to provide some warm savoury snacks for them to eat when they arrive. As your school is part of the Healthy Schools Achievement Program, the organisers realise that any snack food they provide must be healthy and must meet the recommendations of the *Australian Guide to Healthy Eating*. The warm savoury snacks must contain some vegetables and must be wrapped or cut into portions so that they can be eaten with the fingers. The organisers are asking students to upload suggested recipes for a healthy savoury snack food that can be served warm onto their #appsandgames webpage, along with a photograph.

Investigating and defining

1 Develop five design criteria that cover the solution requirements and constraints outlined in the design brief to evaluate your finished product.
2 Use the internet, recipe books or food magazines to research a variety of recipes that could be served as a warm savoury snack.
3 Following your research, identify five recipes that could be suitable designed solution options for your warm savoury snack.
4 Compete a table similar to the one below to record the vegetables used, method of serving and appealing features of each recipe.

	Recipe idea	Vegetables included	Wrapped or cut into portions	Appealing features of the recipe
1				
2				
3				
4				
5				

Generating and designing

1 Select the two designed solution options you think would best meet the solution requirements and constraints in the design brief. Construct a decision table. Select the designed solution option you would prefer and explain your choice.
2 Write up your new recipe ready for production.

Planning and managing

1 Complete a food order.
2 Before producing your warm savoury snack, write up a production plan, noting any safe work practices to be followed and identifying the major processes to be used.

Producing and implementing

1 Produce your preferred option.
2 Style and photograph your warm savoury snack ready to be uploaded onto the coding club's webpage.

Evaluating

1 Respond to your five design criteria in detail to determine the success of your product.
2 Describe the sensory properties – appearance, aroma, flavour, texture and sound – of your savoury snack food.
3 Share your savoury snack food with two other people and record their comments.
4 What modifications would you make to the recipe if you were to make it again?
5 Classify the ingredients of your warm savoury snack on a diagram of the *Australian Guide to Healthy Eating*. Comment on how well it meets the recommendations of this food selection model.

14

RECIPES

Sesame-crusted Fish with Baked Potato Wedges

INGREDIENTS

Baked Potato Wedges

3 medium red-skinned potatoes

olive oil spray

Sesame-crusted Fish

1 tablespoon plain flour

salt and pepper

1 egg

1 tablespoon milk

⅔ cup panko crumbs

3 tablespoons sesame seeds

2 firm white fish fillets, approximately 150 grams each

olive oil spray

METHOD

Baked Potato Wedges

1. Preheat oven to 220°C. Line an oven tray with baking paper.
2. Scrub the potatoes thoroughly. Cut them in half lengthwise, then cut each half into three wedges.
3. Place the wedges in a medium saucepan, cover with cold water and bring to the boil. Turn the heat to medium and cook the wedges for 6 minutes.
4. Drain thoroughly and place on paper towel to dry.
5. Place the wedges on the oven tray and lightly spray with olive oil. Bake in the oven for 30–35 minutes or until they are golden brown.

Sesame-crusted Fish

1. Once the wedges are cooked, remove them from the oven and reduce the oven temperature to 200°C. Line an oven tray with baking paper.
2. Place the flour on a sheet of paper towel and season with salt and pepper.
3. Beat the egg and milk together in a small bowl. Place on a flat plate.
4. Combine the breadcrumbs and sesame seeds in a small bowl. Place on a sheet of paper towel.
5. Coat each fish fillet in the seasoned flour, then brush with the egg mixture and coat in the breadcrumb mixture. Press the crumbs on firmly.
6. Rest the crumbed fish in the refrigerator for 10 minutes. This allows the moisture to be evenly distributed in the crumbing mixture and helps to prevent it from falling off the food during cooking.
7. Lightly spray the fish with olive oil. Place on the lined baking tray and bake in the oven for 10 minutes.
8. Remove from the oven and allow to stand for 2 minutes before serving. Return the potato wedges to the oven to rewarm while the fish is resting.

9780170495714

EVALUATION

1 Describe the sensory properties – appearance, aroma, flavour, texture and sound – of your Sesame-crusted Fish with Baked Potato Wedges.
2 Explain why it is important to parboil the potato wedges before baking them in the oven.
3 Identify and describe the changes that occur to the chemical and physical properties of the fish when it is baked in the oven.
4 Classify the ingredients for the Sesame-crusted Fish with Baked Potato Wedges on a diagram of the *Australian Guide to Healthy Eating*. Write a paragraph to explain how well this meal meets the recommendations of this food selection model.
5 Justify why the Sesame-crusted Fish with Baked Potato Wedges is a healthier meal option than battered fish and fried chips.

Mark Fergus Photography

Cornish Pasties

INGREDIENTS

Glaze
1 small egg
2 tablespoons milk

Shortcrust Pastry
1 cup plain flour
1 cup self-raising flour
pinch salt
125 grams butter
¼ teaspoon lemon juice
⅓–½ cup water

Filling
250 grams lean minced beef
1 medium onion, finely chopped
1 medium potato, grated
½ carrot, grated
2 tablespoons frozen peas
2 teaspoons parsley, chopped
salt and pepper

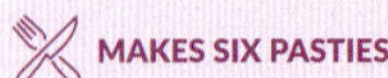

METHOD

Glaze
Prepare the glaze by beating the egg and milk together. This mixture can also be shaken in a lidded jar.

Shortcrust Pastry
1 Sift the flours and salt into a large bowl.
2 Chop the butter into small pieces and rub into the flour, using the fingertips, until the mixture resembles breadcrumbs.
3 Make a well in the centre of the mixture and add the lemon juice and sufficient water to make a firm dough. Use a spatula or knife to carry out this process.

Note: if time is short, steps 1, 2 and 3 can be completed in a food processor.

4 Place the dough on a bench sprinkled with flour and knead lightly until smooth.
5 Press the dough into the shape of a disc and cover with plastic wrap until the filling is complete.

To assemble and cook the dish
1 Preheat oven to 200°C and line a baking tray with baking paper.
2 Break up the minced beef with a fork in a large bowl and add the vegetables, parsley, salt and pepper.
3 Divide the pastry into six equal portions and roll out each one to the size of a saucer.
4 Divide the filling into six portions and place one portion in the centre of each pastry circle.
5 Using a pastry brush, brush the edge of one half of each pastry circle with the egg and milk glaze.
6 Pick up the edges of each pastry circle and draw up so that they meet at the top. Pinch the edges together to create a seal, then place on the lined baking tray.
7 Prick the side of each pasty with a fork to form a small vent. Using a pastry brush, glaze with more of the egg and milk mixture, avoiding the frilled edges.
8 Bake at 200°C for 10 minutes, then reduce the oven to 180°C and bake for a further 30–35 minutes.

9780170495714

EVALUATION

1 How will you know when the butter is rubbed into the flour sufficiently? Explain what could happen if too much water is added to the flour.
2 Why is it important to knead the pastry lightly?
3 Identify and describe the two processes that cause the Cornish Pasties to brown when baked in the oven.
4 Explain how you would test the Cornish Pasties to tell if they were cooked.
5 Classify the ingredients for the Cornish Pasties on a diagram of the *Australian Guide to Healthy Eating* and comment on the health rating you would give this dish.

Mark Fergus Photography

Sausage Rolls

INGREDIENTS

1 slice day-old bread, crumbled

2 tablespoons milk

200 grams sausage mince

½ onion, finely diced

½ medium carrot, grated

½ zucchini, grated

1 sheet puff pastry, defrosted in refrigerator

2 tablespoons flour

1 tablespoon milk, for glazing

MAKES 6 LARGE OR 12 SMALL SAUSAGE ROLLS

METHOD

1. Preheat oven to 220°C and line a baking tray with baking paper.
2. Soak the bread in the milk for 10 minutes, then squeeze out the excess liquid. Combine the sausage mince, onion, carrot, zucchini and soaked bread. Thoroughly mix.
3. Cut the sheet of pastry in half lengthwise.
4. Sprinkle a board with flour. Roll the sausage mixture into two rolls the same length as the pastry.
5. Place one of the rolls of sausage mince along the edge of one half of the pastry. Moisten the edges of the pastry with water.
6. Roll the pastry over the meat with the fold side underneath. Repeat with the other half of the pastry.
7. Cut the rolls into even-sized pieces and place on the lined baking tray. Score three lines across the top of each sausage roll by lightly pressing down on the pastry with a knife.
8. Using a pastry brush, glaze with the milk. Bake at 220°C for 10 minutes, then at 190°C for a further 10–15 minutes.

Mark Fergus Photography

EVALUATION

1. Describe the sensory properties – appearance, aroma, flavour, texture and sound – of your Sausage Rolls.
2. Discuss the benefits of adding carrot and zucchini to the Sausage Rolls.
3. Explain why the Sausage Rolls are initially cooked at 220°C and then with the temperature reduced for the remainder of the cooking time.
4. Identify and describe the process that causes the pastry to brown when the Sausage Rolls are baked in the oven.
5. Classify the ingredients for the Sausage Rolls on a diagram of the *Australian Guide to Healthy Eating* and comment on how well this dish meets the recommendations of this food selection model. Explain why it is recommended that sausage rolls should be eaten 'only sometimes and in small amounts'.

9780170495714

Vietnamese Spring Rolls with Dipping Sauce

INGREDIENTS

Filling
1 teaspoon peanut oil
1 teaspoon fresh ginger, grated
1 clove garlic, crushed
2 spring onions, chopped
125 grams minced pork
100 grams canned small prawns, drained
1 tablespoon fresh mint, chopped
2 teaspoons fish sauce
1 tablespoon sweet chilli sauce
2 teaspoons lime juice
pinch of sugar

Spring Rolls
2 teaspoons cornflour
2 teaspoons water
15 spring roll wrappers
oil, for deep-frying

Dipping Sauce
2 tablespoons sugar
2 tablespoons hot water
1 tablespoon fish sauce
1 tablespoon lime or lemon juice
1 tablespoon sweet chilli sauce (or 1 small red chilli, sliced)
2 teaspoons white vinegar
2 teaspoons fresh coriander leaves, chopped

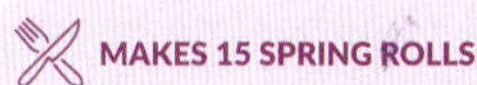
MAKES 15 SPRING ROLLS

Mark Fergus Photography

METHOD

Filling
1. Heat the oil in a wok and add the ginger, garlic and spring onions. Cook until the spring onions are soft.
2. Add the minced pork and cook until the meat is tender and has changed colour. Add the prawns and toss through the meat mixture.
3. Stir in the mint, sauces, lime juice and sugar. Stir and cook for a few minutes until the liquid has reduced. Spread on a plate and allow to cool.

Spring Rolls
1. Combine the cornflour and water in a small bowl to make a paste.
2. Place 2 teaspoons of the filling in the middle of each spring roll wrapper. Fold each side inwards and roll up. Secure the end of the wrapper with some of the cornflour and water paste.
3. Deep-fry the spring rolls in hot oil until golden brown. Drain well on absorbent paper. Alternatively, place the spring rolls on a greased baking tray, lightly spray with oil and bake in a preheated oven at 200°C for approximately 10 minutes or until golden.
4. Serve with the Dipping Sauce.

Dipping Sauce
1. Dissolve the sugar in the hot water in a small saucepan and boil, uncovered and without stirring, for 5 minutes or until the mixture slightly thickens.
2. Stir in the remaining ingredients and allow to cool.
3. Serve in a small bowl.

EVALUATION
1. Why is it important to fry the ginger, garlic and spring onions in step 1 of the filling recipe?
2. Explain why it is recommended that you allow the filling to cool before assembling the spring rolls.
3. Explain how the cornflour and water paste seals the Spring Rolls.
4. What are the advantages of baking the Spring Rolls in the oven, rather than deep-frying them?
5. Classify the ingredients for the Vietnamese Spring Rolls with Dipping Sauce on a diagram of the *Australian Guide to Healthy Eating* and comment on how well this dish meets the requirements of this food selection model. Explain why it is recommended that spring rolls should be eaten 'only sometimes and in small amounts'.

Cinnamon Biscuits

INGREDIENTS

125 grams butter, softened
⅓ cup white sugar
½ cup soft brown sugar
½ egg, lightly beaten
1 teaspoon vanilla extract
1¼ cups plain flour
1 teaspoon bicarbonate of soda
1 teaspoon ground cinnamon
1 teaspoon salt flakes (optional)

MAKES APPROXIMATELY 20 BISCUITS

METHOD

1. Preheat oven to 180°C. Line 2 baking trays with baking paper.
2. In a large mixing bowl, combine the softened butter, white sugar and soft brown sugar. Using a hand-held electric beater, beat the mixture for 5 minutes until it is light and fluffy and looks like whipped cream. Scrape down the bowl several times to ensure that the ingredients are evenly distributed.
3. Add the vanilla to the lightly beaten egg in a small bowl, then add a little of the egg at a time to the butter mixture. Beat well after each addition.
4. Sift the flour, bicarbonate of soda and ground cinnamon together in a medium bowl, then fold into the butter and egg mixture until a soft dough is formed.
5. Shape heaped teaspoons of dough into approximately 20 even-sized balls and place on the lined baking trays. Allow space for the biscuits to spread while in the oven.
6. Press each ball of dough lightly with two fingers to flatten it, then sprinkle with the salt flakes, if using.
7. Bake for approximately 12 minutes. Halfway through the cooking time, rotate the baking trays – front to back and top to bottom – to ensure even cooking.
8. When the biscuits are just beginning to brown around the edges, remove the trays from the oven. Allow to cool for 3–5 minutes before removing from the trays. Finish cooling on a cake rack.

Mark Fergus Photography

EVALUATION

1. Describe the sensory properties – appearance, aroma, flavour, texture and sound – of your Cinnamon Biscuits.
2. Outline the functional roles of bicarbonate of soda in this recipe.
3. Explain how creaming the butter and the sugars together influences the final sensory properties of the Cinnamon Biscuits.
4. Explain why it is important to make the biscuit dough into even-sized portions.
5. Consider the recommendations of the *Australian Guide to Healthy Eating* food selection model and discuss why the Cinnamon Biscuits would be classified into the 'only sometimes and in small amounts' section.

9780170495714

Wholemeal Biscuits

INGREDIENTS

½ cup wholemeal flour

¾ cup plain flour

½ teaspoon bicarbonate of soda

⅓ cup soft brown sugar

⅔ cup rolled oats

½ teaspoon sea salt flakes

90 grams cold unsalted butter, diced

⅓ cup milk

2 teaspoons honey

½ teaspoon vanilla extract

MAKES APPROXIMATELY 16 BISCUITS

METHOD

1 Preheat oven to 180°C. Line a baking tray with baking paper.
2 Place the flours, bicarbonate of soda, soft brown sugar, oats and salt in a food processor and pulse a few times to combine.
3 Add the butter and process until the mixture resembles coarse breadcrumbs.
4 Tip the mixture out into a large mixing bowl and add the milk, honey and vanilla. Use your hands to bring the dough together into a firm ball.
5 Wrap the biscuit mixture in plastic wrap, then flatten it into a disc. Refrigerate for 15 minutes.
6 Lightly flour the workbench, remove the dough from the refrigerator and roll it out to a thickness of about 3 millimetres. Brush off any excess flour.
7 Use a fork to prick holes all over the dough.
8 Cut out circles of dough using a 7-centimetre cutter. Place on the lined baking tray, leaving a little space between the dough circles to allow them to spread. Re-roll the leftover dough and cut more circles until all the dough is used.
9 Bake for approximately 15 minutes or until the biscuits are golden brown and firm to the touch. Allow to cool for 5 minutes on the baking tray before transferring them to a cake rack.

Note: Wholemeal Biscuits are delicious with cheese, or they can be lightly coated with dark or milk chocolate to serve as a sweet treat.

Mark Fergus Photography

EVALUATION

1 Describe the sensory properties – appearance, aroma, flavour, texture and sound – of your Wholemeal Biscuits.
2 Explain why it is important to use cold butter for rubbing into the dry ingredients.
3 Why is the biscuit mixture refrigerated in step 5?
4 Describe one safety rule you observed when using the oven to bake the biscuits.
5 Compare the ingredients for the Wholemeal Biscuits to those for the Anzac Biscuits (see page 91). Which biscuit could be included as part of a healthy diet? Justify your decision with reference to the ingredients used in each.

9780170495714

Vanilla Biscuits

INGREDIENTS

½ cup plain flour

½ cup self-raising flour

60 grams butter, at room temperature

¼ cup caster sugar

1 egg, lightly beaten

1 teaspoon vanilla extract

MAKES 12–18 BISCUITS, DEPENDING ON THE SIZE

METHOD

1. Preheat oven to 160°C. Grease an oven tray.
2. Sift the flours together in a medium bowl.
3. Cream the butter and sugar in another medium bowl until light and creamy in colour.
4. Add the egg and vanilla extract to the butter and sugar mixture and beat well.
5. Add the sifted flours and mix to a firm dough.
6. Lightly flour the bench and knead the dough until it is smooth.
7. Roll out the dough to a thickness of 3–5 millimetres and cut it into shapes. Re-roll the leftover dough and cut more shapes until all the dough is used.
8. Bake for 10–15 minutes or until a pale golden colour.
9. Allow the biscuits to cool on the tray for 5 minutes, then place them on a cake rack until cold.

Variations

1. *To make chocolate biscuits* – replace 1½ tablespoons of plain flour with 1½ tablespoons of cocoa. Sift the cocoa with the flours.
2. *To make lemon and coconut biscuits* – add 1 teaspoon of grated lemon rind and 1 tablespoon of desiccated coconut to the mixture in step 4.
3. *To decorate with coloured sprinkles* – lightly whisk half an egg white in a small bowl. Remove the biscuits from the oven after 10 minutes. Brush the egg white over the biscuits in a thin layer and sprinkle with ¼ cup of hundreds and thousands. Return the biscuits to the oven and continue baking until the bases are a pale golden colour.

Mark Fergus Photography

EVALUATION

1. What tools and small pieces of electrical equipment can be used to cream butter and sugar?
2. Why is the biscuit dough kneaded in step 6?
3. Identify and describe the process that causes the biscuits to brown when they are baked in the oven.
4. How can you test the Vanilla Biscuits to check if they are cooked?
5. Explain why health professionals recommend eating an apple as a snack rather than a Vanilla Biscuit.

9780170495714

RECIPE INDEX

Recipe available on Nelson MindTap

9780170495714

9780170495714

T

V

W

Y

Z

9780170495714

GLOSSARY

aeration the process of incorporating air into food products to increase their volume and create a light, airy texture

aquaculture the breeding, rearing and harvesting of fish and shellfish in coastal marine waters, open oceans and fresh-water systems

Australian Guide to Healthy Eating a pictorial representation of Guideline 2 of the *Australian Dietary Guidelines*: 'Enjoy a wide variety of foods from the five food groups each day'

baking a method of cooking food in an oven without the addition of fat or oil

biodiversity the vast array of living organisms that inhabit the planet and the interactions between them

boiling a method of cooking food in water at boiling point (100°C)

breakfast the first meal eaten soon after waking up from a night's sleep

caramelisation the process that occurs when sugar (sucrose) begins to decompose when it is exposed to high temperatures (190°C) of dry heat

cardiovascular disease a general term used to describe a range of diseases, including heart disease, stroke and blood vessel disease

casein the protein present in the curds of milk

cheese a product made from casein, the protein present in the curds of milk that are separated from the water and lactose, or whey

citrus fruit a fruit family that includes oranges, mandarins and lemons; their peel has a distinctive aroma and their flesh is divided into segments; their flavour ranges from acidic to sweet

coagulation a form of denaturation that occurs when there is a permanent change in the protein from a liquid into a thick mass as a result of heat or the addition of acids

coeliac disease a disease of the small intestine associated with permanent intolerance or hypersensitivity to gluten

conduction when heat is transferred from one molecule to another by collision or movement

connective tissue the tissue in meat that links and holds together the muscles

constraints factors in the design brief with which the product must comply

convection when the molecules in liquids or gases move from a warmer area to a cooler one

cooking the transfer of energy from a heat source to food

cross-contamination the transfer of harmful bacteria from uncooked food to food that has been cooked or prepared

danger zone the temperature – between 5°C and 60°C – at which bacteria can multiply very quickly

denaturation the permanent structural change of the protein molecules in food

dextrinisation the process that occurs when a starch is exposed to dry heat; the starch is broken down to dextrin, resulting in a change in colour to golden brown

design brief a statement that defines the need or opportunity to be resolved and identifies the users of the product; it contains information about the type of product, any constraints, the resources needed and a time frame

design criteria criteria drawn from the constraints and solution requirements in the design brief; they are used to determine if the designed solution is appropriate

design process the process of investigating and defining, generating and designing, planning and managing, producing and implementing, and evaluating

designed solution a product created for a specific purpose that has been outlined or defined in the design brief

diabetes a disease where the pancreas is unable to produce sufficient insulin to enable the glucose produced during digestion to be absorbed into the bloodstream

digestion the breakdown of large pieces of food into smaller components that can be absorbed into the bloodstream

discretionary foods foods that are high in kilojoules and do not provide any of the essential nutrients needed for good health

electric oven an oven that uses radiant and convection heat produced by electricity to cook food

endosperm the main body of the grain, which is composed almost entirely of starch

enzymatic browning a process that occurs when the enzymes in cut or peeled fruit are exposed to oxygen in the air and cause browning

enzymatic hydrolysis a chemical digestive process that breaks down food by breaking the molecular bonds that hold the food together

Fairtrade a movement developed to assist farmers, including cocoa farmers, to produce their crops in a sustainable manner and to have equity in trading their products

fast a period of time during which we eat nothing

feedlot production where cattle are kept in pens for 180–360 days and fed a grain-based, high-energy diet of wheat, barley, sorghum and canola seed

fermentation the conversion of some of the sugar and starch molecules in a yeast dough to alcohol and carbon dioxide as the yeast ferments

fire blanket an insulated blanket used to extinguish small fires in the kitchen

fire-stick farming the practice, undertaken by Australia's First Nations peoples, of using fire to burn vegetation so as to make animal hunting easier and to reorganise the composition of the plants and animals in the area

FODMAP an acronym that stands for fermentable oligosaccharides, disaccharides, monosaccharides and polyols

food any substance that we eat or drink that provides the body with chemical substances called nutrients

food allergy an abnormal immunological reaction to food

food citizenship based on the idea that individuals are not merely consumers at the end of the food chain, but active participants (or citizens) in the food system as a whole; food citizenship encourages individuals to support the environmental sustainability of the food system and the human rights of food producers through the food they buy and consume

food insecurity where an individual or family does not have access to sufficient food, or food of a nutritious quality, to meet their basic needs

food poisoning an illness caused by eating food that has been contaminated with harmful bacteria

food security the state that exists when all people, at all times, have physical, social and economic access to sufficient, safe and nutritious food that meets their dietary needs and food preferences for an active and healthy life

food selection model a guide that visually represents the proportion of different food groups that should be consumed each day

food sovereignty the right of individuals and communities to have control over their own food systems, challenging the dominance of multinational companies, with a particular focus on ensuring that food is produced ethically and sustainably, with a deep consideration for social justice, environmental sustainability and the wellbeing of local communities

food spoilage the deterioration in the physical, sensory and chemical properties of food over time; however, the food will not usually be harmful to eat

food system a complex series of activities that enables food to move from farm to consumer, and includes the growing, harvesting, processing, transporting, manufacturing, consuming, disposing and recycling of food

free sugar any sugar that is added to ultra-processed foods such as confectionery, fruit juices, sugary drinks, cakes and pastries

frying a method of cooking food by total or part immersion in fat or oil that is heated to temperatures between 150°C and 220°C

gas oven an oven that uses radiant and convection heat produced by gas to cook food

gelatinisation the process that occurs when starch granules absorb liquid in the presence of heat, and thicken the liquid, forming a gel

gluten the main protein in wheat flour

gluten intolerance a condition where the gluten proteins present in wheat, oats, barley, triticale and rye cause a person to suffer intestinal discomfort after eating food that contains gluten

glycaemic index (GI) a ranking of carbohydrate foods based on the immediate effect they have on blood sugar levels

glycogen energy stored in the muscle, tissue and liver

grain (cereal) foods edible seeds of certain grasses, including wheat, oats, rice, rye, barley, millet, quinoa and corn

grilling a fast, dry method of cooking food that uses intense heat radiated by an electrical element, a gas flame, glowing charcoal or an open wood fire

gut microbiome the complex ecosystem of microbiota that live in the gut

Health Star Rating a front-of-pack labelling system that rates the overall nutritional profile of packaged food

homogenisation a process that breaks the globules of fat within milk into minute particles so that the cream does not rise to the surface

in season the time of year when a fruit or vegetable has its best sensory properties

kneading the process that makes gluten – the protein in wheat flour – stronger and more elastic so that it can begin to stretch and capture the carbon dioxide bubbles in a dough

lactose the sugar (carbohydrate) present in milk

lactose intolerance the reduced ability to digest milk sugars, due to insufficient amounts of the enzyme lactase

legumes the seeds from the *Leguminosae* family; these vegetables are eaten in their immature form as green peas and beans, and in their mature form as dried peas, beans, lentils and chickpeas

Maillard reaction a reaction that turns food brown and creates pleasant, volatile, aromatic compounds when food is exposed to dry heat; it is a reaction between amino acids in protein and reducing sugars

marbling the even distribution of deposits of fat cells in red muscle tissue

metric measuring tools measuring spoons, cups, jugs and scales that have been calibrated to accurately measure ingredients by volume or weight using the metric system

microbiota the microscopic living organisms such as bacteria, fungi and viruses that live in the small intestine

modified atmosphere packaging (MAP) a method of packaging that causes change in the levels of gases inside a package in order to extend the shelf life of a product

monounsaturated fats fats found in olives, olive oil, avocados and nuts that have been shown to reduce blood cholesterol levels

muscle fibres cells that are bound into thin sheets of connective tissue; these bundles then form groups to create muscles

no-till or minimum tillage farming a farming system that involves planting the new crop by directly drilling in between the rows of the previous crop without tilling the soil

nutrients chemical substances in food that are broken down during digestion, including protein, carbohydrates, fats, vitamins and minerals

organic farming an environmentally sustainable food production system; it produces crops and animals without the use of artificial chemicals, using natural systems instead

9780170495714

osteoporosis a medical condition that occurs when calcium is lost from the bones, making them very fragile and easily broken

pasture-fed cattle beef cattle that spend their lives grazing on pasture

pasteurisation a process that destroys pathogenic, or disease-causing, bacteria and also extends the shelf life of milk

poaching a method of cooking delicate foods in liquid at a temperature just below simmering point (85°C)

pome fruit fruit that have crisp, juicy flesh surrounding a core that contains seeds; for example, apples and pears

processed breakfast cereals grains such as corn, wheat and rice that have been softened by pre-cooking and then dried; most are fortified, meaning they have had vitamins and minerals added during processing

protected cropping a farming practice that involves growing food crops under, or sheltered by, an artificial structure, such as a greenhouse or polytunnel

proving a process in which a yeast dough is rested to allow time for fermentation to take place

qualitative or sensory analysis the evaluation of the sensory properties of food, such as appearance, aroma, flavour, texture and sound

quantitative measures ways to measure the physical, chemical or nutritional properties of food

radiation the transmission of heat energy in the form of rays

recipe a list of ingredients and instructions for preparing food

reducing sugars include all monosaccharides and some polysaccharides; they react with amino acids to create the Maillard reaction

roasting a method of cooking food in an oven using a minimal amount of fat or oil

saturated fats fats found mainly in foods of animal origin (such as meat, cheese and butter) that are linked to raised cholesterol levels; coconut oil and palm oil are also high in saturated fats

sensory properties the appearance, aroma, flavour, texture and sound of food

small appliances pieces of equipment such as toasters, food processors, hand-held beaters or blenders

solanine a toxin that develops when potatoes are exposed to light, resulting in a green colour on exposed surfaces

steaming a method of cooking food in the steam from boiling water

stubble the stalks of the grain crop, which are left once the heads have been cut off during harvesting

sustainable diets accessible, affordable, equitable and acceptable diets to society that have a low environmental impact

sustainable farming farming practices that maintain the land's productivity so that it will be available for future generations

sustainable fishing the practice of leaving enough fish in the ocean so that the fish population can remain productive and healthy

trans fats 'bad' fats that can lead to serious health concerns and should be avoided; they are found mainly in hydrogenated vegetable oil used by food manufacturers in processed and fast foods

UHT milk a type of milk that has undergone ultra-high temperature processing of heating the milk to approximately 135°C for 2–3 seconds

wild native foods the huge variety of edible native Australian herbs, spices, mushrooms, fruits, flowers, vegetables, animals, birds, reptiles and insects

INDEX

9780170495714

C

9780170495714

F

G

H

I

9780170495714

N

O

P

9780170495714

Q

R

S

T

U

V

W

9780170495714

Z